THE LIQUID CRYSTALS B

SMECTIC AND COLUMNAR LIQUID CRYSTALS

CONCEPTS AND PHYSICAL PROPERTIES ILLUSTRATED BY EXPERIMENTS

THE LIQUID CRYSTALS BOOK SERIES

Edited by
G.W. GRAY, J.W. GOODBY and A. FUKUDA

The Liquid Crystals book series publishes authoritative accounts of all aspects of the field, ranging from the basic fundamentals to the forefront of research; from the physics of liquid crystals to their chemical and biological properties; and from their self-assembling structures to their applications in devices. The series will provide readers new to liquid crystals with a firm grounding in the subject, while experienced scientists and liquid crystallographers will find that the series is an indispensable resource.

PUBLISHED TITLES

Introduction to Liquid Crystals: Chemistry and Physics
By Peter J. Collings and Michael Hird

The Static and Dynamic Continuum Theory of Liquid Crystals:
A Mathematical Introduction
By Iain W. Stewart

Crystals That Flow: Classic Papers from the History of Liquid Crystals
Compiled with translation and commentary by Timothy J. Sluckin, David A. Dunmur, and Horst Stegemeyer

Nematic and Cholesteric Liquid Crystals: Concepts and Physical Properties
Illustrated by Experiments
By Patrick Oswald and Pawel Pieranski

Alignment Technologies and Applications of Liquid Crystal Devices
By Kohki Takatoh, Masaki Hasegawa, Mitsuhiro Koden, Nobuyuki Itoh, Ray Hasegawa, and Masanori Sakamoto

Adsorption Phenomena and Anchoring Energy in Nematic Liquid Crystals
By Giovanni Barbero and Luiz Roberto Evangelista

THE LIQUID CRYSTALS BOOK SERIES

SMECTIC AND COLUMNAR LIQUID CRYSTALS

CONCEPTS AND PHYSICAL PROPERTIES ILLUSTRATED BY EXPERIMENTS

Patrick Oswald
Pawel Pieranski

TRANSLATED BY Doru Constantin

With the support of Merck KgaA (LC division), Rolic Research Ltd, and the Laboratoire de Physique des Solides d'Orsay

CRC Press
Taylor & Francis Group
Boca Raton London New York

CRC Press is an imprint of the
Taylor & Francis Group, an **informa** business

Cover Illustration: "Arch-like" texture in a free-standing smectic A film. There is a difference of one layer (or sometimes more) between two areas with different colors. The separation lines correspond to edge dislocations (elementary, most of the time), centered in the thickness of the film (courtesy F. Picano).

CRC Press
Taylor & Francis Group
6000 Broken Sound Parkway NW, Suite 300
Boca Raton, FL 33487-2742

First issued in paperback 2019

© 2006 by Taylor & Francis Group, LLC
CRC Press is an imprint of Taylor & Francis Group, an Informa business

No claim to original U.S. Government works

ISBN-13: 978-0-8493-9840-7 (hbk)
ISBN-13: 978-0-367-39160-7 (pbk)

Library of Congress Cataloging-in-Publication Data

Catalog record is available from the Library of Congress

Visit the Taylor & Francis Web site at
http://www.taylorandfrancis.com

and the CRC Press Web site at
http://www.crcpress.com

CONTENTS

Smectic and Columnar Liquid Crystals

Chapter C.II Continuum theory of smectics A hydrodynamics

Chapter C.III Dislocations, focal conics, and rheology of smectics A

Chapter C.IV Ferroelectric and antiferroelectric mesophases

Chapter C.VI Hexatic smectics

Chapter C.VII The smectic B plastic crystal

IX

Chapter C.VIII   Smectic free films

CONTENTS

XI

Chapter C.X Growth of a columnar hexagonal phase

Nematic and Cholesteric Liquid Crystals

(volume 1)

PART A: OVERVIEW

PART B: MESOPHASES WITH AN ORIENTATIONAL ORDER

To our daughters, Magalie, Séverine, and Natalia

Preface to the English edition

After having dedicated a first volume (titled "Nematic and Cholesteric Liquid Crystals") to mesophases exhibiting an orientational order of the nematic and cholesteric type, in this second volume we focus on the study of smectic and columnar phases exhibiting, beside an orientational order, a one-, two- or three-dimensional translational order.

The number of these mesophases being significant (several dozens), we have limited ourselves to the study of the most representative among them, not only from the point of view of the structures, but also from that of concepts and physical properties. For pedagogical reasons, we strived to illustrate as systematically as possible our theoretical developments using numerous experiments. We also included a large number of diagrams that, it is well known, are often more enlightening than a long discourse.

The present volume starts with the description of the smectic A phase (SmA), the simplest one from a structural point of view. In this phase, the molecules, organized in fluid layers sliding viscously over one another, are on the average perpendicular to these layers. As noted by de Gennes in 1973, this phase is described by a lamellar order parameter, formally equivalent to the order parameter in superconductors. This analogy is presented in chapter C.I, dealing mainly with the nematic-SmA transition. In chapter C.II, we analyze the consequences of lamellar order on the mechanical properties of the phase. In particular, we describe its hydrodynamic properties, insisting on the concept of permeation, discovered by W. Helfrich in 1969, and which describes flow across the layers. In chapter C.III, we show that the viscoelastic properties of the SmA phase are intimately related to the presence of defects. Among them, the best known are the focal conics, whose study can be traced back to the beginning of the 20th century (F. Grandjean and G. Friedel, 1922), and dislocations, the existence of which was predicted by Sir F.C. Frank in 1958. We will first describe these defects in detail, and then analyze their interactions with the flow, in order to understand the origin of the exceptional lubricating properties of the SmA phase. We also describe a cascade of helical instabilities of screw dislocations, explaining the "stepwise" creep observed in compression normal to the layers, as well as the beautiful experiments of the Bordeaux group (D. Roux, O. Diat et al., 1992) on smectics under shear and the nucleation of onions, which are spherical assemblies of stacked lamellae. Finally, in the appendix to chapter C.III, we return to the analogies noticed by P.-G. de Gennes between the structure of a screw dislocation and that of a superconductor vortex, as well as between molecular chirality and a magnetic field.

At this point, we could have continued with the study of twist-grain boundary (TGB) phases, which are the "smectic" counterparts of type II superconductors. We preferred however to start by presenting the SmC* and SmC$_A$ phases, which are the helielectric and antiferroelectric precursors of the TGB phases. In these two phases, the chiral molecules are stacked in fluid layers. Their positioning in these structures is different from that adopted in SmA phases in that they are tilted with respect to the layer normal, forming more or less regular helical structures; this spontaneously generates a locally non-zero electric polarization. This phenomenon, predicted theoretically by R.B. Meyer in 1975 on the basis of symmetry arguments, was demonstrated experimentally together with L. Liébert, L. Strzelecki, and P. Keller who synthesized the same year the first SmC* liquid crystal (named DOBAMBC). This discovery was followed by the observation of several other mesophases, such as the anticlinic SmC$_A$ phase and the ferrielectric phases, each distinguished by a particular relative stacking of their electric dipole moments. All these phenomena are described in chapter C.IV, mainly centered on the experimental discovery of these structures.

It is in chapter C.V that we describe the model of Renn and Lubensky for TGB phases. These chiral phases exhibit locally a lamellar structure of the SmA (TGB$_A$) or SmC (TGB$_C$) type. The former were discovered in 1989 by the Bell group (J.W. Goodby, R. Pindak et al.) and the latter by the Bordeaux group (H.T. Nguyen, L. Navailles, Ph. Barois et al.) in 1992. As we already pointed out, these phases show strong analogies with type II superconductors, apart from the fact that the screw dislocations form walls, while the vortices are uniformly distributed in superconductors. This particular arrangement leads to the formation of a helical structure composed of stacked slabs, separated by walls of screw dislocations. It is noteworthy that the size of the slabs and the pitch of the helix can be commensurate or not. We describe in the same chapter the smectic Blue Phases, which are more complicated versions of the TGB phases and of the Blue Phases described in the first volume. Predicted theoretically by R. Kamien in 1997, they were observed very recently by the Bordeaux (M.H. Li, H.T. Nguyen et al.) and Orsay (B. Pansu and E. Grelet) groups; the smectic order is here rendered compatible with a three-dimensional twist by a new arrangement of the screw dislocations.

The dislocations also play a significant role in the structure of hexatic phases presented in chapter C.VI. We recall that the concept of hexatic order, as well as that of the TGB phase, was purely theoretical in origin. Fortunately, as in the case of TGB phases, it was possible to observe hexatic mesophases. These phases, termed SmB$_{hex}$, SmI, or SmF, were successively discovered by the Bell group (R. Pindak, J.W. Goodby et al.) in 1981, and then by the Orsay group (F. Moussa, J.-J. Benattar, and C. Williams) in 1982. After a short reminder on such fundamental concepts as Peierls disorder and quasi-order, the chapter describes the X-ray scattering, light scattering, microscopy (optical and electron) and rheology experiments by which these phases were characterized. In particular, we show how the observation of five-arm stars ("Pindak scars") in polarized microscopy in the textures of the SmI phase allowed its hexatic order parameter to be revealed.

The last three chapters deal with more personal topics, which are, therefore, less extensively covered in other books.

Thus, we start chapter C.VII with the study of the SmB phase. Its structure, first studied by A.-M. Levelut, J. Doucet, and M. Lambert (1976), has long been a controversial topic, but we know now, after detailed high-resolution X-ray studies, that it is a true plastic crystal, with an ABAB... or other stackings, exhibiting long-range three-dimensional positional order (R. Pindak et al., 1979; P.S. Pershan et al., 1982). This phase is sometimes termed SmB_{cryst}, in contrast with the SmB_{hex} phase, which only has short-range positional order. This hexagonal symmetry phase deserves to be called *plastic*, since it has the astonishing characteristic of flowing under an extremely small applied stress in shear parallel to the layers. This property stems from its very strong elastic anisotropy, the shear modulus parallel to the layers being much lower than the other elastic constants (M. Cagnon and G. Durand, 1980). We also show that this phase can serve as a model system for the study of equilibrium crystal shapes in the presence of facets or "forbidden orientations", as well as for the study of the Herring instability, developing when a facet is forced to grow with a forbidden orientation. The chapter ends by a comparison of the cell morphologies observed in directional growth in the rough, facetted and "forbidden" cases (Mullins-Sekerka instability).

Chapter C.VIII, rather large, is dedicated to free-standing smectic films and deals mostly with the stability of films, their thinning transitions, phase transitions and membrane vibration.

The first part starts with a description of films and their optical and mechanical properties: for instance, their number of molecular layers can be controlled within one unit, between two and a few thousand layers. On the other hand, the films are always attached to the frame supporting them by a meniscus. The films would not exist without this matter reservoir, and we explain why in the beginning of the chapter, dealing in particular with the equilibrium between a film and its meniscus and the profile of the latter, and with the topic of step mobility after a change in film thickness (thinning transitions). We then analyze the influence of the two free surfaces on the thermodynamical properties of films. We show that: in materials that exhibit a SmA-Nematic transition, the films continue to exist for temperatures well above the bulk transition temperature; that, in materials exhibiting a first-order SmA–SmC transition, this transition can simply "disappear" beyond a "critical point" if the films are thin enough; finally, that phase transitions that do not appear in the bulk can appear in the individual surface layers. We conclude that the thickness of a film becomes a relevant thermodynamical parameter, which needs to be taken into account in the case of thin films.

In the second part, the films are used as ideal vibrating membranes, owing to their exceptional mechanical properties (perfectly controlled thickness, high resilience, surface tension constant over the whole surface, etc.). For this reason, they can be used in the study of certain problems concerning the vibrations of flat or curved drums. Among the questions we approached is the Kac problem (1966) formulated thus: "Can one hear the shape of a drum?" After sketching the mathematical solution of this problem, we show how the smectic films can be used to validate it experimentally. The vibrations of fractal- or catenoid-shaped drums, as well as quantum billiards and their diabolic points, are also treated explicitly. The chapter continues with the description of a few nonlinear phenomena in vibrating films that can lead to meniscus destabilization. We also describe the beautiful experiments of C.D. Muzny and N.A. Clark on the Brownian motion of a disclination, as well as those of P. Cladis et al. on backflow effects and topological flow in SmC films. The chapter ends with a description of the molecular diffusion anomalies in films (J. Bechhoefer et al., 1997).

The last two chapters deal with columnar phases. Although they have been known for a long time in lyotropic mixtures and in anhydrous soaps, their discovery in thermotropic systems of disk-like molecules is relatively recent, dating from 1977 (S. Chandrasekhar et al.). A large number of columnar phases is known by now; most of them were synthesized by H.T. Nguyen and J. Malthête. They are briefly described in the beginning of chapter C.IX. We then focus on the simplest one, with hexagonal symmetry, denoted by D_h (D for disk-like and h for hexagonal). As the SmA phase was representative for systems crystalline in one dimension and liquid in two dimensions, the D_h phase is typical for systems that are crystalline in two dimensions and fluid in one dimension. We analyze in this chapter the consequences of this new structure on the elastic and hydrodynamic properties of the phase. We also describe its defects, which can be localized, such as the dislocations, or much more extended, like the developable domains (Y. Bouligand and M. Kléman, 1980), representing the counterpart of the focal domains in smectics. Finally, chapter C.X is dedicated to the study of the free growth of the D_h phase from its isotropic liquid. In this chapter, we treat the problems related to morphological transitions, to dendrite selection and to growth in the kinetic regime.

Looking back over this second volume, we must admit that our endeavour has only encompassed a small part of the current knowledge on liquid crystals; they represent a vast area of condensed matter, still very lively and rich in new discoveries, as indicated by the large number of books recently published on this topic, complementing the great classics already cited in the first volume and recalled as references [1–4]. Among the most recent, one must cite the works by S. Lagerwall [5] and I. Musevic, R. Blinc and B. Žekš [6], giving a very thorough treatment of the very active domain of ferro- and antiferro-electric liquid crystals. Let us also mention a collective volume edited by H.-S. Kitzerow and Ch. Bahr [7] in honor of Professor G. Heppke, presenting in a very pedagogical manner a wealth of recent information on chiral mesophases. Finally, we should not forget an older, but still relevant book, by P.S. Pershan [8], where the reader will find very useful complements on the structure and physical properties of smectic liquid crystals, as well as a collection of the best articles on this topic.

We also regret that we were not able to do more justice to liquid crystal chemistry. In particular, we could not insist on the subtle relations between the anatomy of the molecules and their mesogenic properties. Nevertheless, it is by juggling with the molecular parameters that the chemists discovered all the new mesophases described in this volume. The case of DOBAMBC, discussed in chapter C.IV, or that of disk-like molecules, presented in chapter C.IX., are two typical examples. There are still many others, such as certain diamides that assemble by hydrogen bonds and form long mesogenic filaments, or the phasmid- or banana-shaped molecules that order into very distinctive columnar or lamellar mesophases. To mitigate these shortcomings, we recommend to the reader two works by specialists in this field. The first one, by G.W. Gray and J.W. Goodby [9] contains, among other things, a beautiful collection of texture photos allowing the identification of the main mesophases. The second one was recently edited by D. Demus, J.W. Goodby, G.W. Gray, H.W. Spiess and B. Vill [10].

The translation of this second volume into English is again the work of Doru Constantin to whom we address especially warm thanks for the accomplishment of this hard task.

Our thanks also go to all those who helped us to complete this second volume. First of all, we would like to thank Jacques Friedel, who took the time to read the entire manuscript and who honored us by writing the foreword to the first volume. We also address our thanks to John Bechhoefer, Frédéric Picano and Yves Bouligivand for their careful reading of some chapters and for their (always very useful) comments. We had the help of Jeanne Crassous, who drew the diagrams of several mesogenic molecules, and of Jacques Malthête, who reread the French version of the manuscript very carefully and corrected countless chemistry and many errors that had found their way into the text.

The final proof was made by Prof. G.W. Gray, Editor of the Taylor & Francis series on Liquid Crystals. We are grateful to him for numerous helpful corrections.

We also thank warmly Ms. S.M. Wood who helped us prepare the PDF printer-ready final files of this book.

Finally, we have a thought for all those, too numerous to be cited here by name, with whom we had the pleasure to collaborate. May they find here the expression of our deepest gratitude.

BIBLIOGRAPHY

[1] de Gennes P.-G., Prost J., *The Physics of Liquid Crystals*, Oxford University Press, Oxford, 1993.

[2] Chandrasekhar S., *Liquid Crystals*, Cambridge University Press, Cambridge, 1992.

[3] Kléman M., *Points, Lines and Walls in Liquid Crystals, Magnetic Systems, and Various Ordered Media*, John Wiley & Sons, Chichester, 1983.

[4] Chaikin P.M., Lubensky T.C., *Principles of Condensed Matter Physics*, Cambridge University Press, Cambridge, 1995.

[5] Lagerwall S.T., *Ferroelectric and Antiferroelectric Liquid Crystals*, Wiley-VCH Verlag, Weinheim, 1999.

[6] Musevic I., Blinc R., Žekš B., *The Physics of Ferroelectric and Antiferroelectric Liquid Crystals*, World Scientific, Singapore, 2000.

[7] *Chirality in Liquid Crystals*, Eds. Kitzerow H.-S. and Bahr Ch., Springer-Verlag, Heidelberg, 2001.

[8] Pershan P.S., *Structure of Liquid Crystal Phases*, World Scientific, Singapore, 1988.

[9] Gray G.W., Goodby J.W., *Smectic Liquid Crystals. Textures and structures*, Leonard Hill, Glasgow, 1984.

[10] *Handbook of Liquid Crystals*, Eds. Demus D., Goodby J.W., Gray G.W., Spiess H.W., and Vill B., Wiley-VCH Verlag, Weinheim, 1998.

Part C

SMECTIC AND COLUMNAR
LIQUID CRYSTALS

Chapter C.I

Structure of the smectic A phase and the transition toward the nematic phase

In the previous volume, we recalled the existence of a wide variety of smectic phases, the simplest example being the smectic A. This phase is only distinguished from the uniaxial nematic phase by the presence of positional order along the direction of average molecular alignment (still denoted by the director **n**). This lamellar structuring confers remarkable properties upon the smectic A phase, the most apparent being the elastic behavior exhibited when the layers are compressed or dilated. This novel property is however not the only one, as we shall see throughout this chapter, mostly dedicated to the smectic A-nematic phase transition. It will be shown that, despite its apparent simplicity, this transition poses a multitude of questions, some of them still waiting for a satisfying solution.

The chapter is structured as follows. After having presented some of the experimental evidence for the lamellar structure of this phase (section I.1), we perform a first simplified analysis of the nematic-smectic A phase transition (section I.2). It will be shown that, in agreement with the experimental data, this transition can be first or second order. We then perform a more "rigorous" analysis of the transition using the de Gennes formalism (section I.3) and showing the close analogy between this transition and the one between the normal and superconducting states of a metal at low temperature. In particular, this description predicts the existence of twisted smectic phases (called TGB) that we shall discuss in detail in chapter C.V. Some pretransitional effects observed in the nematic phase close to the transition will then be described (section I.4), as well as the fact that director fluctuations can change the order of the transition (Halperin, Lubensky and Ma theory, section I.5). Finally, we touch upon the Helfrich theory, according to which the transition is induced by defect proliferation (section I.6).

I.1   Lamellar structure of the smectic A phase

Of all smectic phases, the smectic A (Fig. C.I.1) is the simplest one. Its point symmetry is the same as in the nematic phase, the molecules being perpendicular to the plane of the layers and free to turn around the z axis, parallel to the director $\mathbf{n}_0$ and perpendicular to the layers. The smectic A phase is hence optically uniaxial. However, **the translation symmetry is broken** along z, while the molecules are free to move inside each layer, as in a two-dimensional liquid.

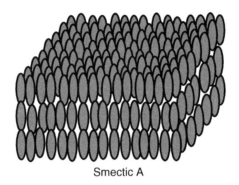

Smectic A

Fig. C.I.1 Schematic depiction of the smectic A phase: the molecules, perpendicular to the layers, move within each layer as in a liquid. They can also jump from one layer to the next.

The lamellar order of the smectic A phase has several macroscopic effects, some of them known since the beginning of the 20th century (Friedel and Grandjean, [1–3]).

For instance, a millimetre-sized free drop of smectic A, placed on a glass slide treated to ensure homeotropic anchoring, does not adopt a hemispherical shape, as in the case of an isotropic liquid or a nematic, but rather a "terrace" structure (Fig. C.I.2) (Grandjean [2]). As for very small droplets (a few tens of micrometers in diameter), they tend to become facetted at the top, as seen in figure C.I.3 [4a–f].

On the other hand, smectics confined between slide and cover slip form "focal conics" textures, easily recognizable using the polarizing microscope. Each time they are observed, these defect patterns, described by Friedel [3], and more recently by Bouligand [5], clearly hint of a lamellar structure (Fig. C.I.4) (for more details, see chapter C.III dedicated to the defects).

Finally, smectics have a creamy consistency, distinct from the much more fluid nematic phase.

To this day, these three "Friedel indices" are used quickly to identify a smectic phase. X-ray analysis is then needed to determine its type (A, B, C, etc.) and to measure its structural parameters.

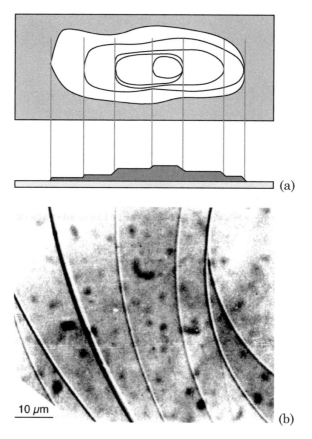

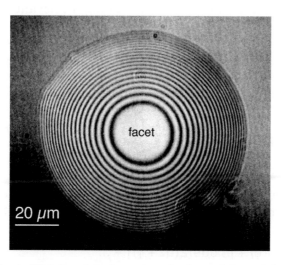

Fig. C.I.2 a) "Terrace" structure of a smectic drop; b) observation of the terraces in unpolarized light (photograph by C. Williams, 1976).

Fig. C.I.3 Observation in Michelson interferometry of a facetted drop of the liquid crystal 8CB placed on a glass slide treated for homeotropic anchoring (photograph by J. Bechhoefer, 1991).

13

Fig. C.I.4 "Focal conic" texture, characteristic of the smectic A phase.

We shall see however in chapter C.VIII that, even without using X-rays, one can determine the layer thickness and obtain some information on the inter- and intramolecular order, for instance by measuring the optical reflectivity of a free-standing smectic film.

I.2 The smectic A-nematic transition: a simplified approach

In the first volume, we successfully treated the nematic-isotropic phase transition using the Landau formalism (section I.2 of chapter B.I, volume 1). We shall see that this approach is also useful here.

As always, the initial problem is finding a relevant order parameter, which describes the symmetry change at the transition. In the present case, the only **broken symmetry** is translation symmetry. Indeed, while the nematic phase is invariant with respect to any translation **t**, including continuous translations along the director $\mathbf{n}_o$,

$$\mathbf{t} = t\,\mathbf{n}_o \qquad (t,\ \text{real}) \qquad\qquad (C.I.1)$$

the smectic phase, due to its lamellar structure, is only invariant with respect to discrete translations

$$\mathbf{t} = l\,a_o\,\mathbf{n}_o \qquad (l,\ \text{integer}) \qquad\qquad (C.I.2)$$

where a_o is the thickness of the smectic layers. As such, the center of mass density ρ in the smectic phase must exhibit along the normal to the layers z a periodical variation of the type:

$$\rho = \rho_o + \rho_1 \cos(qz + \phi) + \rho_2 \cos[2(qz + \phi)] + \dots \qquad\qquad (C.I.3)$$

with ρ_o the average density of the medium, $q = 2\pi/a_o$ the wave vector of the modulation and ϕ an arbitrary phase. In the nematic phase, the parameters ρ_1, ρ_2,... are zero. On the other hand, we know that the energy must be translationally invariant, and consequently independent of the phase ϕ. Thus, the energy can only depend on ρ_1, ρ_2,... that we can choose as **order parameters**. As a general rule, the modulation in center of mass density is fairly different from the electronic density modulation measured using X-rays. More precisely, the X-ray scattering experiments of D.J. Tweet et al. [6a] show that, if the electronic density profiles are always close to a sinusoid (yielding only one Bragg peak), this is not the case for the density profiles of the centers of mass (or mass density). In general, one must then retain several harmonics in order to describe the latter correctly [6b].

The situation is considerably simpler when approaching a second-order (or weakly first-order) phase transition, in this case the mass density modulation being close to a sinusoid. For this reason, one can consider only the dominant ρ_1 term and designate it as the **scalar order parameter** that describes the smectic A-nematic transition.

The next step is to expand the free energy as a power series in ρ_1. Clearly, only even powers are allowed, since changing ρ_1 to $-\rho_1$ is equivalent to changing the phase by π or, equivalently, to an $a_0/2$ translation of the smectic sample along z, which cannot change its energy. The most general development is then:

$$f_L = \frac{1}{2} A \rho_1^2 + \frac{1}{4} B \rho_1^4 + \frac{1}{6} C \rho_1^6 + \dots \qquad \text{(C.I.4)}$$

where, in a mean field theory, A can be written under the simple form:

$$A = A_0 (T - T^*) \qquad \text{(C.I.5)}$$

with T* corresponding to the spinodal temperature of the nematic phase.

The order of the transition depends on the sign of B. It is immediately obvious that:

1. If B is positive, the transition is second order ($T_c = T^*$), with:

$$\rho_1 = \sqrt{\frac{A_0(T^* - T)}{B}} \qquad \text{(C.I.6)}$$

close to the transition.

2. If B is negative, on the other hand, the transition is first order. In this case, one must consider in the expansion C.I.4 the sixth-order saturation term, which must be positive to have a minimum for a ρ_1 value different from zero.

As opposed to the nematic-isotropic transition, which is always first order due to the presence of a third order term in the expansion of the free energy, **the nematic-smectic transition can be first or second order**.

This conclusion remains valid when taking into account the next terms in ρ_2, ρ_3, in the Fourier series expansion of the density (eq. C.I.3). To prove this, let us consider the influence of the ρ_2 term. In this case, the most general expansion is:

$$f_L = \frac{1}{2} A \rho_1^2 + \frac{1}{4} B \rho_1^4 + \ldots - \gamma \rho_1^2 \rho_2 + \frac{1}{\kappa} \rho_2^2 + \ldots \tag{C.I.7}$$

In this expression, the $\rho_2 \rho_1^2$ term is not forbidden as changing the phase by π does not change ρ_2 (while ρ_1 is changed to $-\rho_1$). On the other hand, one must choose $\gamma > 0$ and $\kappa > 0$ so that the energy has a minimum in ρ_2:

$$\rho_2 = \frac{\kappa \gamma}{2} \rho_1^2 \tag{C.I.8}$$

Substituting in eq. C.I.7 yields a new expression for the energy:

$$f_L = \frac{1}{2} A \rho_1^2 + \frac{1}{4} B' \rho_1^4 + \ldots \tag{C.I.9a}$$

with

$$B' = B - \kappa \gamma^2 < B \tag{C.I.9b}$$

Thus, there is no qualitative change with respect to the expansion C.I.4, the main effect of the coupling with the first harmonic ρ_2 being a slight decrease of the constant B. This coupling can only be relevant for small values of B, when it can lead to a change in the order of the transition.

However, this coupling is not the only one possible. Indeed, the quadrupolar order parameter S of the nematic phase varies at the N-SmA transition. This variation δS of the quadrupolar order parameter:

$$\delta S = S - S_{Nem} \tag{C.I.10}$$

is also a distinctive feature of the transition, although its role is less important than that of ρ_1, as it does not directly reflect the symmetry breaking. As for ρ_2, however, it can be taken as a secondary order parameter and the energy can be expanded in a power series of ρ_1 and δS. It can be easily shown that the most general expansion is, as before, of the type:

$$f_L = \frac{1}{2} A \rho_1^2 + \frac{1}{4} B \rho_1^4 + \ldots - \gamma \rho_1^2 \delta S + \frac{1}{\kappa} \delta S^2 + \ldots \tag{C.I.11}$$

Note that the linear term in δS is allowed, since positive and negative values of δS are not equivalent. On the other hand, experiment (birefringence and NMR measurements) shows that the director fluctuates less in the smectic A phase than in the nematic. Let us then take γ and κ positive, so that, at the energy minimum,

$$\delta S = \frac{\kappa \gamma}{2} \rho_1^2 \tag{C.I.12}$$

is positive. Replacing in eq. C.I.11 yields:

$$f_L = \frac{1}{2} A \rho_1^2 + \frac{1}{4} B' \rho_1^4 + \ldots \tag{C.I.13a}$$

with

$$B' = B - \kappa \gamma^2 < B \tag{C.I.13b}$$

This coupling (as the coupling with ρ_2) can lead to a first-order transition if it is strong enough. A priori, this could arise when the temperature range of the nematic phase is narrow as, in this case, δS and hence the product $\kappa\gamma$, must be large. B can thus go down to negative values, driving the transition first order.

Let us now compare these predictions with the experimental results. To this effect, we show in figure C.I.5 the phase diagram of the 8CB-10CB mixture and, in figure C.I.6, the latent heat measurements at the smectic A-nematic transition as a function of the molar concentration in 10CB, x [7]. One can see that the latent heat goes to zero – or, more precisely, is no longer detectable using a classical micro-calorimeter, for $x < x^* \approx 0.33$. The concentration x^* defines in the phase diagram the position of the apparent tricritical point (TCP).

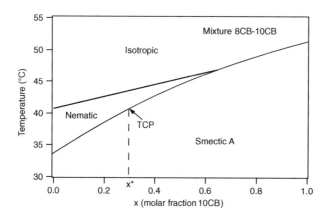

Fig. C.I.5 Phase diagram of the 8CB-10CB mixture. The apparent tricritical point is denoted by TCP (from ref. [7]).

These measurements confirm the model presented by demonstrating the existence of concentration values where the transition is first order, and others for which it appears to be second order.

Furthermore, the transition becomes first order when the temperature range of the nematic phase is sufficiently small, reinforcing the hypothesis that the coupling with the quadrupolar order parameter is driving the transition first order at high 10CB concentration.

In the following, we shall denote by B the coefficient of the fourth-order term in the expansion, after renormalization by the coupling with the order parameter δS and the harmonics $\rho_2, \rho_4, \ldots$.

To test the limitations of the model we proposed, let us try to go further in the analysis of the transition close to the tricritical point (we will return to this problem at the end of the chapter). At this point, the coefficient B must go to zero, it being positive for $x < x^*$, and negative as soon as $x > x^*$; close to x^* one can then write:

$$B = \bar{B} \, (x^* - x) \quad \text{with} \quad \bar{B} > 0 \qquad \text{(C.I.14)}$$

It is then interesting to determine the latent heat (or the entropy jump ΔS) as a function of x. Since $S = -\partial f_L / \partial T$, one has:

$$\Delta S = \frac{1}{2} A_o \, \rho_1^2(T_c) \qquad \text{(C.I.15)}$$

where T_c is the transition temperature. One can easily check that at this temperature in the smectic phase:

$$\rho_1^2(T_c) = \frac{3}{4} \frac{\bar{B}}{C} (x - x^*) \qquad \text{(C.I.16)}$$

whence, by replacing in eq. C.I.15:

$$\Delta S = \frac{3}{8} \frac{A_o \bar{B}}{C} (x - x^*) \qquad \text{(C.I.17)}$$

This formula shows that the latent heat must increase linearly with the concentration above the TCP. The prediction is not verified experimentally, at least close to the TCP, where it should be the most accurate (Fig. C.I.6).

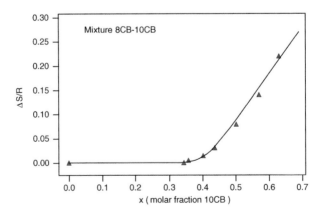

Fig. C.I.6 Latent heat for the 8CB-10CB mixture as a function of the molar concentration in 10CB (from ref. [7]).

This finding prompts us completely to revise the theoretical approach to the transition. In particular, we shall see in section I.5 that the coupling with the director fluctuations, which we ignored so far, must be taken into account for a correct interpretation of the latent heat data close to the TCP.

I.3 De Gennes theory of the smectic A-nematic transition: analogy with superconductivity

I.3.a) Complex order parameter

Before writing down the free energy, let us consider again the choice of the order parameter. We saw that a modulation of the mass density $\rho(\mathbf{r})$ develops along $\mathbf{n}_o$ (parallel to z). On the other hand, we know that close to the transition toward the nematic phase, the (mass or electron) density can be simply written as:

$$\rho(\mathbf{r}) = \rho_o + \text{Re}\{\rho_1 \exp(i\mathbf{q}_o.\mathbf{r} + i\phi)\} \qquad (C.I.18)$$

the higher-order harmonics being negligible. In this expression, ρ_o is the average density, ρ_1 the density modulation, ϕ an arbitrary phase and $\mathbf{q}_o$ the wave vector of the modulation:

$$\mathbf{q}_o = \frac{2\pi}{a_o} \mathbf{n}_o \qquad (C.I.19)$$

In the previous section, we chose the scalar ρ_1 as order parameter. This choice resulted in neglecting two essential properties of the system:

– first, the modulation amplitude can vary over large length scales (with respect to the thickness of the layers). To account for this variation one must consider ρ_1 as a function of the space coordinates $\rho_1(\mathbf{r})$;

– the second effect to be considered is the displacement of the layers (for instance, they can undulate as a result of thermal fluctuations or of an imposed deformation). An easy way of taking this feature into account is to assume that the phase ϕ also depends on the space coordinates: $\phi(\mathbf{r})$.

We can then write:

$$\rho(\mathbf{r}) = \rho_o + \text{Re}\{\rho_1(\mathbf{r}) \exp[i\phi(\mathbf{r})] \exp[i\mathbf{q}_o.\mathbf{r}]\} \qquad (C.I.20)$$

and take as order parameter the **complex amplitude**:

$$\Psi(\mathbf{r}) = \rho_1(\mathbf{r}) \exp[i\phi(\mathbf{r})] \qquad (C.I.21)$$

In this expression, the function $\rho_1(\mathbf{r})$ represents the real amplitude of the complex order parameter and $\phi(\mathbf{r})$ its phase.

We emphasize that $\Psi(\mathbf{r})$, and hence $\rho_1(\mathbf{r})$ and $\phi(\mathbf{r})$, are slowly varying functions of $\mathbf{r}$ (i.e., they vary over distances much larger than the layer thickness a_o).

I.3.b) Physical meaning of the phase of the order parameter and the "Meissner effect" in smectics

As the function Ψ becomes non-zero in the smectic phase, a density modulation $\rho(\mathbf{r})$ develops, with a wave vector $\mathbf{q}_0$. The amplitude $|\Psi| = |\rho_1|$ of the Ψ function sets the "depth" (the amplitude) of the modulation, while the phase ϕ gives the position of this space modulation.

Indeed, a translation $\delta\mathbf{r}$ of the lamellar structure results in a phase variation

$$\delta\phi = -\mathbf{q}_0.\delta\mathbf{r} \tag{C.I.22}$$

as shown in figure C.I.7, and vice versa. More generally, a variation in ϕ along z gives a compression or dilation of the layers.

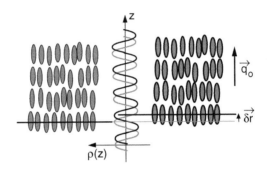

Fig. C.I.7 The translation of the lamellar structure by $\delta\mathbf{r}$ is reflected by a $-\mathbf{q}_0.\delta\mathbf{r}$ variation in the phase of the order parameter.

On the other hand, a variation of the phase ϕ in the (x,y) plane describes a layer undulation, as shown in figure C.I.8. In this case, the vector field of the layer normal shows a "splay" deformation.

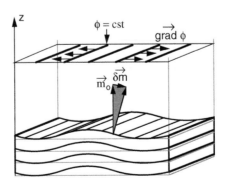

Fig. C.I.8 Relation between the deformation of the lamellar structure, the phase ϕ of the order parameter and its gradient.

Let us now show that if the molecules (described by their average orientation **n**) remain perpendicular to the layers, which we consider of constant thickness, then one must have **curl n** = 0.

Let **m** be the normal to the layers (a unit vector). In an ideal structure ϕ = cst and **m** = **m**$_o$, with **m**$_o$ // **q**$_o$. In the presence of a phase gradient $\phi(x,y)$ that preserves the layer thickness (to first order in the deformation), the smectic layers are tilted, and **m** varies by

$$\delta\mathbf{m} = \mathbf{m} - \mathbf{m}_o = \frac{a_o}{2\pi}\mathbf{grad}\,\phi \qquad\qquad (C.I.23)$$

Assume that the director n remains perpendicular to the layers and let $\delta\mathbf{n}(\mathbf{r})$ be the deviation of **n** with respect to the unperturbed structure:

$$\delta\mathbf{n}(\mathbf{r}) = \mathbf{n}(\mathbf{r}) - \mathbf{n}_o \qquad\qquad (C.I.24)$$

It ensues that:

$$\delta\mathbf{n}(\mathbf{r}) = \delta\mathbf{m}(\mathbf{r}) \qquad\qquad (C.I.25)$$

yielding, from eq. C.I.23:

$$\mathbf{curl}\,\delta\mathbf{n} = 0 \qquad\qquad (C.I.26)$$

In conclusion, the bend and twist deformations of the director field $\vec{n}$ are forbidden in a smectic if the director remains perpendicular to the layers and if these are incompressible.

This is a first similarity with superconductors, which expel the magnetic field **B** = **curl A** (Meissner effect) in the same way that the twist and bend deformations are forbidden in a smectic. In a certain way, the quantities T=**curl**δ**n** and δ**n** play the role of the magnetic field **B** and of the vector potential **A** from which it is derivated.

In the next two sections, we shall further discuss this analogy and draw its consequences.

I.3.c) Landau-Ginzburg-de Gennes free energy

As usual, let us expand the free energy as a power series of the order parameter Ψ. It is important to note that Ψ is complex, while the energy is a real quantity. On the other hand, the energy is necessarily invariant with respect to translations, and thus independent of the phase ϕ of the order parameter. The energy must only depend on the product $\Psi\Psi^* = |\Psi|^2$ yielding, to lowest order:

$$f_L = \frac{1}{2}\alpha\,|\Psi|^2 + \frac{1}{4}\beta\,|\Psi|^4 + \dots \qquad\qquad (C.I.27)$$

In this expression, β takes into account for the coupling with the quadrupolar order parameter discussed in the previous section. Let us also write (in the mean field approximation):

$$\alpha = \alpha_o (T - T^*), \quad \alpha_o > 0 \tag{C.I.28}$$

The formula does not take into account the possible space variations of the order parameter. This is done by including the Ginzburg terms depending on the gradient of the order parameter.

Consider for now that the director remains aligned along the z direction. These terms are:

$$f_G = \frac{1}{2M_{//}} \left| \frac{\partial \Psi}{\partial z} \right|^2 + \frac{1}{2M_\perp} \left[\left| \frac{\partial \Psi}{\partial x} \right|^2 + \left| \frac{\partial \Psi}{\partial y} \right|^2 \right] \tag{C.I.29}$$

where the two stiffness constants $M_{//}$ and $M_\perp$ are different, a priori, because of the uniaxial anisotropy of the medium.

The director **n** does in fact fluctuate, so that a rigorous treatment must include the gradients of the order parameter along **n** and in the perpendicular direction. To second order, only the $M_\perp$ term is changed, because

$$\mathbf{grad}_{//} = \mathbf{\nabla}_{//} \tag{C.I.30a}$$

and

$$\mathbf{grad}_\perp = \mathbf{\nabla}_\perp - iq_o\delta\mathbf{n}_\perp \tag{C.I.30b}$$

where $\delta\mathbf{n}_\perp \approx \delta\mathbf{n} = \mathbf{n} - \mathbf{z}$ (this vector is perpendicular to **n**, as $|\mathbf{n}| = 1$). The correct expression of the Ginzburg terms, including the director fluctuations is as follows:

$$f_G = \frac{1}{2M_{//}} \left| \frac{\partial \Psi}{\partial z} \right|^2 + \frac{1}{2M_\perp} \left| (\mathbf{\nabla}_\perp - iq_o\delta\mathbf{n}_\perp)\Psi \right|^2 \tag{C.I.31}$$

Finally, the complete formula for the free energy f_{NA} must contain the Frank free energy f_F. By putting together all these contributions, one has [8]:

$$f_{NA} = \frac{1}{2}\alpha |\Psi|^2 + \frac{1}{4}\beta|\Psi|^4 + \dots + \frac{1}{2M_{//}} \left| \frac{\partial \Psi}{\partial z} \right|^2 + \frac{1}{2M_\perp} \left| (\mathbf{\nabla}_\perp - iq_o\delta\mathbf{n}_\perp)\Psi \right|^2$$

$$+ \frac{1}{2}\tilde{K}_1 (\mathrm{div}\, \delta\mathbf{n}_\perp)^2 + \frac{1}{2}\tilde{K}_2 (\mathbf{z}.\mathrm{curl}\, \delta\mathbf{n}_\perp)^2 + \frac{1}{2}\tilde{K}_3 \left(\frac{\partial}{\partial z}\delta\mathbf{n}_\perp \right)^2 \tag{C.I.32}$$

This formula can be simplified somewhat by operating an anisotropic scale change $(z \rightarrow \sqrt{M_\perp/M_{//}}\, z)$ to yield:

$$f_{NA} = \frac{1}{2}\alpha |\Psi|^2 + \frac{1}{4}\beta |\Psi|^4 + \dots + \frac{1}{2M_\perp} \left| (\mathbf{\nabla} - iq_o\delta\mathbf{n}_\perp)\Psi \right|^2$$

$$+ \frac{1}{2}\tilde{K}_1' (\mathrm{div}\, \delta\mathbf{n}_\perp)^2 + \frac{1}{2}\tilde{K}_2' (\mathbf{z}.\mathrm{curl}\, \delta\mathbf{n}_\perp)^2 + \frac{1}{2}\tilde{K}_3' \left(\frac{\partial}{\partial z}\delta\mathbf{n}_\perp \right)^2 \tag{C.I.33}$$

Written this way, the free energy bears a strong resemblance to the Landau-Ginzburg functional f_{SC} describing the transition between the normal and the superconducting states:

$$f_{SC} = \frac{1}{2}\alpha \,|\Psi|^2 + \frac{1}{4}\beta\,|\Psi|^4 + \dots + \frac{1}{2m}\,|(\hbar\nabla - iq\mathbf{A})\,\Psi|^2 + \frac{B^2}{2\mu_o} \qquad\qquad \text{(C.I.34)}$$

In this formula, Ψ is the "superconductive" order parameter, $\mathbf{A}$ the vector potential associated to the magnetic field $\mathbf{B} = \mathbf{curl\ A}$, m and q the mass and the charge of the superconductive particles (Cooper pairs) and h the Planck constant (with $\hbar = h/2\pi$). Note however that the K_1 term has no counterpart in superconductors.

The two formulae lead to the following equivalence [8] (see appendix 1):

Smectics	Superconductors
Smectic Ψ	Superconductive Ψ
$\delta n_\perp$	Vector potential $\mathbf{A}$
$\mathbf{T} = \mathbf{curl}\ \delta n_\perp$	Magnetic field $\mathbf{B} = \mathbf{curl\ A}$
Stiffnesses M	Mass m of the superconducting pairs
Frank constants $\tilde{K}_i$	Reciprocal magnetic permeability $1/\mu_o$
Ginzburg energy terms	Kinetic energy
Frank energy	Magnetic energy

This analogy shows that a smectic phase must expel the bend and twist deformations of the director field (for which $\mathbf{T} \neq 0$), as a superconductor expels the magnetic field $\mathbf{B}$ (Meissner effect). We find once again the result proven in the previous section. It can be also shown that this effect is accompanied by divergence of the Frank constants K_2 and K_3 in the nematic phase at the approach of the smectic phase (section I.5).

On the other hand, one can look in the smectic phase for the counterpart of the coherence length ξ and of the penetration length λ in superconductors. We recall that they are defined by:

$$\xi = \frac{\hbar}{\sqrt{|\alpha|m}} \qquad\qquad \text{(C.I.35a)}$$

and

$$\lambda = \sqrt{\frac{m\beta}{\mu_o q^2 |\alpha|}} \qquad\qquad \text{(C.I.35b)}$$

In a smectic phase, the situation is more complicated due to the anisotropy of the medium and to the presence of the K_1 term, so that one can construct:

    1. Two coherence lengths of the order parameter:

$$\xi_{//,\perp} = \frac{1}{\sqrt{2|\alpha|\, M_{//,\perp}}} \tag{C.I.36a}$$

    2. Four penetration lengths of the quantity $\mathbf{T} = \mathbf{curl}\,\delta\mathbf{n}$ corresponding to the magnetic field $\mathbf{B}$:

$$\lambda_{//,\perp}^{(2,3)} = \sqrt{\frac{\tilde{K}_{2,3}\beta\, M_{//,\perp}}{|\alpha|\, q_0^2}} \tag{C.I.36b}$$

    3. Two additional length scales, which have no counterpart in superconductors, and which are related to the penetration length of a splay deformation (see chapter C.II):

$$\lambda_{//,\perp}^{(1)} = \sqrt{\frac{\tilde{K}_1\beta\, M_{//,\perp}}{|\alpha|\, q_0^2}} \tag{C.I.36c}$$

The physical signification of these lengths is illustrated in figures C.I.9 and C.I.10.

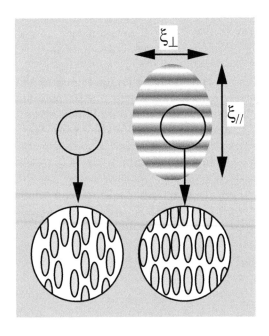

Fig. C.I.9 Cybotactic groups and coherence lengths in the nematic phase.

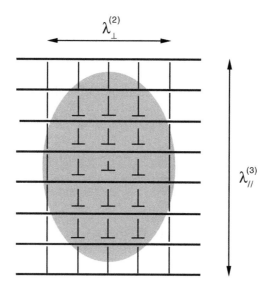

Fig. C.I.10 By locally tilting the director with respect to the layers, one induces a distortion of the smectic phase. This distortion penetrates over a distance $\lambda_{//}^{(3)}$ along the layer normal and over $\lambda_{\perp}^{(2)}$ in the plane of the layers.

The coherence lengths $\xi_{//}$ and $\xi_{\perp}$ determine the size of the cybotactic groups in the nematic phase and the typical distance over which the order parameter varies close to a dislocation core. As to the $\lambda_{//,\perp}^{(2)}$, for instance, they give the penetration length in the nematic phase of a twist deformation.

Once these length scales defined, we can follow de Gennes in a comparison between the two models. To this effect, let us remember a fundamental result from superconductor theory, namely the existence of two types of superconductors:

1. Those with $\lambda\sqrt{2} < \xi$, or "type I" superconductors, in which a perfect Meissner effect is observed below a critical field B_c and normal behavior above this value.

2. Those with $\lambda\sqrt{2} > \xi$, or "type II" superconductors, where two critical values of the magnetic field, B_c^1 and B_c^2, with $(B_c)^2 = B_c^1 B_c^2$, can be defined. For $B_c^1 < B < B_c^2$, the system is penetrated by vortex lines. Below B_c^1, the Meissner effect is perfect, while above B_c^2 behavior becomes normal.

Based on these results, de Gennes conjectured in 1972 that the same kind of behavior must exist in smectics, which should belong in one of two families:

1. Those of "type I," which melt directly to the nematic phase when subjected to a strong enough twist or bend deformation (most smectics, including all those known until 1985 are of this type).

2. Those of "type II," which, instead of directly melting to the nematic phase when subjected to such a deformation, will try to "adapt" to it by creating

edge or screw dislocations similar to the vortices of the superconductors (see appendix 1).

Twenty years had to go by before the first smectic of this type was found [9]. The "forbidden" deformation in this case is the twist, which is not imposed from the outside by mechanical action, but rather from the inside, due to the presence of a non-racemic mixture of chiral molecules. A new phase called "twist-grain boundary" (TGB) appears, similar to the vortex phase in type II superconductors, characterized by the presence of screw dislocations organized in regularly spaced twist walls. The TGB phases will be discussed in detail in chapter C.V.

I.4   Pretransitional effects in the nematic phase

The Landau-Ginzburg-de Gennes theory predicts several pretransitional effects in the nematic phase. They are all related to the presence of smectic fluctuations close to the transition. First, we shall explain why these fluctuations scatter X-rays and show how the associated correlation lengths can be measured. Then we shall analyze their consequences (some of them quite surprising) on the elastic behavior of the nematic phase.

I.4.a) X-ray scattering

W.L. McMillan has the merit of being the first to understand that smectic-type fluctuations developing close to the phase transition must strongly scatter X-rays [10]. He was also the first to measure the intensity of the scattered radiation as a function of the scattering vector $\mathbf{q}$. His results give a satisfactory confirmation of the theoretical predictions we shall now discuss.

Let us first recall that the scattered intensity, or more precisely the structure factor $S(\mathbf{q})$ of a medium with electron density $\rho(\mathbf{r})$ is proportional to the Fourier transform of the density-density correlation function, defined by:

$$C_{\rho\rho}(\mathbf{r}_1, \mathbf{r}_2) = <\rho(\mathbf{r}_1)\rho(\mathbf{r}_2)> \qquad \text{(C.I.37a)}$$

Due to the translation invariance of the nematic phase, this thermodynamic average only depends on $\mathbf{r} = \mathbf{r}_1 - \mathbf{r}_2$, such that

$$C_{\rho\rho}(\mathbf{r}_1, \mathbf{r}_2) = C_{\rho\rho}(\mathbf{r}) = <\rho(\mathbf{0})\rho(\mathbf{r})> \qquad \text{(C.I.37b)}$$

Under these conditions, the structure factor is given by the following simplified formula (see appendix 2):

$$S(q) = \int_V C_{\rho\rho}(\mathbf{r}) e^{-i\mathbf{q}\cdot\mathbf{r}} d^3\mathbf{r} \qquad\qquad (C.I.38)$$

Before computing this integral, let us first try to describe the correlation function $C_{\rho\rho}(\mathbf{r})$ qualitatively, starting from the diagram in figure C.I.9. Indeed, one need only copy this diagram, shift it by $\mathbf{r}$, superpose it to the original and compute the average $<\rho(\mathbf{0})\rho(\mathbf{r})>$. The result is depicted in figure C.I.11.

For an analytical calculation of the correlation function $C_{\rho\rho}(\mathbf{r})$, one must first develop the complex order parameter $\Psi(\mathbf{r})$ over its Fourier components (see appendix 2 of chapter B.II):

$$\Psi(\mathbf{r}) = \frac{1}{V} \sum_{\mathbf{q}} \Psi_{\mathbf{q}} \exp(i\mathbf{q}.\mathbf{r}) \qquad\qquad (C.I.39)$$

Since the modes are statistically independent and since (by definition):

$$\rho(\mathbf{r}) = \rho_o + Re[\Psi(\mathbf{r}) e^{iq_o z}] \qquad\qquad (C.I.40)$$

eq. C.I.37b yields:

$$C_{\rho\rho}(\mathbf{r}) = \rho_o^2 + \frac{1}{2V^2} \sum_{\mathbf{q}} <|\Psi_{\mathbf{q}}|^2> \cos(\mathbf{q}.\mathbf{r} + q_o z) \qquad\qquad (C.I.41a)$$

After replacing the sum by an integral and the cosine function by a sum of integrals, the previous expression becomes:

$$C_{\rho\rho}(\mathbf{r}) = \rho_o^2 + \frac{1}{4V(2\pi)^3} \int <|\Psi_{\mathbf{q}'}|^2> (e^{i(\mathbf{q}'.\mathbf{r}+\mathbf{q}_o.\mathbf{r})} + e^{-i(\mathbf{q}'.\mathbf{r}+\mathbf{q}_o.\mathbf{r})}) d^3\mathbf{q}' \qquad (C.I.41b)$$

with $\mathbf{q}_o = q_o\,\mathbf{z}$.

The structure factor is computed from its definition C.I.38, yielding:

$$S(\mathbf{q}) = (2\pi)^3 \rho_o^2\, \delta(\mathbf{q}) + \frac{1}{4V} \int <|\Psi_{\mathbf{q}'}|^2> [\delta(\mathbf{q}'+\mathbf{q}_o- \mathbf{q}) + \delta(\mathbf{q}'+\mathbf{q}_o+\mathbf{q})] d^3\mathbf{q}'$$

$$\qquad\qquad (C.I.42a)$$

The first term in $\delta(\mathbf{q})$ can be discarded since it cannot be observed experimentally, being completely drowned, at $\mathbf{q} = \mathbf{0}$, in the primary beam transmitted through the sample. Finally, one has:

$$S(\mathbf{q}) \sim <|\Psi_{\mathbf{q}-q_o z}|^2> + <|\Psi_{\mathbf{q}+q_o z}|^2> \qquad\qquad (C.I.42b)$$

up to a constant factor.

The averages $<|\Psi_{\mathbf{q}}|^2>$ are determined using the equipartition theorem and the simplified form C.I.29 of the free energy:

$$\langle|\Psi_{\mathbf{q}}|^2\rangle = \frac{k_B T \Omega}{\alpha + \dfrac{1}{2M_{//}} q_z^2 + \dfrac{1}{2M_\perp}(q_x^2 + q_y^2)} \tag{C.I.43}$$

and finally, using eq. C.I.42 and the definition C.I.36a of the correlation lengths $\xi_{//,\perp}$:

$$S(\mathbf{q}) \sim \frac{k_B T}{\alpha\,[\,1 + \xi_{//}^2\,(q_z - q_o)^2 + \xi_\perp^2\,(q_x^2 + q_y^2)\,]}$$

$$+ \frac{k_B T}{\alpha\,[\,1 + \xi_{//}^2\,(q_z + q_o)^2 + \xi_\perp^2\,(q_x^2 + q_y^2)\,]} \tag{C.I.44}$$

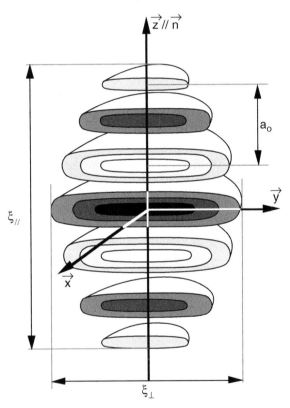

Fig. C.I.11 Sketch of the correlation function $\langle\rho(\mathbf{0})\rho(\mathbf{r})\rangle$ in the nematic phase at the approach of the smectic phase. Only the density fluctuations due to the appearance of the smectic phase are taken into account here.

A perspective view in reciprocal space of the function $S(\mathbf{q})$ is depicted in figure C.I.12. First of all, one notices its revolution symmetry about the **n** axis due to the $D_{\infty h}$ symmetry of the nematic phase.

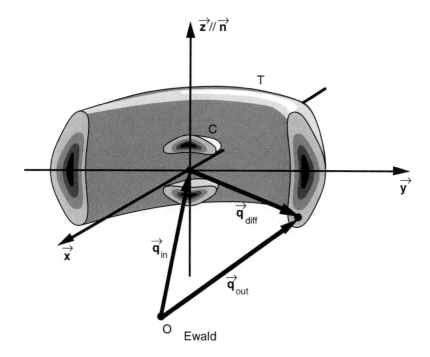

Fig. C.I.12 Perspective view in reciprocal space of a cut through the S(**q**) function, indicating the short-range order in the nematic phase close to the N-SmA transition.

Furthermore, S(**q**) has maxima in certain regions of reciprocal space, namely:

1. The two spherical caps C revealing the existence of smectic correlations around $\mathbf{q}_{\text{diff}} = \pm\, q_o\mathbf{z}$; these caps have a Lorentzian profile given by formula C.I.44. It should also be noticed that $S(q_o\mathbf{z})$ diverges at the transition since $\alpha = 0$ at T = T*. At this temperature, the two caps are replaced by the Bragg peaks of the lamellar structure.

2. The torus T (that we did not compute) signalling the presence of lateral correlations between molecules; these correlations correspond to a short-range order typical of a liquid. The radius of this torus is thus of the order of $2\pi/c_o$, where c_o is the average distance between molecules in the plane normal to the director.

Let us now see how the function S(**q**) can be experimentally determined. In principle, one only needs to perform the two experiments described in figures C.I.13 and 14.

In the setup of figure C.I.13, the incident X-ray beam is parallel to the director. In this case, the photosensitive film records the section of torus T cut by the Ewald sphere, namely a ring of typical radius $2\pi/c_o$. In this geometry, the sphere passes between the caps C of function S(**q**), so they leave no trace on the photographic film.

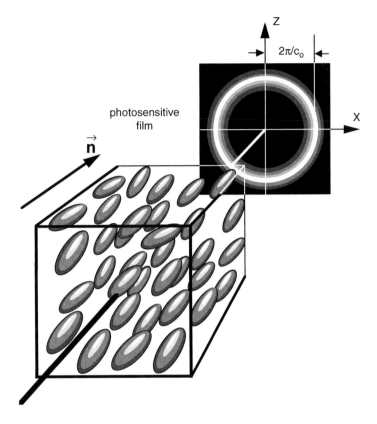

Fig. C.I.13 X-ray scattering when the nematic is oriented along the incident beam. c_o represents the average distance between two molecules in the plane perpendicular to the director.

In order to detect the caps, one must then change the configuration and perform the experiment in figure C.I.14. Now, the incident beam is perpendicular to the director. The cuts of the torus by the Ewald sphere, as well as the two caps appear on the photographic film.

To quantify these measurements, McMillan used a point detector; its position with respect to the incident beam and the sample defines the scattering vector. By changing the orientation of the sample (using a magnetic field) and the position of the detector, McMillan measured $S(\mathbf{q})$ about the position $\mathbf{q} = q_o \mathbf{z}$, along $\mathbf{n}$ and in the plane perpendicular to $\mathbf{n}$. The results of his measurements are shown in figure C.I.15. McMillan's results clearly confirmed the divergence of smectic fluctuations close to the N-SmA transition. However, he found out that the temperature variation of the correlation lengths $\xi_{//}$ and $\xi_{\perp}$ were not compatible with the mean field theory we just presented (which predicts a $(T - T^*)^{-1/2}$ dependence).

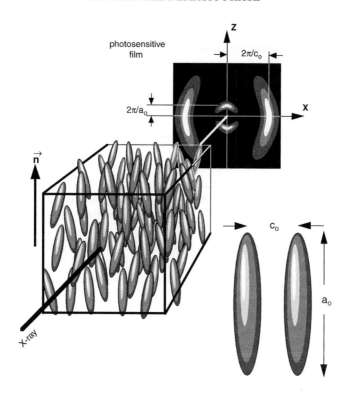

Fig. C.I.14 X-ray scattering when the nematic is perpendicular to the incident beam.

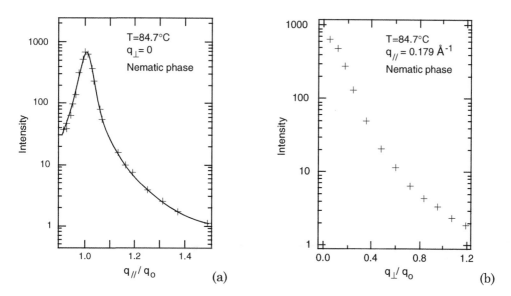

(a)

(b)

Fig. C.I.15 Variation of the scattered intensity close to $q = q_0 z$ along (a) and perpendicular (b) to the director **n**. The material is CBAOB. Solid curves were computed from eq. C.I.44 convolved with the experimental resolution function (from ref. [10]).

This unexpected result triggered a long series of theoretical and experimental studies, which are well beyond the scope of this presentation (see Garland and Nounesis [11] for a review). Finally, let us note that the cybotactic groups are anisotropic, with $\xi_{//} \approx 5\xi_{\perp}$, typically.

I.4.b) Divergence of the elastic constants K_2 and K_3

As we have seen, the lamellar structure of smectics is incompatible with twist and bend deformations of the director field $\mathbf{n}$. Such behavior suggests that the Frank constants K_2 and K_3 of the nematic phase diverge upon approaching the transition.

To prove this result, let us neglect the anisotropy of the medium and assume that the correlation lengths are equal ($\xi_{//} = \xi_{\perp} = \xi$), as are the Frank constants ($K_1 = K_2 = K_3 = K$). In the nematic phase, the Frank free energy is reduced to:

$$f = \frac{1}{2} K (\nabla . \mathbf{n})^2 \qquad \qquad (C.I.45)$$

In this expression, K is the macroscopic bending constant of the nematic. It is different from the bending constant $\tilde{K}$ that the nematic phase would have in the absence of smectic correlations. Indeed, the nematic becomes stiffer as soon as the cybotactic groups appear, as can be seen from the drawing in figure C.I.16.

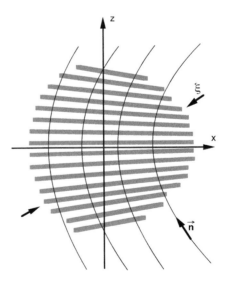

Fig. C.I.16 The appearance of smectic fluctuations in a nematic increases the distortion free energy. For a bend deformation of the director field, the smectic layers (represented as thick gray streaks) cannot remain parallel.

As a matter of fact, the cybotactic groups are elastically deformed when the field lines of the director are distorted. This occurs whenever the smectic layers cannot remain parallel, i.e., when the deformations of the director field are of the twist or bend types (the last situation is shown in figure C.I.16). More energy is then needed to distort the nematic.

Let us determine the order of magnitude of this excess energy using the Landau-Ginzburg-de Gennes theory; one can easily see of eq. C.I.33 that the correction δf to the Frank free energy $1/2.\overline{K}(\mathbf{V}.\mathbf{n})^2$ is:

$$\delta f \approx \frac{q_o^2}{M} |\Psi|^2 \delta \mathbf{n}^2 \qquad (C.I.46)$$

The phenomenon taking place at the scale of a cybotactic group, it is convenient to take $\delta \mathbf{n} = \xi \mathbf{V}.\mathbf{n}$ and to determine the average of $|\Psi|^2$ at this scale. This yields, using eq. C.I.43 and with L the size of the system ($\Omega = L^3$):

$$<|\Psi|^2>_\xi = \frac{1}{\Omega^2} \sum_{q=2\pi/L}^{q=2\pi/\xi} |\Psi_q|^2 \approx \frac{1}{\Omega^2} \int_{2\pi/L}^{2\pi/\xi} \frac{L^3}{(2\pi)^3} \frac{k_B T \Omega}{\alpha(1+\xi^2 q^2)} 4\pi^2 q^2 dq \qquad (C.I.47)$$

or

$$<|\Psi|^2>_\xi \approx \frac{k_B T}{\alpha \xi^3} \qquad (C.I.48)$$

Replacing in eq. C.I.46 yields:

$$\delta f \approx \frac{q_o^2}{M} \frac{k_B T}{\alpha \xi^3} \xi^2 (\mathbf{V}.\mathbf{n})^2 = \frac{q_o^2}{M} \frac{k_B T}{\alpha \xi} (\mathbf{V}.\mathbf{n})^2 \qquad (C.I.49)$$

and, knowing that $1/\alpha = 2M\xi^2$:

$$K = \tilde{K} + \delta K \qquad (C.I.50a)$$

with

$$\delta K \approx q_o^2 k_B T \xi \qquad (C.I.50b)$$

This important formula shows that the correction δK to the Frank constant $\tilde{K}$ diverges with the correlation length at the transition. This divergence is effective for a second-order transition (i.e., for $\beta > 0$). The formula remains however valid for a weakly first-order transition. In this case, the constant K remains finite at the experimentally observed transition temperature T_c (higher than T^*).

This simplified calculation was revisited and extended to the anisotropic case by Jähnig and Brochard [12]. With

$$K_i = \tilde{K}_i + \delta K_i \tag{C.I.51a}$$

these authors find:

$$\delta K_1 = 0 \tag{C.I.51b}$$

$$\delta K_2 = \frac{k_B T q_0^2}{24\pi} \frac{\xi_\perp^2}{\xi_{//}} \tag{C.I.51c}$$

$$\delta K_3 = \frac{k_B T q_0^2}{24\pi} \xi_{//} \tag{C.I.51d}$$

As one would expect, only the K_2 and K_3 constants exhibit critical behavior, K_1 remaining finite at the transition.

Jähnig and Brochard also showed that certain viscosity coefficients also diverge at the transition. For instance, the rotational viscosity γ_1 must behave as

$$\gamma_1 = \tilde{\gamma}_1 + \delta\gamma_1 \tag{C.I.52a}$$

with

$$\delta\gamma_1 \approx \frac{k_B T q_0^2 \tau_m}{16\pi \xi_{//}} \tag{C.I.52b}$$

where τ_m is a microscopic relaxation time varying as $(T - T^*)^{-1}$.

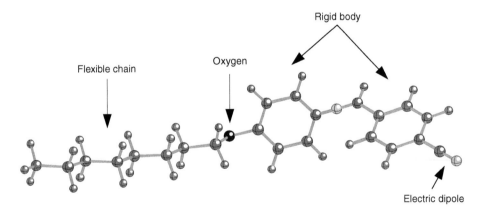

Fig. C.I.17 CBOOA molecule. It is similar to the 8OCB molecule, except for the presence of a different linkage between the two benzene rings. This linkage is $\phi - N = CH - \phi$ instead of $\phi - \phi$ for 8OCB. This molecule exhibits the following phase sequence [13]:

T(°C)	73.2		82.8		107.5	
Crystal	→	Smectic A	→	Nematic	→	Isotropic liquid
ΔS/R	9.1		< 0.02		0.26	

The N-SmA transition is experimentally second order, the latent heat being undetectable.

These critical effects were observed experimentally for the first time by Chung, Meyer and Gruler [14a] in CBOOA (4-cyanobenzylidene-4'-octyloxy-aniline), a substance similar to 8OCB (Fig. C.I.17) and that undergoes a second-order SmA-N phase transition at 82.8°C [13]. These authors measured the bending constants K_1 and K_3 by studying the Frederiks instability in a homeotropic sample in a horizontal magnetic field (Fig. C.I.18). In this geometry, the value of the critical field B_c gives directly K_3/χ_a since

$$\frac{K_3}{\chi_a} = \frac{B_c^2 d^2}{\pi^2 \mu_o} \qquad\qquad (C.I.53)$$

where d is the (known) sample thickness, while a birefringence measurement above the instability threshold yields the K_1/χ_a ratio from the formula:

$$\frac{K_1}{\chi_a} = \frac{B_c^3 d^3}{\pi^2 \lambda} \frac{n_o(n_e^2 - n_o^2)}{n_e^2} \frac{\partial \Delta\psi}{\partial(B^{-1})} \qquad\qquad (C.I.54)$$

with λ the wavelength of the He-Ne laser employed ($\lambda = 6328$ Å), n_o and n_e the ordinary and extraordinary indices of the nematic and $\Delta\psi$ the phase shift between the ordinary and extraordinary rays.

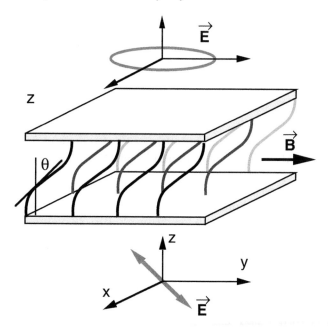

Fig. C.I.18 Frederiks transition in homeotropic geometry. The tilt angle $\theta(z)$ (independent of x and y in the experiment described by Chung et al. [14a]) induced by the magnetic field is determined by measuring the apparent birefringence of the sample. When the incident light is linearly polarized at 45° with respect to the $(\mathbf{B}, \mathbf{z})$ plane, the transmitted light is elliptically polarized, as the ordinary and extraordinary rays acquire a relative phase shift.

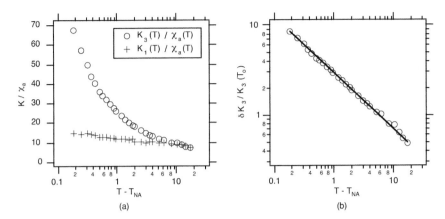

Fig. C.I.19 a) K_1 and K_3 as a function of ΔT in CBOOA; b) Anomalous component of K_3 as a function of temperature in a log-log plot. The slope of the line is –0.635 (from ref. [14a]).

Figure C.I.19a represents the elastic constants measured by Chung et al. as a function of the temperature shift $\Delta T = T - T_{NA}$ [14a]. In agreement with theory, the K_3 coefficient diverges close to the N-SmA transition, unlike K_1.

To determine the critical exponent v of the anomalous part of K_3

$$\delta K_3 \sim (T - T_{NA})^{-v} \qquad (C.I.55)$$

one only needs to subtract from K_3 the "normal background" $\tilde{K}_3$ measured at high temperature and to plot this difference δK_3 against ΔT in log – log representation (Fig. C.I.19b). The result is a straight line of slope

$$v = 0.635 \qquad (C.I.56)$$

It must however be said that this critical exponent is not always the same in all materials, and that it depends on the temperature range of the nematic phase [14b,c].

Finally, the divergence of the K_2 constant leads to the divergence of the cholesteric pitch P at the cholesteric-smectic transition [15]. Indeed, the Frank free energy for a cholesteric is of the form (see eq. B.II.22, volume 1):

$$f_{chol} = k\, \mathbf{n}.\mathbf{curl\,n} + \frac{1}{2} K_2 \,(\mathbf{n}.\mathbf{curl\,n})^2 + \dots \qquad (C.I.57)$$

where k is a constant related to molecular chirality and insensitive to the presence of smectic correlations, unlike K_2. In the absence of an external stress, the cholesteric twist $q = 2\pi/P$ is

$$q = \frac{k}{K_2} \qquad (C.I.58)$$

and it must therefore go to zero in the smectic phase. In other words, the cholesteric pitch P must diverge at the transition like K_2. This predicted effect is easy to observe experimentally [15].

I.4.c) Unexpected pretransitional effects

Two years after the results of Chung et al. were published, Cladis and Torza [16a] also took up the study of the Frederiks transition in CBOOA. Using the same geometry (Fig. C.I.18), they approached the smectic phase even closer in temperature and noticed the formation, above the Frederiks threshold, of a periodic structure with wave vector perpendicular to the field **B** (Fig. C.I.20). The simple theory of the Frederiks transition as used by Chung et al. was no longer valid, as the main assumptions of the calculation, namely:

      1. The director **n(r)** remains in the (**B**, **z**) plane,

      2. The tilt angle $\theta(\mathbf{r})$ only depends on z,

were no longer fulfilled.

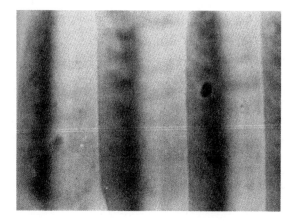

Fig. C.I.20 Periodic structure observed using the polarizing microscope by Cladis and Torza in CBOOA in the presence of a magnetic field and extremely close to the SmA-N transition (from ref. [16a]).

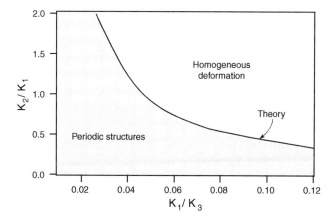

Fig. C.I.21 Theoretical phase diagram showing that for certain values of the elastic anisotropy (below the solid lines), periodic structures are more favorable than homogeneous deformations (from ref. [17]).

The explanation of these stripes was given twelve years later by Allender, Hornreich and Johnson [17], who revised the theory of the Frederiks transition by systematically searching for periodical configurations of lower energy than a homogeneous deformation. The answer to this question is affirmative, in agreement with the observations of Cladis and Torza. Their calculations, summarized in figure C.I.21, show the presence of two separate domains in the parameter plane $(K_2/K_1, K_1/K_3)$:

1. The lower domain (below the solid curve), where the periodical structures are more favorable as soon as the Frederiks threshold is exceeded.

2. The upper domain where deformations are homogeneous (invariant under a translation in the plane of the sample).

Similar effects appear in polymers (see chapter B.II, volume 1), which are much more anisotropic than ordinary liquid crystals due to the elongated shape of their molecules.

Cladis and Torza also worked with nematic samples with "hybrid" boundary conditions (homeotropic on one glass plate and planar on the other). As shown in figure C.I.22, these anchoring conditions impose a bend deformation of the director field, a deformation that is forbidden in the smectic phase. Complications are to be expected. In fact, the experiment shows that the sample does not even need to go into the smectic phase to exhibit "strange" effects since, at the approach of the transition, the nematic locally acquires a very complex deformation. After the transition to the smectic phase, a focal conic texture develops.

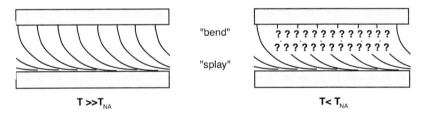

"bend"

"splay"

$T >> T_{NA}$ $T < T_{NA}$

Fig. C.I.22 How does the "forbidden" deformation evolve when approaching the smectic phase?

This experiment shows – once again – that if a "forbidden" distortion is imposed, the system will try to escape it by all possible means, including the formation of complex and defect-containing configurations.

Another example is shown in the photos of figure C.I.23. These spiral configurations are formed by slowly heating, from the smectic A phase, a cholesteric sample contained in a capillary tube with the interior wall treated to ensure homeotropic anchoring. These textures, resulting from the pretransitional effects at the SmA-Cholesteric transition (divergence of the cholesteric pitch and of the elastic constants K_2 and K_3) were theoretically analyzed by Kléman and Lequeux [18a–c].

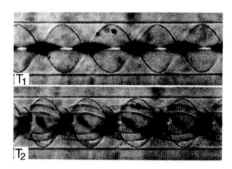

Fig. C.I.23 Mixture of CBOOA with 0.77% of the chiral product CN (cholesteryl nonanoate) contained in a cylindrical capillary with diameter 133 μm, treated for homeotropic anchoring. These complicated helical structures are formed by slowly heating the sample from the smectic phase ($T_2 > T_1$). They are a result of the cholesteric pitch divergence close to the smectic phase (from ref. [16b]).

I.5 Influence of the fluctuations in director orientation on the transition order: the theory of Halperin, Lubensky, and Ma

In 1974, Halperin, Lubensky and Ma showed that the transition between the normal and the superconducting states of a type I material must always be first order [19a,b]. As the N-SmA transition is described by a similar formalism, this conclusion must also apply.

To simplify the derivation, let us **assume that the smectic phase is of type I** (small $\tilde{K}$) and thus that the **orientation fluctuations of the director** (Gaussian in this case) **dominate over the fluctuations of the lamellar order**. Physically, this should be the case close to a tricritical point where $\beta \approx 0$. In this limit, one can consider that $|\Psi|$ is constant so that its space derivatives vanish. The Landau-Ginzburg-de Gennes free energy assumes the following simplified form:

$$f_{NA} = \frac{1}{2}\alpha\,|\Psi|^2 + \frac{1}{4}\beta\,|\Psi|^4 + \frac{1}{2M_\perp}q_0^2(\delta\mathbf{n}_\perp)^2|\Psi|^2 + \frac{1}{2}\,\tilde{K}\,(\nabla.\delta\mathbf{n}_\perp)^2 \qquad \text{(C.I.59)}$$

To account for the fluctuations of $\delta\mathbf{n}_\perp$, Halperin, Lubensky and Ma [19a,b] define the effective Landau energy $f(\psi)$ in the following manner (widely used in statistical mechanics):

$$e^{-f(\Psi)/k_B T} = \int D\{\delta\mathbf{n}_\perp\}\; e^{-f_{NA}(\Psi,\,\delta\mathbf{n}_\perp)/k_B T} \qquad \text{(C.I.60)}$$

where the functional integral is taken over all possible $\delta\mathbf{n}_\perp$ configurations. The derivative of this expression with respect to $|\Psi|$ yields:

$$\frac{df(\Psi)}{d|\Psi|} = \frac{\displaystyle\int \frac{df_{NA}(\Psi)}{d|\Psi|} e^{-f_{NA}(\Psi,\, \delta\mathbf{n}_\perp)/k_B T} D\{\delta\mathbf{n}_\perp\}}{\displaystyle\int e^{-f_{NA}(\Psi,\delta\mathbf{n}_\perp)/k_B T} D\{\delta\mathbf{n}_\perp\}} \tag{C.I.61}$$

whence, by way of eq. C.I.59:

$$\frac{df(\Psi)}{d|\Psi|} = \alpha\,|\Psi| + \beta\,|\Psi|^3 + \frac{1}{M_\perp}q_0^2 <(\delta\mathbf{n}_\perp)^2>|\Psi| \tag{C.I.62}$$

To determine the thermal average $<(\delta\mathbf{n}_\perp)^2>$, it is enough to decompose $\delta\mathbf{n}_\perp$ over the Fourier modes and use the equipartition theorem, assuming that all modes are independent. This yields:

$$(\delta\mathbf{n}_\perp)^2 = \frac{1}{(2\pi)^3\Omega}\int [\delta\mathbf{n}_\perp(\mathbf{q})]^2\, d^3\mathbf{q} \tag{C.I.63}$$

then (from eq. C.I.59)

$$<[\delta\mathbf{n}_\perp(\mathbf{q})]^2> = \frac{k_B T\Omega}{\dfrac{q_0^2|\Psi|^2}{M_\perp} + \tilde{K}q^2} \tag{C.I.64}$$

and finally

$$<(\delta\mathbf{n}_\perp)^2> = \frac{k_B T}{(2\pi)^3}\int_0^{q_{max}} \frac{4\pi q^2 dq}{\dfrac{q_0^2|\Psi|^2}{M_\perp} + \tilde{K}q^2} \approx \frac{k_B T\, q_{max}}{2\pi^2\,\tilde{K}} - \frac{k_B T q_0}{4\pi\, M_\perp^{1/2}\,\tilde{K}^{3/2}}|\Psi| \tag{C.I.65}$$

In this final calculation, the divergent part of the integral was cut off at q_{max} and the other piece integrated from 0 to ∞. It should be noted that the calculation is very similar to the one leading to eq. C.I.43.

Replacing $<(\delta\mathbf{n}_\perp)^2>$ by its expression in eq. C.I.62, one gets the following expression for the Landau free energy:

$$f = \frac{1}{2}\alpha'\,|\Psi|^2 - \frac{1}{3}\tau\,|\Psi|^3 + \frac{1}{4}\beta'\,|\Psi|^4 + \dots \tag{C.I.66a}$$

with

$$\alpha' = \alpha + \frac{k_B T q_0^2 q_{max}}{2\pi^2 \widetilde{K} M_\perp} \tag{C.I.66b}$$

and

$$\tau = \frac{k_B T q_0^3}{4\pi M_\perp^{3/2} \widetilde{K}^{3/2}} \tag{C.I.66c}$$

The important result is that the expansion contains a negative third-order term which, let it be said once more, is not forbidden by the symmetry. It does however break the analyticity of the free energy, but this is not a problem as long as we are dealing with an effective free energy. The main consequence of the cubic term is that the transition **must always be first order**. Another conclusion is that the fluctuations change the transition temperature according to relation C.I.66b. This effect is not negligible, the shift being of the order of $k_B T q_{max}/\widetilde{K} \approx 1\,K$.

Then, Anisimov et al. [20] performed the analysis of the latent heat data of Marynissen et al. [7] close to the TCP (figures C.I.4 and 5) taking into account the cubic term of the HLM theory. At this point, with concentration $x = x^*$, the quartic term goes to zero but the latent heat L^* (or the entropy jump ΔS^*) is finite due to the cubic term. By writing the free energy under the form:

$$f_{NA} = \frac{1}{2}\alpha_o (T - T^*)|\Psi|^2 - \frac{1}{3}\tau|\Psi|^3 + \frac{1}{4}\bar{\beta}(x - x^*)|\Psi|^4 + \frac{1}{6}\gamma|\Psi|^6 + \ldots \tag{C.I.67}$$

it is easily shown that at the tricritical point,

$$\Delta S^* = \frac{1}{2}\alpha_o \left(\frac{\tau}{2\gamma}\right)^{2/3} \tag{C.I.68}$$

and that, close to this point, ΔS is related to ΔS^* and to the concentration difference $x - x^*$, by a relation of the type:

$$\frac{\Delta S}{\Delta S^*} - \sqrt{\frac{\Delta S^*}{\Delta S}} = \frac{3\alpha_o \bar{\beta}}{8\gamma\Delta S^*}(x - x^*) \tag{C.I.69}$$

This relation yields a perfect fit for the experimental microcalorimetry data, as illustrated by the solid line in figure C.I.6. The best fit obtained by adjusting the free parameters x^*, ΔS^* and $3\alpha_o\bar{\beta}/8\gamma$, gives for the 8CB-10CB mixture:

$x^* = 0.424$
$\Delta S^*/R = 0.0261$
$3\alpha_o\bar{\beta}/8\gamma = 0.993$

These results were confirmed by independent measurements of the surface tension at the N-SmA interface [21].

Let us point out that the HLM theory completely neglects the fluctuations of the smectic order parameter (Ψ is assumed constant in the calculation). These fluctuations are however very important, especially since they lead to the divergence of the elastic constants K_2 and K_3. For this reason, Herbut et al. extended the HLM calculation by including the smectic fluctuations [22a]. Their results show that, in certain cases, such as for instance in pure 8CB, the amplitude of the discontinuity at the NA transition is ten times larger than the predictions of the HLM theory. This theoretical prediction is in good agreement with the experimental data [22b, c].

Finally, before concluding this section on the order of the N-SmA transition, let us emphasize that the HLM theory is not applicable in the case of type II smectics, the fluctuations in director orientation being no longer dominant.

Indeed, consider the extreme case of a type II smectic for which $\widetilde{K}$ is very large: in this limit $\delta n = 0$ and the free energy can be written under the simplified de Gennes-McMillan form given in section I.1:

$$f_{NA} = \frac{1}{2} \alpha \, |\Psi|^2 + \frac{1}{4} \beta \, |\Psi|^4 + \frac{1}{6} \gamma \, |\Psi|^6 + \frac{1}{2M_{//}} \left| \frac{\partial \Psi}{\partial z} \right|^2$$

$$+ \frac{1}{2M_\perp} \left[\left| \frac{\partial \Psi}{\partial x} \right|^2 + \left| \frac{\partial \Psi}{\partial y} \right|^2 \right] \dots \tag{C.I.70}$$

where β is a coefficient renormalized by the coupling with the quadrupolar order parameter S. In this case, the transition is second order if $\beta > 0$. To confirm this result, Dasgupta, Halperin, and Bartholomew [23a–c] determined numerically the partition function of a type II smectic by the Monte-Carlo method as the two fields Ψ and δn can no longer be decoupled. They found that the transition can effectively become second order if the smectic is "of strong enough type II character."

I.6   An alternative theory

In 1978, Helfrich [24] suggested that the nematic-smectic transition could be induced by the spontaneous proliferation of dislocation loops (see chapter C.III). This idea was subsequently developed by Nelson and Toner [25a,b], who showed in 1981 that the phase obtained in this way effectively behaved as a nematic to the extent that it exhibits the same director-director correlations and the same elastic constants as a nematic.

This mechanism, similar to the one proposed by Kosterlitz and Thouless for two-dimensional phase transitions [26a,b], is based on the existence of a critical temperature above which the free energy of a dislocation loop becomes negative. One can show it starting from the free energy of a loop:

$$F_{loop} = E L - k_B T \frac{L}{b} \ln(p) \tag{C.I.71}$$

In this expression, E is the line energy, b its Burgers vector, L its length and p a coefficient larger than 1 (in a lattice model, p would be the coordination number of the lattice). The second term is negative and expresses the entropy gained by the system when the loop is created. Clearly, F_{loop} becomes negative above a critical temperature

$$T_{NA} = \frac{E b}{k_B \ln(p)} \tag{C.I.72}$$

corresponding to the smectic-nematic transition temperature. With $E \approx 10^{-6}$ dyn (see chapters C.III and C.VIII), $b = a_0 \approx 3\times10^{-7}$ cm and $k_B = 1.38\times10^{-16}$ erg/K, one has $Eb/k_B \approx 2200$ K! yielding an extremely high transition temperature. If, however, one takes $E \approx 10^{-7}$ dyn, the transition temperature becomes reasonable. The Helfrich scenario can thus be relevant in certain materials.

To date, however, there is no direct evidence of this continuous transition mechanism (the transition would be of order ∞ in the Ehrenfest classification). One way of answering this question was suggested by Holyst and Oswald [27] and consists in measuring the variation in the N-SmA transition temperature (T_{NA}) in free-standing films as a function of their thickness H. Indeed, T_{NA} must increase when the film gets thinner, but following very different laws depending on the type of the transition. Thus, the Helfrich model [28a,b] predicts:

$$\frac{T_{NA}(H) - T_{NA}(\infty)}{T_{NA}(\infty)} \propto \frac{1}{\sqrt{H}} \tag{C.I.73a}$$

while for a first-order phase transition Poniewierski and Sluckin [29] predict:

$$\frac{T_{NA}(H) - T_{NA}(\infty)}{T_{NA}(\infty)} \propto \frac{1}{H} \tag{C.I.73b}$$

Finally, in the case of a second-order phase transition, the prediction is (see chapter C.VIII):

$$\frac{T_{NA}(H) - T_{NA}(\infty)}{T_{NA}(\infty)} \propto \frac{1}{H^{1/\nu}} \tag{C.I.73c}$$

where ν is the critical exponent of the correlation length along the director (equal to $1/2$ in a Landau mean field theory, and to 0.68 for the X-Y model). This model is indeed the most relevant in the case of 8CB, where the N-SmA transition is almost second order ($T_c - T^* \approx 3\text{-}4$ mK, [22a–c]).

BIBLIOGRAPHY

[1] Friedel G., Grandjean F., *Bull. Soc. Fr. Min.*, **33** (1910) 409.

[2] Grandjean F., *Bull. Soc. Fr. Min.*, **39** (1916) 164.

[3] Friedel G., "Mesomorphic states of matter," *Annales de Physique*, **18** (1922) 273. (This consistent work contains a very detailed and very thorough discussion of the properties and the structure of the mesophases known at that time).

[4] a) Bernal J.D., Crowfoot D., *Trans. Faraday Soc.*, **29** (1933) 1032.
b) Chandrasekhar S., *Mol. Cryst.*, **2** (1966) 71.
c) Chandrasekhar S., Madhusudana N.V., *Acta Crystallogr. A*, **26** (1970) 153.
d) Bechhoefer J., Oswald P., *Europhys. Lett.*, **15** (1991) 521.
e) Bechhoefer J., Lejcek L., Oswald P., *J. Phys. II (France)*, **2** (1992) 27.
f) Lejcek L., Bechhoefer J., Oswald P., *J. Phys. II (France)*, **2** (1992) 1511.

[5] Bouligand Y., *J. Physique (France)*, **33** (1972) 525.

[6] a) Tweet D.J., Holyst R., Swanson B.D., Stragier H., Sorensen L.B., *Phys. Rev. Lett.*, **65** (1990) 2157.
b) Holyst R., *Phys. Rev. A*, **44** (1991) 3692.

[7] Marynissen H., Thoen J., van Dael W., *Mol. Cryst. Liq. Cryst.*, **124** (1985) 195.

[8] de Gennes P.G., "On the analogy between superconductors and Smectics A," *Sol. State. Comm.*, **10** (1972) 753.

[9] Goodby J.W., Waugh M.A., Stein S.M., Chin E., Pindak R. and Patel J.S., "Characterization of a new helical smectic liquid crystal," *Nature*, **337** (1989) 449. (The first article evidencing the smectic A* phase analogous to a superconductor of the second kind containing a vortex lattice).

[10] McMillan W.L., *Phys. Rev. A*, **7** (1973) 1419.

[11] Garland C.W., Nounesis G., *Phys. Rev. E*, **49** (1994) 2964.

[12] Jähnig F., Brochard F., *J. Physique (France)*, **35** (1974) 301.

[13] Demus D., "100 Years of Liquid Crystal Chemistry," *Mol. Cryst. Liq. Cryst.*, **165** (1988) 45.

[14] a) Chung L., Meyer R.B., Gruler H., *Phys. Rev. Lett.*, **31** (1973) 349.

b) Mahmood R., Brisbin D., Khan I., Gooden C., Baldwin A., Johnson D.L., Neubert M.E., *Phys. Rev. Lett.*, **54** (1985) 1031.
c) Gooden C., Mahmood R., Brisbin D., Baldwin A., Johnson D.L., *Phys. Rev. Lett.*, **54** (1985) 1035.

[15] Pindak R., Huang C.-C., Ho J.T., *Phys. Rev. Lett.*, **32** (1974) 43.

[16] a) Cladis P.E., Torza S., *J. Appl. Phys.*, **46** (1975) 584.
b) Cladis P.E., White A.E., Brinkman W.F., *J. Physique (France)*, **40** (1979) 325.

[17] Allender D.W., Hornreich R.M., Johnson D.L., *Phys. Rev. Lett.*, **59** (1987) 2645.

[18] a) Kléman M., *J. Physique (France)*, **46** (1985) 1193.
b) Lequeux F., "On some local elastic instabilities in cholesterics," Ph.D. Thesis, University of Paris XI, Orsay, 1988.
c) Lequeux F., Kléman M., *J. Physique (France)*, **49** (1988) 845.

[19] a) Halperin B.I., Lubensky T.C., Ma S.K., *Phys. Rev. Lett.*, **32** (1974) 292.
b) Halperin B.I., Lubensky T.C., *Solid State Comm.*, **14** (1974) 997.

[20] Anisimov M.A., Cladis P.E., Gorodetskii E.E., Huse D.A., Podneks V.E.,Taratuta V.G., van Saarloos W., Voronov V.P., *Phys. Rev. A*, **41** (1990) 6749.

[21] Tamblyn N., Oswald P., Miele A., Bechhoefer J., *Phys. Rev. E*, **51** (1995) 1525.

[22] a) Herbut I.F., Yethiraj A., Bechhoefer J., *Europhys. Lett.*, **53** (2001) 3.
b) Mukhopadhyay R., Yethiraj A., Bechhoefer J., *Phys. Rev. Lett.*, **83** (1999) 4796.
c) Yethiraj A., Bechhoefer J., *Phys. Rev. Lett.*, **84** (2000) 3642.

[23] a) Dasgupta C., Halperin B.I., *Phys. Rev. Lett.*, **47** (1981) 1556.
b) Bartholomew J., *Phys. Rev. B*, **28** (1983) 5378.
c) Dasgupta C., *Phys. Rev. Lett.*, **55** (1985) 1771.

[24] Helfrich W., *J. Physique (France)*, **39** (1978) 1199.

[25] a) Nelson D.R., Toner J., *Phys. Rev. B*, **24** (1981) 363.
b) Toner J., *Phys. Rev. B*, **26** (1982) 462.

[26] a) Kosterlitz J.M., Thouless D.J., *J. Phys. C*, **6** (1973) 1181.
b) Kosterlitz J.M., *J. Phys. C*, **7** (1974) 1046.

[27] Holyst R., Oswald P., *Intern. J. Modern Phys. B*, **9** (1995) 1515.

[28] a) Holyst R., *Phys. Rev. Lett.*, **72** (1994) 4097.
b) Holyst R., *Phys. Rev. B*, **50** (1994) 4097.

[29] Poniewierski A., Sluckin T.J., *Liq. Cryst.*, **2** (1987) 281.

Appendix 1   SUPERCONDUCTING VORTICES AND SCREW DISLOCATIONS

1. Meissner effect: experiment

"Take a piece of tin and cool it down (using liquid helium): at a temperature $T_c = 3.7°K$ an anomaly of the specific heat can be seen. Below T_c, tin is in a new thermodynamic state. What has happened?"

Thus begins de Gennes' book on superconductivity [1].

Today, after the discovery by Bednorz and Müller of high critical temperature (T_c) superconductors, we would rather say:

"Take a piece (a tablet 2 cm in diameter) of YBaCuO ceramic and cool it down using liquid nitrogen. Then take a small permanent magnet and bring it close to the ceramic. You will notice that the magnet and the ceramic strongly repel each other, as if the latter were magnetized. However, by removing the magnet and approaching with a piece of soft iron instead, you will find that the ceramic has no permanent magnetic moment. One must therefore conclude that below a temperature $T_c \approx 70°K$, the ceramic behaves as a perfectly diamagnetic material. What has happened?"

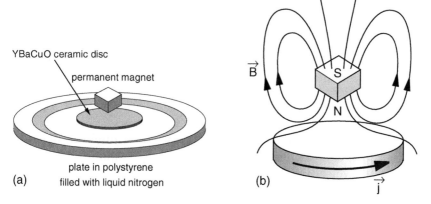

YBaCuO ceramic disc

permanent magnet

plate in polystyrene
filled with liquid nitrogen

(a)

(b)

Fig. A.1.1 Meissner effect; a) magnetic levitation due to the Meissner effect in superconducting YBaCuO ceramic; b) diagram of the magnetic field created by the small magnet and perturbed by the presence of the tablet in its superconducting state. The current **j** induced in the superconductor creates a field that opposes that of the small magnet. The resulting repulsion leads to the levitation of the small magnet.

In a "normal" material, diamagnetism is a result of the induction, under the influence of a variation in magnetic flux, of persistent currents inside the molecules composing the material (see chapter B.II.5, volume 1). These induced currents generate a field opposing the one that created them. In the case of superconductors, diamagnetism is perfect, i.e., as an effect of the induced currents **the magnetic field is entirely expelled from the super-conductor**. This can be written as

B = **0** inside the superconductor. (1.1)

The consequence, as shown in the diagram of figure A.1.1b, is that the magnetic field created by the magnet is perturbed, leading to an additional energy that increases as the magnet approaches the superconductor. The resulting repulsive force leads to the levitation phenomenon.

It should be noted that in superconductors, contrarily to the case of molecular diamagnetism, the permanent currents are not induced inside the molecules, but at the macroscopic scale, meaning that the macroscopic resistance of superconductors must be zero.

2. Meissner effect: theoretical interpretation

The Meissner effect and the presence of macroscopic permanent currents imply the quantum coherence of at least some of the free electrons in the material. The electrons taking part in these currents are all associated in Cooper pairs (for the BCS theory of superconductivity see for instance the book of de Gennes [1], or the one by Martin and Rothen [2]). Since each pair can be considered as a boson, the wave function of the whole system is symmetric with respect to pair exchange. At low temperature, almost all pairs are in the same fundamental quantum state. The amplitude Ψ_0^2 of the corresponding wave function gives the pair density in the fundamental state and plays the role of order parameter for the normal-superconductor transition:

$$\Psi = \Psi_o \exp(i\phi) \qquad (1.2)$$

Formally, this order parameter is the same as in superfluids or in smectic A phases. When the phase ϕ of the wave function is not uniform, depending on the position **r**, one can calculate its gradient and, by analogy with superfluids, associate with it a velocity $\mathbf{v}_\phi$ of expression:

$$\mathbf{v}_\phi = \frac{\hbar}{m} \nabla\phi \qquad (1.3)$$

Is this the velocity of the electron pairs involved in the Meissner effect? The answer is "no" because, as opposed to helium atoms, the BCS electron pairs are charged and, consequently, one must replace in the Schrödinger equation the operator $(\hbar/i).\nabla$ by $(\hbar/i).\nabla - q\mathbf{A}$ [3]. As such, the current $\mathbf{j} = \rho\mathbf{v}$ is not only given by the phase gradient; it also depends on the vector potential **A**:

$$\mathbf{j} = \frac{q}{m}\left[\Psi^*\left(\frac{\hbar}{i}\nabla - q\mathbf{A}\right)\Psi - \Psi\left(\frac{\hbar}{i}\nabla + q\mathbf{A}\right)\Psi^*\right] \qquad (1.4)$$

which can be written as

$$\mathbf{j} = q\,\Psi_0^2\,\frac{\hbar}{m}\left(\nabla\phi - \frac{q}{\hbar}\mathbf{A}\right) \tag{1.5}$$

What is the current distribution when a piece of YBaCuO ceramic is placed in the magnetic field? In the permanent regime, charge conservation requires that:

$$\text{div}\,\mathbf{j} = 0 \tag{1.6}$$

and thus, calculating the divergence of $\mathbf{j}$ from equation 1.5:

$$\Delta\phi - \frac{q}{\hbar}\,\text{div}\,\mathbf{A} = 0 \tag{1.7}$$

In the Coulomb gauge, the $\text{div}\,\mathbf{A}$ term is zero and we are left with:

$$\Delta\phi = 0 \tag{1.8}$$

In a simply connected domain and with the boundary condition $\phi = 0$ at the surface, one finally obtains

$$\phi = 0 \tag{1.9}$$

We then conclude that

$$\mathbf{j} = -\rho_s\,\frac{q^2}{m}\mathbf{A} \qquad \text{where} \qquad \rho_s = \Psi_0^2 \tag{1.10}$$

This equation was proposed for the first time by H. London and F. London in 1935.

To determine the current distribution inside the superconductor, one must use the Maxwell equation

$$\textbf{curl B} = \mu_0\,\mathbf{j} \tag{1.11}$$

which becomes, in Coulomb gauge:

$$\Delta\mathbf{A} = -\mu_0\,\mathbf{j} \tag{1.12}$$

This formula, combined with the London equation, yields:

$$\Delta\mathbf{A} = \lambda^{-2}\,\mathbf{A} \tag{1.13}$$

where λ is a characteristic length, termed "penetration length," defined by:

$$\lambda^{-2} = \rho_s q^2\,\frac{\mu_0}{m} \qquad \text{(with } \mu_0\varepsilon_0 = c^{-2}\text{)} \tag{1.14}$$

In a similar manner, one can find:

$$\Delta\mathbf{j} = \lambda^{-2}\,\mathbf{j} \tag{1.15}$$

whence, after applying the **curl** operator to this equation:

$$\Delta \mathbf{B} = \lambda^{-2}\, \mathbf{B} \tag{1.16}$$

Consider the very simple case of a superconducting slab of thickness 2d exposed to the magnetic field $\mathbf{B} = (0, B_o, 0)$ (Fig. A.1.2).

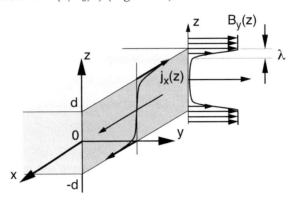

Fig. A.I.2 Superconducting slab of thickness 2d exposed to a magnetic field parallel to its surface. Note that the superconducting current circulates in the immediate proximity of the surface, inside two layers of thickness λ penetrated by the magnetic field. Due to the presence of these currents, the magnetic field is expelled from the inside of the slab.

The slab is perpendicular to the z axis. Owing to the translation invariance in the (x, y) plane, eq. 1.16 simplifies and becomes:

$$\frac{d^2 B_y(z)}{dz^2} = \lambda^{-2}\, B_y(z) \tag{1.17}$$

Integrating it yields the "profile" of the magnetic field inside the superconductor:

$$B_y(z) = B_o \frac{ch(z/\lambda)}{ch(d/\lambda)} \tag{1.18}$$

Knowing the magnetic field distribution one can determine the current $j_x(z)$:

$$j_x(z) = \frac{1}{\mu_o} \frac{dB_y(z)}{dz} \tag{1.19}$$

which gives explicitly:

$$j_x(z) = \frac{B_o}{\mu_o \lambda} \frac{sh(z/\lambda)}{ch(d/\lambda)} \tag{1.20}$$

Note that the superconducting current circulates in the immediate proximity of the surface, inside two layers of thickness λ penetrated by the magnetic field. Due to the presence of these currents, the magnetic field is expelled from the inside of the slab.

3. Superconductors of the first and second kind

If the applied magnetic field is strong enough, the energy cost of totally expelling it from the superconductor material becomes comparable with the cost of suppressing the superconducting state itself. A sort of phase coexistence ensues and the material separates in normal and superconducting domains. Due to the presence of normal domains, the magnetic field is only partially expelled.

Obviously, the residual deformation of the magnetic field is lower when the division into normal and superconducting domains is finer. In superconductors of the first kind, this division can take the form of stripes (Fig. A.1.3a) of greater or lesser width. In the case of superconductors of the second kind, the division reaches the ultimate – quantum – limit, as we shall see later. A 2D triangular lattice of very thin tubes (radius ξ) is then formed, as shown in figure A.1.3b.

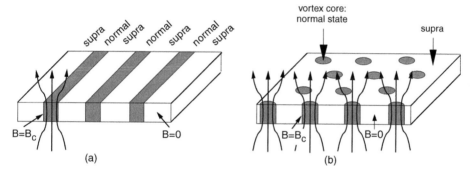

Fig. A.1.3 Coexistence of the superconducting state with the normal state in the presence of a magnetic field; a) in superconductors of the first kind, the normal domains appear as macroscopic stripes; b) in superconductors of the second kind, a lattice of vortices appears.

The choice between these two configurations depends on the energy cost of the interfaces between the normal and the superconducting domains as, at a given volume fraction of one of the phases, finer division implies more interface. It can be shown that the energy cost of the interface is small when the penetration length λ is large compared to the coherence length ξ. Consequently, the relevant parameter for distinguishing the two types of superconductors is the ratio $\kappa = \lambda/\xi$. It can be shown that, for $\kappa > \sqrt{2}$, the superconductor is of the second kind. If, on the contrary, $\kappa < \sqrt{2}$, the superconductor is of the first kind.

4. Structure of a superconducting vortex

In the presence of tube-like normal domains (Fig. A.1.3b), the space taken by the superconducting state is no longer simply connected. Therefore, when looking for the solutions of eq. 1.8, one can consider wave functions Ψ of phase $\phi(\mathbf{r})$ exhibiting singularities similar to those presented in figure A.1.4a).

In general, the integral of the phase ϕ on a closed contour C must be quantized

$$\int_C d\phi = N2\pi \qquad \text{with} \qquad N = 0, \pm 1, \pm 2, \qquad\qquad (1.21)$$

as the wave function must have a well-defined (complex) value in each point of the superconducting state.

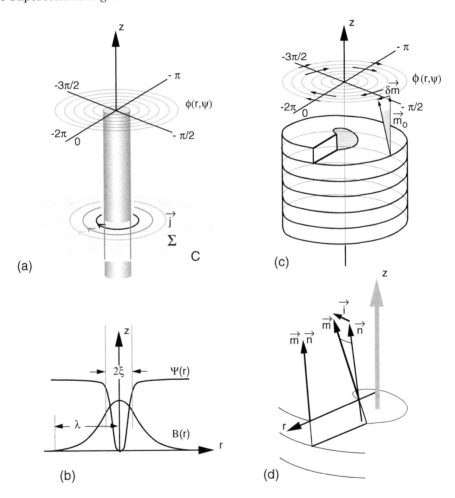

(a)
(b)
(c)
(d)

Fig. A.1.4 Analogy between a superconducting vortex and a screw dislocation in the SmA phase. a) Structure of a superconducting vortex. b) Variation in the order parameter amplitude and in magnetic field at the vortex core. c) Structure of a screw dislocation. d) Close to the dislocation core, for $r < \lambda$, the director **n** is tilted with respect to the layer normal **m**. The difference $\mathbf{i} = \mathbf{m} - \mathbf{n}$ is similar to the permanent current **j**, discussed later.

This condition leads to the quantization of the magnetic flux Φ, defined by:

$$\Phi = \int_{\Sigma} \mathbf{B.dS} = \int_{\Sigma} \mathbf{curl\ A.dS} \tag{1.22}$$

or, from the Stokes theorem:

$$\Phi = \int_{\Sigma} \mathbf{curl\ A \cdot dS} = \int_{C} \mathbf{A.dl} \tag{1.23}$$

On the other hand, eq. 1.5 yields, in the absence of the current $\mathbf{j}$ (i.e., far away from the vortex core):

$$\mathbf{A} = \frac{\hbar}{q}\mathbf{grad}\,\phi(r) \tag{1.24}$$

so that one can write:

$$\Phi = \frac{\hbar}{q}\int_{C} d\phi = N\,\Phi_0 \tag{1.25}$$

In this expression,

$$\Phi_0 = \frac{h}{q} \tag{1.26}$$

is the magnetic flux quantum, also termed **fluxoid**.

Experimental data show that the fluxoid is not equal to h/e, but rather to $h/2e$. This is interpreted in terms of the formation of electron pairs known as "Cooper pairs," of charge 2e. Quantitatively, the magnetic flux quantum is:

$$\Phi_0 = \frac{h}{2e} = 23\,10^{-15}\,\text{T.m}^2 = 23\,10^{-7}\,\text{G.cm}^2 \tag{1.27}$$

This means that, in the presence of a vortex numerical density of the order of $10^6/\text{cm}^2$ (corresponding to an average distance between vortices of the order of 10 µm), the magnetic field is of the order of 0.2 G.

Permanent magnets or electromagnets can very easily produce fields of the order of 2000 G. Under such a field, the vortex density should be 10^4 times larger and the distance between vortices below 0.1 µm. One must then consider the size of the vortex core and the interaction between vortices.

There is an additional – and more basic – reason for considering the structure of the vortex core. Indeed, the vortex carries a magnetic flux, but this flux must be localised in the core, since everywhere else $\mathbf{B} = 0$ as $\mathbf{A}$ is proportional to $\mathbf{grad}\,\phi(r)$. It ensues that the magnetic field rapidly increases on approaching the vortex axis (over a distance of the order of λ), from zero to a maximal value in the center. For a core section of the order of $10^{-10}\,\text{cm}^2$ (corresponding to a diameter of 1000 Å), one finds that a magnetic field of the order of 2000 G is needed to obtain a flux of $2\times10^{-7}\,\text{G.cm}^2$.

This "brutal" space variation of the field shows that $\mathbf{curl\,B} \neq 0$. As a result, from the Maxwell equation $\mathbf{curl\ B} = \mu_0 \mathbf{j}$, a current $\mathbf{j}$ must circulate in the core region.

For a better understanding of the magnetic field and permanent current distribution close to the core, rewrite eq. 1.5 as:

$$\mathbf{j} = q\ \Psi_0^2 \frac{\hbar}{m} \left(\nabla\phi - \frac{q}{\hbar}\mathbf{A} \right) \tag{1.28}$$

Far from the vortex core, where there is no permanent current, the amplitude of the wave function is constant, but the phase turns by 2π on any circle of radius r. The gradient of the phase ϕ is thus:

$$\nabla_\theta \phi = \frac{2\pi}{2\pi r} = \frac{1}{r} \tag{1.29}$$

which, knowing that $\mathbf{j}=0$ and using eq. 1.29, yields:

$$A_\theta = \frac{\hbar}{2e}\nabla_\theta\phi = \frac{\hbar}{2e}\frac{1}{r}, \qquad A_r = A_z = 0 \tag{1.30}$$

On the contrary, in the core region the vector potential is not equal to the gradient of the phase (because of the field $B \neq 0$), a fact that results in a difference between the two contributions and thus in the presence of a finite current $\mathbf{j}$.

5. Structure of a screw dislocation

The structure of a screw dislocation is completely similar to that of a superconducting vortex. Here the smectic phase is replaced by the nematic phase in the core of the dislocation, shaped as a tube with radius ξ. In the center of the tube, at $r = 0$, the director $\mathbf{n}$ is parallel to the z axis. For $0 < r < \xi$, the director is tilted along the orthoradial direction θ and its components in cylindrical coordinates (r, θ, z) are:

$$\mathbf{n} \approx \left(0, \frac{1}{2}Tr,\ 1 \right) \tag{1.31}$$

where $T = \mathbf{n.curl\ n}$ is the local twist. This director field, inside the nematic tube, exhibits all the characteristics of the **double twist** cylinders considered in chapter B.VIII.3 dedicated to Blue Phases. In $r = \xi$, at the interface between the nematic tube and the smectic phase, the orientation of the director is "not yet" completely orthogonal to the smectic layers. A difference i appears between the layer normal $\mathbf{m}$ and the director $\mathbf{n}$. This difference persists in the region $\xi < r < \lambda$ where

$$n = m - i = \frac{a_o}{2\pi} \text{grad}\, \phi - i = \frac{a_o}{2\pi} \left(0, \frac{1}{r}, 1\right) - i \tag{1.32}$$

such that the twist

$$T = n.\text{curl}\, n = -\, n.\text{curl}\, i \tag{1.33}$$

does not abruptly go to zero at the boundary of the smectic phase, decreasing instead progressively in the smectic phase over the penetration length λ. One recognizes that the vector field i plays the same role here as the permanent currents which determine that the magnetic field does not immediately go to zero at the boundary of the normal domain.

BIBLIOGRAPHY

[1] de Gennes P.-G., *Superconductivity of Metals and Alloys*, W.A. Benjamin, New York, 1966.

[2] Martin P.A., Rothen F., *The N-Body Problem and Quantum Fields* (in French), Presses Polytechniques et Universitaires Romandes, Lausanne, 1990.

[3] Gabay Marc, private communication, 1999.

Appendix 2

X-RAY SCATTERING AND THE DETERMINATION OF THE STRUCTURE FACTOR

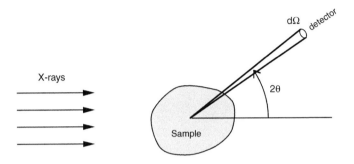

Fig. A.2.1 Typical geometry of an X-ray scattering experiment.

The basic configuration for an X-ray scattering experiment is shown in figure A.2.1. An X-ray beam (plane monochromatic wave with wavelength values between 0.5 and 1.5 Å, usually) is directed at the sample. This wave excites the electrons in the medium which, in turn, emit a scattered wave. If the scattering process takes place at the same wavelength (one speaks of elastic scattering, as the scattered photon has the same energy as the incident one), all the electrons in the medium form a collection of **coherent sources** and the waves they emit will **interfere**. This interference is extremely important and represents the basis of all radiocrystallography. In the following, we shall only be concerned with **coherent scattering**. However, the scattered signal also contains some photons that have undergone inelastic collisions, which have the effect of changing their wavelength λ and thus their frequency $\nu = \omega/2\pi$. For these photons, there is no phase relation, and it is their intensities that must be summed: one speaks of **incoherent** or **Compton scattering**. In practice, one measures the entire scattered intensity, coherent+incoherent. Information on the coherent (and incoherent) fraction of the scattered signal is provided by quantum mechanical calculations. These estimates show that at small scattering angles 2θ (practically, as long as $2(\sin\theta)/\lambda < 0.5 \text{ Å}^{-1}$ for the carbon-based materials considered here), most of the intensity is coherently scattered. As most of the interesting phenomena occur in this region, Compton scattering can be safely ignored.

Let us see how to determine the differential scattering cross section σ_d, which is the experimentally accessible quantity (detected by a photo plate or a CCD detector).

By definition:

$$\sigma_d = \frac{1}{I}\frac{dP_d}{d\Omega} \qquad (2.1)$$

where I is the incident beam intensity (energy transported per unit time and surface) and $dP_d/d\Omega$ the scattered intensity (energy scattered per unit time and solid angle). With this definition, σ_d is expressed in units of surface per solid angle (hence the name). In the language of particle physics,

$$\sigma_d = \frac{\text{number of scattered photons per unit solid angle and unit time in a given direction}}{\text{flux of the incident beam}} \qquad (2.2)$$

In the following, we denote by $\mathbf{k}_o$ the wave vector of the incident wave, by $\mathbf{k}$ the wave vector of the scattered wave and by $\mathbf{q} = \mathbf{k} - \mathbf{k}_o$ the scattering vector. Note that if 2θ is the scattering angle (Fig. A.2.2), then $q = (4\pi/\lambda)\sin\theta$, with λ the wavelength in vacuum ($k = k_o = 2\pi/\lambda$).

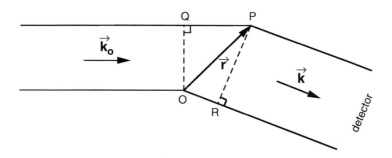

Fig. A.2.2 Determination of the path difference δ in the case of two scatterers.

To calculate the scattered intensity, one must sum the complex amplitudes of all scattered waves. Consider, as a first step, the case of two scatterers placed at the origin O and at point P defined by a position vector $\mathbf{r}$ (Fig. A.2.2). As scattering is coherent, the phase difference $\Delta\phi$ between the two scattered waves as they reach the detector originates in the path difference δ between the two beams, such that:

$$\Delta\phi = \frac{2\pi\delta}{\lambda} \qquad (2.3)$$

where

$$\delta = QP - OR = \mathbf{k}_o.\mathbf{r} - \mathbf{k}.\mathbf{r} = -\mathbf{q}.\mathbf{r} \qquad (2.4)$$

With A_o the amplitude of the incident wave, the (spherical) waves reemitted by the scatterers have as amplitudes $A_o b$ and $A_o b e^{i\Delta\phi}$, respectively. Here b is a length appearing for dimensional reasons, A_o^2 a measure of the power per unit surface in the incident wave, while $A_o^2 b^2$ gives the power of the scattered wave per unit angle. The scattering length b depends on the nature of the scatterer (in this case, the electrons) and on the polarization state of the incident beam. Its determination will be given later for the case of a non-polarized beam. For the time being, consider b known. The detector receives a wave with total amplitude:

$$A(\mathbf{q}) = A_0 b \left(1 + e^{i\Delta\phi} \right) = A_0 b \left(1 + e^{-i\mathbf{q}\cdot\mathbf{r}} \right) \tag{2.5}$$

This calculation is readily extended to the case of N scatterers, yielding:

$$A(\mathbf{q}) = A_0 b \sum_{i=1}^{N} e^{-i\mathbf{q}\cdot\mathbf{r}_i} \tag{2.6}$$

where $\mathbf{r}_i$ gives the position of the ith scatterer with respect to an arbitrary origin. In practice, it is convenient to replace this discrete sum by an integral, introducing the electron density $\rho(\mathbf{r})$ (such that $\rho(\mathbf{r})d^3\mathbf{r}$ gives the number of electrons in the volume element $d^3\mathbf{r}$):

$$A(q) = A_0 b \int_V \rho(\mathbf{r}) \, e^{-i\mathbf{q}\cdot\mathbf{r}} d^3\mathbf{r} \tag{2.7}$$

Written this way, it becomes apparent that the amplitude of the total scattered wave is equal to the Fourier transform of the electron density. One should keep in mind though that this formula is only applicable if each incident photon diffuses at most once in the sample (kinematic theory). More precisely, this amounts to saying that the scattering mean free path of the photons is much larger than the sample thickness. If this condition is not fulfilled, multiple scattering occurs and the calculations become very involved (dynamical X-ray theory). This aspect shall not be dealt with here.

We can now calculate the differential scattering cross section σ_d (in simple and coherent scattering):

$$\sigma_d(\mathbf{q}) = b^2 \int_V \rho(\mathbf{r}) e^{-i\mathbf{q}\cdot\mathbf{r}} d^3\mathbf{r} \int_V \rho(\mathbf{r}') e^{i\mathbf{q}\cdot\mathbf{r}'} d^3\mathbf{r}' \tag{2.8a}$$

or

$$\sigma_d(\mathbf{q}) = \int_V \int_V b^2 \rho(\mathbf{r}) \rho(\mathbf{r}') \, e^{-i\mathbf{q}\cdot(\mathbf{r}'-\mathbf{r})} d^3\mathbf{r} d^3\mathbf{r}' \tag{2.8b}$$

This formula gives the instantaneous scattered intensity. In practice, however, only the time average of this quantity is determined, and this quantity can be replaced by a thermodynamic ensemble average if the system is ergodic. One finally has:

$$\sigma_d(\mathbf{q}) = b^2 \int_V \int_V <\rho(\mathbf{r})\rho(\mathbf{r}')> e^{-i\mathbf{q}\cdot(\mathbf{r}'-\mathbf{r})} d^3\mathbf{r} d^3\mathbf{r}' \tag{2.8c}$$

where <...> designates a thermodynamic average (in the Boltzmann sense).

Consequently, the scattering cross section is proportional to the Fourier transform of the density-density correlation function $C_{\rho\rho}(r, r') = <\rho(r)\rho(r')>$.

This formula simplifies for a translationally invariant system. This is, for instance, the case of an isotropic liquid or of the nematic phase. With $u = r' - r$ and using the fact that $C_{\rho\rho}(r, r') = C_{\rho\rho}(u) = <\rho(0)\rho(u)>$, one obtains:

$$\sigma_d(q) = b^2 V \int_V <\rho(0)\rho(u)> e^{-iq.u} d^3u \qquad (2.9)$$

In this case, the scattered intensity is proportional to the volume V of the sample.

One can notice that, until now, the fact that the incoming beam consisted of (X-ray) photons was never used. This information is in fact contained in the scattering length b. This formula is thus applicable to all kinds of radiation, including neutrons, electrons or even heavy ions. On the other hand, it only holds for simple and coherent scattering, which can pose a problem when it comes to neutrons, and even more so for heavy ions, which interact much more strongly with matter (due to the fact that an X-ray photon with wavelength $\lambda = 1$ Å has a typical energy of about 10^4 eV, while a thermal neutron of the same wavelength has an energy of the order of 0.03 eV, which is comparable to $k_B T$). Regardless of these caveats, the entire information on the structure of the material is contained in the integral that, for this reason, bears the name of **structure factor**:

$$S(q) = \frac{\sigma_d(q)}{b^2 V} \qquad (2.10)$$

where $\sigma_d(q)$ is given by the general formula 2.8c. In the particular case of a translationally invariant system, one simply has:

$$S(q) = \int_V <\rho(0)\rho(r)> e^{-iq.r} d^3r \qquad (2.11)$$

which is none other than the Fourier transform of the correlation function $<\rho(0)\rho(r)>$.

For completeness, let us determine the scattering length b in the case of X-rays. One must calculate the energy scattered by a free electron placed at the origin (Fig. A.2.3). If the incident beam, propagating along the z axis, is not polarized, which is the case for an X-ray tube, one only needs to perform the calculation for the (y, z) plane, the result being independent of the azimuthal angle φ. Under the action of the field E_{ox} of the incident beam, the electron starts to oscillate under the influence of an alternative acceleration with amplitude $E_{ox}e/m$ (e being the electron charge and m its mass).

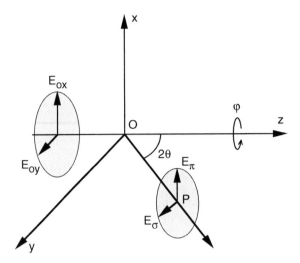

Fig. A.2.3 Scattering of a wave by a free electron.

One knows from classical electromagnetism that an accelerated electron emits electromagnetic radiation which is given at point P by an electrical vector:

$$E_\pi = E_{ox} \frac{e^2}{mc^2} \frac{1}{R}$$

(2.12)

with R the OP distance. In the same way, the E_{oy} component of the incident beam produces at point P the electric field:

$$E_\sigma = E_{oy} \frac{e^2}{mc^2} \frac{1}{R} \cos(2\theta)$$

(2.13)

where 2θ is the scattering angle. Consequently, the average energy flux per unit solid angle and unit time reaching point P is:

$$\left(\frac{dP_d}{d\Omega} \right)_e = (\overline{E}_\pi^2 + \overline{E}_\sigma^2)$$

(2.14)

where the overbar represents a time average. The incident beam being depolarized, its orientation varies randomly with the time, such that:

$$I = E_0^2 = \overline{E}_{oy}^2 + \overline{E}_{ox}^2$$

(2.15a)

with

$$\overline{E}_{oy}^2 = \overline{E}_{ox}^2 = \frac{1}{2} E_0^2$$

(2.15b)

One obtains the Thomson formula, yielding the differential scattering cross section for a free electron in the case of a depolarized incident beam:

$$(\sigma_d)_e = \left(\frac{e^2}{mc^2}\right)^2 \frac{1+\cos^2(2\theta)}{2} \qquad (2.16)$$

This expression contains the classical electron radius $r_e = e^2/mc^2$ (with a value of 2.183×10^{-15} m). Since – by definition – $(\sigma_d)_e = b^2$, one obtains the scattering length b for an electron in a depolarized beam:

$$b = r_e \left(\frac{1 + \cos^2(2\theta)}{2}\right)^{1/2} \qquad (2.17)$$

It should be said that the atomic nuclei also scatter the incident radiation. However, since their mass is much larger than that of the electron, b(nucleus) << b(electron), so that their effective contribution to the overall scattered energy is totally negligible.

A last remark: the Thomson formula contains the entire scattered intensity, elastically as well as inelastically. Eq. 2.17 is thus only valid in coherent scattering at small scattering angles. In this limit, $b \approx r_e$ and b can be taken as a constant. Hence, the experimentally determined scattered intensity is a direct determination (up to a prefactor) of the structure factor defined above in a rigorous manner.

BIBLIOGRAPHY

[1] Guinier A., *X-Ray Diffraction in Crystals, Imperfect Crystals, and Amorphous Bodies*, Freeman, San Francisco, 1963.

[2] James R.W., *The Optical Principles of the Diffraction of X-rays*, Ox Bow Press, Woodbridge, CN, 1982.

[3] Ryong-Joon Roe, *Methods of X-Ray and Neutron Scattering in Polymer Science*, Oxford University Press, New York, 2000.

[4] Chaikin P.M., Lubensky T.C., *Principles of Condensed Matter Physics*, Cambridge University Press, Cambridge, 1995.

Chapter C.II

Continuum theory of smectic A hydrodynamics

The smectic A is the simplest lamellar phase, as the layers are fluid and the molecules are perpendicular to the layers. This phase has the same point symmetry as the usual nematic phase (same rotations and reflections); it is therefore optically uniaxial, with optical axis parallel to the director **n** normal to the layers. On the other hand, its translation symmetry is lower than for a nematic. This symmetry reduction has very important consequences on the nature of the defects one can encounter in this phase (see chapter C.III) and on its rheological behavior. Indeed, the layers can slide one over the other, but will resist compression and dilation. A smectic phase is thus essentially **viscoelastic**, behaving as an usual viscous fluid in the plane of the layers and as an elastic solid under compression or dilation normal to the layers.

To start with, we describe the static properties of a smectic phase (section II.1). This allows us to continue by the study of layer buckling under dilation, and then under a magnetic field (section II.2). Afterward, we generalize the Navier-Stokes equation of ordinary viscous fluids to the smectic case (section II.3) before describing the hydrodynamic modes (first and second sound) (section II.4), as well as some simple flow geometries, with or without obstacles (section II.5). In this chapter, we ignore the numerous topological defects, dislocations or focal conics, that usually proliferate in the samples. Their properties, as well as their role in the very specific rheological behavior of these phases, are treated in the next chapter.

II.1 Static description

II.1.a) Distortion free energy

In this chapter, we take as base state a smectic monodomain consisting of planar and equidistant layers of thickness a_o (Fig. C.II.1a). The z axis is taken along the normal to the layers. Let us begin by searching for the static distortions capable of changing its free energy. As in classical elasticity, we denote by **u** the displacement vector of the molecules.

Clearly, shearing a smectic does not modify its free energy, because the molecules can locally rearrange (at a microscopic scale) in order to remain in the same equilibrium state after deformation. This becomes apparent when the smectic phase is sheared parallel to the layers (Fig. C.II.1b). If the shear is normal to the layers, the molecules must tilt locally about an axis perpendicular to their direction to restore the initial local order (Fig. C.II.1c).

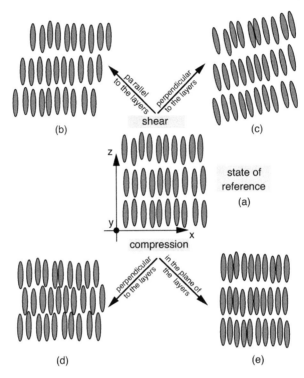

Fig. C.II.1 Different types of possible deformations in a smectic A. After shear, the energy of the smectic phase remains unchanged because the initial local order can be restored due to microscopic rearrangements. In contrast, the average environment of each molecule and the energy of the smectic system change after compression.

The situation is manifestly different when the layer thickness or the density of the medium is changed, in this case the average environment of each molecule being changed after the deformation (Fig. C.II.1d–e). We define ρ_o as the density in the reference state and ρ as its value after deformation. In classical elasticity [1], one shows that

$$\text{div }\mathbf{u} = -\frac{\delta\rho}{\rho_o} \qquad \text{with} \qquad \delta\rho = \rho - \rho_o \qquad\qquad \text{(C.II.1)}$$

In a smectic phase, the density can be changed in two ways:

1. Either by changing the layer thickness, keeping constant the density within the layers (Fig. C.II.1d):

$$\frac{\partial u_z}{\partial z} \neq 0, \qquad \frac{\partial u_x}{\partial x} + \frac{\partial u_y}{\partial y} = 0 \qquad\qquad\text{(C.II.2a)}$$

2. Or by modifying the layer density at constant layer thickness (Fig. C.II.1e):

$$\frac{\partial u_z}{\partial z} = 0, \qquad \frac{\partial u_x}{\partial x} + \frac{\partial u_y}{\partial y} \neq 0 \qquad\qquad\text{(C.II.2b)}$$

One should note that the two quantities $\partial u_z / \partial z$ and $(\partial u_x / \partial x + \partial u_y / \partial y)$ are independent and invariant with respect to the symmetry operations of the smectic phase. The most general expression of the distortion free energy per unit volume is then, to second order in the strain:

$$\rho f_e = \frac{a}{2}\left(\frac{\partial u_x}{\partial x} + \frac{\partial u_y}{\partial y}\right)^2 + \frac{b}{2}\left(\frac{\partial u_z}{\partial z}\right)^2 + c\,\frac{\partial u_z}{\partial z}\left(\frac{\partial u_x}{\partial x} + \frac{\partial u_y}{\partial y}\right) \qquad\text{(C.II.3)}$$

where a, b and c are three elastic constants fulfilling the conditions:

$$a > 0, \quad b > 0 \quad \text{and} \quad ab > c^2 \qquad\qquad\text{(C.II.4)}$$

for stability reasons (the quadratic form must be positive definite).

This choice of variables is not the only one possible; one can use any linearly independent combinations of these quantities. In the following, we shall use as variables:

$$\gamma = \frac{\partial u_z}{\partial z} = \frac{\delta a}{a_o} \qquad\qquad\text{(C.II.5a)}$$

and

$$\theta = \frac{\partial u_x}{\partial x} + \frac{\partial u_y}{\partial y} + \frac{\partial u_z}{\partial z} = \text{div } \mathbf{u} = -\frac{\delta\rho}{\rho_o} \qquad\qquad\text{(C.II.5b)}$$

The first parameter represents the relative variation in layer thickness and the second, the volume dilation. From now on, $u_z = u(x, y, z)$ will represent the vertical **layer** displacement (rather than that of molecules which, as will be shown later, can move between layers by a permeation process).

With this choice of variables, the free energy can be written as [2,3]:

$$\rho f_e = \frac{1}{2} A\,\theta^2 + \frac{1}{2} B\,\gamma^2 + C\,\theta\gamma \qquad\qquad\text{(C.II.6)}$$

where the constants A, B and C also fulfill the stability conditions:

$$A > 0, \quad B > 0 \quad \text{and} \quad AB > C^2 \qquad\qquad\text{(C.II.7)}$$

In fact, the director field **n** is also deformed when the layers are bent, as **n** is normal to the layers, or at least we shall make this assumption throughout this chapter, albeit without giving it a solid foundation. In particular, we shall see in the next section that it is no longer valid close to a SmA-SmC transition or in the regions where the layers are strongly curved (close to a defect core, for instance). The Frank energy associated with deformations of the field **n** is written in the usual form:

$$\rho f_c = \frac{1}{2}\tilde{K}_1\,(\text{div}\mathbf{n})^2 + \frac{1}{2}\tilde{K}_2\,(\mathbf{n}.\text{curl }\mathbf{n})^2 + \frac{1}{2}\tilde{K}_3\,(\mathbf{n}\times\text{curl }\mathbf{n})^2$$

$$+ \frac{1}{2}\tilde{K}_4\,\text{div}\,(\mathbf{n}.\text{div }\mathbf{n} + \mathbf{n}.\text{curl }\mathbf{n}) \qquad\qquad (C.II.8)$$

In this expression, the $\tilde{K}_2$ term disappears as soon as we assume that the director stays along the layer normal. Indeed, one can prove that the existence of a family of surfaces perpendicular to the director imposes a strictly null twist:

n.curl n = 0 $\qquad\qquad$ (C.II.9)

On the contrary, the $\tilde{K}_3$ term cannot be eliminated in general, as it is present each time the layer thickness varies (as a matter of fact, we proved in the previous chapter that **curl n** = 0 only if the layer thickness remains constant). This important observation shows that the $\tilde{K}_3$ term is **inseparable** from a variation in layer thickness (Fig.C.II.2a). In general, it is much more difficult to change the layer thickness than to bend the director field; the $\tilde{K}_3$ term will then be **unobservable**, being much smaller than the layer compressibility term associated with it. We shall therefore neglect it in the following. One should however keep in mind that this term can become important in highly curved areas as, for instance, close to the core of an edge or screw dislocation (see chapter C.III).

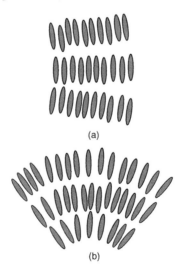

(a)

(b)

Fig. C.II.2 a) A bend deformation of the director field **n** induces necessarily a variation in layer thickness in a smectic phase; b) on the contrary, the layer thickness can remain constant under a splay deformation of the field **n**.

The $\tilde{K}_1$ and $\tilde{K}_4$ terms are completely different in this respect because they can be present even at constant layer thickness (Fig. CII.2b). From a geometrical point of view, the first one is associated with the total curvature of the layers (see chapter B.II, volume 1)

$$\text{div } \mathbf{n} = \frac{1}{R_1} + \frac{1}{R_2} \qquad\qquad (\text{C.II.10})$$

while the second is related to their Gaussian curvature:

$$\text{div } (\mathbf{n}.\text{div } \mathbf{n} + \mathbf{n} \times \mathbf{curl } \mathbf{n}) = \frac{1}{R_1 R_2} \qquad\qquad (\text{C.II.11})$$

Here, R_1 and R_2 are the two principal curvature radii at the considered point. We know that the Gaussian term in div(...) only contributes a surface term to the total free energy of the system. It is generally neglected, as it does not appear in the bulk equilibrium equations. Furthermore, its contribution to the total energy is rigorously zero if the layers remain topologically equivalent to a plane (from the Gauss-Bonnet theorem), which we shall assume in this chapter. This term does however play a role in the energy of focal domains (see chapter C.III on defects), where the layers have the topology of the torus or of the sphere.

From now on, we shall simply write $\tilde{K}_1 = K$ and $\tilde{K}_4 = \bar{K}$ and express the curvature free energy (per unit volume) under the form:

$$\rho f_c = \frac{1}{2} K \left(\frac{1}{R_1} + \frac{1}{R_2} \right)^2 + \bar{K} \frac{1}{R_1 R_2} \qquad\qquad (\text{C.II.12})$$

If the layers are only slightly distorted, one can ignore the Gaussian term completely and express the total curvature of the layers as a function of the displacement u because $n_x = -\partial u/\partial x$ and $n_y = -\partial u/\partial y$ with $n_z = 1$ to first order in the strain:

$$\text{div } \mathbf{n} = \frac{1}{R_1} + \frac{1}{R_2} = \frac{\partial^2 u}{\partial x^2} + \frac{\partial^2 u}{\partial y^2} \qquad\qquad (\text{C.II.13})$$

By putting all the relevant terms together, one obtains the following expression for the distortion free energy:

$$\rho f = \rho f_e + \rho f_c = \frac{1}{2} A \theta^2 + \frac{1}{2} \left(\frac{\partial u}{\partial z} \right)^2 + C \theta \frac{\partial u}{\partial z} + \frac{1}{2} K \left(\frac{\partial^2 u}{\partial x^2} + \frac{\partial^2 u}{\partial y^2} \right)^2 \qquad\qquad (\text{C.II.14})$$

In the static case, two situations must be distinguished:

1. Either the total volume of the smectic phase is fixed, assuming that it is contained in a rigid and closed container. In this rather unrealistic case, $\theta = 0$ and B is the compression modulus of the layers.

2. Or the volume of the smectic can vary freely, the pressure remaining constant inside the sample; this is the case in almost all experiments, when the sample is not sealed on the sides. Then:

$$P - P_{atm} = -\frac{\partial \rho f}{\partial \theta} = -A\,\theta - C\gamma = 0 \qquad (C.II.15)$$

where $\gamma = \partial u/\partial z$ is the layer dilation. This relation yields the equilibrium value of θ as a function of γ:

$$\theta = -\frac{C}{A}\gamma \qquad (C.II.16)$$

and allows further simplification of the free energy, which becomes:

$$\rho f = \frac{1}{2}\overline{B}\left(\frac{\partial u}{\partial z}\right)^2 + \frac{1}{2}K\left(\frac{\partial^2 u}{\partial x^2} + \frac{\partial^2 u}{\partial y^2}\right)^2 \qquad (C.II.17)$$

with

$$\overline{B} = B - \frac{C^2}{A} > 0 \qquad (C.II.18)$$

$\overline{B}$ is the experimentally determined layer compression modulus. In practice, $A \gg C$ and B, so that $\overline{B} \approx B$.

In the static case (and more generally in the dynamical one, when the hydrodynamic velocity is much smaller than the speed of sound), one can ignore the variable θ (given by eq. C.II.16) and use the simplified form C.II.17 of the free energy. Note that here ρf is the energy per unit volume and f the specific free energy (i.e., per unit mass).

II.1.b) Connection with the Landau-Ginzburg-de Gennes theory

The elasticity theory we have just presented is not incompatible with the expression for the free energy that we used in the previous chapter to describe the nematic-smectic transition. To show the relation between these two theories, let us consider a system far enough from the transition: in this case, the absolute value of the order parameter Ψ is constant, and only its phase ϕ (related to the displacement u) can vary. The Landau-Ginzburg-de Gennes free energy can then be written under the simplified form (eq. C.I.32):

$$\rho f = \frac{|\Psi|^2}{2M_{//}}\left(\frac{\partial \phi}{\partial z}\right)^2 + \frac{|\Psi|^2}{2M_{\perp}}|\nabla_{\perp}\phi - q_o\delta\mathbf{n}_{\perp}|^2 + \rho f_c \qquad (C.II.19)$$

where ρf_c is the usual Frank energy. In this expression, the first term corresponds to the layer compressibility term, since

$$\frac{\partial \phi}{\partial z} = - q_o \frac{\partial u}{\partial z}$$ (C.II.20)

from eq. C.I.22 (with $\mathbf{u} = \delta\mathbf{r}$). One can then identify:

$$B = \frac{|\Psi|^2 q_0^2}{M_{/\!/}}$$ (C.II.21)

This important result shows that the B modulus goes to zero at the SmA-N transition when this latter is second order.

As to the second term in eq. C.II.19, it describes a deformation we deliberately ignored so far, representing a director tilt with respect to the layer normal $\mathbf{m}$. Indeed, from eq. C.I.23 we know that

$$\nabla_\perp \phi = q_o \, \delta\mathbf{m}_\perp$$ (C.II.22)

which yields immediately:

$$\nabla_\perp \phi - q_o \delta\mathbf{n}_\perp = q_o \, (\mathbf{m} - \mathbf{n})$$ (C.II.23)

The second term in eq. C.II.19 is then of the type:

$$\frac{1}{2} D \, (\mathbf{n} - \mathbf{m})^2 \qquad \text{with} \qquad D = \frac{|\Psi|^2 q_0^2}{M_\perp} \approx B$$ (C.II.24)

We can now justify our approximation ($\mathbf{n} = \mathbf{m}$). Indeed, the C.II.24 term must be balanced, at equilibrium, by the Frank free energy, yielding:

$$\frac{1}{2} D \, (\mathbf{n} - \mathbf{m})^2 \approx \frac{1}{2} K \, (\nabla.\mathbf{n})^2 \approx \frac{1}{2} \frac{K}{R^2}$$ (C.II.25)

where R is the typical radius of curvature of the director field lines. The difference in angle $\mathbf{n} - \mathbf{m}$ is thus negligible if

$$R \gg (K/D)^{1/2}$$ (C.II.26)

We shall see that $(K/D)^{1/2} \approx a_o$. The preceding inequality is thus almost always satisfied, except very close to a defect core.

II.2 Layer undulation instability

II.2.a) Penetration length λ

The constants K and $\overline{B}$ are in units of energy per unit length and energy per unit volume, respectively. Using these two constants, one can define a length scale λ by:

$$\lambda = \sqrt{\frac{K}{B}} \qquad\qquad\qquad\qquad \text{(C.II.27)}$$

This characteristic length of the material is of the same order of magnitude as the layer thickness a_o, because the layers are not correlated. This length is also related to the penetration distance of a perturbation with wavelength L (Fig. C.II.3) [4]. This can be shown by assuming that the surface layer of a smectic monodomain, placed at $z = 0$, remains in contact with a sinusoid of wave vector $q = 2\pi/L$ and amplitude u_o:

$$u(z = 0) = u_o \cos qx \qquad\qquad\qquad \text{(C.II.28)}$$

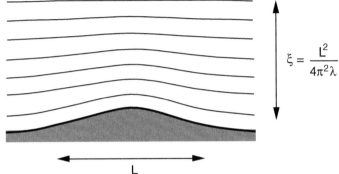

$$\xi = \frac{L^2}{4\pi^2 \lambda}$$

Fig. C.II.3 Penetration length of a modulation with wavelength L.

Minimizing the free energy C.II.17 yields the following equation:

$$K\frac{\partial^4 u}{\partial x^4} - B\frac{\partial^2 u}{\partial z^2} = 0 \qquad\qquad\qquad \text{(C.II.29)}$$

with the obvious solution

$$u = u_o \cos qx \exp\left(-\frac{z}{\xi}\right) \qquad \text{where} \qquad \xi = \frac{1}{\lambda q^2} = \frac{L^2}{4\pi^2\lambda} \qquad \text{(C.II.30)}$$

The length ξ represents the penetration distance of the perturbation. While in a nematic $\xi \approx L$, this distance is much larger in a smectic phase due to the L/λ factor, which is usually much larger than 1.

II.2.b) Anharmonic correction

Before describing the layer undulation instability, one must add a correction to the expression C.II.17 of the smectic free energy. Indeed, the free energy written in this form is not invariant under rotation. This shortcoming must be addressed by including an anharmonic correction beside the layer dilation term $\partial u / \partial z$.

This correction can be determined by considering a deformation of type $u = \beta x$ (Fig. C.II.4). As a result of this displacement, the layer thickness does not change to first order in the strain, since

$$\frac{\partial u}{\partial z} = 0 \qquad\qquad \text{(C.II.31)}$$

It does however incur a second-order variation, by

$$\frac{\delta a}{a_o} = \cos\beta - 1 \approx -\frac{\beta^2}{2} = -\frac{1}{2}\left(\frac{\partial u}{\partial x}\right)^2 \qquad\qquad \text{(C.II.32a)}$$

and more generally by a quantity

$$\frac{\delta a}{a_o} = -\frac{1}{2}\left(\frac{\partial u}{\partial x}\right)^2 - \frac{1}{2}\left(\frac{\partial u}{\partial y}\right)^2 \qquad\qquad \text{(C.II.32b)}$$

for a generic displacement $u(x, y)$.

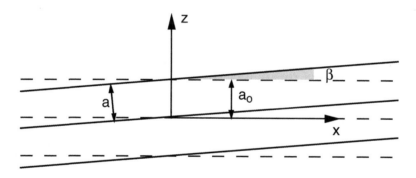

Fig. C.II.4 Imposing the displacement $u = \beta x$ to the layers results in a slight change in thickness.

This thickness change results in an energy increase $(1/2)(\delta a / a_o)^2\, \overline{B}$ that we neglected so far.

To take it into account, one must add the **anharmonic correction** C.II.32b to the layer dilation $\partial u / \partial z$ in the general expression of the free energy. We thus obtain a new expression:

$$\rho f = \frac{1}{2}\overline{B}\left[\frac{\partial u}{\partial z} - \frac{1}{2}\left(\frac{\partial u}{\partial x}\right)^2 - \frac{1}{2}\left(\frac{\partial u}{\partial y}\right)^2\right]^2 + \frac{1}{2}K\left(\frac{\partial^2 u}{\partial x^2} + \frac{\partial^2 u}{\partial y^2}\right)^2 \qquad\qquad \text{(C.II.33)}$$

which has the advantage of being **invariant under rotation** to second order in the deformation, since to this order, $\dfrac{\partial u}{\partial z} - \dfrac{1}{2}\left(\dfrac{\partial u}{\partial x}\right)^2 - \dfrac{1}{2}\left(\dfrac{\partial u}{\partial y}\right)^2 = 0$ for a "solid" rotation.

In the following, we shall use this expression to describe the layer undulation instability.

II.2.c) Undulation instability

It was observed experimentally that a simple layer undulation appears spontaneously when a homeotropic sample of thickness d is slightly (but suddenly) dilated (Fig. C.II.5).

The critical thickness variation δ_c above which this instability appears is of a few tens of Å, explaining why it is very often observed upon rapidly cooling the sample. In this particular case, the dilation is produced by a decrease in layer thickness with temperature. This instability was discovered concurrently in 1973 by Delaye, Ribotta, Durand, and Litster in France [5a–d] and by Clark and Meyer in the United States [6]. It can be detected by simple polarizing microscopy observation (Fig. C.II.5b), but its quantitative study close to the threshold is quite delicate. In particular, the use of piezoelectric ceramics is almost indispensable for imposing (and, most important, for measuring) the required dilations, while a light scattering setup is very useful for detecting the undulation close to the threshold, where it is practically invisible in the microscope (Fig. C.II.6).

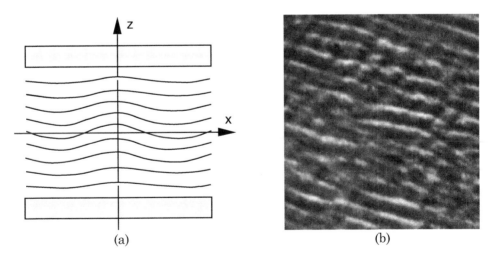

(a)   (b)

Fig. C.II.5 a) Above a critical thickness jump δ_c, the smectic layers start undulating in the center of the sample; b) photograph of this undulation on polarized microscopy, far from the threshold.

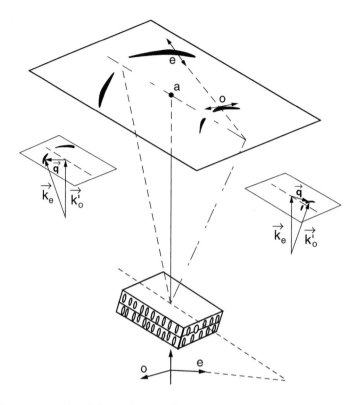

Fig. C.II.6 Geometry employed for evidencing the static layer undulations using light scattering. The scattering pattern appears as two bright crescents, each one corresponding to an eigenvector of the polarization; "a" is the trace of the transmitted beam (from ref. [5b]).

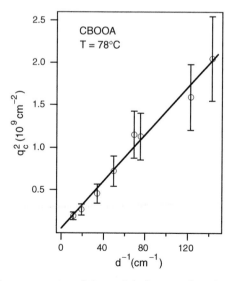

Fig. C.II.7 Variation of the wave vector of the undulation as a function of total sample thickness (from ref. [5a]).

This undulation manifests itself by the sudden appearance on the observation screen of two scattering crescents, symmetric with respect to the plane of incidence. These crescents disappear, usually in a few seconds, even if the dilation is maintained, by a plastic relaxation mechanism to be discussed in chapter C.III. The distance between each diffraction spot and the trace of the incident laser beam gives the wave vector q_c of the undulation. It was shown experimentally that q_c^2 varies linearly with the reciprocal sample thickness (Fig. C.II.7). We shall see that this measurement yields a precise estimate of the penetration length λ. Let us however start by a qualitative explanation of the origin of layer undulations. In figure C.II.8, we have represented two possible mechanisms for relaxing the imposed dilation.

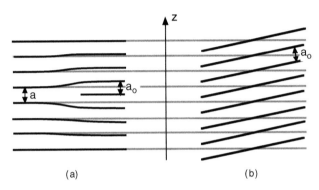

(a)   (b)

Fig. C.II.8 Two possible mechanisms for relaxing the layer dilation. In (a) one layer is inserted, while in (b) the layers are tilted.

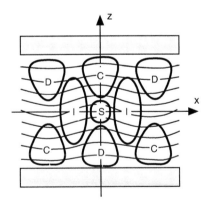

Fig. C.II.9 Sketch of the deformation distribution in the sample: in D and C, the layers are dilated and compressed, respectively; in I, they are tilted, but only slightly deformed (leading to an energy gain); in S, they are curved (which costs energy).

We can see that one must either add new layers, or tilt them if their number is maintained constant. In the experiment we just presented, the dilation is too sudden for the system to be able to adjust the number of layers. (This slow mechanism of plastic relaxation will be described in chapter C.III). The system has thus no other choice but to tilt the layers with respect to the plates, which can only happen in the central region of the sample, the layers being strongly anchored at the surfaces. This fast process (which takes only a few ms, as the layers slide easily over one another) leads to a layer undulation for which we sketched the deformation distribution in figure C.II.9. This diagram shows that the system gains in dilation energy where the layers are tilted (areas denoted by I), but loses in curvature energy in the areas where the layers are strongly curved (denoted by S). This competition drives the instability.

The instability threshold for a simple undulation of wave vector **q** parallel to the x axis can be analytically determined from the Euler equation of layer equilibrium, including the anharmonic terms [7]. Using eq. C.II.33, this yields:

$$K\frac{\partial^4 u}{\partial x^4} - \bar{B}\left[\frac{\partial^2 u}{\partial z^2}\right] + \bar{B}\left[\frac{\partial u}{\partial z}\frac{\partial^2 u}{\partial x^2} + 2\frac{\partial u}{\partial x}\frac{\partial^2 u}{\partial x \partial z} - \frac{3}{2}\left(\frac{\partial u}{\partial x}\right)^2\frac{\partial^2 u}{\partial x^2}\right] = 0 \qquad \text{(C.II.34)}$$

The absolute instability threshold can be found by looking for a deformation of the form

$$u = \alpha z + g(z)\cos qx \qquad \text{with} \qquad \alpha = \frac{\delta}{d} \qquad \text{(C.II.35)}$$

This deformation is the sum of two terms. The first one, αz, describes a homogeneous dilation of the layers; it represents the basic solution to the problem. The second one, unknown, corresponds to a simple layer undulation of wavelength $\Lambda = 2\pi/q$.

After replacing in eq. C.II.35, and only keeping the linear terms, one has:

$$\bar{B}\frac{d^2 g}{dz^2} - Kq^4 g + \bar{B}\alpha q^2 g = 0 \qquad \text{(C.II.36)}$$

Introducing the penetration length λ, this equation for g can be rewritten as:

$$\frac{d^2 g}{dz^2} - (\lambda^2 q^4 - \alpha q^2) g = 0 \qquad \text{(C.II.37)}$$

The strong anchoring conditions on the plates impose

$$g = 0 \quad \text{in} \quad z = 0 \quad \text{and} \quad z = d + \delta \approx d \qquad \text{(C.II.38)}$$

To determine the critical dilation α_c, let us multiply equation C.II.37 by $g' = dg/dz$ and integrate from 0 to d, taking into account the boundary conditions. After integration by parts, we have:

$$\int_0^d g'^2 dz + (\lambda^2 q^4 - \alpha q^2)\int_0^d g^2 dz = 0 \qquad \text{(C.II.39)}$$

yielding

$$\alpha = \lambda^2 q^2 + \frac{1}{q^2} \frac{\displaystyle\int_0^d g'^2 dz}{\displaystyle\int_0^d g^2 dz} \tag{C.II.40}$$

This expression shows that α must necessarily be positive for a solution of the considered type to exist. The instability only occurs under **dilation** normal to the layers (in compression, nothing is gained by tilting the layers). The lowest value of α for which the instability appears can be obtained by minimizing eq. C.II.40 with respect to q^2. The calculation gives the result:

$$q_c = \frac{1}{\lambda} \sqrt{\frac{\displaystyle\int_0^d g'^2 dz}{\displaystyle\int_0^d g^2 dz}} \tag{C.II.41}$$

$$\alpha_c = 2\lambda \sqrt{\frac{\displaystyle\int_0^d g'^2 dz}{\displaystyle\int_0^d g^2 dz}} \tag{C.II.42}$$

Since in the vicinity of the threshold $g = g_o \sin(\pi z / d)$ (we only consider here the first Fourier mode, the most unstable, satisfying the boundary conditions), we find, after replacing in the preceding formulae:

$$q_c^2 = \frac{\pi}{\lambda d} \qquad \text{and} \qquad \alpha_c = 2\pi \frac{\lambda}{d} \tag{C.II.43}$$

q_c **is the critical wave vector of the undulation and** α_c **the critical dilation. The corresponding variation in thickness:**

$$\delta_c = 2\pi\lambda \tag{C.II.44}$$

is thus independent of the sample thickness. This result is well verified experimentally, as well as the q_c dependence with sample thickness (Fig. C.II.7). The measurement of the critical wave vector q_c by light scattering is thus a direct estimate of the penetration length λ. In thermotropic liquid crystals, λ is of the order of the layer thickness (in CBOOA, Delaye et al. [5a] find for instance that $\lambda \approx 22 \pm 3$ Å at 78°C). This also applies in most lyotropic lamellar phases [8, 9], but there are some systems rich in thermodynamical defects (pores or dislocations) where λ is very small, of the order of a few Å [10].

The nature of the bifurcation can be determined by exploring the weakly nonlinear regime. The solution to eq. C.II.34 can be sought for as a series:

$$u = \alpha z + u_o \sin\left(\frac{\pi z}{d}\right) \cos(q_c x) + \sum_{m=2}^{\infty} \sum_{n=0}^{\infty} \left[u_{mn} \sin\left(m \frac{\pi z}{d}\right) \cos(n q_c x) \right] \quad \text{(C.II.45)}$$

After replacing, one obtains [6]:

$$u_o = \frac{8}{3} \lambda \sqrt{\frac{\alpha - \alpha_c}{\alpha_c}} \, , \qquad u_{20} = -\frac{u_o^2}{4\lambda} \qquad \text{(C.II.46)}$$

Close to the threshold, the amplitude varies as $(\alpha - \alpha_c)^{1/2}$. The bifurcation is thus **supercritical** (or normal). It should be noted that this law is only valid very close to the threshold, while $u_o \ll \lambda$ (i.e., as long as $(\alpha - \alpha_c)/\alpha_c < 0.05$). In this regime, the elastic energy per unit sample surface decreases as the square of the distance to the threshold:

$$F = \frac{1}{2} \bar{B} \alpha^2 - \frac{2}{9} \bar{B}(\alpha - \alpha_c)^2 \qquad \text{(C.II.47)}$$

In fact, simple undulation is not the most favorable solution; it is preferable to form a square undulation lattice. This result was shown for the first time by Delrieu [11], in agreement with observations performed for perfectly parallel plates. In the opposite case, experiment shows that a simple undulation develops first, parallel to the wedge, confirming the previous predictions. The calculations are completely analogous in the case of a square lattice. For a deformation field of the form:

$$u = \alpha z + \frac{u_o}{\sqrt{2}} \sin(q_z z) [\cos(qx) + \cos(qy)] \qquad \text{(C.II.48)}$$

it can be shown that the critical dilation α_c and the wave vector at the threshold q_c remain unchanged, being given by eqs. C.II.43. On the other hand, the energy above the threshold decreases faster than for a simple undulation because:

$$F = \frac{1}{2} \bar{B} \alpha^2 - \frac{4}{15} \bar{B} (\alpha - \alpha_c)^2 \qquad \text{(C.II.49)}$$

with

$$u_o = 8 \sqrt{\frac{2}{15}} \lambda \sqrt{\frac{\alpha - \alpha_c}{\alpha_c}} \qquad \text{(C.II.50)}$$

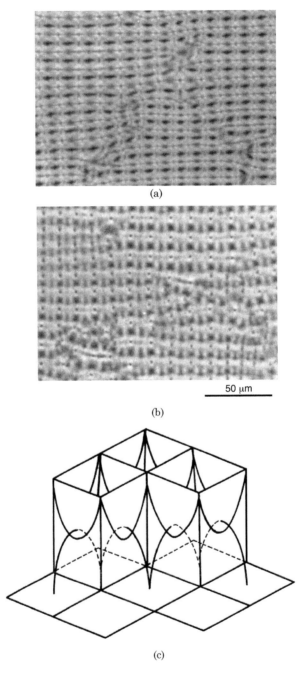

(a)

(b)

50 μm

(c)

Fig. C.II.10 Lattice of focal parabolae in a homeotropic sample (d = 100 μm); a) focus on the center of the sample, where the two parabolae cross (without touching); b) focus on the meeting points of the parabolae; c) schematic representation of the lattice of focal parabolae. Note that in these photos and those of the previous figure, the parabolae form a very regular lattice, as they were obtained by dilating the sample under shear parallel to the layers, which aligns the undulations (from ref. [14]).

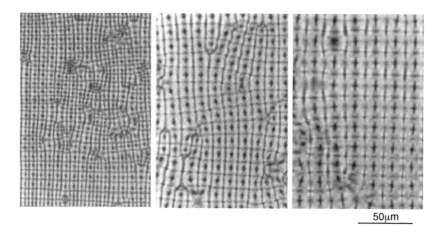

50μm

Fig. C.II.11 Appearance of the lattice of focal parabolae at saturation in samples of varying thickness. From left to right, d = 140, 280 and 420 μm, respectively (from ref. [15]).

Far from the threshold, the other harmonics, as well as the increase of the wavelength must be taken into account. This regime, recently studied by numerical simulations [12], is far from being understood, especially since experimental observation shows that the layers break rapidly when farther away from the threshold, presumably forming dislocation loops (see the next chapter). This mechanism could explain why the applied stress relaxes in two steps, very fast in the beginning and then more slowly, when $\delta > \delta_c$ [10, 13] (see the end of chapter C.III). If the jump in thickness exceeds 1.6 δ_c, layer breaking continues and leads to the appearance of focal parabolae (Fig. C.II.10) [14]. These extended defects, easily seen by polarizing microscopy, form a fairly regular lattice with a parameter depending both on the total sample thickness and on the imposed dilation. More precisely, the lattice parameter increases with the dilation α and saturates when α exceeds a value close to 10^{-3}. At saturation, the size of the focal domains strongly depends on the sample thickness, as illustrated by the photographs in figure C.II.11 [15]. The nucleation mechanisms of these defects are unknown at this time, as are most of their dynamical properties. Their topology and relative arrangement, described in the next chapter, are well understood.

II.2.d) Undulation instability induced by a magnetic field (Helfrich-Hurault)

In principle, it is possible to induce the layer undulation instability by submitting a homeotropic sample to a horizontal magnetic field oriented along the x axis, for instance (Fig. C.II.12a) [16, 17]. Indeed, as in nematics, the molecules tend to orient parallel to the magnetic field if the material exhibits a positive magnetic anisotropy χ_a. Clearly, the layer undulation will decrease the magnetic energy, but it will also create elastic distortions of a high energy cost.

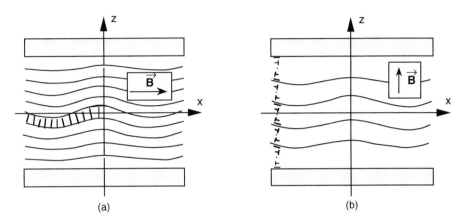

(a)   (b)

Fig. C.II.12 Helfrich-Hurault instability in a smectic (a) and in a cholesteric (b).

This competition leads to the presence of a critical field, which can be determined by minimizing the total free energy of the system, of expression (by neglecting the anharmonic term):

$$F = \int \left[\frac{1}{2} \bar{B} \left(\frac{\partial u}{\partial z} \right)^2 + \frac{1}{2} K \left(\frac{\partial^2 u}{\partial x^2} \right)^2 + \frac{1}{2} \frac{\chi_a}{\mu_o} (\mathbf{B.n})^2 \right] d^3\mathbf{r} \tag{C.II.51}$$

As before, we make the simplifying assumption that the deformation at the threshold is a simple undulation of wave vector $\mathbf{k}$ in the (x,z) plane, of the type

$$u = u_o \sin(k_z z) \sin(k_x x) \tag{C.II.52}$$

with $k_z = \pi z/d$, in order to satisfy the boundary condition of strong homeotropic anchoring. After replacing in eq. C.II.51 and integrating in the thickness of the sample, one gets by averaging in the horizontal plane:

$$<F>_{\text{unit surface}} = Cst + \frac{u_o^2}{8} \bar{B} d \left[\left(\frac{k_z}{k_x} \right)^2 + k_x^2 \lambda^2 \right] - \frac{u_o^2}{8} \frac{\chi_a}{\mu_o} B^2 d \tag{C.II.53}$$

The critical field B_c is the lowest value of B above which the u_o^2 term becomes negative. To calculate it, one must first minimize with respect to k_x the elastic part, always positive. This yields:

$$k_x^2 = \frac{k_z}{\lambda} = \frac{\pi}{\lambda d} \tag{C.II.54a}$$

or equivalently

$$\Lambda = 2\sqrt{\pi\lambda d} \qquad\qquad (C.II.54b)$$

We have obtained the same wavelength Λ as for the classical undulation instability. The critical field can be obtained by putting to zero the u_o^2 term in expression C.II.53:

$$B_c = \sqrt{2\pi \frac{\mu_0}{\chi_a} \frac{\bar{B}\lambda}{d}} \qquad\qquad (C.II.55)$$

A simple numerical estimate shows that this critical field is very high in a smectic, since with $B = 10^8$ erg/cm^3, $\chi_a \approx 10^{-6}$ (in SI units) – note that, although χ_a is dimensionless, χ_a (SI) $= 4\pi\chi_a$ (CGS), the CGS system not being rationalized –, $d = 1$mm, and $\lambda = 20$ Å, one finds $B_c \approx 9$ Tesla. For this reason, the instability was never observed in an ordinary smectic A.

It is, however, very easy to obtain in cholesterics (see Fig. B.VII.59, volume 1) [18], these materials behaving like very soft smectics as long as the sample thickness is much larger than their natural pitch. The geometry of the equivalent experiment is sketched in figure C.II.12b. Here the magnetic field must be applied in the vertical direction in order to disorient the cholesteric layers, because the molecules are parallel to the plates. The critical field can be calculated by noticing that a cholesteric of natural twist $q = 2\pi/p$ behaves, in the low deformation limit, as a smectic with effective susceptibility and elastic modules

$$\chi_a^{eff} \approx \chi_a, \qquad \bar{B}^{eff} \approx K_2 q^2 \qquad \text{and} \qquad K^{eff} \approx K_3 \qquad (C.II.56a)$$

The second relation shows that a change in the pitch increases the twist energy of the cholesteric phase. Note that here $\lambda \approx p$. An exact calculation (to be found in the book by Chandrasekhar [19]) shows that the appropriate values are:

$$\bar{B}^{eff} = K_2 q^2, \qquad K^{eff} = \frac{3}{8}K_3 \qquad \text{and} \qquad \chi_a^{eff} = \frac{\chi_a}{2} \qquad (C.II.56b)$$

for a complete analogy with the smectic A. Using these results and eq. C.II.55, the critical field needed to destabilize the cholesteric field is found as:

$$B_c \,(\text{chol}) = \sqrt{\pi q \frac{\mu_0}{\chi_a} \frac{\sqrt{6K_2 K_3}}{d}} \qquad\qquad (C.II.57)$$

and the wavelength Λ of the undulation:

$$\Lambda \,(\text{chol}) = \left(\frac{3}{2}\frac{K_3}{K_2}\right)^{1/4} \sqrt{pd} \approx \sqrt{pd} \qquad\qquad (C.II.58)$$

These theoretical predictions are in good agreement with the experimental data. Let us just mention that in the photograph of figure B.VII.59 (volume 1) two undulations with perpendicular wave vectors appeared simultaneously, forming a square lattice. We recommend the paper by Delrieu [11] for a detailed discussion of the choice between simple and two-dimensional undulation.

II.3 Equations of smectic hydrodynamics

II.3.a) The conservation equations

Let us begin by establishing the law of mass conservation. To this end, we take the convective (or substantive) derivative of the well-known relation $\theta = \operatorname{div} \mathbf{u}$ (valid to first order in the strain, with $\mathbf{u}$ the displacement of the molecules). Since $\mathbf{v} = D\mathbf{u}/Dt$, one has directly:

$$\frac{D\theta}{Dt} - \operatorname{div} \mathbf{v} = 0 \tag{C.II.59a}$$

This relation is equivalent (to lowest order in the strain) to the well-known mass conservation law

$$\frac{D\rho}{Dt} + \rho \operatorname{div} \mathbf{v} = 0 \tag{C.II.59b}$$

We recall that the convective derivative with respect to time is expressed as (only in Cartesian coordinates):

$$\frac{D}{Dt} = \frac{\partial}{\partial t} + \mathbf{v}.\mathbf{grad} \tag{C.II.60}$$

The second conservation law is the one concerning **momentum**. It is demonstrated as for the nematic phase and can be written as:

$$\rho \frac{D\mathbf{v}}{Dt} = \operatorname{div} \underline{\sigma} + \mathbf{F} \tag{C.II.61}$$

where $\underline{\sigma}$ is the total stress tensor and $\mathbf{F}$ the external bulk forces applied on the system (weight, for instance). The stress tensor is the sum of three terms:

1. The **hydrostatic pressure** $-P\underline{I}$ associated to the bulk dilation θ.
2. The **elastic stress** $\underline{\sigma}^E$ conjugate to the layer displacement u.
3. The **viscous stress** $\underline{\sigma}^v$ coupled to the velocity gradient.

The pressure P and the elastic stress $\underline{\sigma}^E$ can be determined from the bulk free energy ρf, knowing that, for an isothermal transformation:

$$d(\rho f) = -P\, d\theta + \sigma^E_{zi}\, d\frac{\partial u}{\partial x_i} + \operatorname{div}(\ldots) \tag{C.II.62}$$

These three terms represent respectively the work due to the pressure P, to the elastic stress $\underline{\sigma}^E$, and to the surface torque $\underline{C}\boldsymbol{v}$ acting on a surface element of normal vector $\boldsymbol{v}$. This torque is associated with the Frank free energy $(1/2)K_1(\mathrm{div}\,\boldsymbol{n})^2$. On the other hand, considering ρf as an explicit function of variables θ, $\partial u/\partial x_j$ et $\partial^2 u/\partial x_i \partial x_j$:

$$d\rho f = \frac{\partial \rho f}{\partial \theta}\,d\theta + \frac{\partial \rho f}{\partial \dfrac{\partial u}{\partial x_i}}\,d\,\frac{\partial u}{\partial x_i} + \frac{\partial \rho f}{\partial \dfrac{\partial^2 u}{\partial x_i \partial x_j}}\,d\,\frac{\partial^2 u}{\partial x_i \partial x_j}$$

$$= \frac{\partial \rho f}{\partial \theta}\,d\theta + \frac{\partial \rho f}{\partial \dfrac{\partial u}{\partial x_i}}\,d\,\frac{\partial u}{\partial x_i} - \frac{\partial}{\partial x_j}\left(\frac{\partial \rho f}{\partial \dfrac{\partial^2 u}{\partial x_i \partial x_j}}\right)d\,\frac{\partial u}{\partial x_i} + \mathrm{div}\,(\dots) \tag{C.II.63}$$

By identifying with expression C.II.62, we obtain:

$$P = -\frac{\partial \rho f}{\partial \theta} \tag{C.II.64}$$

$$\sigma_{zi}^E = \frac{\partial \rho f}{\partial u_{,i}} - \frac{\partial}{\partial x_j}\frac{\partial \rho f}{\partial u_{,ij}} \tag{C.II.65}$$

where $u_{,i} = \dfrac{\partial u}{\partial x_i}$ and $u_{,ij} = \dfrac{\partial^2 u}{\partial x_i \partial x_j}$.

One can equally show that the two non-vanishing components of tensor $\underline{C}$ are worth:

$$C_{xy} = \frac{\partial \rho f}{\partial u_{,yy}} \qquad\qquad C_{yx} = -\frac{\partial \rho f}{\partial u_{,xx}} \tag{C.II.66}$$

Using the general expression of the free energy C.II.14, these general formulae yield:

$$P = -A\,\theta - C\frac{\partial u}{\partial z} \tag{C.II.67}$$

$$\sigma_{zz}^E = B\frac{\partial u}{\partial z} + C\theta \tag{C.II.68a}$$

$$\sigma_{zx}^E = -K\frac{\partial}{\partial x}(\Delta_\perp u) \tag{C.II.68b}$$

$$\sigma_{zy}^E = -K\frac{\partial}{\partial y}(\Delta_\perp u) \qquad \text{with} \qquad \Delta_\perp = \frac{\partial^2}{\partial x^2} + \frac{\partial^2}{\partial y^2} \tag{C.II.68c}$$

All other components of the elastic stress tensor $\underline{\sigma}^E$ are null.

As for nematics, the stress tensor **is not symmetric**. One can thus associate with it a bulk elastic torque $\Gamma_i^E = - e_{ijk}\sigma_{jk}^E$ of non-vanishing components $\Gamma_x^E = \sigma_{zy}^E$ and $\Gamma_y^E = - \sigma_{zx}^E$. It is then straightforward to check that the equilibrium equation for bulk torques:

$$\operatorname{div}\underline{C} + \Gamma^E = 0 \tag{C.II.69}$$

is identically fulfilled. This should not come as a surprise, since the displacement u and the director **n** are not independent, being related by the geometrical expressions $n_x = - \partial u/\partial x$ and $n_y = - \partial u/\partial y$. The situation would be different if the molecules were free to tilt with respect to the layer normal; we would then have an additional variable, and the extra equation associated with it would be precisely eq. C.II.69 of torque equilibrium [20].

We conclude this section by separating the viscous and elastic terms in the momentum conservation equation:

$$\rho\frac{Dv}{Dt} = - \mathbf{grad}\,P + \operatorname{div}\underline{\sigma}^v + \mathbf{G} \tag{C.II.70}$$

In this expression, div $\underline{\sigma}^v$ is the viscous force, while

$$\mathbf{G} = \operatorname{div}\underline{\sigma}^E \tag{C.II.71}$$

is the elastic force conjugate to the layer displacement u. Oriented along z, it can be expressed to linear order as:

$$\mathbf{G} = (0, 0, G) \quad\text{with}\quad G = -\frac{\delta\rho f}{\delta u} = B\frac{\partial^2 u}{\partial z^2} + C\frac{\partial\theta}{\partial z} - K\Delta_\perp(\Delta_\perp u) \tag{C.II.72}$$

II.3.b) Constitutive laws: irreversible entropy production

Let e be the internal energy per unit **mass** (and not per unit volume, as we do not assume incompressibility). As for nematics, it is easily shown that the energy equation reads:

$$\rho\frac{De}{Dt} = \sigma_{ij}\frac{\partial v_i}{\partial x_j} - \operatorname{div}\mathbf{q} + \operatorname{div}(\dots) \tag{C.II.73}$$

where **q** is the heat flux vector. In this formula the last term, not written down explicitly, corresponds to the work of the surface torque $\underline{C}\mathbf{v}$.

The entropy equation is obtained using the thermodynamic relation:

$$de = T\,ds + dW \tag{C.II.74}$$

where s is the entropy per unit mass. During an elementary reversible transformation, the viscous forces produce no work. We can then write:

$$dW = -\frac{P}{\rho}d\theta + \frac{\sigma_{zi}^E}{\rho}d\frac{\partial u}{\partial x_i} + div\,(...) \tag{C.II.75}$$

where the div (...) term represents the work of the surface torque $\underline{C}\underline{\upsilon}$ acting on a volume of unit mass.

From eqs. C.II.73–75 one immediately obtains:

$$\rho T\frac{Ds}{Dt} = \rho\frac{De}{Dt} + P\frac{D\theta}{Dt} - \sigma_{zi}^E\frac{\partial\dot{u}}{\partial x_i} - div\,(...) \tag{C.II.76}$$

where we used $\dot{u} = \dfrac{Du}{Dt}$.

It should be noted that this relation, calculated along a reversible path, remains valid for a real irreversible transformation, with e and s being state variables. This assumes, however, that the thermodynamic quantities (P, T, e...) remain well defined locally, a condition that is fulfilled if the mean free path of the molecules is very small compared with the typical distances over which these quantities vary. Using the energy equation C.II.73, one gets from eq. C.II.76:

$$\rho T\frac{Ds}{Dt} = \sigma_{ij}^v\frac{\partial v_i}{\partial x_j} + P\left(\frac{D\theta}{Dt} - div\,\mathbf{v}\right) + \sigma_{zi}^E\frac{\partial}{\partial x_i}(v_z - \dot{u}) - div\,\mathbf{q} \tag{C.II.77a}$$

then, using eqs. C.II.59a and C.II.71:

$$\rho T\frac{Ds}{Dt} = \sigma_{ij}^v\frac{\partial v_i}{\partial x_j} + G(\dot{u} - v_z) + div\,[\sigma_{zi}^E(v_z - \dot{u})] - div\,\mathbf{q} \tag{C.II.77b}$$

On the other hand, we know that, according to the second principle of thermodynamics, the entropy production $\rho T\dot{s} = \rho T Ds/Dt + T div(\mathbf{q}/T)$ must be non-negative (see appendix 3 to chapter B.III in volume I). From the previous formula, it can be written as:

$$\rho T\dot{s} = \sigma_{ij}^v\frac{\partial v_i}{\partial x_j} + G(\dot{u} - v_z) + div\,[\sigma_{zi}^E(v_z - \dot{u})] - \frac{\mathbf{q}}{T}\cdot\mathbf{grad}\,T \tag{C.II.78}$$

The first term of this expression should be familiar, as it corresponds to the energy dissipated in the bulk when the smectic phase is sheared. The second one is new and describes **bulk permeation** (flow across the layers), a concept first introduced by Helfrich in relation to cholesterics. The third term is a divergence one. It corresponds to the energy dissipated at the surface by permeation. One can thus separate bulk dissipation from surface dissipation and write:

$$\rho T\dot{s}\,(bulk) = \sigma_{ij}^v\frac{\partial v_i}{\partial x_j} + G\,(\dot{u} - v_z) - \frac{\mathbf{q}}{T}\cdot\mathbf{grad}\,T \tag{C.II.79a}$$

$$\rho T \mathring{s} \text{ (surface)} = - (\dot{u} - v_z) \, \sigma_{zi}^E \, \upsilon_i \qquad \text{(C.II.79b)}$$

In this last equation, υ is the unit vector normal to the surface and pointing out of the smectic volume.

These two equations yield the constitutive laws of the smectic phase. Treating σ_{ij}^v, G, σ_{zi}^E and $-(\mathbf{grad}\,T)/T$ as thermodynamic fluxes and $\partial v_i/\partial x_j$, $\dot{u} - v_z$ and $\mathbf{q}$ as thermodynamic forces, and after noticing that the dissipated energy must vanish in the case of "solid" rotation of the system (which requires that σ_{ij}^v be a function of the shear rates $A_{ij} = 1/2\,(\partial v_i/\partial x_j + \partial v_j/\partial x_i)$), one obtains for the general case of a smectic A with uniaxial symmetry [3]:

$$\sigma_{ij}^v = 2\eta_2\,A_{ij} + 2\,(\eta_3 - \eta_2)\,(\delta_{iz}A_{zj}+\delta_{jz}A_{zi}) + (\eta_4 - \eta_2)\,\delta_{ij}A_{kk} +$$
$$(\eta_1 + \eta_2 - 4\eta_3 - 2\eta_5 + \eta_4)\,\delta_{iz}\delta_{jz}A_{zz} + (\eta_5 - \eta_4 + \eta_2)\,(\delta_{ij}A_{zz} + \delta_{iz}\delta_{jz}A_{kk}) \quad \text{(C.II.80a)}$$

In the incompressible case, three viscosities (instead of five) are enough and the viscous stress tensor can be written under the simplified form:

$$\sigma_{xy}^v = 2\eta_2 A_{xy}$$

$$\sigma_{xz}^v = 2\eta_3 A_{xz}, \qquad\qquad \sigma_{yz}^v = 2\eta_3 A_{yz}$$

$$\sigma_{xx}^v = 2\eta_2 A_{xx}, \qquad\qquad \sigma_{yy}^v = 2\eta_2 A_{yy} \qquad\qquad \text{(C.II.80b)}$$

$$\sigma_{zz}^v = (\eta_1 - \eta_5)\,A_{zz}$$

With these notations (which were introduced by the Harvard group), η_2 is the shear viscosity in the plane of the layers and η_3 the shear viscosity parallel to the layers.

The other constitutive relations are as follows:

$$\dot{u} - v_z = \lambda_p\,G - \frac{\mu}{T}\frac{\partial T}{\partial z} \qquad\qquad \text{in the bulk} \qquad\qquad \text{(C.II.81a)}$$

$$\dot{u} - v_z = -\zeta\,\sigma_{zi}^E \upsilon_i \qquad\qquad \text{at the surface} \qquad\qquad \text{(C.II.81b)}$$

$$q_z = \mu\,G - \kappa_{//}\frac{\partial T}{\partial z} \qquad\qquad \text{in the bulk} \qquad\qquad \text{(C.II.82a)}$$

$$q_i = -\kappa_\perp\frac{\partial T}{\partial x_i}\quad(i = x, y) \qquad\qquad \text{in the bulk} \qquad\qquad \text{(C.II.82b)}$$

In these expressions, λ_p and ζ are the bulk and surface permeation coefficients, respectively, while $\kappa_{//}$ and $\kappa_\perp$ represent the thermal conductivities along the director and perpendicularly. Finally, μ describes the (never yet observed) coupling between heat flux across the layers and the stress gradient.

The irreversible entropy production being positive, one must have

$$\kappa_{//},\,\kappa_\perp,\,\lambda_p \text{ and } \zeta > 0 \qquad\qquad \text{and} \qquad\qquad \mu^2 < T\lambda_p\kappa_{//} \qquad \text{(C.II.83a)}$$

as well as:

86

$$\eta_1 \eta_4 \geq \eta_5^2 , \quad \eta_1 \geq 0 , \quad \eta_2 \geq 0 , \quad \eta_3 \geq 0 , \quad \eta_4 \geq 0 \tag{C.II.83b}$$

One also notices that the viscous stress tensor is symmetric. The bulk dissipation for an isothermal process can then be rewritten under the equivalent form:

$$\rho T \overset{\circ}{s} = \sigma_{ij}^v A_{ij} + G(\overset{\cdot}{u} - v_z) \tag{C.II.84}$$

II.4 Sound propagation in smectics

In a solid, there are three acoustic propagation modes for each wave vector **q** (one longitudinal and two transverse). In a liquid, one acoustic mode (density oscillation) and two damped transverse modes, controlled by the viscosity, are found. What happens in smectics? We shall see that several cases must be distinguished, depending on the orientation of the wave vector **q** with respect to the optical axis.

Start by considering the general case $q_x q_y q_z \neq 0$. To simplify the calculations, we assume that permeation, viscosity, and bending elasticity are negligible, which amounts to setting:

$$\lambda_p = \eta_i = K = 0 \tag{C.II.85}$$

The linearized equations of motion reduce to the simplified expressions:

$$\rho \frac{\partial \mathbf{v}}{\partial t} = - \mathbf{grad}\,P + \mathbf{G} \tag{C.II.86a}$$

$$\mathrm{div}\,\mathbf{v} - \frac{\partial \theta}{\partial t} = 0 \tag{C.II.86b}$$

$$v_z = \frac{\partial u}{\partial t} \tag{C.II.86c}$$

The pressure is given by the relation

$$P = - \frac{\partial \rho f}{\partial \theta} = - A\theta - C\gamma \quad \text{where we defined} \quad \gamma = \frac{\partial u}{\partial z} \tag{C.II.87}$$

Identifying the pressure in eqs. C.II.86a–c yields:

$$\rho \frac{\partial v_x}{\partial t} = A \frac{\partial \theta}{\partial x} + C \frac{\partial \gamma}{\partial x} \tag{C.II.88a}$$

$$\rho \frac{\partial v_y}{\partial t} = A \frac{\partial \theta}{\partial y} + C \frac{\partial \gamma}{\partial y} \tag{C.II.88b}$$

$$\rho \frac{\partial v_z}{\partial t} = (A + C)\frac{\partial \theta}{\partial z} + (B + C)\frac{\partial \gamma}{\partial z} \qquad \text{(C.II.88c)}$$

$$\frac{\partial v_x}{\partial x} + \frac{\partial v_y}{\partial y} + \frac{\partial v_z}{\partial z} = \frac{\partial \theta}{\partial t} \qquad \text{(C.II.88d)}$$

$$\frac{\partial v_z}{\partial z} = \frac{\partial \gamma}{\partial t} \qquad \text{(C.II.88e)}$$

The following step involves searching for eigenmodes of the type

$$v_i = v_i^0 \exp[i(\mathbf{q}.\mathbf{r} - \omega t)], \quad \theta = \theta^0 \exp[i(\mathbf{q}.\mathbf{r} - \omega t)], \quad \gamma = \gamma^0 \exp[i(\mathbf{q}.\mathbf{r} - \omega t)] \quad \text{(C.II.89)}$$

After replacing in equations C.II.88, one is left with a system of homogeneous equations, with the unknowns v_i^0, θ^0 and γ^0. This system only has non-trivial solutions if the following secular equation is fulfilled:

$$\omega \left\{ [\rho\omega^2 - (B + C)q_z^2] \, [\rho\omega^2 - Aq_\perp^2] - (A+C)(\rho\omega^2 + Cq_\perp^2)q_z^2 \right\} = 0 \qquad \text{(C.II.90)}$$

where $q_\perp^2 = q_x^2 + q_y^2$.

Five distinct modes are associated with this fifth-order equation, as many as the unknowns v_i, θ and γ.

1. The first one is the neutral mode $\omega = 0$. For this solution, $u = \theta = 0$ and the velocity is perpendicular to the optical axis and to $\mathbf{q}$. Taking into account the viscosity, it can be shown that this transverse mode involving shear parallel to the layers is damped, and the dispersion relation reads:

$$\rho\omega = - i\eta_3 q^2 \qquad \text{(C.II.91)}$$

2. The other modes are more interesting, as they describe the acoustic modes. They can be written as:

$$\omega_1 = \pm c_1(\Phi) \, q \qquad \text{(C.II.92a)}$$

$$\omega_2 = \pm c_2(\Phi) q \qquad \text{(C.II.92b)}$$

where Φ is the $(Oz, \mathbf{q})$ angle. Setting $x = \rho c^2$, one gets ρc_1^2 and ρc_2^2 by solving the following equation:

$$x^2 - x \, [A \sin^2\Phi + (A+B+2C)\cos^2\Phi] + \sin^2\Phi \cos^2\Phi \, (AB - C^2) = 0 \qquad \text{(C.II.93)}$$

Note that one of the velocities goes to zero for $\Phi \to \pi/2$ ($\mathbf{q}$ parallel to the optical axis) and for $\Phi \to 0$ ($\mathbf{q}$ perpendicular to the optical axis). These two limiting cases will be discussed later.

In general, $A \gg B$ and C and eq. C.II.93 assumes the simplified form:

$$x^2 - Ax + \sin^2\Phi\cos^2\Phi(AB - C^2) = 0 \qquad \text{(C.II.94)}$$

Within the same approximation, this equation has solutions:

$$x = A \quad \text{and} \quad x = \frac{AB - C^2}{A} \sin^2\Phi \cos^2\Phi = \bar{B} \sin^2\Phi \cos^2\Phi \qquad \text{(C.II.95)}$$

They correspond respectively to:

1. The **first sound** (isotropic) of velocity

$$c_1 = \sqrt{\frac{A}{\rho}} \qquad \text{(C.II.96a)}$$

2. The **second sound** (anisotropic) of velocity

$$c_2 = \sqrt{\frac{\bar{B}}{\rho}} \sin\Phi \cos\Phi \qquad \text{(C.II.96b)}$$

The first sound (two propagative modes moving in opposite directions) corresponds to usual sound. It is associated with density variations and corresponds to a longitudinal compression-dilation wave. Within the considered approximation, its velocity is independent of the propagation direction (Fig. C.II.13).

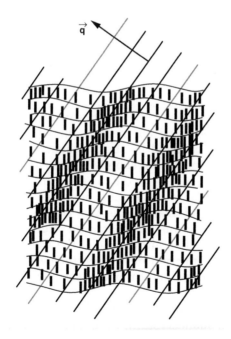

Fig. C.II.13 Schematic representation of first sound ("ordinary" sound).

The second sound (two propagative modes) is more original. It corresponds to a variation in layer thickness at constant density (Fig. C.II.14). Its name comes from superfluid helium, where two phonon branches are also present. Second sound is associated with fluctuations in the phase of the complex order parameter.

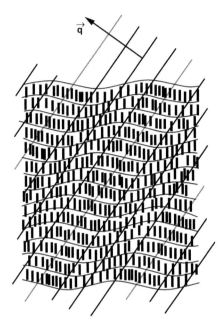

Fig. C.II.14 Second sound in a smectic.

To test these theoretical predictions, Liao, Clark and Pershan [21] studied Brillouin light scattering by the sound waves.

This experiment is based on the fact that a light wave of frequency Ω and wave vector $\mathbf{k_i}$ is scattered by the propagative modes (longitudinal or transverse) thermally excited in the samples. Let $\mathbf{q}$ be their wave vector and $\pm \omega$ their frequencies, such that $\omega = \pm cq$ ($\pm$ as these modes can propagate either way). One can show that these modes engender two scattered light waves of wave vector $\mathbf{k_s} = \mathbf{k_i} + \mathbf{q}$ and frequencies $\Omega \pm \omega$ (inelastic scattering) (Fig. C.II.15). Thus, the spectrum of scattered light contains two lines, symmetrically shifted by $\pm \omega$ on either side of the frequency Ω. By measuring this frequency shift as a function of the scattering vector $\mathbf{q}$, the dispersion relation of sound waves can be determined experimentally and then compared with the theoretical predictions.

The experimental setup of Liao, Clark and Pershan is shown in figure C.II.16. It consists of a monocrystalline sample, illuminated by a monomode laser with wavelength $\lambda = 5145$ Å and a power of less than 200 mW. The scattered light is collected along $\mathbf{k_s}$, and then analyzed using a high-resolution spectrometer (double-pass Fabry-Perrot). By turning the sample, one can acquire different spectra $I(\omega)$ as a function of the angle Φ between the scattering vector $\mathbf{q}$ and the normal to the layers $\mathbf{n}$. In agreement with theory, these spectra possess two pairs of peaks, symmetrical with respect to the

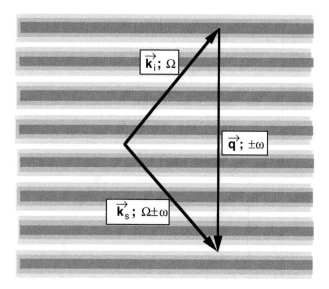

Fig. C.II.15 Principle of Brillouin light scattering by acoustical waves.

frequency Ω of the laser (Fig. C.II.16). The closest two correspond to second sound and the two others to first sound.

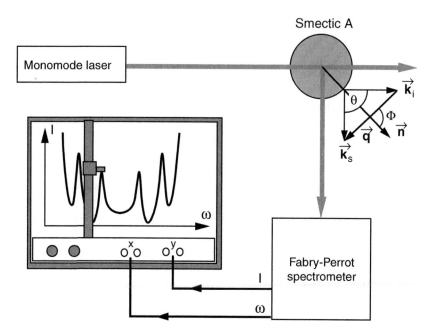

Fig. C.II.16 Diagram of the experiment by Liao, Clark and Pershan [21] on Brillouin scattering in smectics.

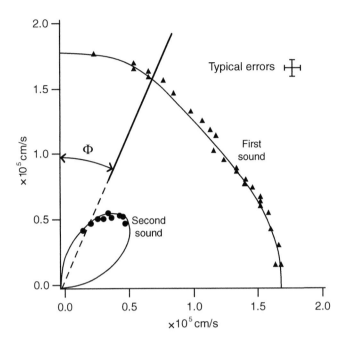

Fig. C.II.17 Polar diagrams of the propagation velocities for the first and second sounds. The experimental points were obtained for three different values (150°, 90°, and 60°) of the angle $\theta = (\mathbf{k_i}, \mathbf{k_s})$. The solid curves are theoretical predictions. (from ref. [21]).

In this way, Liao, Clark and Pershan measured the velocities of the first and second sounds as a function of the angle Φ. Their experimental results are shown in the polar diagram of figure C.II.17. The agreement with classical smectic theory is excellent.

These experiments yielded the three elastic moduli (in the GHz frequency range), as well as their temperature dependence. As an order of magnitude, $A \approx 3 \times 10^{10}$ erg/cm^3, $B \approx 10^{10}$ erg/cm^3, $C \approx 2 \times 10^9$ erg/cm^3. The stiffness constant A is larger than the two others, justifying, in a first approximation, the decoupling between first and second sound. Let us also mention that the value of B obtained by this method is much larger than the low-frequency values determined by mechanical measurements ($B \approx 10^7$–10^8 erg/cm^3 in this limit).

In a similar way, one can study the two extremes corresponding to $\Phi = \pi/2$ and $\Phi = 0$. The first sound is not modified, but the second sound is replaced by damped modes.

Thus, for $\Phi = \pi/2$, corresponding to a scattering vector $\mathbf{q}$ parallel to the layers, one has:

1. The first sound.
2. A slow layer undulation mode, detected by Ribotta et al. [22] in quasi-elastic Rayleigh scattering:

$$\omega \approx -\frac{iKq^2}{\eta_3} \tag{C.II.97}$$

3. A strongly damped shear mode of velocity $\mathbf{v}$ in the plane of the layers and perpendicular to $\mathbf{q}$:

$$\omega \approx -\frac{i\eta_2 q^2}{\rho} \tag{C.II.98}$$

4. A shear mode coupled to sound.

For $\Phi = 0$, i.e., $\mathbf{q}$ perpendicular to the layers, one finds:
1. The first sound.
2. A permeation mode, difficult to observe experimentally:

$$\omega = -i\,\bar{B}\,\lambda_p q^2 \tag{C.II.99}$$

3. Two coupled viscous modes.

II.5 Continuous flow

At low frequency, the smectic phase behaves as an **incompressible** fluid. One has then $\mathrm{div}\,\mathbf{v} = 0$. This condition determines the pressure P, which loses its thermodynamical significance in this limit since the relation C.II.64 becomes useless. The types of flow we shall describe in the following give a very good illustration of the fundamental differences between smectics and ordinary liquids.

II.5.a) Permeation flow

One speaks of permeation when the molecules flow across the layers. This concept was invented by Helfrich [23] for the case of cholesterics, but it also applies to smectics [24]. To induce this flow, let us impose a constant pressure gradient perpendicular to the layers (Fig. C.II.18):

$$\frac{\partial P}{\partial z} \neq 0 \tag{C.II.100}$$

If the layers cannot slide on the surfaces confining the sample, the molecules will flow across the layers as they would in a porous medium. To simplify the calculations, we assume that the smectic sample is infinite along y and that the flow takes place in the (x, z) plane. Let $\mathbf{v} = (0, 0, v_z(x, z))$ be the macroscopic flow velocity in the stationary regime.

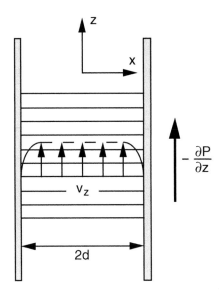

Fig. C.II.18 Permeation flow in a smectic. The velocity is constant, except for two permeation boundary layers of molecular thickness.

The equation of mass conservation

$$\frac{\partial v_z}{\partial z} = 0 \qquad\qquad\text{(C.II.101)}$$

immediately yields:

$$-\frac{\partial P}{\partial x} = 0 \qquad\qquad\text{(C.II.102a)}$$

$$-\frac{\partial P}{\partial z} + \eta_3 \frac{\partial^2 v_z}{\partial x^2} + G = 0 \qquad\qquad\text{(C.II.102b)}$$

$$v_z = -\lambda_p G \qquad\qquad\text{(C.II.102c)}$$

whence:

$$-\frac{dP}{dz} + \eta_3 \frac{\partial^2 v_z}{\partial x^2} - \frac{v_z}{\lambda_p} = 0 \qquad\qquad\text{(C.II.103)}$$

The characteristic length

$$l_p = \sqrt{\eta_3 \lambda_p} \qquad\qquad\text{(C.II.104)}$$

will be called **permeation length**. We shall see later that it is comparable to the layer thickness in thermotropic smectics (see chapter C.III).

Eq. C.II.103 is easily integrated with the boundary conditions $v_z(x = \pm d) = 0$ at the walls:

$$v_z = -\lambda_p \frac{dP}{dz}\left[1 - \frac{ch(x/l_p)}{ch(d/l_p)}\right] \qquad\qquad (C.II.105)$$

This formula describes a so-called "plug flow." The velocity, constant and equal to $-\lambda_p(dP/dz)$ in the center of the sample, goes to zero over two permeation boundary layers of thickness l_p, which is usually much smaller than the thickness $2d$ of the sample. The λ_p coefficient plays here the same role as the Darcy coefficient in porous media. The novelty here is that the smectic medium acts as the fluid and as the porous medium at the same time.

To find the order of magnitude of the permeation length l_p, let us express the permeation coefficient and the viscosity as a function of the self-diffusion coefficient D of the molecules in the liquid crystal. We make here the simplifying assumption that $D_{//} = D_{\perp}$, which is accurate up to a factor of 2 or 3 [25]. Let Ω be the molecular volume; in the presence of a stress gradient $\mathbf{G}$, each molecule feels the elastic force $\mathbf{F} = \Omega\mathbf{G}$ and diffuses across the layers with an average velocity given by the Nernst-Einstein law [26]:

$$<v> = \frac{DF}{k_BT} \qquad\qquad (C.II.106a)$$

This equation is none other than the bulk permeation equation, since:

$$\dot{u} - v_z = <v> = \frac{\lambda_p}{\Omega}\Omega\, G = \frac{\lambda_p}{\Omega}F \qquad\qquad (C.II.106b)$$

This analogy yields the sought-after relation [27a,b]:

$$\lambda_p = \frac{D\Omega}{k_BT} \qquad\qquad (C.II.107)$$

As to the viscosity η, it is related to the diffusion coefficient by the Stokes-Einstein relation:

$$\eta \approx \frac{k_BT}{Da_o} \qquad\qquad (C.II.108)$$

where a_o is a molecular length. Consequently,

$$l_p = \sqrt{\lambda_p\eta} \approx a_o \qquad\qquad (C.II.109)$$

in a thermotropic liquid crystal. As an order of magnitude, $\eta \approx 1$ poise and $a_o \approx 30$ Å, such that $\lambda_p \approx 10^{-13}$ cm^2/poise.

The situation is quite different in lyotropic lamellar phases, as the presence of water hinders diffusion across the layers. In these phases, the intrinsic permeation coefficient must be much lower than in thermotropics. On the other hand, some defects such as screw dislocations, but also pores and passages between the membranes, can contribute very effectively to the permeation process [27b], at least in some lyotropic phases where high defect densities are encountered [27c].

To conclude this section, let us determine the layer deformation induced by the permeation flow. Their tilt $\theta = \partial u/\partial x$ is given by the permeation equation C.II.102c. Taking into account the anharmonic terms and the bending term, and neglecting the variation of v_z across the two permeation boundary layers (of molecular thickness), one has:

$$\frac{3}{2} \bar{B} \theta^2 \frac{\partial \theta}{\partial x} - K \frac{\partial^3 \theta}{\partial x^3} = \frac{dP}{dz} = P' \qquad \text{(C.II.110)}$$

The strong anchoring conditions on the plates impose $u = \theta = 0$ in $x = \pm d$, while on symmetry grounds, $\theta = 0$ in the middle of the sample. This non-linear equation can be solved piecewise in the interval $0 < x < d$:

1. Close to the center ($0 < x < x_o$), the curvature term is dominant and the solution is given by

$$\theta = -\frac{P'}{6K} x^3 + Ax \qquad \text{(C.II.111)}$$

Due to the symmetry of the problem, the solution is odd in x.

2. Outside this region and far enough from the solid wall, ($x_o < x < d - x_1$), the anharmonic term dominates. The solution has the form

$$\theta = 3 (2P'/3 \bar{B})^{1/3} x^{1/3} \qquad \text{(C.II.112)}$$

The constants x_o and A are obtained by connecting the solutions C.II.111 and C.II.112 as well as their derivatives in $x = x_o$. This yields:

$$-\frac{P'}{6K} x_o^3 + Ax_o = 3 (2P'/3 \bar{B})^{1/3} x_o^{1/3} \qquad \text{(C.II.113a)}$$

$$-\frac{P'}{2K} x_o^2 + A = (2P'/3 \bar{B})^{1/3} x_o^{-2/3} \qquad \text{(C.II.113b)}$$

whence:

$$x_o = (144 K^3/ \bar{B} P'^2)^{1/8} \qquad \text{(C.II.114a)}$$

$$A = (256 P'^2/9 K\bar{B})^{1/4} \qquad \text{(C.II.114b)}$$

3. Close to the solid wall ($d - x_1 < x < d$), the angle θ decreases again and goes to zero at the surface if the anchoring is strong enough. In this bend boundary layer, of thickness x_1, the curvature term is again important and the equation is more difficult to solve. However, one can easily estimate x_1 by

writing that, in this region, the curvature and the non-linear terms are of the same order of magnitude:

$$\bar{B}\frac{\theta^3}{x_1} \approx K\frac{\theta}{x_1^3} \qquad \text{(C.II.115a)}$$

This yields:

$$x_1 \approx \frac{\lambda}{\theta} \qquad \text{(C.II.115b)}$$

The angle θ is obtained by putting $x = d - x_1 \approx d$ in equation C.II.112 and replacing in the previous expression

$$x_1 \approx \frac{\lambda}{3}\left(\frac{\bar{B}}{P'd}\right)^{1/3} \qquad \text{(C.II.116)}$$

In figure C.II.19, we plotted the displacement $u(x)$ obtained by integrating equations C.II.111 and C.II.112 for a sample of thickness $2d = 200$ μm, vertically placed and acted upon by its own weight. In this case, $P' = \rho g \approx 10^3$ dyn/cm^2. With $B \approx 10^8$ erg/cm^3, $K \approx 10^{-6}$ dyn and $\lambda \approx 10$ Å, one obtains from eqs. C.II.112, C.II.114a, and C.II.116 a maximum tilt angle $\theta_{max} \approx 0.7°$, as well as $x_o \approx 1.8$ μm and $x_1 \approx 0.07$ μm. The solution is thus very well described in almost the entire sample by the simplified formula C.II.112. As to the permeation velocity v_z, it is extremely low. With $\lambda_p \approx 10^{-13}$ cm^2/poise, one finds indeed that $v_z \approx 1$ μm/day. Experimentally, such a velocity is hardly detectable. One might consider increasing the pressure gradient and hence the flux. Unfortunately, one is very soon confronted with the problem of layer slipping at the walls. This wall slippage could be due to the presence of a high density of edge dislocations in the layers. For this reason, a direct determination of the permeation coefficient by use of this method is not really feasible as the velocity of the layers, taken as zero in the model, is not necessarily so in the experiment.

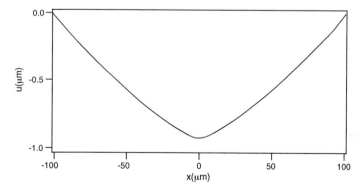

Fig. C.II.19 Displacement u of the layers due to a permeation flow between two parallel walls separated by a distance of 200 μm. The pressure gradient is only due to the gravity. Note the scale difference between the horizontal and vertical axes.

II.5.b) Flow around an obstacle

i) Sphere moving parallel to the layers

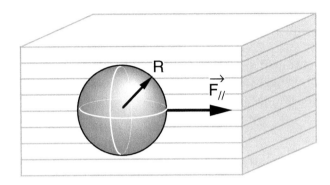

Fig. C.II.20 Hard sphere moving parallel to the smectic layers. The calculation assumes that the sphere does not deform the layers in its vicinity.

In this case (Fig. C.II.20), the flow is mainly in the plane of the layers, permeation being very difficult. De Gennes calculated the friction force $F_{//}$ within this approximation [24]. Up to a prefactor, the result is the same as for an isotropic fluid. With V the velocity of the sphere and R its radius, de Gennes finds (assuming that $\eta_2 = \eta_3 = \eta$):

$$F_{//} = 8\pi\eta RV \tag{C.II.117}$$

One recognizes the classical Stokes formula, with the 6π prefactor replaced by 8π.

ii) Sphere moving perpendicular to the layers

Nobody has solved the equations for this situation. However, dimensional analysis indicates that the friction force is proportional to the volume of the sphere and inversely proportional to the λ_p coefficient, this movement only requiring permeation:

$$F_\perp \approx \frac{4}{3}\pi\frac{R^3}{\lambda_p}V \tag{C.II.118}$$

This friction force is considerable, even for spheres of mesoscopic size (μm). Let us illustrate it by giving the order of magnitude of the settling velocity for a spherical dust particle in a homeotropic sample. By equating the friction force and the weight, one immediately obtains:

$$V_{\text{settling}} \approx \lambda_p \, \Delta\rho \, g \qquad\qquad\qquad\qquad\qquad \text{(C.II.119)}$$

where g is the gravity acceleration and $\Delta\rho$ the difference in density. This yields $V \approx 10^{-11}$ cm/s, which is ridiculously small and impossible to measure.

iii) Flow around a ribbon perpendicular to the layers

We shall study this flow configuration in detail, as it will be useful in the next chapter for calculating the mobility of an edge dislocation moving parallel to the layers (climb). This geometry was first studied by Clark [28].

Consider first the case of a thin semi-infinite obstacle perpendicular to the layers and placed in $z < 0$ and $x = 0$. The system being taken as infinite along y, one has $\partial/\partial y = 0$. The velocity far away from the obstacle **V** is assumed constant, parallel to the layers and perpendicular to the obstacle (Fig. C.II.21). In the stationary regime, the equations of movement write:

$$\frac{\partial P}{\partial z} = -\frac{v_z}{\lambda_p} \qquad\qquad\qquad\qquad\qquad \text{(C.II.120a)}$$

$$\frac{\partial P}{\partial x} = \eta_3 \frac{\partial^2 v_x}{\partial z^2} \qquad\qquad\qquad\qquad\qquad \text{(C.II.120b)}$$

$$\frac{\partial v_x}{\partial x} + \frac{\partial v_z}{\partial z} = 0 \qquad\qquad\qquad\qquad\qquad \text{(C.II.120c)}$$

In the first equation, we neglected the viscous term as compared to the permeation term. We equally assumed that $\partial/\partial x \ll \partial/\partial z$, which can be verified a posteriori.

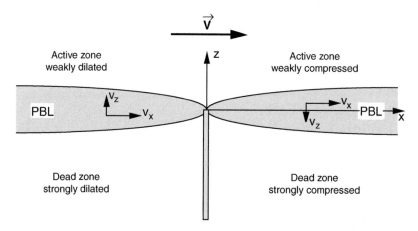

Fig. C.II.21 Flow around a semi-infinite plane perpendicular to the layers. PBL signifies permeation boundary layer. The components of the velocity are shown in these two regions; on the left of the obstacle, the v_z component points upwards, while it points downwards on the right side.

The flow being two-dimensional, we introduce the stream function ψ, such that

$$V_x = \frac{\partial \psi}{\partial z}, \qquad V_z = -\frac{\partial \psi}{\partial x} \qquad\qquad\text{(C.II.121)}$$

Eliminating the pressure between eqs. C.II.120a–c leads to an equation for ψ:

$$\frac{\partial^4 \psi}{\partial z^4} - \frac{1}{l_p^2}\frac{\partial^2 \psi}{\partial x^2} = 0 \qquad\qquad\text{(C.II.122)}$$

where $l_p = \sqrt{\lambda_p \eta_3}$ is the permeation length already introduced. The boundary conditions on the plate and at infinity are:

$$\psi(x, z\to\infty) = Vz \qquad\qquad\text{(C.II.123a)}$$

$$\psi(0, z<0) = 0 \qquad\qquad\text{(C.II.123b)}$$

$$\frac{\partial \psi}{\partial x}(0, z<0) = 0 \qquad\qquad\text{(C.II.123c)}$$

where V is the velocity of the smectic at infinity.

To solve equation C.II.122, we introduce the following two operators A^+ and A^- (this method was proposed by de Gennes in a slightly different context [24]):

$$A^- = \frac{\partial^2}{\partial z^2} - \frac{1}{l_p}\frac{\partial}{\partial x} \qquad\qquad\text{(C.II.124a)}$$

$$A^+ = \frac{\partial^2}{\partial z^2} + \frac{1}{l_p}\frac{\partial}{\partial x} \qquad\qquad\text{(C.II.124b)}$$

In the half-space $x > 0$, let h be a solution to the diffusion equation $A^- h = 0$ such that $h \to 0$ for $|z| \to \infty$. ψ must then fulfill the equation

$$A^+ \psi = h \qquad\qquad\text{(C.II.125)}$$

with solution

$$\psi(x, z) = \frac{1}{2}\int_{-\infty}^{z} dz_1 \int_{-\infty}^{z_1} h(x, z_2)\, dz_2 \qquad\qquad\text{(C.II.126)}$$

The diffusion equation $A^- h = 0$ has the solution (Green function for this operator)

$$h = \frac{Q}{\sqrt{x}}\exp\left[-\frac{1}{4}\frac{z^2}{l_p x}\right] \qquad\qquad\text{(C.II.127)}$$

where Q is a constant to be determined. From relations C.II.121, C.II.126, and C.II.127, one obtains the velocity components:

$$V_x = Q\sqrt{l_p} \int_{-\infty}^{z/(4l_px)^{1/2}} e^{-u^2} du \qquad\qquad (C.II.128a)$$

$$V_z = -\frac{l_p h}{2} = -\frac{l_p Q}{2\sqrt{x}} \exp\left[-\frac{z^2}{4l_px}\right] \qquad\qquad (C.II.128b)$$

Using the boundary conditions at infinity on the velocity immediately yields:

$$Q = \frac{V}{\sqrt{\pi l_p}} \qquad\qquad (C.II.129)$$

The same reasoning can be applied for $x < 0$. The solution is of the same type and can be obtained by changing x to $-x$ and v_z to $-v_z$. Finally, we can rewrite the solution under the following compact form, valid overall:

$$V_x = \frac{V}{2}\left[1 + \text{erf}\left(\frac{1}{2}\frac{z}{\sqrt{l_p|x|}}\right)\right] \qquad\qquad (C.II.130a)$$

$$V_z = -\frac{V}{2\sqrt{\pi}}\sqrt{\frac{l_p}{|x|}}\,\text{sgn}(x)\exp\left[-\frac{1}{4}\frac{z^2}{l_p|x|}\right] \qquad\qquad (C.II.130b)$$

where erf is the error function:

$$\text{erf } Y = \frac{2}{\sqrt{\pi}} \int_{-\infty}^{Y} \exp(-u^2)\, du \qquad\qquad (C.II.131)$$

These two equations yield the velocity profiles in the two **permeation boundary layers** spreading on either side of the obstacle and delimited by two parabola branches of equation

$$z^2 \le 4\, l_p\, |x| \qquad\qquad (C.II.132)$$

Outside these two regions, the permeation velocity v_z is almost zero. Consequently, the velocity in the plane of the layers is close to the one at infinity above the obstacle ($v_x \approx V$ for $z > 2\sqrt{l_p|x|}$) and goes to zero in two "dead areas" situated on either side of the obstacle ($v_x \approx 0$ when $z < -2\sqrt{l_p|x|}$).

It is also interesting to determine the distribution of the normal stress $\sigma = \sigma_{zz}$ on either side of the obstacle. This calculation is trivial, knowing that in the stationary regime $v_z = -\lambda_p G$ with $G \approx \bar{B}\,\partial\sigma/\partial z$. One only needs then to integrate eq. C.II.130b once with respect to z and, imposing $\sigma = 0$ in $z = +\infty$, one finds the following stress field:

$$\sigma = \frac{V}{2}\sqrt{\frac{\eta_3}{\lambda_p}} \, \text{sgn}(x) \left[\text{erf}\left(\frac{z}{2\sqrt{l_p|x|}}\right) - 1 \right] \tag{C.II.133}$$

The stress is strongly inhomogeneous, being almost zero above the obstacle, positive in the dead area upstream of the obstacle and negative in the dead area downstream. This stress distribution can lead to a layer undulation instability downstream of the sample.

To illustrate this inhomogeneous stress distribution, one can perform the experiment described in figure C.II.22 [29]. The homeotropic smectic A sample contains a thread that is thinner than the sample thickness. This thread (a hair) is glued on the lower plate, itself fixed. When the upper plate is set in uniform motion with a velocity V perpendicular to the thread, the approximate velocity profile can be determined by following the movement of dust particles in suspension in the smectic phase. This experiment clearly shows the existence of the two dead areas on either side of the sample, covering the entire width of the sample. This result confirms the existence of two very thin boundary layers separating the dead areas from the flowing medium. Unfortunately, this experiment is not precise enough to allow measurement of the thickness of these layers, estimated at below one micron. It is nevertheless very easy to induce the layer undulation instability by dilating the sample under constant shear. If the shear rate $\tau = V/d$ is not too high (a few s^{-1} to a few tens s^{-1} for this experiment), it can be shown that the critical stress which induces the instability is not noticeably changed. A perturbation calculation shows that the critical stress under shear becomes [30]:

$$\sigma_c \approx 2\pi \, \bar{B} \frac{\lambda}{d}\left(1 + \text{Cst} \frac{V^2 \eta_3^2 \, d \, \lambda}{K^2} \cos^2\theta\right) \quad \text{with} \quad \text{Cst} \approx 8.6 \times 10^{-4} \tag{C.II.134}$$

where θ is the angle between the wave vector of the undulation and the velocity. This correction is negligible at low shear rate. On the other hand, the "permeation stress" generated by the flow around the thread can change the undulation instability and its position in the thickness of the sample, as we can see by comparing the maximum stress C.II.133 generated by the flow, with the critical stress C.II.134. Clearly, the flow around the thread will strongly perturb the undulations as soon as

$$\sigma_{\text{flow}} \approx \frac{V}{2}\sqrt{\frac{\eta_3}{\lambda_p}} \geq \sigma_c \approx 2\pi \, \bar{B} \frac{\lambda}{d} \tag{C.II.135}$$

which is equivalent to

$$V \geq V_c \approx 10\,\bar{B}\frac{\lambda}{d}\sqrt{\frac{\lambda_p}{\eta_3}} \tag{C.II.136}$$

As an order of magnitude, $\bar{B} \approx 10^8$ erg/cm^3, $\eta_3 \approx 1$ poise, $\lambda \approx 20$ Å and $\lambda_p \approx 10^{-13}$ cm^2/poise, yielding $V_c \approx 60$ μm/s for $d \approx 100$ μm.

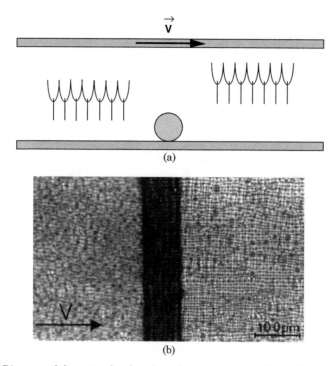

Fig. C.II.22 a) Diagram of the setup for shearing a homeotropic sample in the presence of a thread glued onto one of the plates. b) Photograph of the focal parabola lattice created after dilation under shear. Obviously, the parabolae do not appear at the same height upstream and downstream of the thread. Their relative position in the thickness of the sample is shown in the diagram (from ref. [29]).

This result is qualitatively confirmed by experiments. For instance, the photograph in figure C.II.22b shows clearly that the lattice of focal parabolae does not develop at the same height upstream and downstream of the thread when the sample is dilated under shear. It is easily checked that this distribution of the parabolae is in qualitative agreement with the stress distribution generated by the flow (see Fig. C.II.21).

Consider now the case of flow around a thin ribbon of width 2b, placed perpendicular to the layers (Fig. C.II.23). Let V be the velocity at infinity, assumed parallel to the layers and perpendicular to the obstacle. The relevant equation being linear, the solution is obtained as a superposition of two solutions of the type obtained previously. The first one corresponds to flow around the half-plane $x = 0$ and $z \leq b$, while the second one corresponds to flow around the half-plane $x = 0$ and $z \geq -b$. The permeation velocity is then:

$$V_z = \frac{V}{2\sqrt{\pi}} \sqrt{\frac{l_p}{|x|}} \, \text{sgn}(x) \left[\exp\left(-\frac{(z+b)^2}{4 l_p |x|} \right) - \exp\left(-\frac{(z-b)^2}{4 l_p |x|} \right) \right] \qquad (C.II.137)$$

while the velocity in the plane of the layers is given by:

103

$$v_x = V\left[1 + \frac{1}{2}\,erf\,\frac{z-b}{2\sqrt{l_p|x|}} - \frac{1}{2}\,erf\,\frac{z+b}{2\sqrt{l_p|x|}}\right] \tag{C.II.138}$$

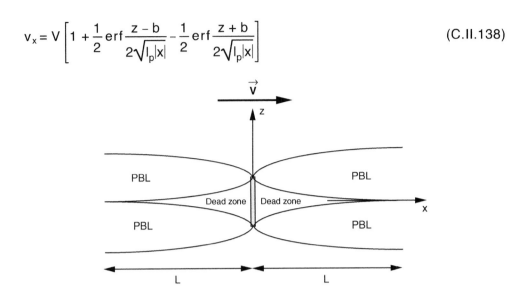

Fig. C.II.23 Flow around a ribbon perpendicular to the smectic layers. The permeation boundary layers merge together at a distance L from the ribbon. Beyond this distance, the flow is practically unchanged and the velocity is again equal to V.

We are at present interested in the friction force acting upon the ribbon. The result will come in handy later, for estimating the mobility of an edge dislocation. This force can be obtained by calculating the total energy dissipation. Preserving only the dominant terms in eq. C.II.84, the dissipation per unit volume is:

$$\phi = \eta_3\left(\frac{\partial v_x}{\partial z}\right)^2 + \frac{v_z^2}{\lambda_p} = 2\,\frac{v_z^2}{\lambda_p} \tag{C.II.139}$$

We used here the fact that $\partial v_x/\partial z = sgn(x)\,v_z/l_p$. Integration over the entire volume yields:

$$\Phi = \int \phi\,dV = \frac{2}{\sqrt{\pi}}\sqrt{\frac{\eta_3}{\lambda_p}}\,b\,V^2\int_0^{+\infty}[1 - \exp(-u)]\frac{du}{u^{3/2}} = 4\sqrt{\frac{\eta_3}{\lambda_p}}\,b\,V^2 \tag{C.II.140}$$

The absolute value of the resistance force per unit length along y is thus:

$$F = 4\sqrt{\frac{\eta_3}{\lambda_p}}\,b\,V \tag{C.II.141}$$

It is proportional to the velocity and to the width of the ribbon. This force can be significant, because the permeation coefficient is very small.

It will be shown at the end of this section that eq. C.II.141 for the viscous force can be found more easily, without explicitly calculating the velocity field and the associated stress field. This approach is however necessary for analyzing the local stability of the flow and to show that the layers are dilated and possibly unstable in the dead area upstream of the obstacle (x < 0). To this end, we need to determine the stress σ normal to the layers, using the permeation equation. Taking it as zero at infinity, we obtain:

$$\sigma = -\frac{1}{\lambda_p} \int_{-\infty}^{0} v_z \, dz \tag{C.II.142a}$$

which gives the final result:

$$\sigma = \text{sgn}(x)\frac{V}{2}\sqrt{\frac{\eta_3}{\lambda_p}}\left[\text{erf}\left(\frac{z-b}{2\sqrt{l_p|x|}}\right) - \text{erf}\left(\frac{z+b}{2\sqrt{l_p|x|}}\right)\right] \tag{C.II.142b}$$

This stress distribution has its maximum on the z = 0 axis and changes sign across the obstacle. It is positive (dilation) upstream (x < 0) and always negative downstream (layer compression). The undulation instability can only develop upstream, provided that σ is high enough. In the dead area upstream, of width b^2/l_p along x, the stress is practically constant, of value

$$\sigma_{\text{Max}} \approx V\sqrt{\frac{\eta_3}{\lambda_p}} \tag{C.II.143}$$

Outside this region, σ vanishes rapidly. The instability will then develop in the dead area ahead of the obstacle if the variation in layer thickness induced by this stress is larger than the critical dilation 2πλ. The instability criterion is thus:

$$2b\frac{\sigma_{\text{Max}}}{B} > 2\pi\lambda \tag{C.II.144}$$

which translates to:

$$V > V_c = \pi \bar{B}\frac{\lambda}{b}\sqrt{\frac{\lambda_p}{\eta_3}} \tag{C.II.145}$$

For an obstacle of width 2b = 10 μm, this critical velocity is of the order of $V_c \approx 60$ μm/s for $l_p = 10$ Å, $K = 10^{-6}$ dyn, $B = 10^8$ erg/cm³ and $\eta_3 = 1$ poise. This instability was observed by Clark in 1978 by compression of a planar sample containing air bubbles [28].

In conclusion of this section, we shall present a different, simpler way of obtaining expression C.II.141 for the friction force on a ribbon in a boundless medium. This approach can even be extended to the case of a sample bound by two free surfaces, which is interesting insofar as it provides a model for the flow around a dislocation line in a free-standing film (see chapter C.VIII).

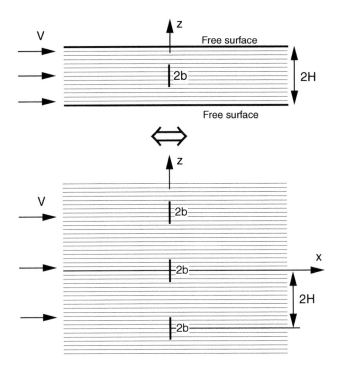

Fig. C.II.24 The flow around a ribbon of width 2b in a free-surface film of thickness 2H is equivalent to that in a boundless medium around an infinite number of parallel ribbons separated by 2H.

For simplicity, assume that the 2b wide ribbon is placed at the center of the sample of thickness 2H. It is easily shown that this problem reduces to that of the flow around an infinite array of parallel ribbons, with spacing 2H (Fig. C.II.24).

The proof starts with the observation that the hydrodynamic equations C.II.120 are identically satisfied by taking:

$$P = c \pm \sqrt{\frac{\eta_3}{\lambda_p}}\, v_x \qquad (C.II.146)$$

If the flow direction is from left to right, as shown in the diagram of figure C.II.24, one must take the "–" sign to the left of the ribbon and the "+" sign to the right. The constant c is then c_l to the left of the wall and c_r to its right, such that:

$$P_l = c_l - \sqrt{\frac{\eta_3}{\lambda_p}}\, v_x(x,z) \quad \text{in} \quad x < 0 \qquad (C.II.147a)$$

$$P_r = c_r + \sqrt{\frac{\eta_3}{\lambda_p}}\, v_x(x,z) \quad \text{in} \quad x > 0 \qquad (C.II.147b)$$

yielding the pressure difference between $x = -\infty$ and $x = +\infty$ where the pressure reaches constant values:

$$P_l - P_r = c_l - c_r - 2\sqrt{\frac{\eta_3}{\lambda_p}}\ V \qquad\qquad (C.II.148)$$

where we used $v_x(-\infty, z) = v_x(+\infty, z) = V$. To find $c_l - c_r$, write that the two pressure fields must be equal in $x = 0$:

$$P_l - P_r = 0 \qquad \text{in} \qquad x = 0 \qquad\qquad (C.II.149)$$

yielding:

$$c_l - c_r = 2\sqrt{\frac{\eta_3}{\lambda_p}}\ v_x(0,z) \qquad\qquad (C.II.150)$$

The velocity v_x being independent of z in $x = 0$, flux conservation results in:

$$v_x(0,z) = \frac{H}{H-b}\ V \qquad\qquad (C.II.151)$$

and thus:

$$P_l - P_r = 2\sqrt{\frac{\eta_3}{\lambda_p}}\ \frac{b}{H-b}\ V \qquad\qquad (C.II.152)$$

One finally obtains the viscous force F_v acting on a ribbon by equating the power injected into the system in a slab of thickness $2H$ with the energy dissipated around one ribbon:

$$(P_l - P_r)\ 2H\ V = F_v\ V \qquad\qquad (C.II.153)$$

which yields the force we have been searching for:

$$F_v = 4\sqrt{\frac{\eta_3}{\lambda_p}}\ \frac{Hb}{H-b}\ V \qquad\qquad (C.II.154)$$

This formula indeed reduces to eq. C.II.141 in the $H \to \infty$ limit. It also shows that the friction force increases in a film of finite thickness, diverging when the width of the obstacle approaches the film thickness.

II.5.c) Anomalies in Poiseuille flow parallel to the layers

We have seen that a plug flow develops when applying a pressure gradient across the layers. This kind of flow has no equivalent in ordinary Newtonian fluids. In this section, we consider the a priori simpler case of a Poiseuille flow between two planar surfaces, when the pressure gradient is applied parallel to the smectic layers. This case might seem trivial at first sight, as the smectic phase behaves as an ordinary Newtonian fluid, with a parabolic velocity profile.

The situation is actually much more complex, for multiple reasons. One of them is the presence of defects (focal conics, but also dislocations) that hinder the flow. We shall return to this subtle problem in chapter C.III. Another fundamental reason, which usually goes unnoticed, is the difficulty in reaching the stationary Poiseuille flow when the layers are parallel to the boundaries of the sample. This can be shown by considering the situation of a smectic phase injected at a constant flow rate Q at one end of a rectangular capillary internally treated to ensure homeotropic anchoring. If the tube is thin but very wide, we can neglect the edge effects and consider the flow as two-dimensional in the (x, z) plane (Fig. C.II.25). The question is at what distance from the entry section is the parabolic velocity profile reached?

In the case of an ordinary Newtonian fluid, the answer is found by searching for the distance L* where the two Blasius boundary layers formed at the walls join together. Knowing that, in an ordinary fluid of viscosity η, the boundary layer thickness is:

$$\delta = \sqrt{\frac{\eta x}{\rho V}} \qquad\qquad (C.II.155)$$

one obtains by setting $\delta(L^*) = d$:

$$L^* \approx \frac{\rho V d^2}{\eta} \qquad\qquad (C.II.156)$$

For a fluid with viscosity η = 1 poise and a capillary thickness of 2d = 200 μm, one finds L* ≈ 1 μm at V = 1 cm/s. Clearly, the parabolic velocity profile given by the Poiseuille law is reached over very short distances.

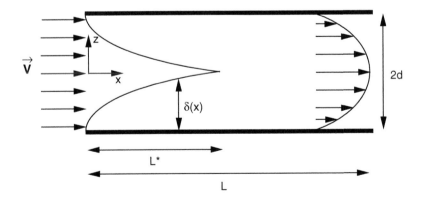

Fig. C.II.25 Flow in a flat capillary of length L. The layers are parallel to the (x, y) plane, and the velocity is considered constant over the entry section. The Poiseuille profile is reached when the two boundary layers meet, after a distance L* that can be much larger than the total length of the capillary.

In smectics, the situation is fundamentally different. Indeed, the thickness of the boundary layer developed in contact with the wall increases much more slowly than in an ordinary fluid due to the difficulty encountered by the molecules in crossing from one layer to the next (permeation). This permeation boundary layer, first studied by de Gennes [24], can be treated as in the previous section. Its thickness varies as

$$\delta \approx \sqrt{l_p x} \tag{C.II.157}$$

The parabolic velocity profile, which is still the solution to the problem far enough from the entry point, is reached after a distance:

$$L^* \approx \frac{d^2}{l_p} \tag{C.II.158}$$

This distance, independent of the velocity, is very large due to the very small value of the permeation length. For instance, $L^* \approx 3$ m for $d = 100$ μm and $l_p = 30$ Å. Consequently, the Poiseuille law must rarely be relevant in experiments and the permeation boundary layers at the walls must be taken into account.

In conclusion, we present the expression of the effective viscosity η_{eff} of a smectic phase when the total capillary length L is much smaller than L^*. This is the viscosity of a Newtonian fluid that, under the same pressure gradient, would give the same flow rate. Writing that the work effected by the pressure is entirely dissipated within the two boundary layers, one finds:

$$\eta_{eff} = \frac{4\sqrt{2}}{3\sqrt{\pi}} \frac{d}{\sqrt{l_p L}} \eta_3 \tag{C.II.159}$$

To our knowledge, this relation was never verified experimentally, although it provides a direct method for measuring the permeation length l_p. It is however possible that this experiment be very difficult as all samples, even the most carefully prepared, always contain a high density of edge or screw dislocations. These defects can interact with the flow and profoundly change the expected rheological behavior, as noticed very early on by Kim et al. [31]. We shall consider these effects in chapter C.III.

BIBLIOGRAPHY

[1]   Landau L., Lifshitz E., *Theory of Elasticity*, Pergamon Press, New York, 1996.

[2]   de Gennes P.G., *J. Physique Colloq. (France)*, **30** (1969) C4-65.

[3]   Martin P.C., Parodi O., Pershan P.S., *Phys. Rev. A*, **6** (1972) 2401.

[4]   Durand G., *C. R. Acad. Sci. (Paris)*, **275B** (1972) 629.

[5]   a) Delaye M., Ribotta R., Durand G., *Phys. Lett.*, **44A** (1973) 139.
b) Ribotta R., Durand G., Litster J.D., *Solid State Commun.*, **12** (1973) 27.
c) Ribotta R., "Etude expérimentale de l'élasticité des smectiques," D.Sc. Thesis, University of Paris XI, Orsay, 1975.
d) Ribotta R., Durand G., *J. Physique (France)*, **38** (1977) 179.

[6]   Clark N.A., Meyer R.B., *Appl. Phys. Lett.*, **22** (1973) 10.

[7]   Ben-Abraham S.I., Oswald P., *Mol. Cryst. Liq. Cryst.*, **94** (1983) 383.

[8]   Safinya C.R., Roux D., Smith G.S., Sinha S.K., Dimon P., Clark N.A., Bellocq A.-M., *Phys. Rev. Lett.*, **57** (1986) 2718.

[9]   Roux D., Safinya C.R., *J. Physique (France)*, **49** (1988) 307.

[10] Oswald P., Allain M., *J. Physique (France)*, **46** (1985) 831.

[11] Delrieu J.M., *J. Chem. Phys.*, **60** (1974) 1081.

[12] Fukuda J., Onuki A., *J. Phys. II (France)*, **5** (1995) 1107.

[13] Oswald P., "Dynamique des dislocations dans les smectiques A et B," D.Sc. Thesis, University of Paris XI, Orsay, 1985.

[14] Rosenblatt Ch.S., Pindak R., Clark N.A., Meyer R.B., *J. Physique (France)*, **38** (1977) 1105.

[15] Oswald P., Behar J., Kléman M., *Phil. Mag. A*, **46** (1982) 899.

[16] Helfrich W., *J. Chem. Phys.*, **55** (1971) 839.

[17] Hurault J.P., *J. Chem. Phys.*, **59** (1973) 2068.

[18] Rondelez F., Hulin J.P., *Solid State Commun.*, **10** (1972) 1009.

[19] Chandrasekhar S., *Liquid Crystals*, Second Edition, Cambridge University Press, Cambridge, 1992.

[20] Kléman M., Parodi O., *J. Physique (France)*, **36** (1975) 671.

[21] Liao Y., Clark N.A., Pershan P.S., *Phys. Rev. Lett.*, **30** (1973) 639.

[22] Ribotta R., Salin D., Durand G., *Phys. Rev. Lett.*, **32** (1974) 7.

[23] Helfrich W., *Phys. Rev. Lett.*, **23** (1969) 372.

[24] de Gennes P.G., *The Physics of Fluids*, **17** (1974) 1645.

[25] Krüger G.J., *Phys. Rep.*, **82** (1982) 229.

[26] Philibert J., *Atom Movements: Diffusion and Mass Transport in Solids*, Les Éditions de Physique, Les Ulis, 1991.

[27] a) Parodi O., *J. Physique Lettres (France)*, **37** (1976) L-143.
b) Oswald P., *C. R. Acad. Sci. (Paris)*, **304** (1987) 1043.
c) See, e.g., Constantin D., Oswald P., *Phys. Rev. Lett.*, **85** (2000) 4297.

[28] Clark N.A., *Phys. Rev. Lett.*, **40** (1978) 1663.

[29] Oswald P., *J. Physique Lettres (France)*, **44** (1983) L-309.

[30] Oswald P., Ben-Abraham S.I., *J. Physique (France)*, **43** (1982) 1193.

[31] Kim M.G., Park S., Cooper M. Sr., Letcher S.V., *Mol. Cryst. Liq. Cryst.*, **36** (1976) 143.

Chapter C.III

Dislocations, focal conics and rheology of smectics A

Smectics A exhibit two types of defects:

1. The **focal conics**, involving large-scale curvature deformations of the layers. These extended defects are distinctly visible using a polarizing microscope and played a historical role in the discovery of liquid crystals. Indeed, it is by studying their topology that G. Friedel was able in 1922 to predict the lamellar nature of smectics, later confirmed by X-ray analysis [1];

2. The **dislocations**, locally breaking the translational order and involving local variations of the layer thickness. These defects are not always visible using polarizing microscopy, but can be revealed by other techniques, for instance by freeze-fracture and electron microscopy. F.C. Frank was the first to direct the physicists' attention to their existence in smectic phases in 1958 [2].

In this chapter, we describe the topology and the energetic properties of these defects. We shall then see that they are very different from the defects encountered in usual solids (e.g., grain boundaries, dislocations, and vacancies) and that they participate in the very special rheological behavior of these phases, situated in between classical hydrodynamics and the plasticity of solids.

The outline of this chapter is as follows. We first discuss some static and dynamic properties of focal conics (section III.1) and of dislocations (section III.2): energy, stress field, interaction with the surfaces, mobility, etc. We then describe some rheological experiments where the defects play a fundamental role. We show the existence of several flow regimes, depending on the orientation of the layers and on the shear rate. Thus, we describe rheology measurements at low shear rate in continuous or oscillating flow, performed on homeotropic, planar or disoriented samples. Particular attention will be given to the lubrication regime, especially important for applications (section III.3). In all these experiments, the anchoring of the layers on the surfaces limiting the sample is important and determines the behavior of the smectic phase. On the other hand, we shall see that, at high shear rates, the orientation of the layers depends more on the flow than on the boundary conditions, allowing "orientation diagrams" similar to phase diagrams to be established (section III.4). We end this chapter with the description of a microplasticity experiment in compression and in dilation normal to the layers (section III.5). We shall mainly focus on the screw dislocations and their helical instabilities in compression, as well as the nucleation of focal parabolae under dilation.

III.1 Focal conics

III.1.a) Topology of focal domains of types I and II

In a smectic, it is very easy to bend the layers, since they are uncorrelated from each other. For this reason, their bending modulus K is very small (of the order of 10^{-6} dyn). On the other hand, it is much more difficult to compress or dilate them.

An ideal focal domain describes a molecular configuration such that the **layer thickness remains constant**, except on two singular lines.

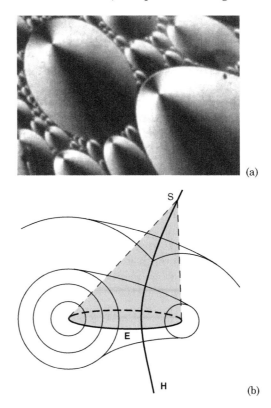

(a)

(b)

Fig. C.III.1 a) Focal conics texture seen between crossed polarizers in the polarizing microscope; b) focal domain of the first type (FD1): the two singular lines are an ellipse E and its conjugated hyperbola H. The physical part of an FD1 is contained inside the cone of vertex S and having as base the ellipse E. The texture seen in a) is an association of such cones.

Two types of domains can be distinguished according to the sign of the Gaussian curvature of the layers [3].

The first, called "focal domains of the first type" (in short, FD1), in which the Gaussian curvature is negative, the two principal radii of curvature (denoted in the following by R_1 and R_2) having opposite signs everywhere. Inside an FD1, the layers are deformed along Dupin's cyclides (Fig. C.III.1), their normals passing through two conics in focal position. This means that they are in two orthogonal planes, the focus of each one passing through the vertex of the other. In general, these two conjugated conics are a hyperbola and an ellipse, the product of their eccentricities being equal to unity: $e_1 e_2 = 1$ (Fig. C.III.1) [4a–c].

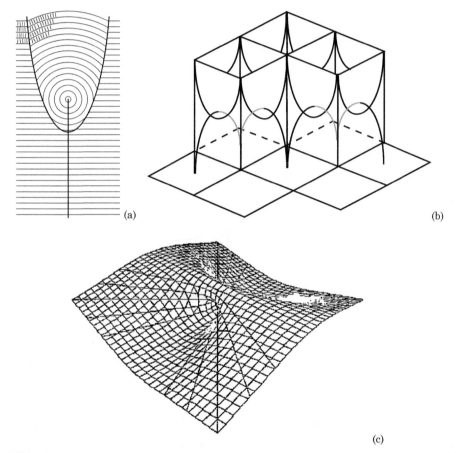

(a)

(b)

(c)

Fig. C.III.2 a) Layer arrangement in a plane containing one parabola of a parabolic domain (FD1); b) lattice of focal parabolae. It can be shown [7] that this arrangement does not fulfil G. Friedel's association rules; c) sketch of a layer seen in perspective (from ref. [5a]).

Two particular cases can be distinguished. The first one corresponds to $e_1 = e_2 = 1$: the two conics are parabolae. Such domains can easily nucleate under dilation normal to the layers and assemble to form a lattice (Figs. C.II.10

and 11) [5–8]. In a "parabolic domain," the layers have a rather peculiar shape, shown in figure C.III.2c [5a,b]. The other extreme case corresponds to $e_1 = \infty$ and $e_2 = 0$, the two conics degenerating in a straight line and a circle, respectively (Fig. C.III.3). Such a structure is called a "toric domain," the layers having the shape of concentrical tori. It should also be noted that, in an FD1, only the segments joining the two conics have physical reality. Otherwise stated, an FD1 is always bound on the outside by a revolution cone, having its vertex on the hyperbola and the generatrices passing through the ellipse. It is interesting to know that the entire space can be filled with such cones in an iterative manner [9–11], provided some association rules [1,4a–c] – definitively stated by G. Friedel in 1922 – are respected. Finally, we note that the only cases in which the domain is compatible with a planar layer structure at infinity are the parabolic domain (Fig. C.III.2) and the toric domain (Fig. C.III.3).

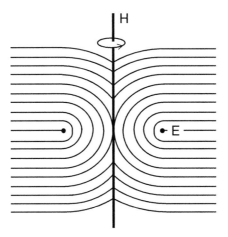

Fig. C.III.3 Toric domain (FD1).

In a focal domain of the second type (FD2), the layers have, on the contrary, a positive Gaussian curvature ($1/R_1 R_2 > 0$). These domains are obtained by considering that the physical parts of the layer normals passing through the two conics are the two half-lines originating on the hyperbola (Fig. C.III.4). In such a domain, with a structure resembling that of an onion, only the hyperbola is singular. In the limiting case when both the ellipse and the hyperbola are reduced to a point, the layers are spheres centered on this point. Such a domain is called a spherulite or an onion.

Experimentally, only focal domains of the first type are encountered in thermotropic liquid crystals. Anyway, they were the only ones considered by G. Friedel. The situation is sometimes different in lyotropics, where the two types of domains can appear. However, the domains of the second type are rarely encountered and only develop under very specific experimental conditions, for instance after prolonged shear of a lamellar lyotropic phase

[12a–e] or upon approaching the transition toward a direct micellar phase [3]. In addition, these defects seem to exist usually under the form of spherulites, wherein the ellipse and the hyperbola are replaced by the same singular point, with a lower energy cost.

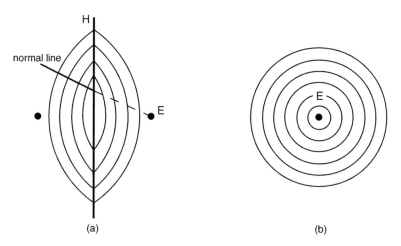

(a)   (b)

Fig. C.III.4 a) Focal domain of the second type when the ellipse is a circle and the hyperbola a straight line. b) Spherulite or onion (the ellipse is reduced to a point).

The appearance of one type of domain rather than the other (FD1 or FD2) in a given material is closely related to the sign and the absolute value of the Gaussian curvature constant $\bar{K}$ of the layers, as shown by the following calculations.

III.1.b) Energy of a focal domain

Kléman et al. [13, 14] calculated the curvature energy of a focal domain in the general case.

Let us start by considering the case of an FD1. Let e be the eccentricity of the ellipse and a its major axis. The energy is obtained by integrating the curvature energy:

$$\rho f_c = \frac{1}{2} K \left(\frac{1}{R_1} + \frac{1}{R_2} \right)^2 + \bar{K} \frac{1}{R_1 R_2} \qquad (C.III.1)$$

over the entire volume of the focal domain, taken as limited by two semi-infinite cylinders, having as basis the ellipse and of generatrices parallel to the two asymptotes of the hyperbola. The result of the calculation reduces to a relatively simple form when the ellipse E is close to a circle (e small) [14]:

$$E_c(DF1) = 2\pi(1 - e^2)K(e^2) \left[2\ln\left(\frac{a}{r_c}\right) + 1 - 2x \right] aK + (2\ln 2 - 5)\pi^2(1 - e^2)aK$$

$$(C.III.2)$$

where $K(e^2)$ is the complete elliptic function of the first kind ($K(0) = \pi/2$). In this expression, $x = \bar{K}/K$ and r_c is a core radius along the ellipse and the hyperbola, of the order of the layer thickness. To the curvature energy, one must add a core energy close to

$$E_{core}(FD1) \approx \xi KL_E \tag{C.III.3}$$

where ξ is a geometric constant and L_E the length of the ellipse ($L_E = 4aK(e^2)$).

These formulae show that the curvature energy of a focal domain of the first type is proportional to the major axis of the ellipse and decreases for $\bar{K} > 0$ ($x > 0$).

Conversely, the FD2 are favored when $\bar{K} < 0$, as shown by calculating the curvature energy of a spherulite with radius R:

$$E_c(\text{spherulite}) = 2\pi(4K + \bar{K})R \tag{C.III.4}$$

Throughout these calculations, we have neglected the layer compressibility. Is this approximation justified? Kléman and Parodi were the first to show that the ideal focal domains do not exactly satisfy the equilibrium equation of bulk forces [15]. This can be shown by adding to the curvature energy C.III.1 the layer compressibility term (we set $\bar{B} = B$ for simplicity in this chapter):

$$\rho f_e = \frac{1}{2} B \left(\frac{a - a_0}{a_0}\right)^2 \tag{C.III.5}$$

and by minimizing the total energy in its covariant formulation:

$$E = \int dV \left\{ \frac{1}{2} K \left(\frac{1}{R_1} + \frac{1}{R_2}\right)^2 + \bar{K} \frac{1}{R_1 R_2} + \frac{1}{2} B \left(\frac{a - a_0}{a_0}\right)^2 \right\} \tag{C.III.6}$$

This calculation, recently simplified by Fournier [16], leads to the following equilibrium equation:

$$G = \text{div}(K \, \mathbf{V}_{//}\Sigma + B\varepsilon \, \mathbf{n}) = 0 \tag{C.III.7}$$

In this expression, $\mathbf{V}_{//}$ is the gradient parallel to the layers $\mathbf{V}_{//} = \mathbf{V} - \mathbf{n}(\mathbf{n}.\mathbf{V})$, Σ the total curvature of the layers ($\Sigma = \text{div} \, \mathbf{n} = 1/R_1 + 1/R_2$) and ε the layer dilation $\varepsilon = (a - a_0)/a_0$. Note that only first-order terms in ε were kept, since $\varepsilon \ll 1$ in classical elasticity. This equation translates the equilibrium of forces normal to the layers originating in layer dilation, on the one side, and in the tangential variation of the curvature torques (expressed by $\mathbf{C} = \mathbf{n} \times K\mathbf{V}_{//}\Sigma$). Fournier also showed that eq. C.III.7 could be rewritten in the equivalent form:

$$G = \mathbf{n}.\text{div}\,\underline{\sigma}^E \qquad\qquad\qquad\qquad\qquad\qquad\qquad (\text{C.III.8})$$

where σ^E is the elastic stress tensor, given in covariant expression by:

$$\underline{\sigma}^E = B\,\varepsilon\,\mathbf{n}\otimes\mathbf{n} + K\,\mathbf{n}\otimes\boldsymbol{\nabla}_{/\!/}\Sigma \qquad\qquad\qquad (\text{C.III.9})$$

with $(\mathbf{a}\otimes\mathbf{b})_{ij} = a_i b_j$. Let us now return to ideal focal domains. Inside them, $\varepsilon = 0$ by definition; on the other hand, it can be shown that

$$\text{div}\,(K\,\boldsymbol{\nabla}_{/\!/}\Sigma) \neq 0 \qquad\qquad\qquad\qquad\qquad (\text{C.III.10})$$

Consequently, the equilibrium equation C.III.7 cannot be strictly satisfied inside a focal domain (except for a spherulite). We conclude that the layers must be slightly compressed (or dilated) inside a domain, as Kléman and Parodi had already pointed out in 1975 [15]. One then needs to calculate the associated energy contribution. This question is not yet solved, in spite of recent progress by Fournier. This author showed the existence around each singular line of a "dilation sheath," well separated from the core, where the elastic stress decreases algebraically [17]. More precisely, he showed that the layers are dilated close to the core, where the stress relaxes as $1/r^2$ while far from the core the layers are compressed, the stress decreasing more slowly, as $1/r$. Unfortunately, this model contains an undetermined parameter, difficult to estimate theoretically unless the boundary conditions at the domain borders are known precisely.

So far, we have only considered isolated domains. Experimentally, these domains can associate and form (sometimes very complex) structures. The lattice of focal parabolae that develops on dilating a homeotropic sample is an example. In the next section, we shall briefly describe other aggregation patterns of conics, frequently observed in samples. We shall assume that the conics are close to the geometrical models of G. Friedel, which is clearly the case in experiments.

III.1.c) Oily streaks and Grandjean walls

In homeotropic geometry, one very often notices in the sample bright lines, sometimes as wide as the sample thickness (Fig. C.III.5). These lines, also visible in cholesterics [18a,b], are very often formed by a regular stacking of focal conics. They were called "oily streaks" by G. Friedel [1] and can be interpreted as "giant" dislocations of very large Burgers vector having undergone a structural transformation [14, 18–20]. This transformation, leading to the appearance of focal conics, is explained in figure C.III.6.

The first diagram (a) represents a dislocation with Burgers vector b, formed by two dislocations with Burgers vectors of opposite sign ($b = b_1 - b_2$). To avoid layer compression or dilation, a discontinuity wall was introduced. If the layers are to join seamlessly along this wall, the additional condition

$$\left|ML_1 - ML_2\right| = \frac{b}{2} \tag{C.III.11}$$

must be fulfilled, where M is a running point on this surface. The equation defines a hyperbolic cylinder of eccentricity

$$e_H = \frac{b_1 + b_2}{b_1 - b_2} \tag{C.III.12}$$

Such a wall being energetically unfavorable, the system will rather form an array of focal domains (Fig. C.III.6b) where each ellipse has an eccentricity $e_E = 1/e_H$, a major axis $a_E = (b_1 + b_2)/2$ and a minor axis $b_E = (b_1b_2)^{1/2}$. As to the conjugated hyperbolae, they inherit the eccentricity e_H of the discontinuity surface that engendered them.

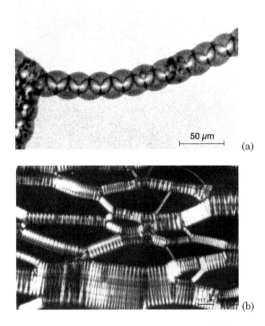

(a)

(b)

Fig. C.III.5 Examples of oily streaks observed in unpolarized light in a homeotropic 8CB sample (a) and, between crossed polarizers, in a homeotropic sample of a lamellar lyotropic phase L_α (b) (from ref. [20]).

The focal domains are also involved in the formation of **walls** separating two disoriented monodomains. Let $2\theta_\infty$ be the misfit angle of the layers across such a wall, assumed to be **symmetric**.

If this angle is small, experiments show that the layers are continuously bent, without breaking. Such a wall is termed a **curvature wall** (Fig. C.III.7).

It is easily shown that the local misfit angle varies with the coordinate x across the wall thickness as $\theta_\infty \tanh(\theta_\infty x/2\lambda)$, and that the wall energy E_c per unit surface is [9, 21]:

$$E_c = 2\sqrt{KB}\left(tg\theta_\infty - \theta_\infty\right) \approx \frac{2}{3}\sqrt{KB}\,\theta_\infty^3 \qquad\qquad \text{(C.III.13)}$$

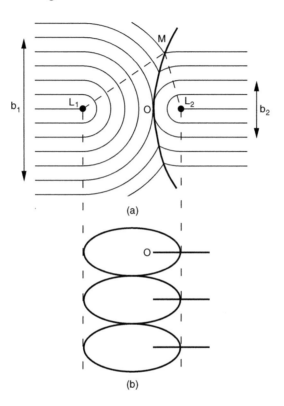

(a)

(b)

Fig. C.III.6 Formation of ellipses along an oily streak composed of two adjoining dislocations of Burgers vectors b_1 and b_2.

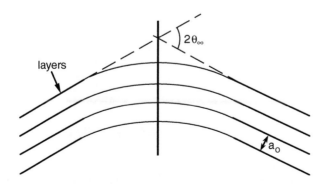

Fig. C.III.7 Bend wall in a smectic A.

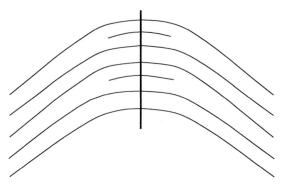

Fig. C.III.8 Mixed wall. The presence of dislocations allows the layer dilation to relax at the center of the wall.

As this energy increases very abruptly with the angle θ_∞, these walls become unfavorable as soon as θ_∞ exceeds a few degrees. It is then preferable to introduce pairs of edge dislocations at regular intervals in order to relax the layer dilation at the center of the wall (Fig. C.III.8). This result was confirmed experimentally by Williams and Kléman [21, 22]. It is then very easily checked that the optimum distance Λ between two successive pairs of elementary dislocations (with a Burgers vector b equal to the layer thickness a_0) is given by $\Lambda = 2a_0/\theta_\infty^2$. The energy of the wall is then mainly due to the dislocations that, as we shall see later, also contribute their own energy, of the order of K. Admitting this result, one finds for a combined wall:

$$E_{comb} \approx \frac{K}{a_0}\theta_\infty^2 \qquad\qquad (C.III.14)$$

In this case, the energy increases less rapidly with θ_∞ than in the previous model. Experimentally, combined walls are obtained for angles θ_∞ between a few degrees and 25°.

Fig. C.III.9 Grandjean walls in a thin planar sample. The layers are perpendicular to the glass plates, but there is no preferred orientation in the horizontal plane (from Demus and Richter [23]).

For even larger angles, Grandjean walls, or **focal domain walls**, appear Such walls are frequently seen in planar samples with degenerate anchoring (Fig. C.III.9).

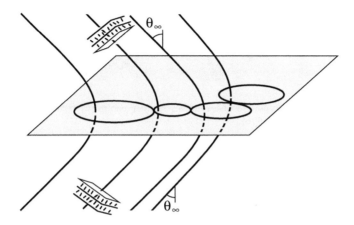

Fig. C.III.10 Schematic representation of a Grandjean wall.

In this case, the symmetry plane of the wall is tiled with ellipses, tangential to each other. At the same time, the asymptotes of the hyperbolae conjugated to the ellipses are parallel to each other and perpendicular to the average direction of the layers far from the wall (Fig. C.III.10). It is easily shown that, for each ellipse, the ratio a/b between the major and the minor axes is a constant, related to θ_∞ by the geometrical relation

$$\mathrm{cotg}^2\theta_\infty = \frac{a^2}{b^2} - 1 \tag{C.III.15}$$

However, the size of each individual ellipse is free to vary.

Fig. C.III.11 "Friedel bâtonnets" observed at the smectic A-isotropic liquid transition (from ref. [23]).

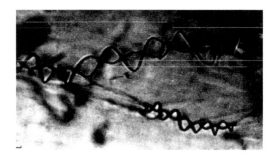

Fig. C.III.12 Double helix structure observed on rapidly cooling a twist line from the nematic phase (photo by C.E. Williams, 1976)

Other types of focal conic association are also possible. The most classical are the "Friedel bâtonnets" [1], which appear when the smectic phase grows in its isotropic liquid (Fig. C.III.11), or the helical structures of C. Williams (simple and double helices, [21, 24]) generated by quenching the twist lines of the nematic phase, often present at higher temperature (Fig. C.III.12).

III.2 Dislocations

III.2.a) Definitions

By definition, these defects break the translational symmetry of the lamellar structure. Let **b** be their Burgers vector and **t** the unit vector tangent to the line, and let us adopt the conventions used in solids for defining the orientation **b**. We recall that for **edge dislocations b⊥t** (Fig. C.III.13), while **b//t** for **screw dislocations** (Fig. C.III.14). Since these defects have very different properties, we analyze them separately.

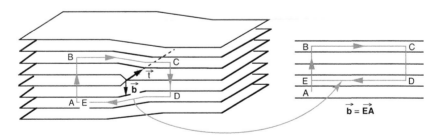

Fig. C.III.13 Edge dislocation in a smectic. To define the orientation of **b** with respect to the unit vector **t** tangent to the line, consider the image in the perfect material of the Burgers circuit enclosing the line. This circuit is oriented using the usual right-hand rule. The image circuit has a closure defect that we take as equal to the Burgers vector: **b** = **EA**.

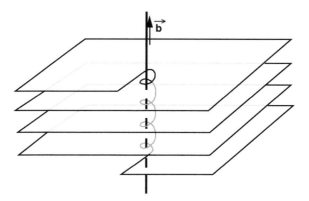

Fig. C.III.14 Screw dislocation in a smectic. The layers have the shape of a director-plane helicoid.

We shall also see that, due to the fluidity of the layers, important differences exist between smectics and usual solids.

III.2.b) Edge dislocations

i) Experimental observation

There are several methods for visualizing these defects. The most elegant one was discovered by Meyer et al [25a] and Lagerwall et al. [25b] and is applicable to materials that exhibit a smectic A-smectic C phase transition.

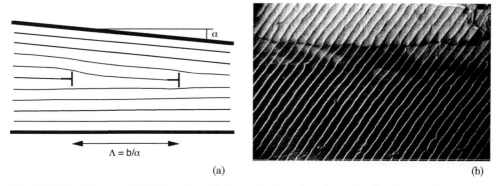

(a)   (b)

Fig. C.III.15 a) Geometrical dislocations that must be introduced to relax the dilation of the layers in homeotropic anchoring when the plates make a small angle α; b) close to the SmA-SmC phase transition, the dislocations become visible, as they locally induce the tilted SmC phase (photo by S. Lagerwall, 1978).

The homeotropic sample is sandwiched between two glass plates making a small angle α. Due to the wedge geometry, edge dislocations must be introduced at regular intervals in order to compensate the variation in sample thickness (Fig. C.III.15a). If the angle α is small, these dislocations are elementary (b = a_o) and the average distance between them is Λ = b/tanα ≈ b/α. In the smectic A phase, these dislocations are difficult to see using the polarizing microscope, since the layer distortions they induce are too small. Nevertheless, they become perfectly visible very close to a smectic A-smectic C second-order phase transition, as they locally compress the layers, favoring the smectic C phase in their vicinity. In this way, the dislocations are "decorated," and the local tilt of the molecules (and hence of the optical axis) is easily detectable in the microscope between crossed polarizers (Fig. C.III.15b).

Another optical method, based on fluorescence, also allows the detection of the lattice formed by the elementary edge dislocations inside a wedge filled with a lamellar lyotropic phase L_α [26]. The material (a phospholipid) is doped with a fluorescent molecule of a similar chemical structure. After prolonged annealing over a few weeks, allowing the dislocations to form a well-ordered lattice (a few minutes to a few hours are usually sufficient for thermotropics), the sample is observed in fluorescence. More precisely, the sample is scanned along a direction perpendicular to the wedge with a very fine laser beam. The beam excites the fluorescent molecules, and the emitted intensity is recorded as a function of position (Fig. C.III.16). The steps appearing on the intensity graph mark the presence of elementary edge dislocations parallel to the wedge.

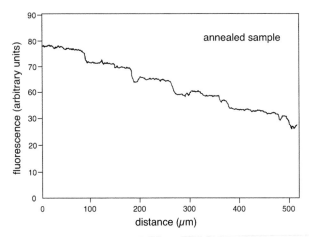

Fig. C.III.16 Record of the fluorescence intensity emitted as a function of position. Each step corresponds to the end of a lamellum and marks the presence of a dislocation. The residual noise represents about 10% of the step height (from ref. [26])

The elementary edge dislocations are also visible in free-standing films [27], as shown in Fig. C.III.17. Here, the observations were made using reflection microscopy in monochromatic light to increase the contrast of the lines (see chapter C.V).

Electron microscopy equally provides a way of observing the elementary edge dislocations. Such an experiment was recently performed on the lamellar phase of a diblock copolymer on a solid substrate [28a]. The copolymer (after being dissolved in a solvent) is spread onto the substrate by centrifugation. In the solid phase initially, it is then heated above its glass transition temperature; the lamellar order appears, with the lamellas parallel to the substrate. Since the quantity of spread material does not in general correspond to an integral number of lamellae, micron-sized circular islets appear at the surface of the layer and can be clearly seen in reflection microscopy. Atomic force microscopy (AFM) studies of the free surface of the film, as well as observation by transmission electron microscopy (TEM) of thin slices cut perpendicular to the film, revealed that the edge of an islet corresponds to an elementary edge dislocation localized in the thickness of the film. These observations, as well as those on free-standing films, show that the dislocations are repelled by a free surface, behavior without any analogy in solids.

Fig. C.III.17 "Arch-like" texture in a free-standing smectic A film. There is a difference of one layer (or sometimes more) between two areas with different gray levels. The separation lines correspond to edge dislocations (elementary, most of the time), centered in the thickness of the film (photo by F. Picano, 2000).

In conclusion, we would like to point out that "giant" dislocations of large Burgers vector (a few tens to a few hundreds of layers) are frequently encountered in smectics. Very conspicuous in the microscope due to the characteristic structure of their cores (see figure C.III.19), they are spontaneously formed by bunching of elementary edge dislocations. This mechanism was first observed by Williams and Kléman by shearing a planar sample perpendicular to the layers [21, 22]. Giant dislocations also appear on violently compressing a homeotropic sample.

ii) Elastic energy and deformation field

Let $\mathbf{t}$ be the unit vector tangent to the line and $\mathbf{b}$ its Burgers vector ($b = na_0$, with n integer). We assume for the moment that the sample is infinite. The strain field u around the dislocation must obey the equilibrium equation in the bulk (see eq. C.II.29):

$$\frac{\partial^2 u}{\partial z^2} - \lambda^2 \frac{\partial^4 u}{\partial x^4} = 0 \quad \text{with} \quad \lambda = \sqrt{\frac{K}{B}} \tag{C.III.16}$$

with the boundary conditions on the cut surface (Fig. C.III.18):

$$u(z = 0, x \leq 0) = 0 \tag{C.III.17a}$$

$$u(z = 0, x > 0) = \text{sgn}(z)\frac{b}{2} \tag{C.III.17b}$$

where sgn(z) represents the sign of z.

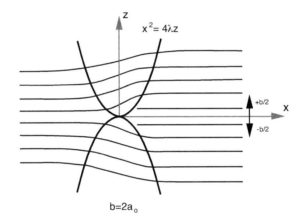

Fig. C.III.18 Edge dislocation in a smectic A. The stress is significant within the two parabolae drawn in heavy line.

It is easy to check that the solution to this problem reads [29–31]

$$u(x, z) = \frac{b}{4}\text{sgn}(z)\left(1 + \text{erf}\frac{x}{2\sqrt{\lambda|z|}}\right) \tag{C.III.18}$$

where erf designates the error function:

$$\text{erf}(t) = \frac{2}{\sqrt{\pi}}\int_0^t e^{-u^2} du \tag{C.III.19}$$

The function goes from -1 to $+1$ as t goes from $-\infty$ to $+\infty$.

This expression yields the tilt angle of the layers $\partial u/\partial x$, as well as the dilation $\partial u/\partial z$:

$$\frac{\partial u}{\partial x} = \frac{\text{sgn}(z)b}{4\sqrt{\pi\lambda|z|}}\exp\left(-\frac{x^2}{4\lambda|z|}\right) \tag{C.III.20}$$

$$\frac{\partial u}{\partial z} = - \frac{b}{8\sqrt{\pi\lambda}} \frac{x}{|z|^{3/2}} \exp\left(-\frac{x^2}{4\lambda|z|}\right)$$

(C.III.21)

One can notice that the strain (and hence the stress) propagates far away in the glide plane perpendicular to the layers and decreases exponentially (as $\exp(-x^2)$) in the plane of the layers. This behavior is very different from that of solids, where the stress decreases isotropically, with a $1/r$ dependence. This very strong anisotropy of the deformation field is due to the fluidity of the layers. More specifically, the deformations are only important within the two parabolae of equations (Fig. C.III.18):

$$x = \pm 2\sqrt{\lambda|z|}$$

(C.III.22)

and which, for this reason, are called "deformation parabolae."

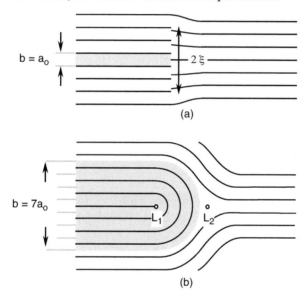

(a)

(b)

Fig. C.III.19 Core model for an edge dislocation. a) Core spread in the glide plane; b) core of a giant dislocation split into two disclination lines of rank 1/2 (L_1) and $-$ 1/2 (L_2).

The stress is calculated from formulae C.III.20–21, as well as the energy of the dislocation per unit length [19, 30–33]:

$$E_{edge} = \frac{Kb^2}{2\lambda\xi} + E_c$$

(C.III.23)

In this calculation, ξ is a cut length along the z direction, quantifying the spread of the core in the glide plane (Fig. C.III.19a) and E_c is the corresponding core energy. Writing that $E_c \approx \gamma_c \xi$, where γ_c is the energy of the cut, minimization with respect to ξ results in a core size proportional to b, more precisely $\xi = b\,(B\lambda/2\pi\gamma_c)^{1/2}$. Substituting in eq. C.III.23 yields:

$$E_{edge} \approx \sqrt{2/\pi} \; B^{1/4} K^{1/4} \gamma_c^{1/4} b \qquad\qquad\qquad (C.III.24)$$

This formula shows that the **energy of an edge dislocation is proportional to its Burgers vector** (and not to b^2 as in solids). This result depends however on the assumption that the core energy is proportional to ξ. We shall see in chapter C.VIII that this prediction is well verified in experiments. Thus, a dislocation with Burgers vector 2b will not tend to split into two dislocations of smaller Burgers vector b, as happens in the solids. We understand better now why dislocations with a large Burgers vector are so frequent in smectics A.

For very large b, Williams and Kléman propose a different core model [22], whereby the dislocation is split into two disclination lines of rank $+1/2$ and $-1/2$, separated by a distance b/2 (Fig. C.III.19b). Within this model, the core energy is given by:

$$E_{core} \approx \frac{\pi K}{2} \ln\left(\frac{b}{2r_c}\right) + E_c \qquad\qquad\qquad (C.III.25)$$

The first term represents the curvature energy of the intermediate layers, r_c (of the order of a_o) is the core radius of the two disclinations and E_c their core energy. This structure is expected for Burgers vectors larger than five layers (typically).

iii) Peach and Koehler force

It is known that, in solids, a stress field $\underline{\sigma}$ exerts on a dislocation a configurational force given by the Peach and Koehler formula [34]:

$$\mathbf{F} = (\underline{\sigma}^t \mathbf{b}) \times \mathbf{t} \qquad\qquad\qquad (C.III.26)$$

In this formula, $\underline{\sigma}^t$ is the transpose of the stress tensor (due to the other defects or to the applied stress), calculated at the position of the line, $\mathbf{b}$ the Burgers vector and $\mathbf{t}$ the unit vector tangent to the line (Fig. C.III.13). In solids (but not in smectics), the stress tensor is symmetric, resulting in the usual formula $\mathbf{F} = (\underline{\sigma}\mathbf{b}) \times \mathbf{t}$. This force is defined such that its work Fx (x being the displacement of the dislocation) is equal to the work of the stress σSl (S is the surface on which the stress is exerted and l the macroscopic displacement that occurs when the dislocation moves by x). Kléman and Williams showed that formula C.III.26 remains valid for smectics [32], provided one takes as stress tensor the elastic stress $\underline{\sigma}^E$ calculated in the previous chapter (eq. C.II.65). The non-zero components of the force acting on an edge dislocation parallel to the y axis ($\mathbf{t} = \mathbf{y}$) and such that $\mathbf{b} = b\mathbf{z}$ (with b positive or negative), are then explicitly, from eqs. C.III.26 and C.II.64:

$$F_x = -\sigma_{zz}^E b = -B \frac{\partial u}{\partial z} b \qquad\qquad\qquad (C.III.27a)$$

$$F_z = \sigma^E_{zx} b = -K\left(\frac{\partial}{\partial x}(\Delta_\perp u)\right) b \qquad\qquad\text{(C.III.27b)}$$

The first component is due to layer compression: this is the **climb** force tending to move the dislocation parallel to the layers. The second component, termed **glide** force, is associated with the curvature of the layers: it induces dislocation glide in a plane perpendicular to the layers.

These two formulae can be used to calculate the force exerted by a dislocation on another one. To simplify, let us assume that the two dislocations are parallel, the first one (of Burgers vector b_1) placed at the origin, and the second one (of Burgers vector b_2) at coordinates (x_0, z_0). The force exerted by the first one on the second is, from eqs. C.III.20, 21, 27a and b:

$$F_x = \frac{Bb_1b_2}{8\sqrt{\pi\lambda}}\frac{x_0}{|z_0|^{3/2}}\exp\left(-\frac{x_0^2}{4\lambda|z_0|}\right) \qquad\qquad\text{(C.III.28a)}$$

$$F_z = \frac{sg(z_0)Kb_1b_2}{16\sqrt{\pi}\lambda^{3/2}|z_0|^{3/2}}\left(2 - \frac{x_0^2}{\lambda|z_0|}\right)\exp\left(-\frac{x_0^2}{4\lambda|z_0|}\right) \qquad\qquad\text{(C.III.28b)}$$

These formulae show that, if the two dislocations have opposite signs, the climb force F_x exerted by the dislocation placed at the origin on the other dislocation is everywhere attractive, while the glide force is repulsive outside the parabola of equation $x^2 = 2\lambda|z|$ and attractive elsewhere (Fig. C.III.20). One can also see that, due to the exponential decay of the stress in the plane of the layers, the elastic interaction between the two dislocations is significant only when they are more or less in the same glide plane. In particular, their elastic interaction vanishes exactly when they are in the same climb plane ($\mathbf{F} = 0$ for $z_0 = 0$).

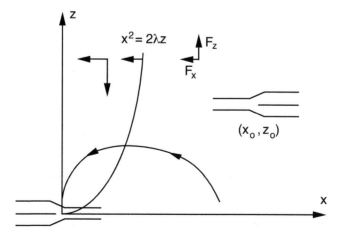

Fig. C.III.20 Schematic representation of the interaction between two parallel dislocations of opposite signs. The arrowed line gives the direction and the orientation of the interaction force exerted by the dislocation at the origin on the other dislocation (from ref. [32]).

iv) Interaction with a free surface or a solid wall

Experiments show that the edge dislocations are repelled by the free surface in contact with the air, unlike the case of solids. To explain this fact, one must take into account the surface tension γ and the excess free energy due to the deformation of the surface [35]. Let us consider an edge dislocation at a distance z_0 from the surface, that we take as parallel to the layers. The smectic phase occupies the half-space $z < 0$. The total energy contains a bulk and a surface contribution given, in the small deformation limit, by:

$$F = \frac{1}{2} \int_V \left[K \left(\frac{\partial^2 u}{\partial x^2} \right)^2 + B \left(\frac{\partial u^2}{\partial z} \right) \right] dxdz + \frac{\gamma}{2} \int_S \left(\frac{\partial u}{\partial x} \right)^2 dx \qquad \text{(C.III.29)}$$

Minimizing the free energy yields the usual bulk equation:

$$B \frac{\partial^2 u}{\partial z^2} = K \frac{\partial^4 u}{\partial x^4} \qquad \text{(C.III.30a)}$$

to which one must add the boundary condition

$$B \frac{\partial u}{\partial z} = \gamma \frac{\partial^2 u}{\partial x^2} \quad \text{in} \quad z = 0 \qquad \text{(C.III.30b)}$$

On the right-hand side of this expression, we recognize the Laplace excess pressure in γ/R, with R the local curvature radius of the interface.

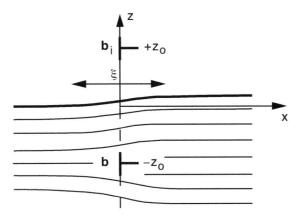

Fig. C.III.21 Interaction of a dislocation with a free surface. The boundary conditions at the free surface can be satisfied by introducing an image dislocation.

To satisfy the stress equilibrium condition at the free surface, a possible calculation method consists of introducing one (or more) image dislocations on the other side of the interface [35]. In the present case, one image dislocation, placed symmetrically with respect to the interface, and of Burgers vector b_i given by

$$b_i = b \frac{\gamma - \sqrt{KB}}{\gamma + \sqrt{KB}} \tag{C.III.31}$$

is sufficient. Note that b_i is not necessarily an integral multiple of a_o, this dislocation being **virtual**. It is easily checked that the deformation field given by this superposition,

$$u(x,z) = \frac{b}{4} sgn(z + z_0) \left(1 + erf \frac{x}{2\sqrt{\lambda|z + z_0|}} \right)$$

$$+ \frac{b_i}{4} sgn(z - z_0) \left(1 + erf \frac{x}{2\sqrt{\lambda|z - z_0|}} \right) \tag{C.III.32}$$

satisfies the equilibrium equations C.III.30a and b. This formula also yields the equation, $z = u(x,0)$, of the free surface deformed by the presence of the dislocation:

$$z = \frac{b}{2} \frac{\sqrt{KB}}{\gamma + \sqrt{KB}} \left(1 + erf \frac{x}{2\sqrt{\lambda|z_0|}} \right) \tag{C.III.33}$$

It can be seen that the surface is deformed over a typical width $\xi \approx 2\sqrt{\lambda|z_0|}$ and tilts with respect to the horizontal by a maximum angle

$$\theta_M = \frac{b}{2} \frac{\sqrt{KB}}{\gamma + \sqrt{KB}} \frac{1}{\sqrt{2\lambda|z_0|}} \tag{C.III.34}$$

As an example, one obtains $\xi \approx 6b$ and $\theta_M \approx 2°$ for $\lambda = b = a_o$, $\gamma/(KB)^{1/2} = 3$ and $z_0 = 10a_o$. We note that the surface profiles determined by AFM in copolymer films are in excellent agreement with this theory [28b].

We would also like to point out that the free surface exerts on the dislocation (via the image dislocation) a glide force F_z that can be estimated from eq. C.III.28b:

$$F_z = - \frac{Kb^2}{16\sqrt{2\pi}(\lambda z_0)^{3/2}} \frac{\gamma - \sqrt{KB}}{\gamma + \sqrt{KB}} \tag{C.III.35}$$

This important formula shows that the interaction is repulsive ($F_z < 0$) for $\gamma > (KB)^{1/2}$, and attractive ($F_z > 0$) for $\gamma < (KB)^{1/2}$. In general, the surface tension in contact with air $\gamma \approx 30$ erg/cm^2 is larger than $\sqrt{(KB)^{1/2}} \approx 10$ erg/cm^2 since, typically, $B \approx 10^8$ erg/cm^3 and $K \approx 10^{-6}$ dyn. The situation must be different at the interface between a smectic phase and its isotropic liquid, γ being much smaller ($\gamma \approx 0.1-1$ erg/cm^2, typically). In this case, the dislocations should be attracted by the interface and form steps. Such steps were observed by Galerne and Liébert at the interface between a smectic O phase (antiferroelectric, also called SmC$_A$, see chapter A.II, volume 1, p. 27 and section C.IV.8) and its isotropic liquid [36].

The preceding calculation can also be applied to the case of a rigid solid wall. It is enough to take γ to infinity in the equations. In particular, eq. C.III.35 shows that a solid wall always repels the dislocations, as was already shown by Pershan in 1974 [37].

The situation gets more complicated when the dislocation is sandwiched between two interfaces. In this case, the two boundary conditions can only be satisfied by introducing an infinity of image dislocations. The general solution to this problem is given in Lejcek and Oswald [35]. One can also cast the solution in the form of an integral over Fourier space, which can be more convenient than an infinite discrete series. This calculation method was proposed independently by Turner et al. [28b] and Holyst and Oswald [38–40]. It is worth noting that the dislocation is in stable equilibrium in the center of the sample when the two limiting interfaces are identical and stiff enough to repel the dislocation (smectic free film, homeotropic sample between two glass plates, etc.). The dislocation can, however, be trapped in a different, metastable, position. This can occur if the glide force acting on the dislocation is much smaller than the force pinning the dislocation in a Peierls valley. This pinning force given by the lattice potential (also termed Peierls-Nabarro force) was calculated by Lejcek in smectics A. It involves the width ξ of the core in the glide plane, defined here as the distance over which the layer displacement varies from 0 to b/4. This force is [41]:

$$F_{P.N.} = \sigma_{P.N.}\, b \qquad \text{with} \qquad \sigma_{P.N.} = \frac{3}{4} Bb \sqrt{\frac{\pi\lambda}{\xi}} \exp\left(-\frac{2\pi\xi}{b}\right) \qquad (C.III.36)$$

The quantity $\sigma_{P.N.}$ is the Peierls-Nabarro stress. As for solids, $F_{P.N.}$ strongly depends on ξ. If the dislocation is not dissociated ($\xi \approx b/2$), $F_{P.N.}$ is considerable (of the order of Bb) and, in general, much larger than the glide force exerted by a wall placed at a distance z_0 from the dislocation (of the order of $0.02\, Bb(b/z_0)^{3/2}$ in the case of a solid wall, eq. C.II.35). The dislocation is then strongly pinned in its Peierls valley. The situation can change if the dislocation core spreads, as can happen close to a second-order smectic A-nematic transition. In this case, the Peierls-Nabarro force "collapses" and the dislocation can easily glide toward its equilibrium position.

To conclude this section, let us give the expression for the free energy of an edge dislocation placed at the center of a free-standing film with N layers and surface tension γ:

$$E_{edge} = E_\infty + \delta E_s \qquad (C.III.37)$$

In this formula, $E_\infty \propto b$ is the energy in an infinite medium (see eq. C.III.24) and δE_s is a surface correction given by the image dislocations. Since there is an infinity of them, this correction can be written as:

$$\delta E_s = \frac{B\lambda b^2}{4\sqrt{\pi\lambda a_0}(N + 1/2)} Li_{1/2}(A) \qquad (C.III.38)$$

where $A = \dfrac{\gamma - \sqrt{KB}}{\gamma + \sqrt{KB}}$ and $Li_{1/2}(A)$ is a special function given by:

$$Li_{1/2}(A) = \sum_{p=1}^{\infty} A^p \frac{1}{\sqrt{p}}$$

(C.III.39)

This function is plotted in figure C.III.22.

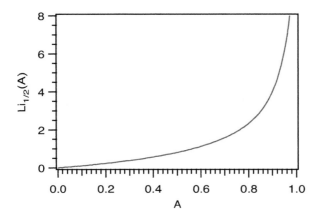

Fig. C.III.22 Polylogarithm function $Li_{1/2}$.

It diverges in $A = 1$. It is also instructive to calculate the correction δE_s in the $B \rightarrow 0$ limit (K and γ remaining finite). This occurs close to the transition temperature N-SmA, when the transition is second order. In this limit, we find:

$$\delta E_s = \frac{B^{1/2}\gamma^{1/2}b^2}{4\sqrt{\pi a_o}(N+1/2)}$$

(C.III.40)

This surface correction decreases "slowly" as $1/\sqrt{N}$ and only becomes negligible in very thick films, of several hundred layers. It also increases as b^2. This dependence explains why it is very difficult to obtain dislocations with large Burgers vectors in thin films, while giant dislocations are frequently encountered in bulk samples (very thick, by definition).

v) Dislocation bunching inside a wedge

We consider again a homeotropic sample limited by two plates making a small angle α. Several experiments showed that the dislocations bunch together as the angle of the wedge increases [42–44]. This phenomenon is of the same nature as the formation of double lines in a Cano wedge filled with a cholesteric phase (section B.VII.4.c in volume 1). To understand this bunching, one must compare the energy of the two configurations drawn in figure C.III.23. In the first one, the dislocations are elementary (of Burgers vector $b = a_o$), while in the second one, we consider that the dislocations regroup two by two and form dislocations of Burgers vector 2b. In the first configuration (Fig. C.III.23a), the energy of the grayed area is:

$$E_1 = 2E_{edge}(b) + \frac{Bb^3}{12\alpha d} \qquad \text{(C.III.41a)}$$

with d the sample thickness. The first term $2E_{edge}(b)$ represents the self-energy of the two dislocations; the second one corresponds to the elastic energy of layer compression and dilation on either side of the dislocations.

A similar calculation yields, for the second configuration (Fig. C.III.23b):

$$E_2 = E_{edge}(2b) + \frac{Bb^3}{3\alpha d} \qquad \text{(C.III.41b)}$$

If $2E_{edge}(b) > E_{edge}(2b)$, the second configuration becomes energetically favorable as soon as the angle α exceeds the critical angle α_c given by

$$\alpha_c = \frac{Bb^3}{4d[2E_{coin}(b) - E_{coin}(2b)]} \qquad \text{(C.III.42)}$$

Experimentally, this angle is close to 10^{-3} rad [42] in thermotropics for a typical thickness of 100 μm.

One can thus conclude that $2E_{edge}(b) - E_{edge}(2b) \approx 10^{-3}$ K, meaning that $E_{edge}(2b) \approx 2E_{edge}(b)$ in agreement with the theoretical predictions (eq. C.III.24). This small energy difference can result from subtle changes in the core of the defects and from the approximations used (for instance, we neglected the anharmonic terms, but this does not entail a qualitative change, as shown by the more rigorous calculation of Brener and Marchenko, in particular in the limit of penetration length λ larger than the layer thickness [45]).

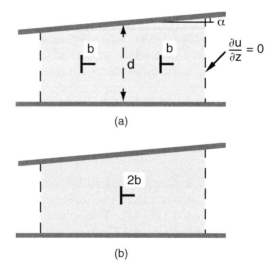

(a)

(b)

Fig. C.III.23 Two possible configurations inside a wedge. In (a), the dislocations are elementary; in (b), they are bunched together two by two.

vi) Friction force on an edge dislocation

The most interesting case from the experimental point of view is the one of **pure climb**. In this case the movement is not conservative, as one must locally "carry" matter by diffusion. Let V be the velocity of the dislocation in the stationary regime, when submitted to the action of an applied stress σ normal to the layers. This stress exerts on the dislocation a Peach and Koehler force equal (in absolute value) to σb. This force is balanced by a viscous friction force F_V, usually written as $F_V = Vb/m$, such that

$$V = m\sigma \qquad\qquad (C.III.43)$$

in the permanent regime. The quantity m, a priori a function of b and V, is the climb **mobility** of the dislocation. One can estimate it either using microscopic models, as metallurgists do for solids, or by solving the problem of permeation flow around the dislocation in the hydrodynamic approximation.

The first method applies to elementary dislocations. It starts from the assumption of an infinity of kinks, of one molecule each, along the line. This approach yields, by analogy with solids [32]:

$$m \approx \frac{D\Omega}{k_B Tl} \approx \frac{Dl^2}{k_B T} \qquad\qquad (C.III.44)$$

where D is an average self-diffusion coefficient, Ω the molecular volume and l a diffusion length close to the molecular size, as the layers are fluid.

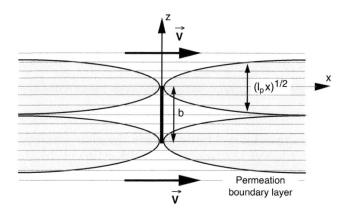

Fig. C.III.24 Hydrodynamical model used to determine approximately the mobility of a dislocation, replaced here by a ribbon of width b.

The hydrodynamical method leads to fairly complicated equations, not yet exactly solved [46]. There is, however, an approximate model that yields a similar result, at least as far as the dimensional dependence is concerned. This model, not very different from the one proposed by de Gennes in 1974 [47], consists of replacing the mobile dislocation by an obstacle of width b perpendicular to the layers (Fig. C.III.24). The solution to this problem was given in the previous chapter (section C.II.5). The essential result is that the friction force F_v is proportional to b and to the velocity V (eq. C.II.141). Consequently, the mobility m of a dislocation is (in the first approximation) independent of its velocity and of its Burgers vector, and its value is of the order of:

$$m = \beta \sqrt{\frac{\lambda_p}{\eta}}$$

(C.III.45)

with β a numerical prefactor of order unity.

This formula is not incompatible with the one previously obtained. To see this, one only needs to write that $\eta \approx k_B T/lD$ and that $\lambda_p \approx D\Omega/k_B T$ (see section C.II.5.a) and to plug these expressions into eq. C.III.45. Formula C.III.44 is then retrieved, up to a numerical coefficient of the order of unity.

The advantage of the hydrodynamical model is its applicability to dislocations of large Burgers vector. It shows that their mobility is independent of their Burgers vector, an unexpected result stemming from the fluidity of the layers. At the end of this chapter, we shall see how to measure the mobility m of a dislocation.

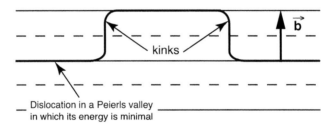

Fig. C.III.25 Glide of an edge dislocation strongly pinned in its Peierls valleys by the formation of thermally activated kinks.

For completeness, let us briefly discuss the glide mobility. In this case, the dislocation moves across the layers under the action of the shear stress σ_{zi}^E. This conservative movement (no diffusion) requires breaking the layers. For this reason, two situations must be distinguished, corresponding to an applied shear stress σ below or above the Peierls-Nabarro stress $\sigma_{P.N.}$ pinning the dislocation in its potential valley. In the first case ($\sigma \ll \sigma_{P.N.}$), it is difficult for the dislocation to glide, as its movement can only occur by thermally activated kink generation (Fig. C.III.25). If, on the other hand, $\sigma \gg \sigma_{P.N.}$, the glide becomes easier, being limited only by the viscosity of the medium. Assuming that the strain field in the reference system of the dislocation is the same as in the static case, the energy dissipated by viscous friction (there is no permeation, the movement being conservative) can be determined, and the mobility M in "easy glide" is given by [48]:

$$M_{edge} = \frac{32\,\xi\sqrt{2\pi\lambda\xi}}{\eta b\left(1 + \dfrac{3}{4}\dfrac{\lambda}{\xi}\right)} \tag{C.III.46}$$

where ξ is the core width (Fig. C.III.19a), λ the penetration length, and η the viscosity. This formula shows that, as for solids, the mobility increases when the dislocation core is dissociated. If the core is not dissociated (which is to be expected), $\xi \approx b/2 \approx \lambda$, yielding $M_{edge} \approx 23\lambda/\eta \approx 10^{-6}$ cm²/poise with $\lambda \approx 10$ Å and $\eta \approx 2$ poise (see sections III.5.b and III.3.c of this chapter, respectively). This mobility is comparable to the climb mobility m, of the order of 5×10^{-7} cm²/poise. This result is in stark contrast with the situation of solids, where glide is usually much easier than climb, but has not yet been verified experimentally.

vii) Trapping of a solid particle by a moving edge dislocation
    (Orowan process)

Experience shows that small solid particles can be dragged along by an edge dislocation, provided the climb velocity of the latter is not too high [48, 49a,b]. This trapping process, also visible in free-standing films, is interesting to the extent that it gives access to the velocity of the dislocations, even when they are not visible under the microscope. During this process, the dislocation is locally deformed (Fig. C.III.26a), so that the motive force it applies to the particle balances the viscous friction force opposing the movement of the particle (section II.5.b.i). In the case of a spherical particle with radius r, this force equilibrium reads:

$$2T\sin\alpha = 8\pi\eta rV \tag{C.III.47}$$

where T is the line tension of the edge dislocation, close to K, α the maximum tilt angle of the dislocation with respect to the direction of movement, V the dragging velocity of the dislocation and η the smectic viscosity (assuming that $\eta = \eta_2 = \eta_3$). This relation can only be satisfied at sufficiently low velocities:

$$V < V^* = \frac{T}{4\pi\eta r} \tag{C.III.48}$$

This relation defines a critical velocity V^* above which the dislocation snaps free from the dust particle leaving around it a dislocation loop (Orowan process, Fig. C.III.26b).

In order of magnitude, one has $V^* \approx 10$ µm/s for a particle of radius r = 1 µm. This result is well verified experimentally [48]. One should keep in mind that this reasoning assumes that the dislocation cannot escape the particle by glide. This seems to be the case experimentally, otherwise the pinning would be ineffective. This observation suggests that the dislocation core is not dissociated, since it can hardly glide.

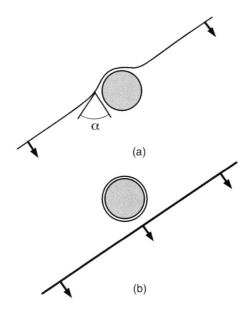

Fig. C.III.26 a) Dust particle trapped by an edge dislocation moving parallel to the layers; b) after release, the particle is surrounded by an edge dislocation loop with the same Burgers vector.

Here lies another essential difference with respect to solids. In the latter, the particles cannot move, so that the material is considerably hardened. In a smectic phase, the hardening is negligible because the particles are easily carried along by the dislocations.

III.2.c) Screw dislocations

i) Experimental observation

Screw dislocations are invisible in polarizing microscopy, both in the smectic A and in the smectic C phase. They can however be observed in homeotropic [50] and planar [51] samples of ferroelectric smectic C* (more precisely, they are wedge-screw dispirations). In the first case, the lines are perpendicular to the plane of the sample and appear in the microscope as black dots on a white background or white dots on a black background, depending on the polarization of light (Fig. C.III.27a). In the second case, one can clearly see the defect lines perpendicular to the layers (Fig. C.III.27b). The screw dislocations were also observed in the lamellar lyotropic phases L_α by freeze-fracture and electron microscopy [52–55]. This experimental technique consists of breaking under vacuum a lamellar phase sample initially quenched in undercooled liquid nitrogen, in order to inhibit crystallization (liquid propane or freon are also used to this effect). If the quench is fast enough (10^3 to 10^4 K/s), the lamellar structure is not destroyed, as the ice crystals do not have time to appear. A replica of the fracture surface (very often a cleavage plane parallel to the layers) is then made by successively evaporating 30 Å of platinum at oblique incidence (45°), and then 200 Å of carbon at normal incidence (replica support).

This replica is then carefully peeled off the surface (by dissolving the surfactant in water) and dried after transfer to an electron microscopy grid. The observation of this replica in TEM reveals the presence of numerous screw dislocations, recognizable by the steps they leave as they pierce the cleavage surface parallel to the layers (Fig. C.III.27). The density of screw dislocations varies strongly from one sample to the next. It varies from $10^6 \, cm^{-2}$ in lecithin [52, 53] to $10^{10} \, cm^{-2}$ in certain non-ionic surfactants such as $C_{12}EO_5$, when approaching the transition toward the isotropic phase [54].

We note that the screw dislocations can gather along hyperbolae of focal conics of the first type [56], and that the helical structures observed by Williams (Fig. C.III.12) are topologically equivalent to screw dislocations with a giant Burgers vector [24].

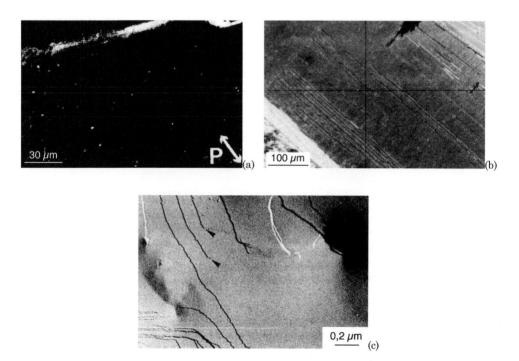

Fig. C.III.27 a) Thick (75 μm) SmC* sample in homeotropic orientation: the screw dislocations appear as white dots on the black background (photo by L. Bourdon, 1980); b) very thin (3 μm) SmC* sample in planar orientation: the lines, perpendicular to the layers, are lengthy and very straight, although some jogs can be detected. The segments of edge dislocations connecting their extremities are less sharp and more difficult to see (from ref. [51]); c) TEM (transmission electron microscopy) observation of the replica of the surface of a lyotropic lamellar phase, frozen and then cleaved parallel to the layers. Each mark (arrow) points to the emergence point of a screw dislocation (photo by M. Allain, 1987).

ii) Elastic energy

We recall that, by definition, the Burgers vector **b** is parallel to the line (of unit vector **t**) and perpendicular to the layers (Fig. C.III.14). The orientation of **b** is given by the Burgers convention described in figure C.III.13.

In the first approximation, the strain of the layers is written, as in the case of solids:

$$u = \frac{b\varphi}{2\pi} = \frac{b}{2\pi} \arctan\left(\frac{y}{x}\right) \quad (\mathrm{mod}\ \pi) \tag{C.III.49}$$

with φ the azimuthal angle in the plane of the layers (x,y). This equation describes a director-plane helicoid (Fig. C.III.14).

It is easily checked that this strain field fulfills the linearized equilibrium equation:

$$B\frac{\partial^2 u}{\partial z^2} - K\left(\frac{\partial^4 u}{\partial x^4} + \frac{\partial^4 u}{\partial y^4} + 2\frac{\partial^2 u}{\partial x^2 \partial y^2}\right) = 0 \tag{C.III.50}$$

One can also verify that the layers have zero total curvature (minimal surfaces) because:

$$\frac{1}{R_1} + \frac{1}{R_2} = \Delta u = 0 \tag{C.III.51}$$

Since, moreover, $\partial u/\partial z = 0$, the elastic energy is zero to lowest order in the strain.

On the other hand, the anharmonic term $(1/8)B[(\partial u/\partial x)^2 + (\partial u/\partial y)^2]^2$ is non-zero, as well as the bend term of the director field $(1/2)\tilde{K}_3(\mathbf{n}\times\mathbf{Rotn})^2$, appearing in the complete expression of the free energy (eq. C.II.8). Calculating these two terms using expression C.III.49, then integrating them over the entire volume, leads to the following expression for the energy (per unit length) of a screw dislocation [57, 58]:

$$E_{screw} = \frac{Bb^4}{128\pi^3 r_c^2} + \frac{\tilde{K}_3 b^4}{64\pi^3 r_c^4} + E_c \tag{C.III.52}$$

Note that the energy does not diverge for an infinite sample, as in the case of solids (where $E_{disl} \propto \ln(R/r_c)$). One still needs, however, to introduce a core radius r_c and a core energy E_c, the elastic energy diverging on the line. Because Bb^2 and $\tilde{K}_3$ are of the same order of magnitude, the two elastic contributions participate equally in the distortion energy. One can also remark that the b^2 dependence of the first term in eq. C.III.52 (assuming $r_c \approx b$) favors the screw dislocations of small Burgers vectors. This result is confirmed by electron microscopy observations showing that most dislocations are elementary. Finally, the elastic energy of a screw dislocation is 1000 times smaller than that of an edge dislocation. This difference can also explain the abundance of screw dislocations in the experimental samples.

The core structure of a screw dislocation is not well known [58a–c]. As for edge dislocations [31], one can reasonably consider it to be nematic, or even isotropic in certain cases.

iii) Line tension

We are dealing with an important concept that becomes fundamental for an anisotropic system, as in the case of smectic A phases.

This parameter was not discussed in the case of edge dislocations, as their energy is independent of the orientation in the plane of the layers. As long as the edge dislocation remains in the same plane, its line tension is equal to its energy:

$$T_{edge} = E_{edge} \approx K \tag{C.III.53}$$

The situation is completely different for a screw dislocation; as soon as it tilts with respect to the layer normal, it acquires an edge component that strongly changes its energy. To evaluate this change in energy, let us assume that the line makes an angle θ with the z axis (Fig. C.III.28). The strain field:

$$u = \frac{b}{2\pi} \text{arctg}\left(\frac{y - \theta z}{x}\right) \tag{C.III.54}$$

correctly describes this dislocation as long as the angle θ is small.

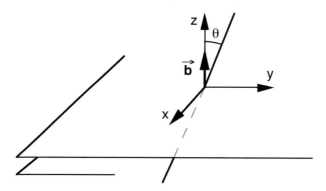

Fig. C.III.28 Tilted (mixed) screw dislocation making an angle θ with the layer normal.

It is then easy to determine the energy of this mixed dislocation and to show that [49]:

$$E(\theta) = E_{screw}(\theta = 0) + \frac{Bb^2}{4\pi}\theta^2 \ln\left(\frac{R}{r_c}\right) \tag{C.III.55}$$

Here, R is an outer cutoff radius needed in the calculation. Its value is typically the average distance between two dislocations of opposite signs. The line tension (the analogue in the case of a line of the surface stiffness of an interface) results immediately from the general formula [59a] (demonstrated in section C.VII.4 for the case of an interface):

$$T_{screw} = \left(E(\theta) + \frac{d^2 E(\theta)}{d\theta^2} \right)_{\theta = 0}$$ (C.III.56a)

and finally, using the approximate expression C.III.55:

$$T_{screw} = E_{screw} + \frac{Bb^2}{2\pi} \ln\left(\frac{R}{r_c}\right)$$ (C.III.56b)

In this expression, the self-energy of the line is negligible with respect to the other term, close to K. One can see that

$$T_{screw} \approx K >> E_{screw}$$ (C.III.57)

This feature is essential for understanding the buckling instability of screw dislocations, to be discussed in section III.5.a. Another important consequence of this result is that a mixed dislocation connecting two points A and B belonging to different layers, will "split" into two perpendicular dislocation segments, of edge and screw character, respectively (Fig. C.III.29).

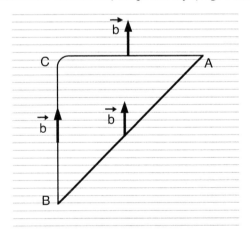

Fig. C.III.29 The mixed dislocation AB has a higher energy than the split structure ACB composed of the edge dislocation segment AC and the screw dislocation segment CB.

iv) Mobility of a screw dislocation

When a screw dislocation glides along its own direction, it locally deforms the layers, inducing shear. One can calculate the viscous dissipation in the same way as for an edge dislocation in easy glide, neglecting the permeation (a reasonable assumption, the movement being conservative) and considering that the strain field u of the layers is the same as for the static case (in the reference system of the line). This calculation yields the following formula [49]:

$$M_{vis} = \frac{8\pi^2 r_c^2}{\eta b} \qquad \text{(C.III.58)}$$

where r_c is the (hydrodynamic) core radius of the dislocation, unknown but of the order of the molecular size, and η a viscosity.

III.2.d) Dislocation crossing

It often happens in plasticity that only some of the dislocations are acted upon by the applied stress. One must then distinguish the mobile dislocations from the others.

During their motion, the mobile dislocations will often encounter the fixed ones. The metallurgists sometimes speak of "trees" and "forest" to designate the family of immobile dislocations [59a,b]. The ease with which the dislocations can cross each other will result in a more or less ductile material.

This problem is encountered in smectics when an edge dislocation with Burgers vector b_2 crosses a screw dislocation with Burgers vector b_1 (Fig. C.III.30). It is easy to see that, in this case, the screw is not modified, being parallel to b_2. On the other hand, b_1 is perpendicular to the edge dislocation, so this latter will bear, after the crossing, a jog of screw character, of a height equal to the Burgers vector of the screw dislocation. The typical energy of the jog is

$$E_{jog} \approx \frac{3Bb_2^4}{128\pi^3 r_c^2} b_1 \approx \frac{3Ba_o^3}{128\pi^3} \qquad \text{(C.III.59a)}$$

if we assume that $b_1 = b_2 = r_c = (\tilde{K}_3/B)^{1/2} = a_o$ (the layer thickness). Experimentally, $B \approx 10^8$ erg/cm^3 and $a_o \approx 30$ Å, yielding

$$E_{jog} \approx 2\times10^{-15} \text{ erg} \approx 6\times10^{-2} \, k_B T \qquad \text{(C.III.59b)}$$

This extremely small energy facilitates thermally activated jumps. The crossing of elementary dislocations (or of small Burgers vector) is thus, in principle, easy in smectic A phases, even at room temperature. At any rate, this result is compatible with microplasticity experiments performed on homeotropic 8CB samples where the stress relaxes almost to zero after suddenly applying a constant compression or dilation deformation (see section III.5).

In contrast, the situation seems to be rather different in materials exhibiting a SmA-SmC phase transition, such as AMC11 (n-undecyl *p*-azoxy-α-methylcinnamate) where the edge dislocations can be visualized in the vicinity of the transition temperature $T_{AC} \approx 79°C$ (see section III.2.b). In this case,

Bourdon states [50] that the edge dislocations are strongly pinned on the screw dislocations in the smectic C phase (Fig. C.III.30b), but not in the smectic A phase, where they seem to cross very easily.

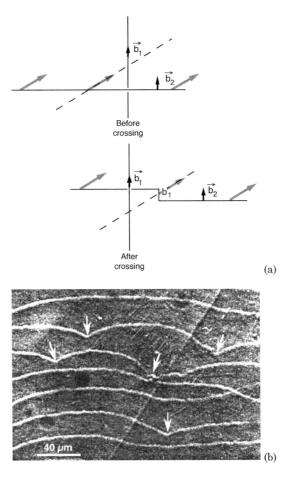

(a)

(b)

Fig. C.III.30 a) Crossing of an edge dislocation (with Burgers vector b_2) with a screw dislocation (with Burgers vector b_1). After crossing, the edge dislocation exhibits a jog of screw character and of height b_1. b) In the smectic C phase, the edge dislocations are easily pinned on the screw dislocations (tagged by arrows) that cross the sample. The photo is taken between slightly uncrossed polarizers. Homeotropic AMC11 sample, 36 μm thick. The tilted straight line is a scratch on one of the glass plates (photo by L. Bourdon, 1980).

These observations are at odds with the much more recent results of Lelidis et al. [60a–c], who used a different material (BDH173 from Merck, which is a fluorinated octyloxyphenyl octyloxybenzoate), also exhibiting a SmA-SmC transition at $T_{AC} \approx 54°C$. Unlike Bourdon, these authors observe a strong pinning of the two types of dislocations in the smectic A phase, slightly above the transition temperature T_{AC}. On the other hand, they estimate (by counting the pinning points) that the density of screw dislocations in their samples varies between 10^5 and 10^8 cm^{-2} depending on the anchoring quality or on the roughness of the glass plates. These values are comparable to those measured by freeze-fracture in samples of lamellar lyotropic phases [52–54]. Finally, these authors notice that the pinning "seems" to disappear when the edge dislocations move rapidly under the influence of a strong applied stress (a fact they explain very easily by calculating the shape of the moving edge dislocations, which straighten up as the velocity increases).

Various mechanisms have been proposed to account for the pinning [60a–c] (or lack of it [50]) between dislocations. As these explanations are contradictory, we shall not discuss them in detail here.

III.3   Rheological behavior at low shear rate and lubrication theory

When sheared parallel to the layers, a smectic phase should behave as an ordinary fluid. We shall show that this ideal behavior is not easy to demonstrate, even at low shear rates, due to the presence of defects. We recall that these can be edge or screw dislocations or focal conics. We shall successively present four experimental situations where the role of each defect type is clearly delineated. In the first one, performed in lubrication geometry, the edge dislocations dominate. This experiment emphasizes the strongly lubricating properties of smectics. In the second one, the lubrication effect is suppressed using a floating-plate setup: in this case, only the screw dislocations are coupled to the flow, resulting in a plug flow distribution and hence in non-Newtonian behavior. In the third experiment, we show that it is possible to "freeze" the motion of edge and screw dislocations under alternating shear and, in this way, to reach the intrinsic smectic viscosity η_3. In the last experiment, we consider the role of focal conics. We shall see that they generate a strong internal stress leading to Bingham behavior:

$$\sigma = \sigma_s + \eta_{app}\dot{\gamma} \qquad\qquad\qquad \text{(C.III.60a)}$$

We shall conclude this section with the description of two additional (and somewhat more complicated) situations. In the first one, the sample, oriented in planar anchoring, is sheared perpendicular to the layers. We shall see that Grandjean walls can appear and accommodate the shear deformation. In the second one, the sample is completely disoriented and forms a focal conic texture with many oily streaks. It will be shown that, in this limit, the smectic phase has a shear-thinning behavior, obeying a power law of the type:

$$\dot{\gamma} = A\,\sigma^{5/3} \qquad\qquad\qquad \text{(C.III.60b)}$$

III.3.a) Lubrication theory in shear parallel to the layers

It is well known that the lamellar phases (e.g., soap, graphite, custard, etc.) are excellent lubricants. Engineers also know that rolling "oils" (obtained from very complex aqueous mixtures) exhibit, under the extreme temperature and pressure conditions they are used in, a locally lamellar structure that increases their lubricating power [61]. Why is that?

The geometry of a lubrication experiment is presented in figure C.III.31 [49, 62]. The lower plate is fixed and makes a small angle α with the upper plate. The surfaces are treated with a silane compound, ensuring perfect alignment of the layers parallel to the plates (strong homeotropic anchoring). The sample thickness at an abscissa x is given by

$$d = d_o + \alpha x \qquad\qquad (C.III.61)$$

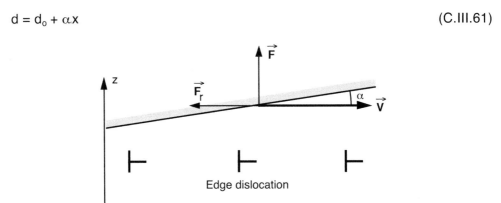

Fig. C.III.31 Geometry of a lubrication experiment.

With this thickness variation is associated a sub-grain boundary composed of regularly spaced edge dislocations. These dislocations being repelled by the plates, we shall consider that they occupy the center of the sample. Let Λ be the distance between two dislocations:

$$\Lambda = \frac{b}{\alpha} \qquad\qquad (C.III.62)$$

These dislocations have been observed in polarizing microscopy (Fig. C.III.15). One should keep in mind that in this ideal (but experimentally feasible) sample, only the equilibrium dislocations imposed by the geometry are present. In particular, one only needs to vary the angle α to change the dislocation density.

At the moment $t = 0$, the upper plate is set in motion with a horizontal velocity V. During its motion, the angle α remains constant. Let us assume for the moment that $V > 0$. At time t and at the position of a fixed point on the lower plate, the thickness d varies by

$$\Delta d = - V t \alpha \qquad \text{(C.III.63)}$$

Because the smectic phase is under compression, it is stable with respect to the layer buckling instability. This stress acts on the bulk dislocations that "climb" together to relax the thickness variation. Let v(t) be their velocity at time t. The relaxed thickness variation is

$$- \frac{b}{\Lambda} \int_0^t v(t)dt = - \alpha \int_0^t v(t)dt \qquad \text{(C.III.64)}$$

such that the smectic sees an effective thickness variation

$$\alpha \left[\int_0^t v(t)dt - Vt \right] \qquad \text{(C.III.65)}$$

inducing a uniaxial stress

$$\sigma_{zz}^E = \sigma = B \frac{\alpha}{d} \left[\int_0^t v(t)dt - Vt \right] \qquad \text{(C.III.66)}$$

This uniaxial stress exerts a pure climb force on each dislocation, to be balanced by the viscous friction force also acting on every dislocation. In the center of the sample, the average flow velocity is V/2 (by symmetry), yielding:

$$v(t) - \frac{V}{2} = - m B \frac{\alpha}{d} \left[\int_0^t v(t)dt - Vt \right] \qquad \text{(C.III.67)}$$

where m is the mobility of an edge dislocation, calculated in the previous section. Integrating this equation is straightforward, resulting in

$$v(t) = V \left[1 - \frac{1}{2} \exp\left(- \frac{t}{\tau} \right) \right] \qquad \text{(C.III.68a)}$$

with

$$\tau = \frac{d}{mB\alpha} \qquad \text{(C.III.68b)}$$

At the moment $t = 0$, the dislocations have the same velocity V/2 as the flow that drags them along (as the time needed to establish the velocity profile is much shorter than τ). They are then accelerated under the action of the stress σ and reach a limiting velocity V equal to that of the upper plate in the stationary regime. The vertical lubrication force opposing the compression of the smectic film is written (up to a $\cos\alpha$ prefactor):

$$F = \int_0^L (P - P_{ext} - \sigma)dx \qquad \text{(C.III.69)}$$

This formula shows that besides the usual term involving the hydrostatic pressure, there is an additional term related to the layer compressibility.

The first term can be calculated as for simple liquids, since within our approximation, $\sigma = Cst$ across the sample. One therefore has $G = 0$ (except close to the core of the dislocations, which only accounts for a negligible volume), and the equation describing the flow is equivalent to the Navier-Stokes equation for usual liquids. In the classical lubrication theory, one has:

$$\int_0^L (P - P_{ext})\,dx = \frac{1}{2}\eta_3 \left(\frac{L}{d}\right)^3 V\alpha \tag{C.III.70}$$

The second term is easily obtained from eqs. C.II.66–68. First, one has

$$\sigma = B\frac{\alpha}{d}\frac{V\tau}{2}\left[\exp\left(-\frac{t}{\tau}\right) - 1\right] \tag{C.III.71}$$

and then, assuming the stationary regime has been reached ($t \gg \tau$)

$$\int_0^L -\sigma\,dx = \frac{B\alpha}{d}\frac{V\tau L}{2} = \frac{VL}{2m} \tag{C.III.72}$$

Putting together eqs. C.III.70 and C.III.72, we find that, in the permanent regime, the lubrication force is

$$F = F_p + F_e = \frac{1}{2}\eta_3\left(\frac{L}{d}\right)^3 V\alpha + \frac{VL}{2m} \tag{C.III.73}$$

These two forces are often very different, as shown by the following numerical application. For $\eta_3 \approx 1$ poise, $d = 10$ μm, $L = 1$ cm, $\alpha = 10^{-4}$ rad and $\lambda_p = 10^{-13}$ cm^2/poise, one has:

$$F_p\,(dyn) = 5 \times 10^4\ V\,(cm/s) \tag{C.III.74a}$$

$$F_e\,(dyn) = 1.5 \times 10^6\ V\,(cm/s) \tag{C.III.74b}$$

One can see that the elastic force is about 100 times larger than the hydrostatic force one would obtain using an isotropic fluid of similar viscosity. Consequently, a smectic phase resists compression much better than a simple fluid, owing to its lamellar structure. It is therefore an excellent lubricant.

It is also instructive to determine the drag force F_r exerted by the smectic on the mobile plate in the stationary regime. It contains the usual viscous term (plus a negligible correction in α^2, stemming from hydrostatic pressure, as for classical lubrication) and a second one, of viscoelastic origin, related to the motion of the dislocations:

$$F_r = -\eta_3\frac{VL}{d} - \frac{\alpha VL}{2m} \tag{C.III.75}$$

Thus, the smectic phase behaves as an ordinary fluid with an apparent viscosity (to first order in α) [49,62]

$$\eta_{app} = \eta_3 + \frac{\alpha d}{2m} \tag{C.III.76}$$

This formula reveals the experimental difficulty of measuring directly, in continuous shear, (for instance, using a Weissenberg rheogoniometer) the intrinsic viscosity η_3. Indeed, an angular misalignment α as small as 2×10^{-4} rad (and, experimentally, one can hardly do any better!) leads to a correction $\alpha d/2m \approx 10$ poise using $m \approx 2 \times 10^{-7}$ cm^{2}sg^{-1} and $d = 200$ μm. The lubrication effect (the thicker the sample, the more important it is) can thus lead to a considerable error in determining the intrinsic viscosity η_3. One must however keep in mind that this calculation only takes into account the grain boundary dislocations, corresponding to an ideal, albeit experimentally achievable, situation. Nevertheless, we shall see in section III.3.d how to generalize this formula to the case of strongly "dislocated" samples.

Let us now briefly consider the case $V < 0$. The layers are now dilated and the – positive – stress is given by eq. C.III.71, as long as it remain below the critical value σ_c above which the layers are unstable and start buckling. It can be shown that, for low shear rates, σ_c is still given by the static formula

$$\sigma_c = \frac{2\pi \sqrt{KB}}{d} \tag{C.III.77}$$

the main effect of the shear being to orient the wave vector of the undulation perpendicular to the velocity [63]. The smectic phase is thus unstable in the stationary regime, if the shear rate $\dot{\gamma}_c = V/d$ exceeds (in absolute value) the critical value

$$\dot{\gamma}_c = \frac{4\pi m \sqrt{KB}}{d^2} \tag{C.III.78}$$

One could be intrigued by the fact that the angle α drops out of this formula, as no instability is expected for $\alpha = 0$. This apparent contradiction is removed if we calculate the time t_c necessary to develop the instability. This time is determined by putting $\sigma = \sigma_c$ in eq. C.III.71:

$$t_c = \tau \ln\left(\frac{\dot{\gamma}}{\gamma - \dot{\gamma}_c}\right) = \frac{d}{mB\alpha} \ln\left(\frac{\dot{\gamma}}{\dot{\gamma} - \dot{\gamma}_c}\right) \tag{C.III.79}$$

This time diverges for $\alpha = 0$: as expected, the smectic phase is stable if the plates are parallel. Note that, in this formula, $\dot{\gamma} < 0$ and $\dot{\gamma}_c > 0$.

Formula C.III.78 is interesting from the experimental point of view, showing that the mobility of edge dislocations can be determined without knowing their density or, equivalently, the angle α between the plates, which is generally very difficult to measure precisely. In 8CB, at room temperature, the experiment yields $\dot{\gamma}_c \approx 0.1$ s^{-1} for a 140 μm thick sample. Knowing that $(KB)^{1/2} \approx 10$ erg/cm^2, we obtain $m \approx 2 \times 10^{-7}$cm^{2}sg^{-1}. Using expression C.III.45 for the mobility m (with $\beta = 1$), one can also estimate $\lambda_p \approx 1.6 \times 10^{-13}$ cm^2/poise with $\eta_3 \approx 4$ poise (see the next section).

III.3.b) Measurement of the intrinsic viscosity by the floating-plate technique: role of the screw dislocations

To measure the intrinsic viscosity of a smectic phase in shear parallel to the layers, the lubrication effect must be eliminated. One way of achieving this goal is to impose a shear stress to the mobile plate (rather than a rigid motion), while leaving it free to move along the vertical axis. The cause of the lubrication effect is thus suppressed, since the smectic layers are no longer compressed (or dilated). The "floating-plate" setup shown in figure C.III.32a satisfies these conditions, with the isotropic liquid opposing a negligible resistance in compression or in dilation [64]. In this experiment, the intermediate plate is held in place by the capillary forces. The upper plate is set in motion with a velocity V and the velocity v of the intermediate plate is measured directly in the microscope. If the lower plate is fixed, the apparent viscosity of the homeotropic phase can be easily calculated in terms of the viscosity η_r of the reference fluid and of the respective thicknesses D and d of the two samples:

$$\eta_{app.} = \eta_r \frac{d}{D}\left(\frac{V}{v} - 1\right) \qquad\qquad (C.III.80)$$

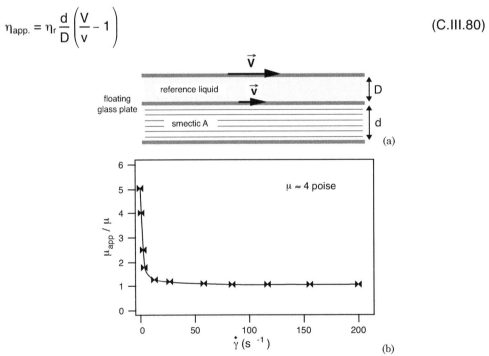

Fig. C.III.32 a) Floating plate setup allowing the lubrication effect to be overcome; b) viscosity measured as a function of the shear rate using this method (8CB at room temperature, d = 100 μm). The solid line corresponds to the best fit with the theoretical prediction in eqs. C.III.81a–c.

The experiment was performed with 8CB, using glycerine as a reference fluid (not necessarily a good idea, since its viscosity strongly depends on its water content). The experimental graph in figure C.III.32b shows that, at high shear rates, the measured viscosity is constant (about 4 poise at 21°C, which is certainly an overestimation, considering the aforementioned property of glycerine). However, the viscosity increases strongly at low shear rates. The lubrication effect cannot explain this non-Newtonian behavior (shear-thinning in this case) in continuous shear.

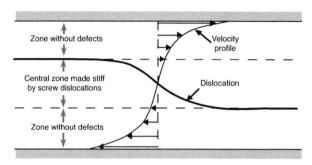

Fig. C.III.33 Typical velocity profile in a sample sheared parallel to the layers ($\dot{\gamma} \approx 1$ s^{-1}). A strong "stiffening" of the flow can be noticed in the center of the sample, no doubt related to the presence of dislocations.

One possible explanation of this anomaly involves the screw dislocations, always present in the samples (see also section III.5.a). It can be shown that these defects hinder the flow at low shear rates, leading to a partial "stiffening" of the velocity profile ("plug"-type flow) in the center of the sample [65]. This effect can be observed in polarizing microscopy by following the motion of microparticles contained in the sample (Fig. C.III.33) [64]. This velocity profile, very different from the expected linear profile, can be estimated, as well as the associated increase in apparent viscosity. The law thus deduced can be written in the approximate form [65]:

$$\frac{\eta_{app}}{\eta_3} \approx \frac{d}{2\varsigma} \coth\left(\frac{d}{2\varsigma}\right) \tag{C.III.81a}$$

with

$$\varsigma = 1 \quad \text{for} \quad \dot{\gamma} < \dot{\gamma}* \quad \text{and} \quad \varsigma = 1\left(\frac{V}{V*}\right)^{1/4} \quad \text{for} \quad \dot{\gamma} > \dot{\gamma}* \tag{C.III.81b}$$

where l is the average distance between two screw dislocations and $\dot{\gamma}* = V*/d$ a critical shear rate:

$$\dot{\gamma}* = 16\frac{Ba_o^2}{\eta_3 d^2} \tag{C.III.81c}$$

above which the screw dislocations are unstable in the direction of the sample thickness and "stretch" along the flow (Fig. C.III.33). The theoretical law C.III.81 provides a correct explanation for the shear-thinning behavior observed under continuous shear as well as for the saturation of the apparent viscosity at high shear rates ($\eta_{app.} \to \eta_3$ for $\dot{\gamma} \to \infty$). The best fit of the experimental curve in figure C.III.32b with the theoretical expression C.III.81a yields $l \approx 20\ \mu m$, a reasonable value in light of the other experiments.

The continuous shear experiments not being very easy to perform (difficulty in maintaining the parallelism, etc.), it can be more convenient to work in oscillating shear. The obvious question is then whether the previous results can be directly transposed to this situation. We shall only give a partial answer, by specifying under which conditions the viscosity measured in oscillating shear tends toward the intrinsic viscosity η_3 of the smectic phase.

III.3.c) Measurement of the intrinsic viscosity η_3 in oscillating shear

The geometry is the same as in figure C.III.31, but this time, the upper plate oscillates around an axis perpendicular to the lower plate, at an angular frequency ω. We also assume that the setup is stiff and that the angle α remains constant. For the measured viscosity be close to the intrinsic viscosity of the smectic phase, the dislocations must move at the velocity of the flow, without generating any additional dissipation. This condition of solidary motion is only fulfilled if the oscillation frequency ω is high enough, and this depends on the type of dislocations under consideration.

For edge dislocations, ω only needs to be much higher than the reciprocal of the time needed to set them in motion in a reference system co-moving with the fluid. From eq. C.III.68b, this condition reads:

$$\omega \gg \frac{1}{\tau} = \frac{mB\alpha}{d} \approx \text{a few Hz} \tag{C.III.82}$$

For screw dislocations, ω must be larger than the reciprocal of the typical time a tilted dislocation needs to return to the vertical position. One can show that, for a screw dislocation crossing the entire sample, this condition is given by [66]:

$$\omega \gg \frac{24K}{\eta_3 d^2} \approx \text{a few Hz} \tag{C.III.83}$$

Thus, it is enough to work at frequencies above roughly 10 Hz in order to "freeze" the relative motion of dislocations and to reach the viscosity coefficient η_3. The experiment, performed using 8CB, fully confirms these predictions, with the measured viscosity decreasing with the frequency and saturating at a value of approximately 4 poise at room temperature (21°C) as soon as $\omega > 5$ Hz. This value is the same as the one obtained under continuous shear by the floating plate method. Note that the viscosity η_3 decreases strongly with the temperature in 8CB, reaching 1 poise at 28°C. It can also be noticed that, at high frequency, the smectic phase exerts on the plates a normal force that oscillates in phase with the strain: its response along this direction is purely elastic, as one would expect once the dislocations are "frozen."

Until now, we have completely ignored the role of focal conics, which is a justified approximation as long as the samples are perfectly homeotropic and only contain very few dust particles. This was indeed the case in the previous experiments.

Nevertheless, focal conics can easily nucleate in homeotropic samples under shear, leading to a different rheological behavior. These effects are presented in the following section.

III.3.d) Rheological behavior under shear parallel to the layers in the presence of focal domains

This experiment was performed by Horn and Kléman [67] in the cone-plate geometry in a Weissenberg rheometer. The material employed is – once again – 8CB. The cone and the plate, both made out of glass, are treated with lecithin, ensuring homeotropic anchoring. The rheometer imposes a fixed shear rate and is equipped with a polarizing microscope, so that the sample can be observed under shear. Two situations can appear: either the sample is homeotropic (and appears black between crossed polarizers) or it contains focal domains (Fig. C.III.34).

In the first case, the behavior is Newtonian between 2 and 80 s^{-1} (curve "a" in figure C.III.35). This (somewhat unexpected) observation deserves a more detailed discussion in light of the previous results.

Indeed, the measured viscosity (1.4 poise) is close to the intrinsic viscosity expected at 25°C. It appears that the lubrication effect is absent here. This result can be explained if the sample contains a high density ρ_m of mobile edge dislocations (some of them created by the screw dislocations stretched by the shear). In this case, eq. C.III.76 is easily generalized and the smectic phase can be shown to exhibit Newtonian behavior with an apparent viscosity:

$$\eta_{app} = \eta_3 + C\frac{\alpha^2}{mb\rho_m} \tag{C.III.84}$$

Here, α is the angle between the cone axis and the normal to the plate, while the numerical factor C (between 1/2 and 1) accounts for the orientation of the dislocation lines with respect to the velocity (C = 1/2 when the line is perpendicular to the velocity and 1 when it is parallel to the velocity). With $\alpha = 5\times10^{-4}$ rad (which is already a very good alignment of the rheometer), $m \approx 3\times10^{-7}$ cm^2/poise (value measured at 25°C, [49]), b = 30 Å, one finds that the correction is negligible (below 0.1 poise) for $\rho_m > 10^7$ cm^{-2}.

These dislocations are also involved in preventing the development of the layer buckling instability in the region of the sample that is dilated under shear. Indeed, it can be checked that the dilation stress stays below the critical value σ_c given by eq. C.III.77 if the shear rate does not exceed:

$$\dot{\gamma}_c = \frac{2\pi\sqrt{KB}}{d}\frac{\rho_m b_m}{A\alpha} \tag{C.III.85}$$

where d is the local thickness of the sample. This expression generalizes eq. C.III.78 given previously. It shows that the risk of developing the layer buckling instability decreases as the concentration of edge dislocations susceptible to release the dilation stress increases. On the other hand, it is clear that the instability will first develop in the thicker areas of the sample. Experimentally, the maximum thickness at the border of the sample is 260 μm (for a cone angle of 1° and a radius of 1.5 cm). Using this maximal value, as well as $(KB)^{1/2} = 10$ erg/cm^2, one obtains $\dot{\gamma}_c \approx 10^{-6}\rho_m$. A minimal edge dislocation density of the order of 10^8 cm^{-2} (corresponding to an average distance between dislocations of 1 μm) is needed so that the instability does not appear at the highest shear rates employed (80 s^{-1}).

These estimates show that the samples used by Horn and Kléman, although homeotropic, must have been strongly "dislocated" (very likely, since they were oriented under oscillating shear at 0.1 Hz, without going into the nematic phase, which is an essential step for annealing the out-of-equilibrium dislocations). The situation is complicated by the appearance of focal conics (focal parabolae in this case) (Fig. C.III.34). These defects are usually generated in the wake of dust particles contained in the sample, by a process similar to the one described in section II.5.b.iii).

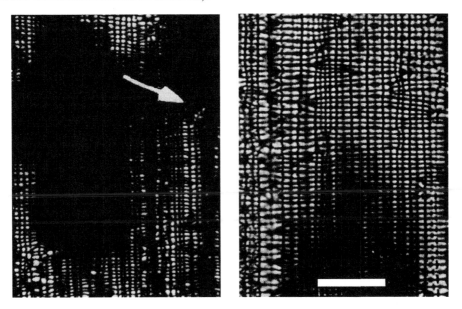

Fig. C.III.34 Appearance of the sample between crossed polarizers. The arrow indicates a dust particle developing in its wake focal conics, of the focal parabola type. These defects form a rectangular lattice and cover more or less extended areas. The bar represents 40 μm, which is also the thickness of the sample in the photo. The shear is along the vertical direction (from ref. [67]).

They can also appear if the sample is dilated beyond the threshold for the appearance of focal parabolae (see section III.5.b). It is therefore possible to produce samples with a variable density of focal conics. One must however keep in mind that this density depends on the history of the sample as much as on the applied shear rate; this means that, for a given $\dot{\gamma}$, one can observe multiple stationary states, with different apparent viscosities and defect densities. On the other hand, for a given shear rate there is a maximal defect density that cannot be exceeded.

To estimate the quantity of defects, Horn measured the intensity transmitted by the sample between crossed polarizers using a photodiode. This estimation can only be qualitative, since there is no obvious relation between the intensity and the number of defects per surface area or their extension in the thickness of the sample. Nevertheless, Horn finds that, for each transmitted intensity, the sample obeys a Bingham law (Fig. C.III.35):

$$\sigma = \sigma_s + \eta_{app}\dot{\gamma} \qquad\qquad (C.III.86)$$

In this expression, σ_s represents a yield stress below which the sample does not flow and η_{app} an apparent viscosity. These two quantities increase more or less linearly with the quantity of defects, as shown in figure C.III.36.

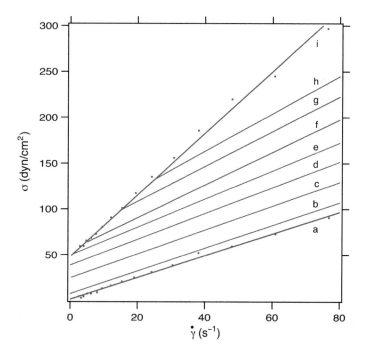

Fig. C.III.35 Rheological behavior in the presence of focal conics. Each curve corresponds to a different defect density, characterized by the intensity I transmitted between crossed polarizers (given in arbitrary units). a) I = 0 (almost homeotropic sample); b) I = 0.5; c) I = 1.5; d) I = 2.5; e) I = 3.5; f) I = 4.5; g) I = 5.5; h) I = 6.5 (corresponding to a state where 80% of the sample surface is covered by a focal conics texture). Curve i) represents the maximum stress measured at each shear rate (from ref. [67]).

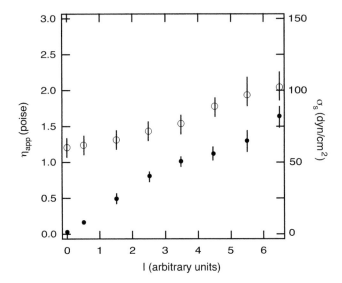

Fig. C.III.36 Yield stress σ_s (solid dots) and apparent viscosity η_{app} (circles) corresponding to each Bingham curve plotted in figure C.III.35 as a function of the light intensity transmitted between crossed polarizers (from ref. [67]).

We suggest that the appearance of the yield stress σ_s could be due to the already cited lubrication effect. Indeed, it is reasonable to assume that the focal domains hinder the motion of edge dislocations by creating an internal stress σ_i of elastic origin, which must be overcome so that the sample may flow. This internal stress, perpendicular to the layers, is related to the experimental values for the yield stress of the Bingham model by the simple geometrical relation

$$\sigma_s \approx \alpha \, \sigma_i \tag{C.III.87}$$

where α is the "parallelism defect" of the system.

To estimate the order of magnitude of the internal stress σ_i, let us imagine that the focal domains are localized obstacles, insurmountable by thermal activation. In this case:

$$\sigma_i \approx \frac{K}{bL} \tag{C.III.88a}$$

where K is the typical line tension of an edge dislocation, b its Burgers vector and L the distance between focal domains. One could also imagine that the focal domains (that often bunch together forming extended areas of focal parabolae) create an oscillating stress field with an amplitude

$$\sigma_i \approx B \frac{u_o}{d} \tag{C.III.88b}$$

where u_o is a distance of the order of the focal distance of a parabolic domain (a few layers to a few tens of layers) and d the sample thickness. With typical values $K = 10^{-6}$ dyn, $B = 10^8$ erg/cm^3, $d = 100$ μm, $L = 3$ μm and $u_o = L^2/d = 1000$ Å (30 layers), one finds for σ_i a value of 10^4 dyn/cm^2 using the first formula and 10^5 dyn/cm^2 from the second one. This latter estimate yields $\sigma_s \approx 50$ dyn/cm^2 if we take $\alpha = 5 \times 10^{-4}$ rad, which is exactly the expected order of magnitude (Fig. C.III.36).

One still needs to explain the increase of the apparent viscosity η_{app} in the presence of focal parabolae. The lubrication effect is not sufficient because the edge dislocations are too numerous (see eq. C.III. 84). On the other hand, the layer deformations induced by the focal parabolae are another source of dissipation, since they locally deform the flow. To quantify this effect, let us assume that the sample is completely covered by focal parabolae (this is almost the case for curve h in figure C.III.35). Under shear, the parabolae aggregate and form a square lattice with a wavelength L (Fig. C.III.34). The associated strain field is complicated, especially in the central region (comprising a few layers to a few tens of layers) where the focal parabolae meet. Outside of this region, however, the strain field is similar to the one created by a square lattice of undulating layers, given by:

$$u = u_o \sin(kz) \sin(qx) \sin(qy) \qquad \text{(C.III.89)}$$

with $k = \pi/d$, $u_o \approx L^2/d$ and $q \approx 2\pi/L$. Permeation being negligible, the molecules do not move across the layers. The dissipation is then easily determined [49]:

$$\Phi = \eta_3 \int_0^d \left[\left(\frac{\partial v_x}{\partial z}\right)^2 + \frac{u_o^2}{4} q^4 [\cos(kz)]^2 v_x^2 \right] dz \qquad \text{(C.III.90)}$$

At low Reynolds number, the system tries to minimize dissipation, yielding the following differential equation for the velocity profile:

$$\frac{d^2 v_x}{dz^2} - \frac{u_o^2}{4} q^4 [\cos(kz)]^2 v_x = 0 \qquad \text{(C.III.91)}$$

The boundary conditions are $v_x = \pm V/2$ in $z = \pm d/2$. This equation has no simple analytical solution but can be solved numerically. In figure C.III.37a, we show the velocity profile calculated with $d \approx 100$ μm, $L = 3$ μm and $u_o \approx 1000$ Å (these values account for the order of magnitude of the experimentally measured internal stress). We can see that the velocity profile is no longer linear, becoming "stiffer" in the central region where the layers undulate the most (this effect is similar to the one described in the presence of screw dislocations). This profile leads to an increase in the apparent viscosity, which we traced as a function of the control parameter $1/4\,u_o^2 d^2 q^4$ in figure C.III.37b.

With the previous values for d, L and u_o, one finds an apparent viscosity of $2.7\,\eta_3$ when the sample is completely covered by focal parabolae. If we admit that the viscosity increase is proportional to the coverage rate X of the focal parabolae, we find $\eta_{app} \approx [1(1-X)+2.7X]\,\eta_3 = [1+1.7X]\,\eta_3$, yielding for the curve h in figure C.III.35 an apparent viscosity of the order of $2.3\,\eta_3$ (the coverage rate is about 80%). This value is compatible with the experimental result, namely $\eta_{app} \approx 2\eta_3$ (see figure C.III.36).

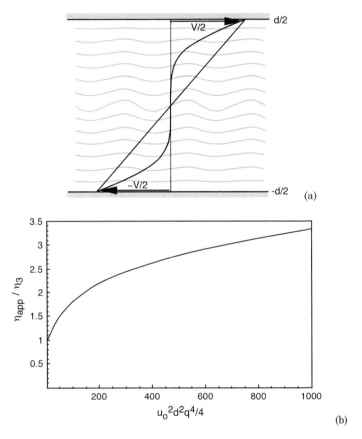

Fig. C.III.37 a) Numerically calculated velocity profile with $d \approx 100$ μm, $L = 3$ μm and $u_o \approx 1000$ Å; b) apparent viscosity as a function of the product $u_o^2 d^2 q^4/4$.

In conclusion, the viscous response of a smectic phase in shear parallel to the layers is quite complex, even at small Reynolds numbers, $Re = \rho Vd/\eta$ (note that this number is always below 0.01 in the experiments we described). In particular, the measured viscosity of a sample, albeit perfectly homeotropic, depends on the density of edge and screw dislocations it contains. This density can vary widely from one sample to another and depends on the quality of the anchoring, on the annealing process and thus, very often, on the "history" of the sample. A clear distinction must be made between the two types of dislocations: the edge dislocations, which increase the apparent viscosity measured in continuous shear by a constant quantity, and the screw dislocations, leading to a non-Newtonian and shear-thinning behavior. In oscillating shear, the motion of the dislocations with respect to the flow can be "frozen," provided the frequency is high enough, typically above a few Hz. In this case, one measures the intrinsic viscosity of the smectic phase. It is also noteworthy that the smectics act as excellent lubricants, resisting compression much better than simple fluids, due to their layered structure. In the presence of focal parabolae, the situation becomes more complicated, these defects creating a stress field that leads to a rheological behavior corresponding to a Bingham fluid. These defects also increase the apparent viscosity of the sample, by hindering layer slipping with respect to one another.

III.3.e) Shear perpendicular to the layers

In continuous shear, Williams and Kléman [21, 22] showed experimentally that the layers break up close to the glass plates and form edge dislocations that bunch together in giant dislocations if the tilt angle of the layers exceeds 20 degrees (Fig. C.III.38).

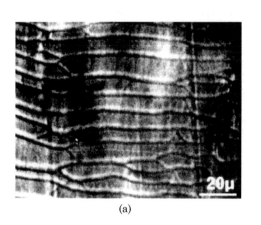
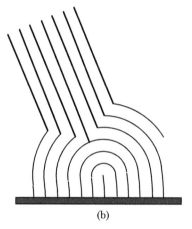

(a)   (b)

Fig. C.III.38 a) Bunching of giant edge dislocations in a planar CBOOA sample of thickness 75µm, sheared perpendicular to the layers. The tilt angle of the layers at the center of the sample is close to 20°. These dislocations have apparent widths of a few µm, corresponding to Burgers vectors of several hundreds of layers (photo by C. Williams,1976). b) Schematic representation of a giant dislocation.

In oscillating shear [68a,b], the situation is more complex due to the appearance of a hydrodynamic instability, even if the tilt angle of the layers remains small. The control parameter of this instability:

$$N = B\frac{\theta_0^2}{\mu\omega} \tag{C.III.92}$$

depends both on the tilt angle of the layers $\theta_0 = z_0/d$ and on the oscillation frequency ω of the upper plate, set in motion according to $z = z_0\sin(\omega t)$ (Fig. C.III.39). Physically, this parameter is the ratio between the elastic force (resulting from the anharmonic term) and the viscous force. Marignan et al. [68a,b] found that, above a critical threshold $N_c \approx 7$, two Grandjean walls parallel to the layers appear in the sample; they oscillate along the vertical at the imposed frequency and with opposite phases (Fig. C.III.39).

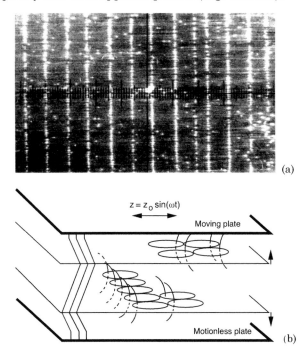

(a)

(b)

Fig. C.III.39 a) Observation under the polarizing microscope of a planar sample undergoing oscillating shear perpendicular to the layers. The band structure is due to the formation of Grandjean walls. The bright points along each line correspond to the vertices of the ellipses. b) Schematic representation of the two Grandjean walls allowing the shear stress to relax (from ref. [68a]).

III.3.f) Continuous shear of disoriented samples

In all the experiments presented so far, the anchoring of the layers onto the surfaces was important, as it determined their orientation with respect to the direction of the shear. Meyer et al. [69] studied the rheology of 8CB samples in the absence of any surface treatment. In this case, the samples are strongly disoriented, containing a high density of focal conics and oily streaks. The experiments, performed at imposed shear stress, were carried out in the cone-plate geometry between 22° and 29°C (i.e., far enough from the transition temperature toward the nematic phase, close to 33°C). Meyer et al. showed that, in this temperature range and for stress values between 100 and 300 dyn/cm^2, the smectic phase is shear-thinning, following a power law of the type:

$$\dot{\gamma} = A_o \exp(- W/k_B T)\, \sigma^m \qquad\qquad\qquad\qquad \text{(C.III.93)}$$

with $m \approx 1.7$, $A_o \approx 9.1 \times 10^{17}$ (in CGS units) and $W \approx 48 k_B T_R$ ($T_R = 298$ K). Note that, in this experiment, the sample is typically 20 times more viscous than if it was perfectly oriented in homeotropic anchoring.

An explanation of this rheological behavior and of the 5/3 exponent, involving the motion of screw dislocations, is given in Meyer et al [69].

III.4 Rheological behavior at high shear rates

In this limit, the rheological behavior of the smectic phase is once again very different. As for the experiments of Meyer et al. [69], the layer anchoring at the surfaces is of secondary importance, as the average orientation of the layers in the permanent regime, seems to depend only on the imposed shear rate and on physical parameters, such as the temperature (or the solvent concentration, in the case of lyotropic mixtures). At least, this is what one can infer from the results of Diat, Roux et al. [12a–e, 70a,b] who gathered all their experimental results in **orientation diagrams**, that we shall exemplify in the following. Similar results were also obtained by Safinya et al. in the United States [71, 72].

All these experiments were performed for cylindrical geometry (between two coaxial cylinders, of radius much larger than the gap between them) or for the cone-plate geometry, where the shear rate is constant throughout the sample. Their innovating character resides in the simultaneous use of various techniques (optical microscopy, conoscopy, scattering techniques using light, neutrons or X-rays, dielectric measurements) that give access to the average orientation of the layers and to the microstructure of the smectic phase as a function of the shear rate $\dot{\gamma}$ or the applied stress σ. A distinction must be made between the behavior of the thermotropic and lyotropic liquid crystals because, even if the orientation diagrams exhibit similar features, the defects and the textures encountered ("leeks" in thermotropics and "onions" in lyotropics) are different. It should be said right away that the these defects appear on micrometric scales, very small compared to the size of the shear cell, in contrast with the convective structures observed in classical hydrodynamics, where the dimensions are fixed by the sample thickness.

III.4.a) Orientation diagram in a thermotropic liquid crystal

The results described in this section were obtained by Panizza et al. [70a,b] using 8CB, in cylindrical geometry, with a gap between cylinders of 0.5 mm.

The orientation diagram (Fig. C.III.40) in the plane of parameters $(\dot{\gamma}, T)$ consists of three distinct regions:

    1. In region I, the layers arrange themselves, on average, in multilayer cylinders ("leek"-like structure) oriented along the velocity.

    2. In region II, corresponding to higher shear rates, the layers are oriented perpendicular to the surfaces of the coaxial cylinders and parallel to the velocity, a result confirmed by Safinya et al. [71, 72].

    3. Finally, in the intermediate ("biphasic") region, the two types of orientation coexist. It should be noted that, in the case of 8CB, region I is progressively reduced as the temperature increases, and disappears completely at the temperature $T_c \approx 31.83°C$. This temperature does not coincide with the transition temperature toward the nematic phase T_{NA}, which is about 1.2°C higher ($T_{NA} \approx 33°C$). It is noteworthy that, below T_c and at **fixed shear rate**, the transition between the states I and II is discontinuous and exhibits hysteresis (Fig. C.III.41a). In these experimental conditions, the coexistence of states I and II cannot be observed, as the system jumps abruptly from one to the other as the stress is increased or decreased. If, on the other hand, the experiment is performed at **fixed shear rate** (Fig. C.III.41b), the measured stress exhibits a plateau over the entire range of shear rates where states I and II coexist. Borrowing from the terminology of phase transitions, one can speak here of a first-order dynamical transition. At T_c, this transition becomes continuous (the width of the plateau vanishes), and then disappears above T_c (which plays the role of a critical point). Finally, the experimental data show that the smectic phase is non-Newtonian in region I, its viscosity varying as $(T_c - T)/\dot{\gamma}^{1/2}$ at low shear rates, while its behavior is Newtonian in region II.

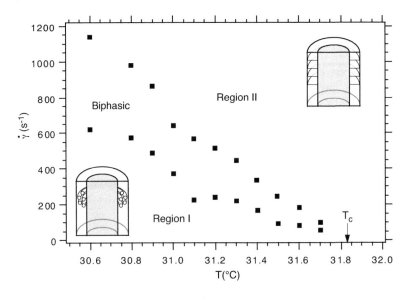

Fig. C.III.40 Orientation diagram of 8CB in the $(\dot{\gamma}, T)$ representation. The smectic phase goes to the nematic phase at 33°C. Region I disappears between T_c and T_{NA} (from ref. [70]).

The viscosity measured in this latter orientation is η_2 (see eq. C.II.80b): it does not vary much with temperature and is typically worth 0.36 poise, a value close to the one measured in the nematic phase.

To date, there is no theory to account for this orientation diagram. It seems though that it results from a competition between elastic, entropic and viscosity anisotropy effects.

In particular, one could think that the smectic phase will orient such as to dissipate the least energy possible, favoring, in the case of 8CB, the planar orientation with respect to the homeotropic one, since $\eta_2 < \eta_3$ (region II of the orientation diagram). An alternative explanation would be that the thermal fluctuations of the layers, becoming coupled to the shear [73a–d], induce the orientation transitions [74].

The underlying idea is that the shear tends to suppress the thermal fluctuations of the layers, decreasing the entropy of the system and, hence, increasing its free energy. This effect depends on the orientation of the layers with respect to the velocity and the velocity gradient, as shown by this formula taken from reference [74]:

$$F = F(\dot{\gamma} = 0) + 0.25 \frac{k_B T}{\lambda^3} \left(\frac{\dot{\gamma}}{B / \eta_3} \cos\theta \right)^{4/3} + \dots \qquad \text{(C.III.94)}$$

Here, F is the "free energy" in the dynamic regime (if such a quantity can be defined), θ is the angle between the layers and the velocity gradient (Fig. C.III.42), B the elastic modulus, η_3 the viscosity in shear parallel to the layers, and λ the penetration length $(\lambda = (K/B)^{1/2})$. This formula shows that F has a minimum for $\theta = \pi/2$, corresponding to the planar orientation observed at high shear rates in 8CB (region II of the orientation diagram in figure C.III.40).

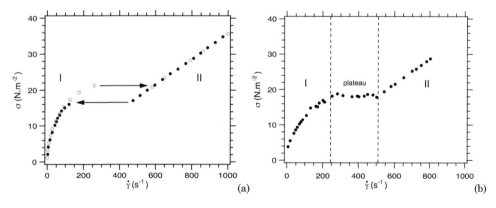

Fig. C.III.41 Stress as a function of the shear rate (8CB, T = 31°C, cone-plate geometry). a) Measurement at a fixed stress; b) measurement at a fixed shear rate. At low shear rate ($\dot{\gamma} < 200$ s^{-1}), the experimental curve is well described by a power law of the type $\dot{\gamma} = A\sigma^2$ (from ref. [70]). Note that the exponent is different from 5/3, which is the value found by Meyer et al. at room temperature [69] (see section III.3e).

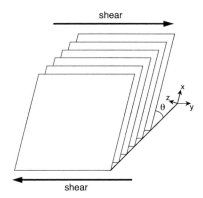

Fig. C.III.42 Definition of the tilt angle θ. In this orientation, the apparent viscosity is given by $\eta_{app} = \eta_3 \cos^2\theta + \eta_2 \sin^2\theta$.

The lubrication effects induced by the cylinders being off-center or by the alignment defects of the cone-plate setup should also have a strong influence, especially in the "leek" regime, but they were never systematically studied.

The shear should also change the smectic A-nematic [75] or smectic A-isotropic [76] transition temperatures, but these effects are always extremely small in thermotropics and difficult to characterize experimentally.

III.4.b) Orientation diagrams and onions in the lamellar phases of lyotropic liquid crystals

A typical example of an orientation diagram is given in figure C.III.43 for a lamellar lyotropic system. Here, the experimental parameters are the shear rate and the volume fraction of membranes (for lyotropics, this thermodynamic variable is more relevant than the temperature). The system is the quaternary mixture used by Diat et al. in their seminal work on the rheology of lamellar lyotropic phases [12a–e]: it consists of water, an ionic surfactant SDS (sodium dodecyl sulphate), oil (dodecane) and a cosurfactant, pentanol. In this system, the smectic repeat distance can be varied over a wide range by swelling with oil the lamellar phase (stabilized here by the Helfrich entropic interactions between membranes [77]).

The orientation diagram of this mixture contains four distinct regions that are traversed successively with increasing shear rate, at fixed membrane concentration (vertical line in the diagram of figure C.III.43):

1. In region 1, the sample remains globally homeotropic, but contains many defects, such as edge dislocations or oily streaks, which are partially carried along by the flow. The sample exhibits Newtonian behavior but remains quite viscous (Fig. C.III.44), no doubt due to the reduced mobility of the edge dislocations (strong lubrication effect). As the shear is stopped, the texture freezes into place without a noticeable change.

2. In region 2, the sample is much more homogeneous at the macroscopic scale than in region 1, but it strongly scatters light. Observation using polarizing microscopy reveals a granular structure on micron scales (Fig. C.III.45a). The size of the grains D can be measured precisely by observing the diffraction ring produced by a laser beam shone through the sample (Fig. C.III.45b).

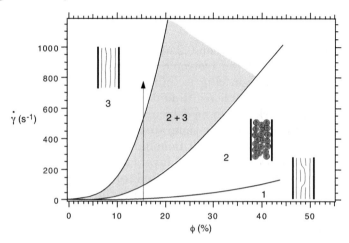

Fig. C.III.43 Typical orientation diagram for the quaternary mixture SDS/water/dodecane/pentanol at a ratio water/SDS = 1.55 (from ref. [12c]).

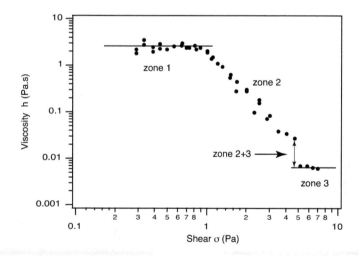

Fig. C.III.44 Typical curve representing the viscosity of the sample as a function of the stress (ϕ = 18%) (from ref. [12a]).

In this case, the diameter is given by the Debye formula:

$$D = \frac{\lambda}{2n} \frac{1}{\sin(\theta/2)} \qquad \text{(C.III.95)}$$

where λ is the wavelength of the laser, n the average refraction index of the lamellar phase and θ the diffraction angle corresponding to the ring. A remarkable result is that the grain size is well-defined (indeed, the size dispersion, given by the angular width of the diffraction ring, is usually lower than 20 %) and decreases as the reciprocal square root of the shear rate:

$$D \propto \frac{1}{\sqrt{\dot{\gamma}}} \qquad\qquad (C.III.96)$$

The typical grain size varies between 0.2 µm and 50 µm depending on the shear rate and the composition of the lyotropic mixture employed. Experimentally, the intensity of the diffraction ring is uniform along all directions, meaning that the grains are randomly stacked (liquid-type amorphous packing). Finally, the fine structure of the grains can be determined by diluting the phase with its own solvent: in this case, the grains separate and swell under the action of the osmotic pressure, revealing their multilamellar structure. The grains are thus vesicles formed by more or less spherical layers stacked one over the other. For this reason, these vesicles are also called **onions**. Electron microscopic observations of frozen sample replicas [78] show that they have a polyhedral shape and form under shear very compact packing arrangements. The onions are also very robust and persist for a long time (several days to several months) after the shear is stopped, yielding a metastable phase. Finally, the sample exhibits shear-thinning behavior in region 2, as shown in figure C.III.44.

3. In region 3, observed at high shear rates, the sample is again oriented in homeotropic anchoring, as in region 1. The sample regains a Newtonian behavior, but the measured viscosity is lower than in region 1 (Fig. C.III.44), certainly because the edge dislocations or the oily streaks have already disappeared (or become aligned along the velocity). If the shear is stopped, the sample destabilizes immediately and forms a texture bearing a strong resemblance to a focal parabolae texture: the homeotropic alignment is partially destroyed (which is not the case in region 1). We can infer that the sample under shear stores much more elastic energy in region 3.

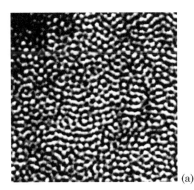
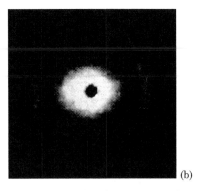

(a)   (b)

Fig. C.III.45 a) Granular appearance of the onion texture observed under the microscope between crossed polarizers; b) typical diffraction pattern given by the onions (from ref. 12a).

Fig. C.III.46 Electron microscopy images of onion replicas obtained by freeze-fracture. The samples (mixture of 60% imidazoline (SOCHAMINE 35 from Witco Chemicals) and 40% water containing 30% glycerine) were sheared at 13 s^{-1} (a) and at 20 s^{-1} (b). The typical sizes of the spherulites are 1.4 and 1 μm, respectively (from ref. [78]).

4. Finally, there is a wide range of shear rates where states 2 and 3 coexist (shaded area in the orientation diagram of figure C.III.43). In this regime, homeotropic areas coexist in the sample with onion-filled areas, arranged in stripes parallel to the shear [12a]. Note that the transition between "onions" (region 2) and homeotropic orientation (region 3) is discontinuous, **of first-order type with a jump in shear** when the experiment is performed at fixed shear stress. (In this case, the system goes directly from the onion structure to the homeotropic state, without coexistence, in contrast to the configuration of fixed shear rate.) This jump in shear is associated with the jump in viscosity on the curve in figure C.III.44. This transition is of the same nature as the one observed between "leeks" and planar orientation in 8CB (Fig. C.III.40). The difference is that this time one deals with onions instead of leeks, and that the lamellas are oriented in homeotropic anchoring rather than in planar anchoring at high shear rates. The reason for this difference is certainly that, in the lyotropic systems under study, $\eta_2 > \eta_3$, unlike the case of 8CB. We point out that the observation of a homeotropic state at high shear rates goes against the model of Goulian and Milner cited previously, which predicts a planar orientation ([74] and eq. C.III.94). We note however that the reduction in layer undulation fluctuations under shear, which is at the foundation of this model, is real and has been observed in lyotropic lamellar systems by Yamamoto and Tanaka [79], and more recently by Al Kahwaji et al. [80].

As for the leeks in 8CB, a convincing model for the formation of onions does not yet exist. Nevertheless, one thing is certain: mechanical energy must be injected into the system in order to destabilize it and form onions. In our opinion, the alignment and centering defects of the cylinders (or of the cone-plate setup), inducing at each rotation (in continuous shear) a compression and then dilation of the sample, are sufficient to induce layer buckling and nucleation of onions (or of leeks in 8CB). This occurs when the stress normal to the layers can no longer be effectively relaxed by the motion of edge dislocations present in the sample. A different model, proposed by Zilman and Granek [81], rests upon the idea that the "natural" repeat distance of the lamellas decreases under shear (due to the reduction of undulation fluctuations). This effect, intrinsic to the system, leads to the same result as a defect in the alignment of the cylinders, namely a local dilation of the layers that can lead to buckling under shear. There is, however, a significant difference between these models: in the first one, the lamellar phase is continuously compressed, and then dilated and can never relax the stress by adjusting the number of layers; in the second model, on the other hand, it is always possible (by propagation of edge dislocations) completely to adjust the number of layers in the sample in order to relax the dilation stress induced by the shear, irrespective of the value of the shear rate. In this case, why form onions? Another shortcoming of the model of Zilman and Granek is that the intrinsic dilation stress must increase linearly with the shear rate. If this stress cannot be relaxed, how can one explain the disappearance of the onions at high shear rate, since the stress should increase monotonically? In the lubrication model, on the other hand, the maximal dilation of the sample is constant, being set once and for all by the centering defect of the cylinders. On the other hand, we know that the critical dilation for the appearance of the layer buckling instability increases with the shear rate [82]. Hence, there must exist a critical shear rate above which the buckling instability disappears. This type of explanation was recently proposed by Wunenburger et al. to explain the disappearance of onions at high shear rates [83].

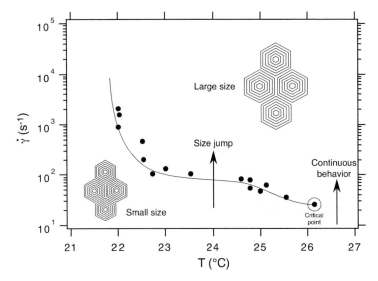

Fig. C.III.47 Transition line for the jump in size and critical point close to 26°C (from ref. [84]).

Even more complex behavior was observed in a mixture of SDS, octanol and brine [84a]. In this system, the onions stack first of all at random for "low" shear rates (as in the previously described system), and then form hexagonally ordered layers above a critical shear rate. Light and neutron scattering studies show that, within the onions, the layers are strongly compressed (the smectic distance decreases by several percent [84a]) and that the resulting structure is a mixture of face-centered cubic (ABCABC... stacking) and hexagonal compact (ABAB... stacking) structures [84b]. At even higher shear rates, a second transition occurs between two ordered states, characterized by a sudden and sharp increase in the size of the onions [84c,d]. In a $(\dot{\gamma}, T)$ diagram, this transition line ends at the equivalent of a critical point, beyond which the change in onion size is continuous (Fig. C.III.47). Finally, the onion structure has a viscoelastic response under oscillating shear. A (still not well understood) result is that, in the linear regime, the storage modulus G' varies as $1/D$ (D being the diameter of the onions) when the onions form an amorphous structure, while in the ordered state, G' is almost independent of D [84b,e].

In conclusion, let us mention that onions are nowadays used for the encapsulation of chemical substances (or biological, such as enzymes or DNA [85a–c]) and can be used as chemical microreactors [85d–f].

III.5 Microplasticity

The first microplasticity experiment was performed by Bartolino and Durand in 1977 [86]. It allowed them to measure directly the compression modulus of the layers B and to detect the plastic relaxation of the applied stress. In their experiment, the deformations are produced by a piezoelectric ceramic; they are therefore extremely small (10^{-5} or less), precluding the study of defect dynamics and of their instabilities. Much larger deformations (up to 10^{-3}) can be reached, for instance by stacking several tens of ceramic elements (40, for instance) [49, 87]. This technique recently revealed an original sequence of relaxation instabilities due to the helical deformation of screw dislocations [88, 89]. In this section, we summarize the main results obtained by this method.

III.5.a) Creep under compression normal to the layers and helical instabilities of screw dislocations

The experiment consists of brutally compressing a homeotropic sample while at the same time measuring the stress. The thickness variation is imposed by a piezoelectric ceramic, or by a stack of ceramics if a large deformation is to be obtained. The stress is measured directly using a second ceramic element [86] or indirectly, by measuring the deformations of the relaxation cell frame [87, 89].

Let $\delta(t)$ stand for the thickness variation of the sample (taken here as positive in compression, Fig. C.III.48). By applying a voltage jump to the ceramics, a rigid strain is imposed

$$\delta(t < 0) = 0 \tag{C.III.97a}$$

$$\delta(t \geq 0) = \delta_o \tag{C.III.97b}$$

In the experiments presented, δ_o varies from a few Å to a few hundred Å. The instantaneous response of the smectic, recorded by an oscilloscope or by a tracing table is elastic, as expected

$$\sigma(t = 0) = \sigma_o = B\frac{\delta_o}{d} \tag{C.III.98}$$

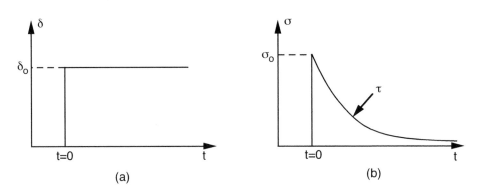

Fig. C.III.48 Schematic representation of the applied thickness variation and of the stress measured during a relaxation test. δ and σ are taken as positive in compression.

The measurement of σ_o yields the B modulus, generally taking values between 10^7 and 10^8 erg/cm^3. The stress (taken as positive in compression, for simplification) then slowly relaxes toward 0 (or a very small value, corresponding to compression by half a layer [49, 90, 91]) (Fig. C.III.48). In general, this relaxation is well described by an exponential law:

$$\sigma(t) = \sigma_o \exp\left(-\frac{t}{\tau}\right) \tag{C.III.99}$$

The relaxation time τ depends on the temperature and the initial applied stress, as shown in figure C.III.49. Its main feature is the stepwise variation (as a function of the stress). The first jump occurs for a value σ_1 of the instantaneous stress and the p-th jump at a value $\sigma_p = p\sigma_1$. More specifically, each jump of order p is characterized by:

    – the number of layers, of a thickness a_o, by which the sample is elastically compressed during a jump p:

$$N_p = pN_1 \tag{C.III.100}$$

with $\sigma_p = BN_pa_o/d$. In this experiment, performed on 8CB, N_1 varies from 3 to 5 depending on the temperature;

    – the relaxation time τ_p defined by the relation

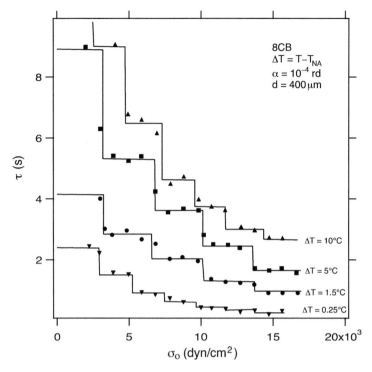

Fig. C.III.49 Experimentally determined relaxation time at different temperatures as a function of the instantaneous stress σ_0. The experimental data were corrected for the parasitic effects related to the finite rigidity of the relaxation cell frame. The solid lines emphasize the stepwise decrease of the relaxation time (from ref. [89]).

$$\frac{1}{\tau} = \frac{1}{\tau_g} + \frac{1}{\tau_p}$$
(C.III.101)

This decomposition implicitly assumes the presence of two independent relaxation mechanisms.

τ_g is the relaxation time measured at very small deformations. Experiments show that it is proportional to the sample thickness and inversely proportional to the angle α:

$$\tau_g = \frac{d}{\alpha}$$
(C.III.102)

This dependence is the clear signature of a relaxation mechanism by climb of edge dislocations from the grain boundary. On the other hand, the experimental data show conclusively that this relaxation time does not depend on the surface area of the sample. These observations rule out the creep model proposed by the Orsay Group on Liquid Crystals [92], whereby the layers (and thus the molecules) can easily leave the glass plates, an assumption that is not confirmed experimentally.

The second time τ_p is independent of α and decreases as $1/p$:

$$\tau_p \propto \frac{d}{p} \qquad\qquad\qquad \text{(C.III.103)}$$

We shall show that screw dislocations are responsible for this "anomalous" relaxation.

Before we tackle this problem, let us return for a moment to the microstructure of the sample. Ever since the work of Meyer et al. [25a,b], we know that a grain boundary composed of edge dislocations exists in the center of the sample. Under compression, these dislocations move in order to relax the stress, as they do in lubrication. Let x be the distance covered by each dislocation at the time t (the geometry is the one depicted in figure C.III.15a). At that moment, the value of the stress is

$$\sigma = \frac{B}{d}(\delta_0 - \alpha x) \qquad\qquad\qquad \text{(C.III.104)}$$

Writing that the velocity of each dislocation is $v = dx/dt = m\sigma$ (we recall that the stress is taken as positive in compression), one obtains an evolution equation for σ by taking the derivative of eq. C.III.104 with respect to time:

$$\frac{d\sigma}{\sigma} = -\frac{dt}{\tau_g} \qquad \text{with} \qquad \tau_g = \frac{d}{mB\alpha} \qquad\qquad \text{(C.III.105)}$$

According to this relation, the stress relaxes exponentially towards 0, with a characteristic time τ_g proportional to d/α. This simple mechanism explains the relaxation observed at low compression. By comparing the theoretical expression C.III.105 with the experiment, one finds for the mobility m a value in good agreement with the one found in the previous section ($m \approx 3 \times 10^{-7}$ cm^{2}sg^{-1} for 8CB at 25°C). One can also see that the mobility m is thermally activated, with an activation energy of typically 1.8 eV in 8CB. This simple mechanism cannot explain the successive jumps in relaxation time. To this end, one would have to admit that the density of dislocations suddenly increases above σ_1, then doubles above σ_2, etc. It is easily shown that the nucleation of edge dislocation loops ex nihilo is impossible, the activation energy required by this process (of the order of $\pi K^2/b\sigma$) being considerable (larger than 1000 eV) and exceeding the typical stress values used in the experiment (of the order of a few thousand dynes). Another, topologically equivalent, possibility is the helical destabilization of screw dislocations. We show in the following that this is spontaneously possible under high enough stress.

To analyze this transformation, let us consider a screw dislocation with Burgers vector b bridging the two plates (usually, $b = a_0$). This dislocation is initially straight, due to its strong line tension. We shall not consider here the case of incomplete segments of screw dislocations, very seldom encountered in experiments (such a segment is present each time an edge dislocation ends abruptly, so that they could be counted at the SmA-SmC transition). We consider that the line assumes a helical form, with a pitch $P = d/p$. If the line is anchored onto the surface, p is necessarily an integer. From a topological point of view, this helix is equivalent to the initial dislocation, encircled by p edge dislocation loops with a Burgers vector b (Fig. C.III.50). Thus, p layers were eliminated from the cylinder bearing the helix of equation $z = d\theta/(2\pi p)$.

To study the dynamics of the line, let us consider the radial forces acting upon it. We find:

1. The Peach and Koehler force:

$$F_1 = \sigma b \sin\Phi = \sigma b \frac{r}{\sqrt{r^2 + d_o^2}} \qquad \text{with} \qquad d_o = \frac{d}{2\pi p} \qquad \text{(C.III.106)}$$

where Φ is the angle made by the helix with the z axis and r the radius of the cylinder bearing the helix.

2. The elastic restoring force associated with the line tension T:

$$F_2 = -\frac{T}{R} = -T\frac{r}{r^2 + d_o^2} \qquad \text{(C.III.107)}$$

with R the curvature radius of the helix.

3. A viscous friction force:

$$F_3 = -\frac{b}{M}\frac{dr}{dt} \qquad \text{(C.III.108)}$$

where M is a mobility given by eq. C.III.58.
The mass of the dislocation being negligible, one has $F_1 + F_2 + F_3 = 0$, yielding:

$$\frac{\sigma b}{\sqrt{d_o^2 + r^2}} - \frac{T}{r^2 + d_o^2} = \frac{b}{M}\frac{1}{r}\frac{dr}{dt} \qquad \text{(C.III.109)}$$

In this expression, σ is a decreasing function of time, from an initial value $\sigma_o = B\delta_o/d$. The helix develops spontaneously if

$$\left(\frac{dr}{dt}\right)_{t=0} > 0 \qquad \text{(C.III.110)}$$

At the initial moment, $r \approx 0$ (up to the effect of thermal fluctuations). The instability is thus absolute if

$$\sigma_o > \frac{T}{bd_o} = p\frac{2\pi T}{bd} = \sigma_p \qquad \text{(C.III.111)}$$

This condition defines the threshold for development of the instability of order p. With the notations employed, we have, using $b = a_o$:

$$N_p = pN_1 = p\frac{2\pi T}{Ba_o^2} \qquad \text{(C.III.112)}$$

This relation is consistent with the experiments. It predicts regularly spaced jumps.

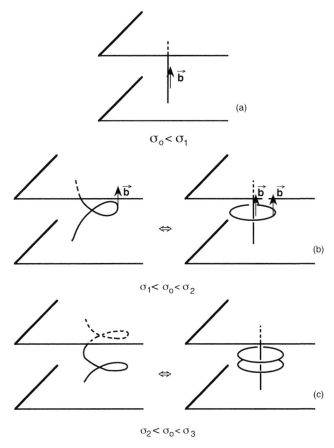

Fig. C.III.50 a) Screw dislocation bridging the two surfaces; b) helical dislocation (p = 1) and equivalent topological diagram; c) idem for p = 2. Note that we have not represented the kink of height b that necessarily exists on each dislocation loop due to the presence of the screw dislocation.

The experimental measurement of N_1 gives access to the line tension T which contains, as we recall, a core energy term and an elastic term of the order of K (see eq. C.III.56b). This elastic term was rigorously calculated by Bourdon et al. [88], who find for a helical dislocation:

$$T = E_c + 0.04\, Ba_o^2 \ln \frac{2\pi d\lambda}{r_c^2} \approx E_c + 0.5 Ba_o^2 \qquad \text{(C.III.113)}$$

The elastic term being known, comparison with the experiment yields the core energy: $E_c \approx 0.1\, Ba_o^2$. This value is close to the one obtained by taking the core as filled with the nematic phase.

To determine the relaxation time τ_p, one must first solve equation C.III.105 taking into account the fact that σ decreases with time. This

176

calculation shows that τ_p is proportional to d/p, in agreement with the experiment [89]. The p dependence should not come as a surprise, since a helix of order p relaxes the stress p times faster than a helix of order 1, the number of removed layers being multiplied by p.

In conclusion, the two mechanisms presented here explain all the experimental features observed. Let us emphasize that, once again, the two kinds of dislocations play distinct roles, without really interacting. This independence is due to the easy crossing of the edge and screw dislocations and to the fact that the interaction between two edge dislocations is negligible unless they are in the same glide plane.

We conclude this section by pointing out that the helical instability of screw dislocations was recently observed by Herke et al. [93a,b] for 8CB using a surface force apparatus. In this experiment, the relaxation dynamics are more complex due to the sphere-plane geometry employed and to the reduced sample thickness (a few μm in the central area).

III.5.b) Creep in dilation normal to the layers and nucleation of a focal parabolae lattice

In this case, the situation is much more complicated due to the layer buckling instability [94] completely dissimulating the helical instability of the screw dislocations (still possible, in principle, since the same dislocation can relax either a compression or a dilation by a simple change in the sign of the helix). On the other hand, the experiments of Bartolino and Durand [42] show that, if the instability threshold δ_c is the one given by the classical elasticity theory as long as the angle between the plates is close to zero ($\alpha < 10^{-3}$ rad):

$$\delta_c = 2\pi\lambda \tag{C.III.114}$$

it increases significantly when the angle α exceeds the critical value α_c above which the dislocations of the grain boundary start to bunch together (see section III.2.b.v). This phenomenon still lacks a clear-cut explanation.

We shall restrict the discussion to the case where α is close to 0. In this limit, experiment shows that, below the buckling threshold, the stress relaxes as under compression − by climb of the edge dislocations in the grain boundary (for well annealed samples). Above the threshold, on the other hand, the relaxation can schematically be separated into two steps:

1. In the first step, the stress decreases very fast from its initial value σ_o to an intermediate value σ_i. This characteristic time τ_i was measured by Bartolino and Durand by light scattering; it is of the order of a few tens of ms [42].

2. In the second step, the stress relaxes slowly toward zero, as it would do under compression, with a characteristic time τ_r of the order of a few seconds, or even much longer if the focal parabolae already appeared (for $\delta_o > 1.6\,\delta_c$) [49a,b] (Fig. C.III.51).

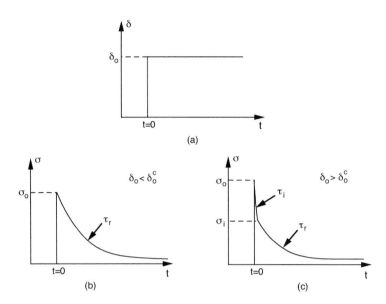

Fig. C.III.51 Schematic representation of the imposed thickness variation (a) and of the stress measured in a relaxation test under dilation. Below the threshold of the layer buckling instability, the stress relaxes slowly, with a characteristic time τ_r (b); above the threshold (c), the stress relaxes in two steps: very fast at first (over a time τ_i) and then slowly (over a time τ_r).

The behavior of the smectic sample immediately above the buckling instability threshold is not easily interpreted quantitatively, as one does not really know what happens during the first step of the relaxation process (even in the regime of undulations). It is nevertheless possible that the layers break very easily, as when they are exposed to oscillating shear perpendicular to their plane (see section III.3.e), forming dislocation loops that facilitate the relaxation. Let us check the hypothesis by comparing these two experiments. Indeed, creating a layer undulation amounts to locally shearing them at a typical angle $\theta_o = q_c u_o$, where q_c is the wave vector of the most unstable mode (very close to the wave vector at the threshold given by eq. C.II.43 [94]) and u_o the saturation amplitude of the undulation (given by eq. C.II.46 in the weakly nonlinear approximation):

$$\theta_o \approx \frac{8}{3}\sqrt{\frac{\delta - \delta_c}{2d}} \qquad\qquad (C.III.115)$$

The growth rate of the most unstable mode, given by the linear theory, is given by [95]:

$$\omega_{max} = (\delta - \delta_c)\frac{B}{\mu d} \qquad\qquad (C.III.116)$$

These two equations yield the control parameter N of the instability (see eq. C.III.92):

$$N = B \frac{\theta_0^2}{\mu \omega_{max}} = \frac{16}{3} \qquad \text{(C.III.117)}$$

We note that this value is independent of the distance to the threshold and close to the critical value for which the layers break [68a,b]. The comparison between the two experiments suggests that the buckling instability is accompanied by layer breaking, which might lead to a partial, but very fast, relaxation of the applied stress.

The plastic relaxation effects are much more significant once the focal parabolae have developed, namely for initial deformations above $1.6\delta_c$ [5, 49, 96]. As in the undulation regime, the stress relaxes in two steps: very fast at first, over a time τ_i equal to the nucleation time of the focal parabolae, then extremely slowly by thermally activated disappearance of parabolic domains. In this regime, the stress σ_i varies little with the initially imposed deformation, as the focal parabolae relax it almost completely from the beginning.

Experimentally, the stress σ_i can be measured by filtering the signal of the stress sensor and preserving only the low frequencies. An ordinary tracing table is perfectly suited, its reaction time $\tau_{react} \approx 0.1$ s being much larger than τ_i and much lower than τ_r. The figure C.III.52 shows the results obtained in this way at 25°C with an 8CB sample (of angle α close to 10^{-4} rad). One can see that the curve $\sigma_i(\delta_0)$ has a downward inflection above 70 Å and exhibits a break in slope for a value close to 110 Å. This break corresponds to the threshold for the appearance of focal parabolae, as shown by simultaneous observation in the polarizing microscope. The first threshold, more difficult to determine experimentally, corresponds to the undulations. This experiment again shows that the focal parabolae appear for a dilation above $1.6 \delta_c$, in agreement with the light scattering experiments [5b]. These mechanical measurements give access to the value of λ, close to 10 Å in 8CB at 25°C. This method was also used to measure λ in the lamellar phase of the lyotropic mixture $C_{12}EO_6/H_2O$ [97].

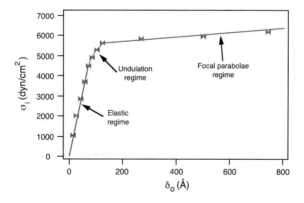

Fig. C.III.52 Stress σ_i as a function of the imposed thickness variation δ_0 (homeotropic 8CB sample at 25°C, of thickness d = 200 μm). The stress increases linearly with δ_0 in the elastic regime (the slope yields $B \approx 1.3 \times 10^8$ erg/cm³); above a critical value δ_c close to 80 Å, the stress σ_i increases much more slowly (plastic domain).

In practice, the precision of the method can be significantly improved by applying a sinusoidal oscillation and using a lock-in amplifier [96]. This method allows discrimination between the elastic and the viscous components of the signal, and better separation of the undulation regime (where little energy is dissipated) from that of focal parabolae (where the dissipation increases brutally, sign of a sort of ductile fracture). Thus, the elastic modulus B and the penetration length λ of 8CB can be measured more precisely (Fig. C.III.53a and b) [98].

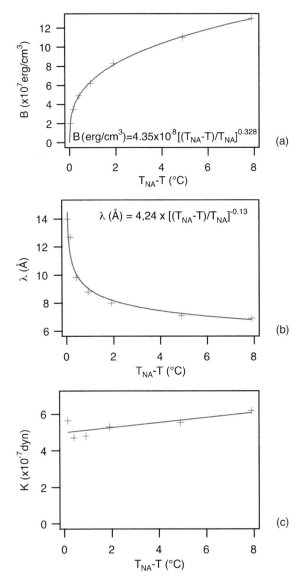

Fig. C.III.53 Compression modulus B, penetration length λ and bending constant K obtained by mechanical measurements on the smectic A phase of 8CB (from ref. [98]).

These two measurements yield the bending constant K of the layers (Fig. C.III.53c), almost constant in temperature and exhibiting no anomaly close to the transition towards the nematic phase (in agreement with the results of chapter C.I). Moreover, the value of K measured at the transition ($\approx 4.5 \pm 1 \times 10^{-7}$ dyn) is close to the value of K_1 measured for the nematic phase, close to the transition to the smectic phase ($\approx 7 \pm 1 \times 10^{-7}$ dyn [99]).

BIBLIOGRAPHY

[1] Friedel G., *Annales de Physique*, **18** (1922) 273.

[2] Frank F.C., *Disc. Faraday Soc.*, **25** (1958) 19.

[3] Boltenhagen P., Kléman M., Lavrentovich O., *C. R. Acad. Sci. (Paris)*, Série II, **315** (1992) 931.

[4] a) Bouligand Y., *J. Physique (France)*, 33 (1972) 525.
b) Friedel G., Grandjean F., *Bull. Soc. Franç. Minér.*, **33** (1910) 409.
c) Bragg W., *Supplement to Nature*, **3360** (1934) 445.

[5] a) Rosenblatt Ch.S., Pindak R., Clark N.A., Meyer B., *J. Physique (France)*, **38** (1977) 1105.
b) Clark N.A., Hurd A.J., *J. Physique (France)*, **43** (1982) 1159.

[6] Benton W.J., Toor E.W., Miller C.A., Fort T. Jr., *J. Physique (France)*, **40** (1979) 107.

[7] Oswald P., Béhar J., Kléman M., *Phil. Mag. A*, **46** (1982) 899.

[8] Stewart I.W., *Liq. Cryst.*, **15** (1993) 859.

[9] Bidaux R., Boccara N., Sarma G., De Seze L., de Gennes P.-G., Parodi O., *J. Physique (France)*, **34** (1973) 661.

[10] Sethna J.P., Kléman M., *Phys. Rev. A*, **26** (1982) 3037.

[11] Lavrentovich O.D., *Sov. Phys. J.E.T.P.*, **64** (1986) 984.

[12] a) Diat O., "Etude du cisaillement sur des phases lyotropes : phase lamellaire et phase éponge," Ph.D. Thesis, University of Bordeaux I, Bordeaux, Order no. 833, 1992.
b) Diat O., Roux D., *J. Phys. II (France)*, **3** (1993) 9
c) Diat O., Roux D., Nallet F., *J. Phys. II (France)*, **3** (1993)1427.
d) Diat O., Roux D., Nallet F., *J. Phys. II (France)*, **3** (1993) 193.
e) Roux D., Nallet F., Diat O., *Europhys. Lett.*, **24** (1993) 53.

[13] Kléman M., *J. Physique (France)*, **38** (1977) 1511.

[14] Boltenhagen P., Lavrentovich O., Kléman M., *J. Phys. II (France)*, **1** (1991) 1233.

[15] Kléman M., Parodi O., *J. Physique (France)*, **36** (1975) 671.

[16] Fournier J.B., *Eur. J. Phys.*, **15** (1994) 319.

[17] Fournier J.B., *Phys. Rev. E*, **50** (1994) 2868.

[18] a) Rault J., *C. R. Acad. Sci. (Paris)*, série B, **280** (1975) 417.
b) Rault J., *Phil. Mag.*, **34** (1976) 753.

[19] Kléman M., "Points, Lines, and Walls in Liquid Crystals, Magnetic Systems, and Various Ordered Media," John Wiley & Sons, Chichester, 1983.

[20] Benton W.J., Miller C.A., *Prog. Colloid & Polymer Sci.,* **68** (1983) 71.

[21] Williams C.E., "Défauts de structure dans les smectiques A," D.Sc. Thesis, University of Paris XI, Orsay, Order no. A.O.12720, 1976.

[22] Williams C.E., Kléman M., *J. Physique Colloq. (France)*, **36** (1975) C1-315.

[23] Demus D., Richter L., *Textures of Liquid Crystals*, Verlag Chemie Weinheim, New York, 1978.

[24] Williams C.E., *Phil. Mag.*, **32** (1975) 313.

[25] a) Meyer R.B., Stebler B., Lagerwall S.T., *Phys. Rev. Lett.*, **41** (1978) 1393.
b) Lagerwall S.T. , Meyer R.B., Stebler B., *Ann. Physique*, **3** (1978) 249.

[26] Chan W.K., Webb W.W., *J. Physique (France)*, **42** (1981) 1007.

[27] Pieranski P., Beliard L., Tournellec J.-Ph., Leoncini X., Furtlehner C., Dumoulin H., Riou E., Jouvin B., Fénerol J.-P., Palaric Ph., Heuving J., Cartier B., Kraus I., *Physica A*, **194** (1993) 364.

[28] a) Maaloum M., Ausserré D., Chatenay D., Coulon G., Gallot Y., *Phys. Rev. Lett.*, **68** (1992) 1575.
b) Turner M.S., Maaloum M., Ausserré D., Joanny J.-F., Kunz M., *J. Phys. II (France)*, **4** (1994) 689.

[29] de Gennes P.-G., *C. R. Acad. Sci. (Paris)*, Série B, **275** (1972) 939.

[30] Kléman M., *J. Physique (France)*, **35** (1974) 595.

[31] Loginov E.B., Terent'ev E.M., *Sov. Phys. Crystallogr.*, **30** (1985) 4.

[32] Kléman M., Williams C.E., *J. Physique Lett. (France)*, **35** (1974) L49.

[33] Kléman M., *Rep. Prog. Phys.*, **52** (1989) 555.

[34] Landau L, Lifshitz E., *Theory of Elasticity*, Second Edition, Pergamon Press, New York, 1981.

[35] Lejcek L., Oswald P., *J. Phys. II (France)*, **1** (1991) 931.

[36] Galerne Y., Liébert L., *Phys. Rev. Lett.*, **66** (1991) 2891.

[37] Pershan P.S., *J. Appl. Phys.*, **45** (1974) 1590.

[38] Holyst R., *Phys. Rev. Lett.*, **72** (1994) 4097.

[39] Holyst R., Oswald P., *Intern. J. Modern Phys. B*, **9** (1995) 1515.

[40] Holyst R., Oswald P., *J. Phys. II (France)*, **5** (1995) 1525.

[41] Lejcek L., *Czech. J. Phys. B*, **32** (1982) 767.

[42] Bartolino R., Durand G., *Mol. Cryst. Liq. Cryst.*, **40** (1977) 117.

[43] Nallet F., Prost J., *Europhys. Lett.*, **4** (1987) 307.

[44] Quilliet C., Fabre P., Veyssié M., *J. Phys. II (France)*, **3** (1993) 1371.

[45] Brener E.A., Marchenko V.I., *Phys. Rev. E*, **59** (1999) R4752.

[46] Dubois-Violette E., Guazzelli E., Prost J., *Phil. Mag.*, **48** (1983) 727.

[47] de Gennes P.G., *Physics of Fluids*, **17** (1974) 1645.

[48] Oswald P., Lejcek L., *J. Phys. II (France)*, **1** (1991) 1067.

[49] a) Oswald P., "Dynamique des dislocations dans les smectiques A et B," D.Sc. Thesis, University of Paris XI, Orsay, 1985.
b) Oswald P., *C. R. Acad. Sci. (Paris)*, Série II, **296** (1983) 1385.

[50] Bourdon L., "Contribution à l'étude des défauts dans les phases smectiques à couches liquides," M.Sc. Thesis, University of Paris XI, Orsay, 1980.

[51] Ishikawa K., Uemura T., Takezoe H., Fukuda A., *Jap. J. Appl. Phys.*, **23** (1984) L666.

[52] Kléman M., Colliex C.,Veyssié M., in *Lyotropic Liquid Crystals*, Ed. S. Freiberg, 1976, p. 71.

[53] Kléman M., Williams C.E., Costello J.M., Gulik-Krzywicki T., *Phil. Mag.*, **35** (1977) 33.

[54] Allain M., *J. Physique (France)*, **48** (1985) 225.

[55] Zasadzinski J.A.N., *Biophys. J.*, **49** (1986) 1119.

[56] Williams C.E., Kléman M., *Phil. Mag.*, **33** (1976) 213.

[57] Kléman M., *Phil. Mag.*, **34** (1976) 79.

[58] a) Pleiner H., *Liq. Cryst.*, **1** (1986) 197.
b) Pleiner H., *Phil. Mag. A*, **54** (1986) 421.
c) Pleiner H., *Liq. Cryst.*, **3** (1988) 249.

[59] a) Friedel J., *Dislocations*, Pergamon Press, Oxford, New York, 1967.
b) Martin J.-L., *Dislocations et Plasticité des Cristaux*, Presses polytechniques et universitaires romandes, Lausanne, 2000.

[60] a) Lelidis I., Kléman M., Martin J.-L., *Mol. Cryst. Liq. Cryst. Sci. Technol.*, Section A, **330** (1999) 457.
b) Lelidis I., Kléman M., Martin J.-L., *Mol. Cryst. Liq. Cryst. Sci. Technol.*, Section A, **351** (2000) 187.
c) Blanc C., Zuodar N., Lelidis I., Kléman M., Martin J.-L., *Phys. Rev. E*, **69** (2004) 011705.

[61] Hollinger S., "Comportement d'un lubrifiant aqueux dans un contact à très hautes pressions," Ph.D. Thesis, Ecole Centrale de Lyon, Ecully, Order no. ECL99-54, 1999.

[62] Oswald P., Kléman M., *J. Physique Lett. (France)*, **43** (1982) L411.

[63] Oswald P., Ben-Abraham S.I., *J. Physique (France)*, **43** (1982) 1193.

[64] Oswald P., *J. Physique Lett. (France)*, **44** (1983) L303.

[65] Oswald P., *J. Physique (France)*, **47** (1986) 1091.

[66] Oswald P., Allain M., *J. Coll. Int. Sci.*, **126** (1988) 45.

[67] Horn R.G., Kléman M., *Ann. Physique*, **3** (1978) 229.

[68] a) Marignan J., Parodi O., *J. Physique (France)*, **44** (1983) 665.
b) Marignan J., Parodi O., Dubois-Violette E., *J. Physique (France)*, **44** (1983) 263.

[69] Meyer C., Asnacios S., Bourgaux C., Kléman M., *Rheol. Acta*, **39** (2000) 223.

[70] a) Panizza P., Archambault P., Roux D., *J. Phys. II (France)*, **5** (1995) 303.
b) Panizza P., "Etude des états stationnaires et transitoires de phases lamellaires sous cisaillement," Ph.D. Thesis, University of Bordeaux I, Bordeaux, Order no. 1603, 1996.

[71] Safinya C.R., Sirota E.B., Plano R.J., *Phys. Rev. Lett.*, **66** (1991) 1986.

[72] Plano R.J., Safinya C.R., Sirota E.B., Wenzel L.J., *Rev. Sci. Instrum.*, **64** (1993) 1309.

[73] a) de Gennes P.-G., *Mol. Cryst. Liq. Cryst.*, **34** (1976) 91.
b) Ramaswamy S., *Phys. Rev. A*, **29** (1984) 1506.
c) Frederickson G.H., *J. Chem. Phys.*, **85** (1986) 5306.
d) Bruinsma R., Rabin Y., *Phys. Rev. A*, **45** (1992) 994.

[74] Goulian M., Milner S.T., *Phys. Rev. Lett.*, **74** (1995) 1775.

[75] Bruinsma R.F., Safinya C.R., *Phys. Rev. A*, **43** (1991) 5377.

[76] Cates M.E., Milner S.T., *Phys. Rev. Lett.*, **62** (1989) 1856.

[77] Helfrich W., *Z. Naturforsch.*, **33a** (1978) 305.

[78] Gulik-Krzywicki T., Dedieu J.C., Roux D., Degert C., Laversanne R., *Langmuir*, **12** (1996) 4668.

[79] Yamamoto J., Tanaka H., *Phys. Rev. Lett.*, **74** (1995) 932.

[80] Al Kahwaji A., Greffier O., Leon A., Rouch J., Kellay H., *Phys. Rev. E*, **64** (2001) 041502.

[81] Zilman A.G., Granek R., *Euro. Phys. J. B*, **11** (1998) 593.

[82] Oswald P., Ben Abraham S.I., *J. Physique (France)*, **43** (1982) 1193.

[83] Wunenburger A.-S., Colin A., Colin T., Roux D., *Euro. Phys. J. E*, **2** (2000) 277.

[84] a) Diat O., Roux D., Nallet F., *Phys. Rev. E*, **51** (1995) 3296.
b) Leng J., "Structure et dynamique de la texture ognon des phases lamellaires lyotropes," Ph.D. Thesis, University of Bordeaux I, Bordeaux, Order no. 2140, 1999.
c) Sierro Ph., "Effet du cisaillement sur la phase lamellaire du système quaternaire : octanol/SDS/eau/NaCl," Ph.D. Thesis, University of Bordeaux I, Bordeaux, Order no. 1386, 1995.
d) Sierro Ph., Roux D., *Phys. Rev. Lett.*, **78** (1997) 1496.

e) Leng J., Nallet F., Roux D., *Eur. Phys. J. E*, **4** (2001) 337.

[85] a) Freund O., Mahy P., Amedee J., Roux D., Laversanne R., *J. Microencapsulation*, **17** (2000) 157.
b) Berheim-Grosswasser A., Ugazio S., Gauffre F., Mahy P., Roux D., *J. Chem. Phys.*, **112** (2000) 3424.
c) Mignet N., Bup A., Degert C., Delord B., Roux D., Helene C., Francois J.C., Laversanne R., *Nucleic Acid Research*, **28** (2000) 3134.
d) Gauffre F., "Utilisation de vésicules préparées par cisaillement d'une phase lamellaire comme microréacteurs chimiques," Ph.D. Thesis, University of Bordeaux I, Order no. 1756, 1997.
e) Roux D., Gauffre F., *European Chemistry Chronicle*, **3** (1998) 17.
f) Gauffre F., Roux D., *Langmuir*, **15** (1999) 3070 and 3738.

[86] Bartolino R., Durand G., *Phys. Rev. Lett.*, **39** (1977) 1346.

[87] Oswald P., Lefur D., *C. R. Acad. Sci. (Paris)*, Série II, **297** (1983) 699.

[88] Bourdon L., Kléman M., Lejcek L., Taupin D., *J. Physique (France)*, **4 2** (1981) 261.

[89] Oswald P., Kléman M., *J. Physique Lett. (France)*, **45** (1984) L319.

[90] Bartolino R., Durand G., *Il Nuovo Cimento*, **3D** (1984) 903.

[91] Galerne Y., "Etude interférométrique de la transition smectique A-smectique C," D.Sc. Thesis, University of Paris XI, Orsay, 1982.

[92] Orsay Group on Liquid Crystals, *J. Physique Colloq. (France)*, **36** C1 (1975) 305.

[93] a) Herke R.A., Clark N.A., Handschy M.A., *Phys. Rev. E*, **56** (1997) 3028.
b) Herke R.A., Clark N.A., Handschy M.A., *Science*, **267** (1995) 651.

[94] Ribotta R., Durand G., *J. Physique (France)*, **38** (1977) 179.

[95] Ben Abraham S.I., Oswald P., *Mol. Cryst. Liq. Cryst.*, **94** (1983) 383.

[96] Picano F., Holyst R., Oswald P., *Phys. Rev. E*, **62** (2000) 3747.

[97] Oswald P., Allain M., *J. Physique (France)*, **46** (1985) 831.

[98] Zywocinski A., Picano F., Oswald P., Géminard J.-C., *Phys. Rev. E*, **62** (2000) 8133.

[99] Madhusudana N.V., Pratibha R., *Mol. Cryst. Liq. Cryst.*, **89** (1982) 249.

Chapter C.IV

Ferroelectric and antiferroelectric mesophases

The history of liquid crystals abounds in discoveries of new phases and original phenomena. If most mesophases were found by accident, characterizing their structure was not always easy. The main difficulty is that, in order to accept the existence of a new phase, one must sometimes reconsider existing ideas. The controversy between Otto Lehmann and Georges Friedel, evoked in chapter A.I (volume I), provides an outstanding example.

The history of the SmC* phase is very different, as the existence of this mesophase, very important for applications (new display technology), was contemplated theoretically before being demonstrated experimentally.

In the first part of this chapter, dealing with the SmC* phase, we recall the symmetry arguments and the flexoelectricity elements which led R.B. Meyer to posit the existence of ferroelectric phases in liquid crystals (section IV.1). On the occasion, we shall also present a subsequent explanation of ferroelectricity in smectics C* due to P.-G. de Gennes and based on an illuminating ichthyological analogy. We shall then see how the chemists (L. Liébert, L. Strzelecki, and P. Keller) brought to life Meyer's ideas, by synthesizing the first ferroelectric mesophase, that they named smectic C* (section IV.2), before we describe (in section IV.3) the first experiment that Meyer performed in order to demonstrate the spontaneous polarization of this new phase. After these theoretical and historical preliminaries, we shall describe the ways of measuring the spontaneous polarization and its use in creating fast displays (section IV.4). We also discuss the hydroelectrical and electromechanical effects, which resemble to some extent the piezoelectric effects encountered in solid crystals (sections IV.5 and IV.6). In conclusion, we give a description of the smectic A-smectic C* transition and of the electroclinic effect (section IV.7).

In the second part of the chapter, we present more recently discovered mesophases, in which the electric dipole moments are also ordered, but in a different manner than in the SmC* phase. In particular, we discuss the antiferroelectric SmC$_A$ phase (at first named SmO, section IV.8), and then the "banana" phases (section IV.9), which can exhibit a chiral structure even though they are not composed of chiral molecules.

IV.1   Do ferroelectric mesophases exist?

This question eventually led to the discovery of the SmC* phase. For a better understanding of its relevance, one should remember that, by definition, a phase is ferroelectric when it exhibits a finite average density of dipole moment. As shown in chapter B.I, mesogenic molecules, considered individually, very often bear an electric dipole. Nevertheless, in the statistical ensemble of the mesophase, the average value of the dipole moment is almost always zero, due to the point symmetry of the phase.

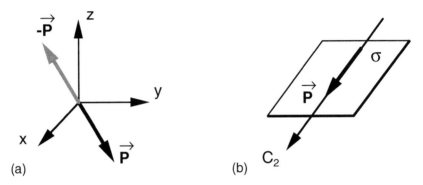

Fig. C.IV.1 Effect of the symmetry operations on the spontaneous polarization: a) under inversion, **P** changes sign; b) on the other hand, **P** remains invariant under the action of the C_2 axis and of the mirror σ.

One symmetry element – inversion – is enough, for instance, to annihilate the spontaneous polarization. We can show it by the following reasoning (see Fig. C.IV.1a):

Let ***P*** *be the polarization of the mesophase. Under inversion,* ***P*** *is transformed into* ***– P***. *If inversion is a symmetry element of the mesophase, this latter transforms in itself under its action, so that* ***P = – P***. *This is only possible if* ***P*** *is zero.*

By a similar argument, one can show that the spontaneous polarization must also be zero in phases with symmetries C_{2h} (the second-order axis being perpendicular to the symmetry plane σ_h) and D_2 (the three second-order axes being orthogonal). On the other hand, spontaneous polarization is allowed in systems with symmetries C_n, C_{nv} (the nth order axis being parallel to the symmetry plane σ_v) or C_s (possessing only one symmetry plane). In the first two cases, it must be directed along the C_n axis, while in the last case, it is necessarily contained in the symmetry plane σ (see Fig. C.IV.1b).

This being said, the question concerning the existence of ferroelectric mesophases can be rephrased as:

"Are there mesophases exhibiting symmetry compatible with ferroelectricity?"

Meyer [1] considered this question in 1975 while investigating the origins of flexoelectricity (see section B.I.4.c, volume 1). The undistorted nematic phase has a $D_{\infty h}$ symmetry, which – as we have just seen – forbids spontaneous polarization. Nevertheless, this symmetry is broken in the presence of a splay or bend deformation of the director field. More precisely, the symmetry is reduced to C_{2v} in the case of two-dimensional distortions (when **n** only depends on two space coordinates: x and y for instance) of the "pure splay" or "pure bend" types (Fig. C.IV.2). In the case of a generic two-dimensional deformation combining splay and bend, only the reflection with respect to the (x, y) plane is preserved. As a consequence of this symmetry breaking, the distorted nematic can acquire a **finite polarization P** contained in the symmetry plane σ.

Having understood this relation between polarization and distortion (which breaks the symmetry) for the case of flexoelectricity, Meyer wondered if one of the known mesophases (at that time) might exhibit one of the desired symmetries, C_2 or C_{2v}. His attention was attracted by the smectic C phase.

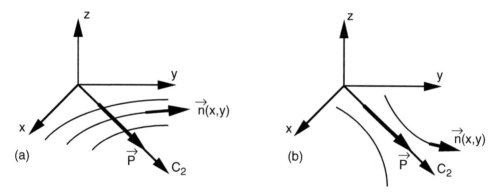

Fig. C.IV.2 Relation between the symmetry of the distorted nematic phase and the existence of a spontaneous polarization: the polarization **P** is directed along the local symmetry axis C_2.

IV.2   Genesis of the smectic C* phase

IV.2.a) Structure of the smectic C phase

In the classification diagram of mesophases (see chapter A.II), the smectic C phase occupies the same position as the smectic A phase, since in the two phases the positional order within the layers is identical, characteristic of a liquid. The difference between the two phases is given by the orientation of the molecules with respect to the layer normal **m**. They are on the average parallel to **m** in the SmA phase and tilted with respect to **m** in the SmC phase.

The difference between the two phases appears even more clearly when considered in terms of point symmetry. We saw in chapter C.I that the point symmetry of the smectic A phase is the same as for a nematic phase, namely $D_{\infty h}$. In particular, the smectic A phase is invariant with respect to rotations about the axis $C_{\infty}//\mathbf{m}$. From a microscopic point of view, this means that all molecular configurations related to each other by rotations around **m** are equivalent. In other words, the molecules are free to turn around **m**. Optically, the symmetry with respect to the C_{∞} axis induces the uniaxial character of the dielectric tensor ε_{ij} in the SmA phase.

In the smectic C phase, however, the axial symmetry C_{∞} is broken. The only remaining symmetry elements are: 1. The reflection in a plane σ containing **m**. 2. The rotation about the C_2 axis, perpendicular to the plane σ. 3. The inversion, which is composed of the two previous operations. Consequently, the symmetry of the SmC phase is C_{2h} (see Fig. C.IV.3).

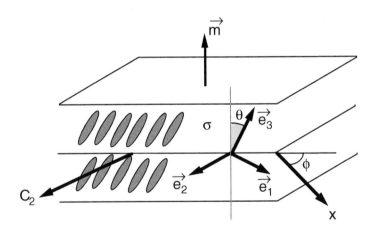

Fig. C.IV.3 Symmetry elements of the smectic C phase: the mirror plane σ, the C_2 axis and the inversion $I = \sigma C_2$. The vectors $\mathbf{e}_1$, $\mathbf{e}_2$ and $\mathbf{e}_3$ define the three eigenaxes of the dielectric tensor. The orientation of the ($\mathbf{e}_1$, $\mathbf{e}_2$, $\mathbf{e}_3$) trihedron is given by the polar and azimuthal angles (θ, ϕ).

The smectic C phase is thus **optically biaxial**, since its dielectric tensor, invariant under transformations belonging to the C_{2h} group, can have three distinct eigenvalues. It should be noted that one of the eigenvectors, say $\mathbf{e}_2$, must be parallel to the C_2 axis, while the two others, $\mathbf{e}_1$ and $\mathbf{e}_3$, are contained in the mirror plane σ. The orientation of the dielectric tensor is thus completely determined by two angles: the azimuthal angle ϕ between the plane σ and the axis **x** (perpendicular to **m**), and the polar angle θ formed by the eigenvector $\mathbf{e}_3$ and the normal to the layers **m** (Fig. C.IV.3).

The difference between the SmA and SmC phases is thus given by the value of the polar angle θ: it is zero in the SmA phase and finite in the SmC phase. As for the azimuthal angle ϕ, it is *uniform* in an undistorted sample (when the smectic layers are flat), but its value is *arbitrary* (at least in the absence of an external field and if the boundary conditions do not impose a preferred orientation).

Concerning the electric polarization, Meyer already knew (see the previous section) that it must go to zero in any system with C_{2h} symmetry, including the SmC phase. A different phase must then be used in order to obtain the sought-after ferroelectric behavior.

IV.2.b) Structure of the smectic C* phase: synthesis of DOBAMBC

Having realized this point, Meyer had the brilliant idea of using optically active molecules to induce the spontaneous polarization, by breaking the mirror symmetry of the smectic C phase. He was inspired in this by the cholesteric phase, which possesses no mirror symmetry and which is obtained either by adding chiral (non-racemic) impurities to an ordinary nematic or by directly using chiral mesogens (again, non-racemic). Meyer chose the second option and asked his chemist friends, Lionel Liébert, Leszek Strzelecki and Patrick Keller, to conceive and synthesize molecules that are at the same time optically active (i.e., in non-racemic form), susceptible to having a smectic C phase, and endowed with a permanent electric dipole.

The first compound synthesized by Liébert, Strzelecki and Keller [1] and fulfilling these specifications is known by the acronym DOBAMBC (standing for (S)-(-)-p'-decyloxybenzylidene p'-amino 2-methylbutyl cinnamate). Its structure is shown in figure C.IV.4. The DOBAMBC molecule contains an asymmetric carbon atom C* rendering it chiral, while a lateral C = O group provides a transverse permanent dipole moment. The aliphatic chain attached to the other end of the molecule by an oxygen atom is relatively long (ten carbons), favoring the SmC phase. The chemists also prepared the racemic version of DOBAMBC (equal parts mixture of molecules with opposite absolute configurations).

As they had expected, Meyer, Liébert, Strzelecki and Keller observed that, below 95°C, the phase sequence of racemic DOBAMBC (Fig. C.IV.4) has a SmC phase and that, in chiral DOBAMBC, this phase (named SmC*) exhibits very strong rotatory power. The molecular structure of the SmC* phase, compatible with this strong rotatory power and proposed by Meyer, Liébert, Strzelecki and Keller, is depicted in figure C.IV.5. It is very similar to that of the usual SmC phase, except that the azimuthal angle ϕ continuously rotates as a function of the coordinate z // **m** (Fig. C.IV.5a):

$$\phi = 2\pi \frac{z}{p} \qquad\qquad (C.IV.1)$$

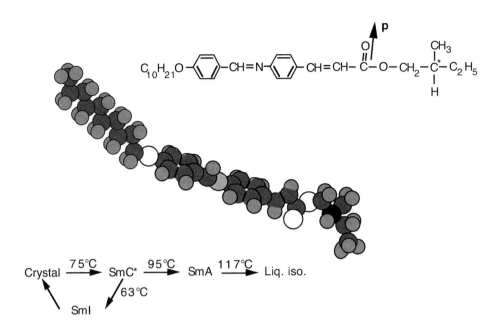

Fig. C.IV.4 DOBAMBC: the first compound showing spontaneous polarization in the SmC* phase. The black atom is the asymmetric carbon.

This rotation of the molecular tilt plane (Fig. C.IV.5) is the sign that the mirror symmetry is broken; only the C_2 axis, orthogonal to **m** and parallel to $\mathbf{e}_2$, is preserved and turns with ϕ. Another way of representing the smectic C* phase is to draw at every point the unit vector tangent to the average direction of the molecules (more precisely, the eigenvector $\mathbf{e}_3$ of the local dielectric tensor). As shown in the diagram of figure C.IV.5b, this vector generates a helicoid when the current point moves across the layers. The deformation of this vector field is of the "bend" type. By analogy with the flexoelectric effect in nematics, Meyer postulated that a polarization **P** is induced by this (spontaneous) distortion. Since this flexoelectric polarization must furthermore be parallel to the C_2 axis, it must turn with it along the z axis. Summarizing, the expression of the spontaneous polarization of a smectic C* phase can only be of the type:

$$P_x = P_o \cos(\phi - \pi/2)$$

$$P_y = P_o \sin(\phi - \pi/2) \tag{C.IV.2}$$

$$P_z = 0$$

according to Meyer's arguments. It should be pointed out that, due to the polarization turning around the z axis, the SmC* phase is not strictly speaking ferroelectric, but rather helielectric.

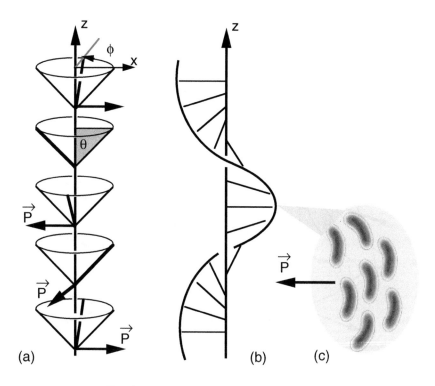

Fig. C.IV.5 Structure of the smectic C* phase.

IV.2.c) Spontaneous polarization and ichthyology

Although edifying and historically important, the explanation of the spontaneous polarization based on the flexoelectric effect in a helical texture is not entirely correct. Indeed, as we shall see in the following, in very thin display cells the helix is unwound due to planar molecular anchoring at the walls but the spontaneous polarization still exists. Furthermore, when mixing molecules with opposite twist power (but not of the same compound), the helix pitch diverges, but without removing the spontaneous polarization.

On the other hand, any explanation based exclusively on symmetry considerations must be true. We present in this section such an explanation, founded on an ichthyological analogy, discovered by P.-G. de Gennes.

To start with, consider a molecule bearing a dipolar moment and a configuration A of this molecule such that the dipolar moment $\mathbf{p}_A$ is orthogonal to the σ plane (Fig. C.IV.6). From the symmetry point of view, this molecule is equivalent to a fish speared by a harpoon (the dipole moment).

If the molecule is not chiral, there exists necessarily another configuration B such that $\mathbf{p}_B = -\mathbf{p}_A$, deduced from the first by reflection in the mirror plane σ. These two configurations being equiprobable by symmetry, the average dipolar moment must be zero. In the case where the molecule is chiral, the fish must be given a dorsal fin, to continue the ichthyological analogy. The two configurations A and B are then no longer equiprobable, since the positions of the fins are different with respect to the plane of the layers in the two cases. As a result, the average value of the dipolar moment must no longer vanish.

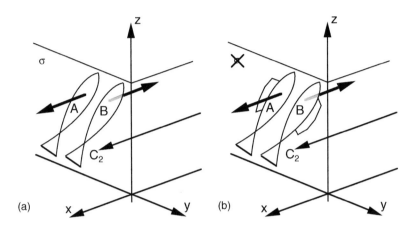

Fig. C.IV.6 In an achiral smectic C, configurations A and B, with opposite polarizations, are related by a reflection in the mirror plane σ. This is no longer possible when the molecules are chiral; indeed, the configurations A and B are different due to the position of the dorsal fin (this ichthyological analogy is due to P.-G. de Gennes).

IV.3 Experimental evidence for the spontaneous polarization

IV.3.a) *Experimentum crucis*

Once the synthesis of the smectic C* phase was achieved, one still had to prove that this phase really had a spontaneous polarization. To do this, Meyer et al. chose the most direct method, namely exposing the SmC* phase to the action of an electric field and verifying that its response is **linear**.

Their experimental setup is schematically presented in figure C.IV.7. The liquid crystal is placed between two glass plates treated for homeotropic anchoring.

In the SmA phase (T > 95°C), the smectic layers are oriented parallel to the glass surfaces. This orientation is preserved when cooling the sample below 95°C to the SmC* phase, yielding a monocrystalline sample of the SmC* phase. The electric field is applied by means of two conducting wires parallel to the **y** axis. Halfway between the two wires, the electric field **E**, albeit non-uniform (its intensity E depends on z), is horizontal and parallel to the **x** axis. The action of the electric field on the liquid crystal is optically detected by monitoring the interference pattern produced by the sample in divergent polarized light (conoscopy, see section B.I.5.f of volume I).

In the absence of the electric field, the interference pattern given by a DOBAMBC sample of thickness d ≈ 200 µm in the smectic phase C* bears a strong resemblance to that of the smectic A phase: it consists of a system of concentric rings centered on the **z** axis.

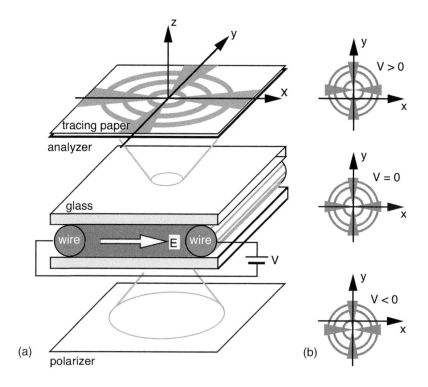

Fig. C.IV.7 Experimental evidence for the spontaneous polarization in DOBAMBC: under the effect of an electric field directed along **x**, the conoscopic figure shifts along **y**. The shift is **linear** in **E**.

Meyer and his collaborators ascertained that the conoscopic figure shifts along the **y** axis when applying an electric field parallel to **x**. More important, they observed that this **shift varies linearly with the applied field**, with the conoscopic figure going through the origin and then shifting in the opposite direction, when the field is reversed.

They concluded from these results that the response to an electric field of chiral DOBAMBC in the SmC* phase is *linear*, in agreement with the conjectured existence of *spontaneous polarization*.

IV.3.b) Interpretation of the experiment

Although the experiment described previously fulfilled its role as crucial experiment, it still raises some questions:

1. How do we explain that, in spite of the smectic C* phase being biaxial, its conoscopic figure in zero field shows uniaxial symmetry?

2. Why does the conoscopic figure shift in a direction perpendicular to the electric field?

3. What is the action of the electric field at the microscopic level?

i) Conoscopic figure of the SmC phase at rest*

Light propagation in the SmC* phase is controlled by the space distribution of the dielectric tensor ε_{ij}. According to the diagram in figure C.IV.5, the eigenaxes of this tensor turn with the angle ϕ about the z axis. Consequently, ε_{ij} is a periodic function of z. In the limit $d \gg p$ (d being the sample thickness and p the helix pitch), the Fourier series expansion of $\varepsilon_{ij}(z)$ yields, in the (x,y,z) reference system:

$$\underline{\varepsilon}\,(z) = \underline{\varepsilon}_0 + \underline{\varepsilon}_1\, e^{-i\phi} + \underline{\varepsilon}_{-1}\, e^{i\phi} + \underline{\varepsilon}_2\, e^{-i2\phi} + \underline{\varepsilon}_{-2}\, e^{i2\phi} \qquad \text{(C.IV.3a)}$$

with

$$\phi = 2\pi z/p \qquad \text{(C.IV.3b)}$$

and

$$\underline{\varepsilon}_0 = \begin{pmatrix} \varepsilon_\perp & 0 & 0 \\ 0 & \varepsilon_\perp & 0 \\ 0 & 0 & \varepsilon_{//} \end{pmatrix}, \quad \underline{\varepsilon}_{\pm 1} = \varepsilon_q \begin{pmatrix} 0 & 0 & 1 \\ 0 & 0 & \pm i \\ 1 & \pm i & 0 \end{pmatrix}, \quad \underline{\varepsilon}_{\pm 2} = \varepsilon_{2q} \begin{pmatrix} 1 & \pm i & 0 \\ \pm i & -1 & 0 \\ 0 & 0 & 0 \end{pmatrix} \quad \text{(C.IV.3c)}$$

The first term in the expansion, $\underline{\varepsilon}_0$, represents the average value of the ε_{ij} tensor over the thickness h. Note that here:

$$\varepsilon_\perp = \frac{\varepsilon_2}{2} + \frac{\varepsilon_1 \cos^2\theta + \varepsilon_3 \sin^2\theta}{2} \qquad \text{(C.IV.4)}$$

and

$$\varepsilon_{//} = \varepsilon_3 \cos^2\theta + \varepsilon_1 \sin^2\theta \qquad \text{(C.IV.5)}$$

where ε_1, ε_2 and ε_3 are the three eigenvalues of the ε_{ij} tensor calculated in the (e_1, e_2, e_3) reference system attached to the molecule (see Fig. C.IV.3). The average tensor is uniaxial, as expected, since it must be invariant under the helical symmetry of the smectic C* phase. This result explains the uniaxial character of the conoscopic figure given by the SmC* phase, which is mostly determined by the average value of the ε_{ij} tensor (in a sample much thicker than the pitch).

ii) Deformation of the SmC* phase by the electric field

In the presence of an electric field, the dipole moments of the molecules, aligned on the average along the C_2 axis (perpendicular to z), feel a torque tending to align them along the field direction (x). With:

$$\mathbf{E} = [E, 0, 0]\, e^{-i\omega t} \tag{C.IV.6}$$

and

$$\mathbf{P} = [\cos(\phi - \pi/2),\, \sin(\phi - \pi/2),\, 0]\, P_o \tag{C.IV.7}$$

the torque per unit volume is:

$$\Gamma = \mathbf{P} \times \mathbf{E} = [0, 0, P_o E \cos\phi]\, e^{-i\omega t} \tag{C.IV.8}$$

directed along z. This torque deforms the structure by inducing deviations of the ϕ angle with respect to the ideal configuration $\phi_o(z) = 2\pi z/p$.

The equation describing the motion of the angle ϕ is obtained by writing the equilibrium of the torques acting on the director:

$$K_{eff} \frac{\partial^2 \phi}{\partial z^2} - \gamma_{eff} \frac{\partial \phi}{\partial t} + P_o E \cos\phi = 0 \tag{C.IV.9}$$

In this equation, K_{eff} and γ_{eff}, respectively, are the coefficients of the elastic and viscous torques opposing the deformation of the helix. The last term represents the "deforming" torque due to the field.

This nonlinear differential equation has a simple solution in the weak field approximation, when the equilibrium helical configuration $\phi(z) = 2\pi z/p = qz$ is only slightly perturbed by the field. In this case, the solution can be written as:

$$\phi(z, t) = qz + \delta\phi\, e^{-i(\omega t + \psi)} \cos(qz) \; ; \qquad \delta\phi \ll 1 \tag{C.IV.10}$$

By linearizing eq. C.IV.9 with respect to $\delta\phi$, one has:

$$\delta\phi = \frac{P_o E}{\gamma_{eff}} \left(\omega^2 + \frac{1}{\tau^2} \right)^{-1/2} \tag{C.IV.11}$$

where

$$\tau = \frac{\gamma_{eff}}{K_{eff}\, q^2} \tag{C.IV.12}$$

is the viscoelastic relaxation time of the structure.

Equation C.IV.11 confirms that the structure deformation is linear in the electric field E. The deformation is strongest in the low frequency limit, when $\omega \ll 1/\tau$ and its value is:

$$\delta\phi = \frac{P_o\,E}{K_{eff}\,q^2} \tag{C.IV.13}$$

When the frequency of the electric field increases, the orientational viscosity impedes the rotation of the molecules and the amplitude $\delta\phi$ decreases with ω as:

$$\delta\phi = \frac{P_o\,E}{\gamma_{eff}\,\omega} \tag{C.IV.14}$$

iii) Conoscopic figure of the deformed SmC*

Let us now estimate the change in the conoscopic figure as a result of a deformation of the helical structure of the smectic C*. Proceeding as before, the average value of the dielectric tensor is:

$$\underline{\varepsilon_o} \approx \begin{pmatrix} \varepsilon_\perp & 0 & 0 \\ 0 & \varepsilon_\perp & \delta \\ 0 & \delta & \varepsilon_{//} \end{pmatrix} \tag{C.IV.15}$$

with

$$\delta \approx \frac{1}{4}(\varepsilon_3 - \varepsilon_1)\,\delta\phi\,\sin(2\theta) \tag{C.IV.16}$$

The presence of off-diagonal terms $\varepsilon_{yz} = \delta$ now indicates that the $\underline{\varepsilon_o}$ tensor underwent a rotation about the **x** axis. The conoscopic figure being related to the $\underline{\varepsilon_o}$ tensor, it must turn in the same way, which provides a qualitative explanation for the experimental observation that the conoscopic figure shifts along the **y** axis. In order to calculate this rotation angle, note that a tensor of the form

$$\begin{pmatrix} \varepsilon_\perp & 0 & 0 \\ 0 & \varepsilon_\perp & 0 \\ 0 & 0 & \varepsilon_{//} \end{pmatrix} \tag{C.IV.17}$$

rotated through a **small** angle ζ about the x axis becomes:

$$\begin{pmatrix} \varepsilon_\perp & 0 & 0 \\ 0 & \varepsilon_\perp & (\varepsilon_{//} - \varepsilon_\perp)\,\zeta \\ 0 & (\varepsilon_{//} - \varepsilon_\perp)\zeta & \varepsilon_{//} \end{pmatrix} \tag{C.IV.18}$$

Comparing expressions C.IV.15 and C.IV.18 yields:

$$\zeta = \frac{\delta}{\varepsilon_{//} - \varepsilon_\perp} \tag{C.IV.19}$$

Replacing δ and $(\varepsilon_{//} - \varepsilon_\perp)$ by their expressions C.IV.16 and C.IV.4–5 in this formula, yields:

$$\zeta \approx \frac{\delta\phi\,\theta}{2} \tag{C.IV.20}$$

in the limit of small polar angles ($\theta \ll 1$) and with the approximation $\varepsilon_1 = \varepsilon_2$.

In summary, when an electric field is applied along a direction parallel to the smectic layers, the molecules are submitted to the torque $\boldsymbol{\Gamma} = \mathbf{P} \times \mathbf{E}$ which tends to align the polarization $\mathbf{P}$ along the direction of the field. Under the action of this torque, which varies as $\cos(qz)$, the helical configuration of the liquid crystal is distorted, leading to a tilt of the average optical axis of the medium and to a shift of its conoscopic figure in the direction perpendicular to the field (Fig. C.IV.8).

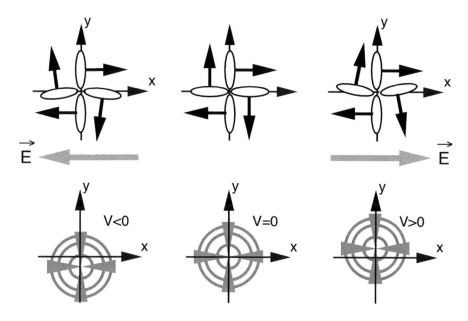

Fig. C.IV.8 Deformation of the helical structure of the smectic C* phase induced by an electric field parallel to the **x** axis (view along the **z** axis) and its effect on the conoscopic figure.

IV.4 Measurement and use of the spontaneous polarization

One of the important features of smectic C* phases is the value of their spontaneous polarization, which must be as large as possible for practical applications in displays. How can one measure this polarization? We have just seen that, in DOBAMBC, the polarization turns with ϕ (eq. C.IV.2), such that, in a sufficiently thick sample ($h \gg p$), the average value of P is zero. Setting aside for the moment the question whether the polarization P must necessarily follow a helix, let us show how to unwind this helix in order to induce a finite average polarization P, which is easier to measure.

IV.4.a) Principle of SmC* displays

As with the cholesteric phase, the SmC* phase can be unwound:

– either by applying a strong enough electric field;

– or by confining it between two surfaces close enough to impose a well-defined molecular orientation (anchoring effect).

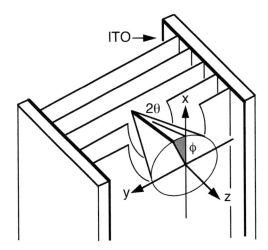

Fig. C.IV.9 Display cell using a smectic C* material.

Here, we shall be interested in the second method, which represents the principle of smectic C* based displays [2–5]. In a display cell (Fig. C.IV.9), the liquid crystal is placed between two semi-transparent electrodes (ITO-covered glass plates) treated to ensure strong planar anchoring of the molecules. When the two electrodes are very close together (d < p), the smectic C* adopts a so-called "bookshelf" configuration in which the smectic layers are perpendicular to the surfaces. Furthermore, the molecules no longer turn inside the cell, the azimuthal angle ϕ assuming one of the two values, 0 or π, corresponding to planar alignment of the molecules at the surfaces.

These two configurations of the unwound smectic C* are **energetically equivalent** in zero field, but possess average polarizations **P** of **opposite signs**.

When a DC voltage V is applied between the two electrodes, the resulting electric field **E** // **y** favors one of the two states according to its direction (Fig. C.IV.10). Reversing the field direction triggers the transition from one state to the other by rotating the molecules through an angle 2θ about the **y** axis. This orientation change being discontinuous, the transition is first order. It can only occur by first nucleating a domain with opposite polarization, then increasing its surface by wall propagation. The dynamics of the transition are thus determined by the nucleation rate of new domains and by the mobility of the walls separating two domains of opposite polarizations.

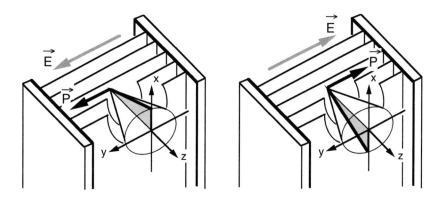

Fig. C.IV.10 When the direction of the electric field is reversed, the smectic layers preserve their orientation, but the azimuthal angle goes from $\phi = 0$ to $\phi = \pi$.

From the optical point of view, the proper axes of the ε_{ij} tensor turn by the same angle 2θ as the molecules during this transition. It is then enough to place the cell between two crossed polarizers, and this rotation leads to a change in the transmitted intensity. This effect is the fundamental principle of smectic C* displays (Fig. C.IV.11). For a more detailed description of the principles, construction, and operation of SmC* liquid crystal displays, we refer the reader to the book by S.T. Lagerwall [3].

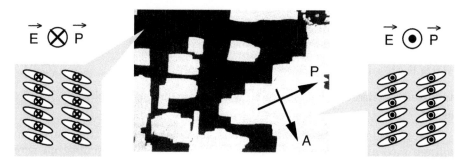

Fig. C.IV.11 The commutation between the two states **P** up and **P** down is the operation principle of smectic C* displays (the photograph is reproduced from ref. [3]).

IV.4.b) Measuring the polarization

The display cell we have just described can be used to measure the spontaneous polarization using the electric circuit shown in figure C.IV.12 [6a,b]. In this setup, the cell is represented by the capacitor C and the resistor R_i is connected in parallel (liquid crystals have finite resistivity). The cell is connected in series with resistor R_s, used as a probe for measuring the polarization charge q_p. When the polarization **P** of the liquid crystal changes sign after reversal of the electric field, the resistor R_s is crossed by the current:

$$i_q = \frac{dQ_p}{dt} \qquad\qquad\qquad (C.IV.21)$$

created by the charges Q_p arriving on the electrodes of the cell to screen the polarization charges. In a cell of surface area S and spontaneous polarization P, the polarization reversal involves a total charge

$$\int i_p\,dt = 2q_p = 2\sigma_p\,S = 2PS \qquad\qquad (C.IV.22)$$

In the circuit of figure C.IV.12, two other currents contribute, besides i_p. One of them:

$$i_c = \frac{dQ_c}{dt} \quad \text{with} \quad Q_c = \frac{V_c}{C} \qquad\qquad (C.IV.23)$$

is due to the intrinsic capacity C of the sample, calculated without taking into account the spontaneous polarization. The other one, due to the finite resistivity R_i of the sample, is:

$$i_r = \frac{V}{R_i + R_s} \qquad\qquad\qquad (C.IV.24)$$

Thus, the voltage recorded by the oscilloscope:

$$V_s = R_s i = R_s(i_p + i_c + i_r) \qquad\qquad (C.IV.25)$$

is composed of three contributions, with only one of which, $R_s i_p$, we are concerned.

To separate these three contributions, let us analyze the time variation of the voltage $V_s(t)$ after reversal of voltage V. Its profile on the oscilloscope screen (Fig. C.IV.12) reveals the presence of:

1. A contribution due to the i_c current, exponentially decreasing with a very short time constant $\tau_c = R_s C$, of the order of $10^4\ \Omega \times 10^{-9}\ F = 10\ \mu s$.

2. A contribution due to current i_r, which, contrary to the previous term, does not vary in time.

3. A contribution due to the i_p current, its behavior reflecting the dynamics of the transition between the two states of the smectic C* sample.

In a cell a few µm thick, the transition takes about 200 µs, the time needed by the walls to cross the entire sample.

The polarization charge density σ_{pol} measured by this method depends on the material. It is typically of the order of 10 nC/cm^2 (or 100 µC/m^2), but it varies strongly as a function of the temperature, especially at the approach of the SmC*→SmA transition. In materials where this transition is second order, the polar angle θ and the polarization P vary as:

$$P \sim \theta \sim (T_{AC} - T)^{\alpha} \qquad \text{with} \qquad 0 < \alpha < 1 \qquad \text{(C.IV.26)}$$

This pretransitional behavior is illustrated by the diagram in figure C.IV.14.

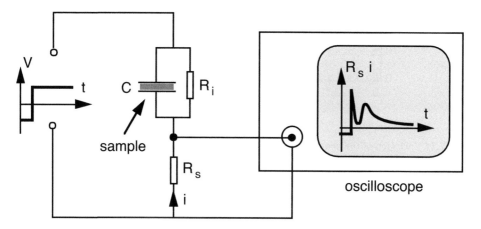

Fig. C.IV.12 Setup for measuring the spontaneous polarization.

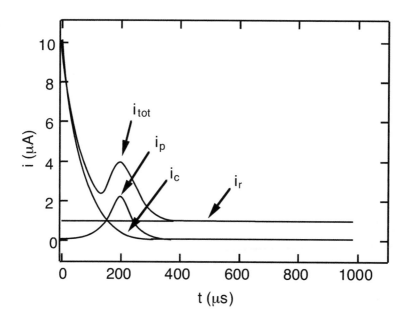

Fig. C.IV.13 Decomposition of the measured signal i_{tot} into three components: i_r = internal resistance of the sample; i_c = sample capacity; i_p = spontaneous polarization.

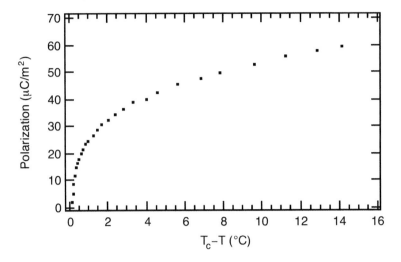

Fig. C.IV.14 Variation of the spontaneous polarization in DOBAMBC as a function of temperature close to a second-order SmA → SmC* transition (from ref. [6b]).

IV.5 Hydroelectric effect

We know that, in thick DOBAMBC samples, the average polarization <P> is zero due to its helical configuration. However, in the presence of an electric field perpendicular to its axis, the **helix is deformed such that an average polarization appears along the direction of the field**. The amplitude $\delta\phi$ of the deformation minimizes the total free energy of the system:

$$F_{tot} = F_{elast} + F_{diel} \tag{C.IV.27}$$

where

$$F_{elast} = \frac{1}{2} K_{eff} (\delta\phi \; q)^2 \int dz \; \cos^2(qz) \tag{C.IV.28}$$

and

$$F_{diel} = - P_0 E \int dz \; \sin[qz + \delta\phi\cos(qz)] \approx - P_0 E \; \delta\phi \int dz \; \cos^2(qz) \tag{C.IV.29}$$

Minimizing with respect to $\delta\phi$ yields the amplitude of the deformation:

$$\delta\phi = P_0 \, E / \, K_{eff} \, q^2 \tag{C.IV.30a}$$

in agreement with expression C.IV.13. As for the average polarization due to this deformation, it is given by

$$<P_x> = P_o(T) <\sin[qz + \delta\phi \cos(qz) - \pi/2]> = P_o(T) \, \delta\phi/2 \qquad \text{(C.IV.30b)}$$

We shall now show that the polarization of the SmC* phase can also be evidenced by purely mechanical means. Indeed, the helix of the SmC* phase can be deformed in other ways than using an electric field, for instance by shearing the sample. This deformation of the helix, demonstrated by conoscopy in reference [7a], must also induce an average polarization, as per Meyer's experiment. One can thus say that one is dealing with a viscoelectric or hydroelectric effect. In order to demonstrate this polarization, the experiment shown in figure C.IV.15 was performed [7a,b]. A DOBAMBC sample in the smectic C* phase is placed between two planar electrodes, parallel and treated to orient the smectic layers along the surfaces (Fig. C.IV.15). One of the plates is fixed, while the other oscillates in the plane along the y direction. Its sinusoidal displacement is given by the equation:

$$y = y_o \sin(\omega t) \qquad \text{(C.IV.31)}$$

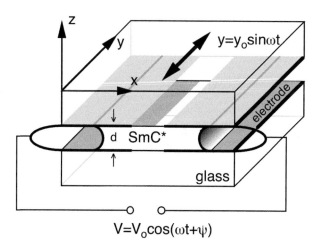

Fig. C.IV.15 Setup used to illustrate the viscoelectric effect.

When the oscillation frequency is low enough, the flow induced in the liquid crystal sample is simple shear, with a velocity:

$$v_y = \omega y_o \frac{z}{d} \cos(\omega t) , \qquad v_x = v_z = 0 \qquad \text{(C.IV.32)}$$

and does not involve permeation.

Due to the shear:

$$s = \frac{\partial v_y}{\partial z} = \frac{\omega y_o}{d} \cos(\omega t) \qquad \text{(C.IV.33)}$$

207

the liquid crystal molecules inside each layer are submitted to a viscous torque of component Γ_z given in the "nematohydrodynamic" approximation (Fig. C.IV.16) by:

$$\Gamma_z = \alpha_{eff}\, s\, \cos[\phi(z)] \tag{C.IV.34}$$

where

$$\alpha_{eff} = \alpha_2\, \sin\theta\, \cos\theta \tag{C.IV.35}$$

This viscous torque tends to align the molecules along the direction of the flow, exactly as the electric field tends to align the polarization along its direction (eq. C.IV.9). Writing that:

$$\phi(z,t) = qz + \delta\phi\, \cos(qz)\, \cos(\omega t + \psi) \tag{C.IV.36}$$

and using this analogy and eqs. C.IV.10–12, the amplitude $\delta\phi$ of the helix deformation is found, without further calculation:

$$\delta\phi = \frac{\alpha_{eff}\, y_o}{\gamma_{eff}\, d}\, \frac{\omega}{\sqrt{\omega^2 + 1/\tau^2}} \tag{C.IV.37}$$

where

$$\tau = \frac{\gamma_{eff}}{K_{eff}q^2} \tag{C.IV.38}$$

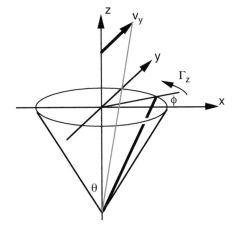

Fig. C.IV.16 Torque exerted by the flow on the smectic C* molecules.

In the "nematohydrodynamic" approximation:

$$\gamma_{eff} = \gamma_1\, \sin^2\theta \tag{C.IV.39a}$$

and

$$K_{eff} = K_3\, \sin^2\theta\, \cos^2\theta + K_2 \sin^4\theta \tag{C.IV.39b}$$

resulting in:

$$\delta\phi = \frac{1}{tg\theta} \frac{\alpha_2}{\gamma_1} \frac{y_o}{d} \frac{\omega}{\sqrt{\omega^2 + 1/\tau^2}}$$ (C.IV.40)

The average polarization induced by this deformation is then:

$$P_x = \frac{1}{d} \int dz\, P_o \sin[qz + \delta\phi(t)\cos(qz)] \approx P_o \delta\phi(t) \int dz\, \cos^2(qz) = P_o \frac{\delta\phi(t)}{2}$$ (C.IV.41)

In the experimental setup of figure C.IV.15, this polarization perpendicular to the shear direction is detected using four electrodes that enclose the sample, forming a sort of capacitor containing the liquid crystal. The voltage variation at the terminals of this capacitor is proportional to the variation of the polarization P_x. Figure C.IV.17 shows a typical record of the amplitude V_o as a function of temperature. In qualitative agreement with this model, the polarization induced by the shear appears below $T_{AC*} = 95°C$ (i.e., in the SmC* phase).

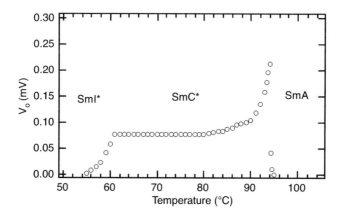

Fig. C.IV.17 Variation of the voltage induced by shear in DOBAMBC as a function of the temperature (from ref. [7a]).

To understand the variation of the signal as a function of the temperature, one must know the variation of the polar angle θ and of the polarization P_o with temperature. Indeed, if θ is small, from eqs. C.IV.40 and 41, the induced polarization:

$$P_x \sim P_o(T)\, \delta\phi[\theta(T)] \approx \frac{P_o(T)}{\theta(T)}$$ (C.IV.42)

depends on the temperature through $P_o(T)$ and $\theta(T)$. If, for small θ, P_o is proportional to θ, the signal should not depend on the temperature. The actual behavior is more complex than that since, as shown in the record of figure

C.IV.17, the measured signal remains more or less constant throughout the temperature range of the SmC* phase, except very close to the SmA → SmC* transition, where it seems to diverge.

The explanation of this experimental fact requires a deeper analysis of the way the liquid crystal behaves close to the transition. In particular, one must check the validity of the approximation according to which the polar angle θ is not perturbed by the shear. Indeed, the susceptibility controlling the molecular tilt must diverge if the SmA → SmC* transition resembles the SmA → SmC transition.

IV.6 Electromechanical effect

Solid piezoelectric materials, such as quartz or barium titanate, become electrically polarized under mechanical deformation and, conversely, become deformed in the presence of an externally applied electric field.

We have just shown that a macroscopic sample in the SmC* phase becomes polarized under shear. By analogy with the behavior of piezoelectric materials, one can thus expect the SmC* phase to exhibit a electrohydrodynamic effect reciprocal to the hydroelectric effect.

This effect was evidenced for the first time by Jakli et al. [8a,b]. In their experiment, described in figure C.IV.18, the liquid crystal (the mixture FK4 [8b], exhibiting a SmC* phase at room temperature) is in the "bookshelf" configuration between two transparent ITO electrodes treated for planar anchoring. In this material, the pitch p of the helix is of the order of 5 μm, such that, for a thickness d of the order of 10 μm, the helix is completely unwound. One of the glass plates is fixed, while the other can slide in the (y, z) plane. The motion of the latter, limited by a spring of stiffness k, is detected by means of a piezoelectric sensor. In another variant of the setup, the mobile plate is mechanically coupled to the membrane of a loudspeaker, so that its oscillation can be directly heard when the frequency ω of the applied field is within the audible domain.

The experiment consists of measuring the voltage $V = V_o\cos(\omega t + \psi)$ delivered by the sensor as a function of the amplitude U_o and the frequency $\omega = 2\pi f$ of the excitation voltage, at different temperatures. The diagram in figure C.IV.19 shows the existence of a **linear** electromechanical effect in the SmC* phase, which disappears in the SmA and SmI* phases. Jakli et al. also proved the linearity of the effect, showing that V_o is proportional to U_o, at constant T and f, in the limit of weak excitation amplitudes. Finally, the measurements at constant T and U_o show that the response V is proportional to the excitation frequency below 1 kHz; at higher frequency, the signal saturates, as the inertia of the mobile plate comes into play. Finally, at very low frequency, there is a π/2 phase shift between the applied voltage and the induced displacement.

In conclusion, these experiments show that, at constant temperature and for low enough frequencies ω and excitation amplitudes, one has:

$$y(T) \approx a(T) \frac{dE}{dt} = a(T) \, i \, \omega \, E \qquad\qquad (C.IV.43)$$

where a(T) is a coefficient depending both on the investigated property of the liquid crystal and on the parameters of the experimental setup. To discriminate between these contributions, one can make use of the fact that, at low frequency, the inertia of the mobile plate can be neglected, such that the ky(t) product gives the force exerted by the liquid crystal on the mobile plate.

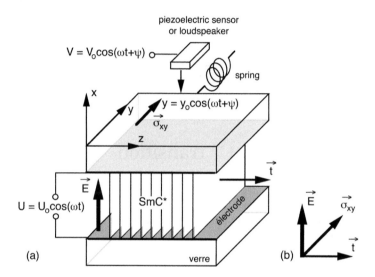

Fig. C.IV.18 Experimental setup used by Jakli et al. [8a,b] to demonstrate the electromechanical effect in the SmC* phase. a) Diagram of the setup; b) demonstration of the chiral nature of the effect.

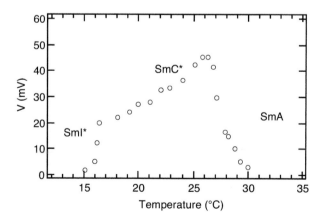

Fig. C.IV.19 Electromechanical effect of Jakli et al. [8b]. Measurements performed on the mixture FK4. The excitation frequency $f = \omega/2\pi$ is 500 Hz and the amplitude U_o is 25 V.

It suffices then to divide this force by the surface S of the sample to obtain the stress exerted by the liquid crystal on the mobile plate:

$$\sigma_{yx} = \frac{ky(t)}{S} = \frac{k}{S}a(T)\frac{dE_x}{dt} = b(T)\frac{dE_x}{dt} \tag{C.IV.44}$$

The b(T) coefficient, relating the rate of variation of the electric field to the measured stress, only depends on the material studied.

It should be noted that the effect demonstrated by Jakli et al. is chiral in nature, since the $(\mathbf{t}, \mathbf{f}, \mathbf{E})$ trihedron formed by the twist axis $\mathbf{t} = \mathbf{n} \times \mathbf{curl}(\mathbf{n})$ of the helix, the stress $\mathbf{f} = \sigma_{xy}\,\mathbf{y}$ and the applied field $\mathbf{E}$ has a well-defined chirality. As a consequence, the b(T) coefficient must change sign when the optical antipode is used.

For the same symmetry reasons, the Jakli effect cannot exist in a racemic mixture or in an achiral material, where the normal $\mathbf{m}$ to the layers, which replaces $\mathbf{t}$, is only defined up to a sign change.

IV.7   The SmA → SmC* transition

IV.7.a) "Landau" model and order parameters θ and P

Throughout all thermodynamical considerations concerning the SmA → SmC* transition, one must keep in mind the following experimental result: the SmC* phase of a chiral material exists over practically the same temperature range as the SmC phase of the racemic mixture of the same substance. Thus, the relevant parameter of this transition is not the spontaneous polarization P, which is only present in the SmC* phase, but rather the tilt angle θ of the molecules, common to the SmC and SmC* phases.

In the following expansion of the free energy close to the SmA → SmC* transition [9]:

$$F = F_A + \frac{1}{2}A(T)\,\theta^2 + B\theta^4 + \frac{1}{2}\frac{1}{\chi_p}P^2 - t\,\theta\,P \tag{C.IV.45}$$

this observation is included by assuming that the A(T) coefficient changes sign at the transition, and not χ_p^{-1}.

Minimizing this expression with respect to P, one obtains a linear relation between the polarization and the molecular tilt:

$$P = (t\chi_p)\,\theta \tag{C.IV.46}$$

This equation defines the trajectory (see figure C.IV.20, $T < T_{AC^*}$) that the system must follow in the (θ, P) space in zero field $(\partial F/\partial P = 0)$ when it crosses over to the SmC* phase. Since the susceptibility χ_p is always positive, it is the coefficient t that determines the sign and amplitude of the polarization. This coefficient is termed by Meyer **piezoelectric**, since it describes the relation between the tilt θ and the polarization P in the absence of electric fields. This coefficient is different from zero in chiral materials and has opposite signs in the two enantiomers of the same substance. On the other hand, t goes to zero in an achiral material (such as for instance a racemic mixture). In a mixture of two enantiomers, the coefficient t varies as a function of the molar concentrations c_+ and $c_- = 1 - c_+$ in the following way (Fig. V.IV.21):

$$t = t_{max}(c_+ - c_-) \qquad (C.IV.47)$$

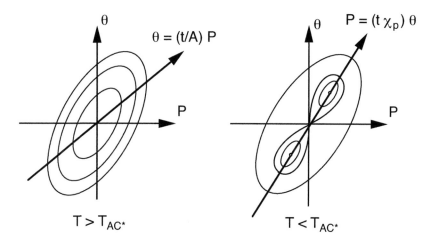

Fig. C.IV.20 Diagram of the free energy landscape as a function of θ and P above (left) and below (right) the SmA $\rightarrow$ SmC* transition.

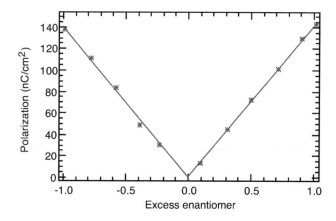

Fig. C.IV.21 Variation of the spontaneous polarization as a function of the excess enantiomer concentration $\Delta c = c_+ - c_-$ (from ref. [10]).

Replacing the polarization by its expression C.IV.46 in eq. C.IV.45, one obtains the expansion of the free energy as a function of only θ:

$$F = F_A + \frac{1}{2} A'(T)\, \theta^2 + B\, \theta^4 \qquad\qquad (C.IV.48)$$

where A'(T) is the renormalized coefficient:

$$A'(T) = A(T) - \chi_p t^2 \qquad\qquad (C.IV.49)$$

Knowing that

$$A(T) = a\,(T - T_{AC}) \qquad (a > 0) \qquad\qquad (C.IV.50)$$

the critical temperature $T_{AC}*$ of the SmA → SmC* transition is determined from the condition A'(T) = 0 (assuming that B > 0 and that the transition is second order):

$$T_{AC^*} = T_{AC} + \chi_p \frac{t^2}{a} \qquad\qquad (C.IV.51)$$

The transition to the SmC* phase is reflected in the right diagram of figure C.IV.20 by the appearance of two minima symmetrically placed with respect to the origin in the (θ, P) plane.

IV.7.b) Experimental findings for enantiomer mixtures when the SmA → SmC* transition is second order

In order to verify this theoretical prediction Bahr, Heppke, and Sabaschus [11] measured $T_{AC}*$ for enantiomer mixtures as a function of the excess concentration $c_{exc} = c_+ - c_-$ (Fig. C.IV.22).

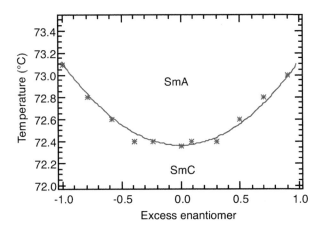

Fig. C.IV.22 Influence of the chirality on the temperature of the SmA → SmC* phase transition (from ref. [11]).

For these mixtures, the theory of Garoff and Meyer predicts that the transition temperature varies as the square of the excess concentration, with the piezoelectric coefficient t being proportional to c_{exc}. The results shown in figure C.IV.22 confirm this theoretical prediction.

IV.7.c) Electroclinic effect

Let us consider the smectic A phase, at a temperature slightly above the transition temperature T_{AC^*}. The AC* transition is considered second order. In the absence of an applied electric field, the polarization is zero. In the presence of an electric field E parallel to the plane of the layers, the experiment shows that a finite polarization P is induced. To calculate this, let us rewrite the free energy taking into account only the dominant terms and the additional term due to the coupling between the electric field and the induced polarization P:

$$F = F_A + \frac{1}{2} A(T) \, \theta^2 + \frac{1}{2\chi_p} P^2 - t \, \theta \, P - PE + \ldots \qquad \text{(C.IV.52)}$$

This finite polarization P leads to a molecular tilt due to the piezoelectric coupling (tθP term). The parameters θ and P being independent, one has at equilibrium:

$$\frac{\partial F}{\partial \theta} = 0 \qquad \text{(C.IV.53)}$$

yielding:

$$\theta = \frac{t}{A} P \qquad \text{(C.IV.54)}$$

This relation defines the trajectory in space of the parameters (θ, P) that the system must follow when the field is applied (see figure C.IV.20). By inserting this equation into eq. C.IV.52, one gets:

$$F = F_A + \frac{1}{2\chi_{eff}} P^2 + PE \qquad \text{(C.IV.55)}$$

where

$$\frac{1}{\chi_{eff}} = \frac{1}{\chi_p} - \frac{t^2}{A} \qquad \text{(C.IV.56)}$$

represents the effective dielectric susceptibility. Indeed, minimization of F with respect to P gives:

$$P = \chi_{eff} \, E \qquad \text{(C.IV.57)}$$

This effective susceptibility diverges for:

$$A = \chi_p t^2 \qquad \text{(C.IV.58)}$$

That is at $T = T_{AC}^*$ (eq. C.IV.49).

Finally, substituting P by its expression C.IV.57 in eq. C.IV.54, we find the relation between the applied field and the induced tilt θ :

$$\theta = \frac{t}{A}\chi_{eff}\, E \qquad\qquad\qquad\qquad (C.IV.59)$$

This equation expresses the **electroclinic** effect, namely that in a smectic A, the molecules (if they are chiral) tilt with respect to the layer normal when an electric field is applied in the plane of the layers.

The electroclinic effect was observed for the first time by Garoff and Meyer in DOBAMBC [9]. It was later studied more precisely by Bahr and Heppke in other materials, with higher spontaneous polarization, where this effect is more easily induced and measured [12]. The geometry used by these authors is described by the diagram in figure C.IV.23. The liquid crystal in the SmA phase is confined between two electrodes imposing a "bookshelf" configuration.

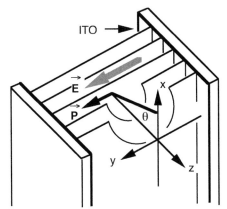

Fig. C.IV.23 Geometry used by Bahr and Heppke for their measurement of the electroclinic effect.

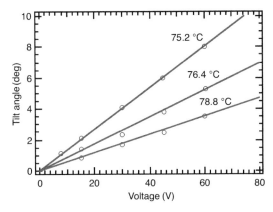

Fig. C.IV.24 Electroclinic effect: measurement of the tilt angle θ as a function of the applied voltage (from ref. [12]).

The angle θ is measured optically by searching for extinction of the texture between crossed polarizers. The diagram in figure C.IV.24 shows the variation of the angle θ as a function of the voltage V applied between the two electrodes. The slope of the straight line $\theta(V)$ increases on cooling the sample and then diverges at the approach of the A-C* transition.

IV.7.d) Induced SmC* phases, eutectic mixtures

The experiment of Bahr et al. [10] on enantiomer mixtures shows that the spontaneous polarization of the SmC* phase is **directly proportional** to the excess concentration $\Delta c = c_+ - c_-$ of one of the enantiomers. Such experiments can be performed using either two pure enantiomers, or a racemic mixture and a pure enantiomer. To determine the excess concentration $\Delta c = c_+ - c_-$ in this second case, note that, if c_r and c_p are the concentrations of the racemic product and of the pure enantiomer "+", respectively, then $c_+ = c_p + c_r/2$, $c_- = c_r/2$ such that $\Delta c = c_p$.

On the other hand, what happens if, instead of mixing the racemic mixture and a pure enantiomer, one uses, for instance, an achiral compound exhibiting the phase SmC and a pure enantiomer, even a non-mesomorphic one? In general, the SmC phase itself becomes chiral under the action of the chiral product (it is thus of the SmC* type) and a **spontaneous polarization P is induced**. Nevertheless, the variation of the induced polarization $P(c_p)$ as a function of the concentration c_p of the chiral product is no longer linear, as shown by the experiments of Stegemeyer et al. [13], performed on a large number of products.

This method allowed the authors to extend the existence domain of the ferroelectric SmC* phase using eutectic mixtures instead of a single achiral product (see section A.III.1c, volume1).

IV.8 Anticlinic mesophases, antiferroelectric and ferrielectric

The discovery of the helielectric SmC* phase in DOBAMBC by R. Meyer, P. Keller, L. Liébert, and L. Strzelecki [1] opened the way for the search of other mesophases, with different arrangements of the electric dipoles. It is impossible, in only one chapter, to present the history of the many discoveries marking this area of liquid crystal research. On this topic, we refer the reader once again to the book by S. Lagerwall [3], as well as to the contribution of Takezoe and Takanishi [14]. As far as we are concerned, we shall only present some crucial experiments demonstrating the essential properties of antiferroelectric and ferrielectric mesophases.

IV.8.a) The concept of an anticlinic structure

In 1977, Michelson, Cabib and Benguigui [15] examined the structures that can be obtained starting from the smectic A phase, through a second-order phase transition, by tilting the molecules with respect to the layer normal.

One of the conclusions of this article is that, aside from the classical SmA → SmC transition (Fig. C.IV.25) characterized by **uniform** molecular tilt in all the layers (θ independent of z), one can have another kind of symmetry breaking, leading to an **anticlinic** structure, in which the layers are still of the smectic C type, but where the molecular tilt angle θ changes sign from one layer to the next (Fig. C.IV.25).

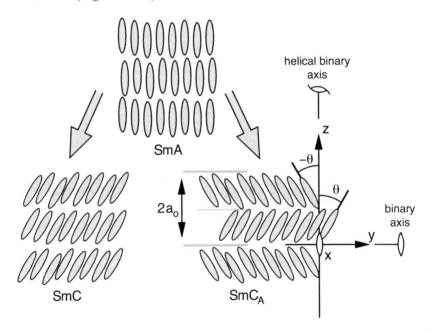

Fig. C.IV.25 Symmetry breaking obtained starting from the SmA phase and tilting the molecules with respect to the layer normal. In the anticlinic SmC$_A$ phase, the tilt angle θ changes sign from one layer to the next.

When contemplating the dielectric properties of this anticlinic phase, one is tempted naively to transpose the "ichthyological" argument developed in the case of the SmC* phase. Thus, when the molecules are chiral and bear a transverse electric dipole $p_\perp$, this argument leads directly to the conclusion that, in the SmC$_A$* phase, the smectic layers considered separately have a spontaneous polarization P_x orthogonal to the tilt plane (y, z), which changes sign from one layer to the next due to the alternating sign of the angle θ (Fig. C.IV.26a). Such a structure, composed of two sub-lattices with opposite polarizations P_1 and P_2 ($P_1 = -P_2$), is, by definition, **antiferroelectric**.

Due to the molecular chirality, and as in the SmC* phase, the anticlinic SmC$_A$*
structure should also be twisted. More to the point, the molecular orientations
should form a **double helix** at a scale much larger than the layer thickness,
instead of a simple helix, as in the SmC* phase.

It is important to understand that **in order to be antiferroelectric,
the anticlinic phase needs not be chiral** [16, 17]. To prove it, one only
needs to assume that the achiral molecules bear a longitudinal dipole $\mathbf{p}_{//}$, and
that this dipole is not placed at the center of the molecules but at one of their
extremities instead.

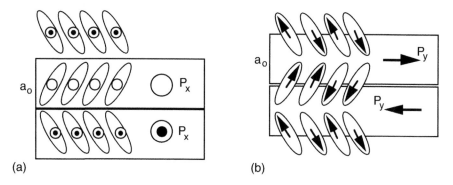

(a)   (b)

Fig. C.IV.26 Antiferroelectric structures of the anticlinic phase: a) chiral molecules bearing a
transverse electric dipole; b) achiral molecules endowed with a longitudinal off-centered dipole.

The diagram in figure C.IV.26b shows that, on the average, in each stripe
of width a_o at the junction of two adjacent smectic layers there appears a
spontaneous polarization P_y of alternating sign from one stripe to the other
along z. This particular antiferroelectric structure is compatible with the
symmetry of the anticlinic and achiral SmC$_A$ phase (see figure C.IV.25). In
particular, it is invariant with respect to reflection in the (y, z) plane and has a
second-order helical axis perpendicular to the layers (rotation by 180° around
the layer normal z, followed by a translation of vector $\mathbf{a_o}$ along the **z** direction).

IV.8.b) First experimental suggestion of an anticlinic phase

The anticlinic structure and its antiferroelectric properties were invoked for the
first time by Beresnev et al. [18] in order to explain the results of a study on
the pyroelectricity of mixtures of cinnamates with two benzene rings and chiral
additives. In this experiment, a static electric field is applied to samples of a
thickness of 200 μm sandwiched between two glass plates with electrodes, and
the polarization variation dP, induced by a heating dT produced by a pulsed
laser beam, is measured. The results of the measurements of the pyroelectric
coefficient $\gamma = dP/dT$ as a function of the applied field are given in figure
C.IV.27 for one such mixture.

Two curves, of different shape according to the temperature, are shown. Let us comment on them.

The first one, recorded at 86°C is, according to its authors, very similar to the curves recorded in the helielectric SmC* phase (at the time, this phase had already been positively identified in other materials). It is characterized by a linear variation of the pyroelectric coefficient with the field in the low field limit and then by saturation at strong fields.

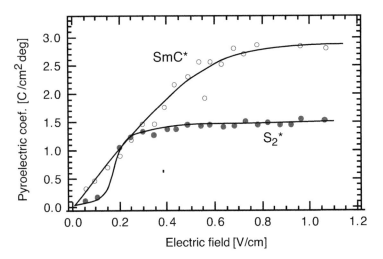

Fig. C.IV.27 Measurements of the pyroelectric effect in a mixture of NOBAPC with 20% of HOBACPC (from ref. [18]).

To understand this behavior, let us first recall (see section C.IV.3b) that, in the SmC* phase, the helical configuration of the polarization is deformed under the action of the electric field E, such that the sample acquires a global polarization <P> proportional to the local polarization in the individual layers:

$$<P> = c(E) \, P_o(T) \qquad\qquad (C.IV.60)$$

As a result, the measured pyroelectric coefficient

$$\gamma = \frac{d<P>}{dT} = c(E) \, \frac{dP_o}{dT} \qquad\qquad (C.IV.61)$$

depends both on the local polarization $P_o(T)$ and on the coefficient c(E) quantifying the deformation of the helix. The analysis of Meyer's experiment (see section C.IV.3) shows that, in the low field limit, the amplitude $\delta\phi$ of the helix deformation varies linearly with the field E. From eqs C.IV.7–10, we can determine the average polarization induced by a static field $\mathbf{E} = (E,0,0)$ perpendicular to the helix, which is oriented along z:

$$<P_x> = P_o(T) \, <\cos[qz + \delta\phi \cos(qz) - \pi/2]> \qquad\qquad (C.IV.62)$$

whence, using eq. C.IV.13:

$$c(E) = \frac{\delta\phi}{2} = \frac{P_oE}{2K_{eff}q^2} \qquad \text{(C.IV.63)}$$

It is noteworthy that the deformation of the helix depends both on the applied field and on the local polarization P_o. Since this polarization is modified by the heating dT, the deformation $\delta\phi$ of the helix should also be affected. Nevertheless, if the heating dT is very brief, the orientational viscosity γ_{eff} strongly reduces this effect. The pyroelectric coefficient of the helielectric SmC* phase must then be linear in E in the limit of weak fields. When the field is strong, however, the helix is completely unwound. Under these conditions, $<P_x> \rightarrow P_o$ so that $c(E) \rightarrow 1$ and $\gamma(E)$ saturates. This behavior of $\gamma(E)$ is perfectly compatible with the curve in figure C.IV.27 obtained using a mixture of NOBAPC (4-nonyloxybenzylidene-4'-amino-pentylcinnamate) with 20 % of HOBACPC (L-4-hexyloxybenzylidene-4'-amino-2-chloropropyl-cinnamate) at 86°C. For this reason, the authors conclude that, at this temperature, the mixture is in the helielectric SmC* phase.

The $\gamma(E)$ curve obtained at 67°C, on the other hand, is quite different, since the pyroelectric response is almost zero in the low-field limit and since the applied field must exceed a certain threshold for it to suddenly become apparent. Based on these observations, the authors propose:

1. That at 67°C, the sample is in a new phase, different from the SmC*.

2. That, given the very typical shape of the $\gamma(E)$ curve, this novel phase is antiferroelectric.

3. That the structure of this phase is anticlinic.

To better understand the reasoning of Beresnev et al. [18], let us analyze more closely the action that a static electric field can have upon the anticlinic SmC$_A$* phase. The torques of opposite sign exerted by the electric field on the polarizations $P_1 = -P_2 = P$ of the two sub-lattices are the strongest when the field is parallel to the (y, z) plane of molecular tilt (Fig. C.IV.28). Under the action of these torques, the azimuthal angle of the molecules is modified, resulting in a net polarization:

$$<P> = P_y \approx P_o\,\delta\phi \qquad \text{(C.IV.64)}$$

where $\delta\phi$ is the change in azimuthal angle between two neighboring layers (see Fig. C.IV.28). This angle $\delta\phi$ can be estimated by equating the electric torque per unit surface P_oEa_o acting on a layer of thickness a_o, with the restoring elastic torque, of the order of $K_{eff}\delta\phi/a_o$. This yields:

$$\delta\phi \approx \frac{P_oE\,a_o^2}{K_{eff}} \qquad \text{(C.IV.65)}$$

The deformation $\delta\phi$ and the induced polarization are consequently very small compared to the analogous quantities in the SmC* phase (eq. C.IV.13), since

$qa_o \ll 1$. Thus, when the electric field increases, the free energy of the anticlinic structure

$$F_A(E) = F_A(0) - E P_o \delta\phi \qquad\qquad (C.IV.66)$$

remains almost constant as $\delta\phi \ll 1$. On the other hand, the free energy of the monoclinic phase (completely unwound SmC* phase of polarization P_o):

$$F_C(E) = F_C(0) - E P_o \qquad\qquad (C.IV.67)$$

decreases rapidly with increasing field. Consequently, above a critical field E_c, defined by the relation

$$E_c P_o(1 - \delta\phi) = F_C(0) - F_A(0) \qquad\qquad (C.IV.68)$$

the slightly deformed anticlinic structure is replaced by the "monoclinic" structure. On symmetry grounds, the $SmC_A{}^*\to SmC^*$ transition is first order, resulting in a hysteresis effect related to the barriers of nucleation of one phase into the other. When the electric field is reversed, the analysis is identical, up to a sign change for P_o and $\delta\phi$.

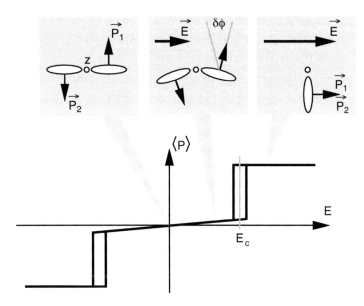

Fig. C.IV.28 Double hysteresis loop characteristic of the behavior of the $SmC_A{}_*$ antiferroelectric phase under electric field.

In conclusion, we predict that the variation of the induced polarization <P> as a function of the applied field exhibits two hysteresis loops, as shown in the diagram of figure C.IV.28. Note that the experimental curve in figure C.IV.27, obtained at 67°C by Beresnev et al. [18], only partially reflects this expected behavior of an antiferroelectric phase. On the other hand, the entire double hysteresis loop predicted for the $SmC_A{}^*$ phase was observed by

Chandani et al. [19] in a different material, the MHPOBC, as shown in the diagram of figure C.IV.29.

The pyroelectric effect measurements by Beresnev et al. [18] indicate that, in constant field, the pyroelectric coefficient depends strongly on the temperature. The shape of the $\gamma(T)$ curves obtained in various mixtures and at different concentrations prompted the authors to propose the existence of at least seven different phases (S_i^* i = 1,...,7) that were antiferroelectric and ferrielectric.

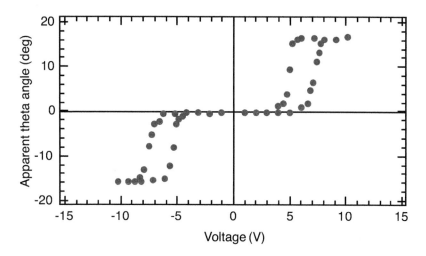

Fig. C.IV.29 Double hysteresis loop measured by Chandani et al. [19] for the SmC$_A^*$ phase of MHPOBC.

IV.8.c) SmO = SmC$_A$

The preparation of terephtalydene-*bis-p*-aminocinnamates by Keller, Liébert and Strzelecki [20] was at the origin of a second trial, leading to an independent discovery of the anticlinic mesophase. Indeed, it is in such a compound, MHTAC (1(methyl)-heptyl-terephtalydene-*bis-p*-aminocinnamate, see Fig. C.IV.30), that Levelut et al. [21] observed two new phases, which they named SmO and SmQ, respectively (the phase sequence is given in Fig. C.IV.30). In the first one, present between 95°C and 130°C, the layers are fluid and the molecules are tilted with respect to the layer normal, as in the SmC phase. However, the X-ray diffraction diagram of SmO exhibits **two pairs of crescents symmetrical** with respect to the layer normal (Fig. C.IV.30b), instead of a unique pair as in SmC (Fig. C.IV.30a). We recall that, in the SmC phase, these crescents are the signature of short-range order in the plane of the layers, while their orientation indicates a uniform tilt of the molecules with respect to the layer normal, which is defined by the Bragg spots. This "anomaly" of the SmO phase indicates that the tilt angle in the layers takes the two values $+\theta$ and $-\theta$ with the same probability.

Seen under the polarizing microscope (Fig. C.IV.30c), the SmO phase resembles the SmC phase, its textures ("schlieren") comprising 2π disclinations characteristic of the director field **c**, but also differs from it by the presence of **± π orientational defects incompatible with the structure of the SmC phase**.

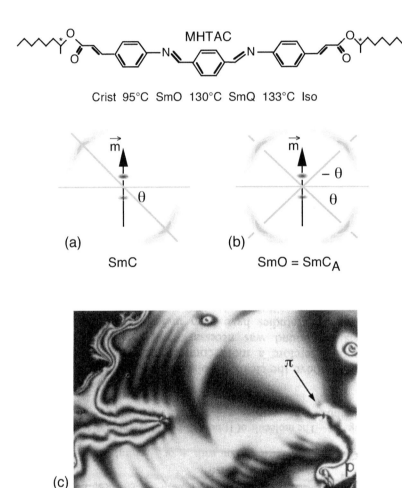

Fig. C.IV.30 Demonstration of the SmC$_A$ (SmO) anticlinic phase in MHTAC [21]. The molecule contains two asymmetric carbons. X-ray diffraction diagrams in the SmC phase (a) and in the SmO phase (b). Texture of the SmO phase seen under the microscope between crossed polarizers (c); a singularity of angle π can be recognized.

To prove that the angle θ changes sign from one layer to the next, as in the anticlinic SmC$_A$ phase, Galerne and Liébert [22] studied optically the free surface of a droplet of the racemic mixture (more exactly, of the + +/− −/+ − mixture) of MHTAC in the isotropic phase (Fig. C.IV.31a). At 164°C, which is about 6°C above the Iso → SmO transition (T_c = 158°C), a presmectic film starts to develop at the surface of the droplet (see chapter C.VIII), consisting of a number N of smectic layers that increases as the temperature approaches T_c. The crucial observation of the authors is that the optical aspect of the presmectic film texture depends on the parity of the layer number N. This is particularly conspicuous when several thicknesses, N, N − 1, N − 2, etc., coexist in the same film. According to Galerne and Liébert, the alternance in the optical aspect of the presmectic film indicates an anticlinic structure.

A similar alternance in the optical aspect of a thin suspended film of an anticlinic phase was observed by Link et al. in a different material − optically active MHPOBC − under the influence of a static electric field parallel to the plane of the layers (Fig. C.IV.31b) [16].

As shown in the photograph of figure C.IV.31c (taken with slightly uncrossed polarizers), the areas of thickness N even are light, while those with N odd are dark. The optical analysis of this image allowed the authors to conclude that, in the areas with N even, the tilt plane of the molecules is parallel to the applied electric field, such that the global polarization of the film is parallel to the plane of molecular tilt. The same optical analysis showed that, in the areas where N is odd, the molecular tilt plane is perpendicular to the applied field, meaning that this time the global polarization is orthogonal to the tilt plane.

To find the orientation of the global polarization of a film, let us investigate the orientation of the electric dipole borne by each molecule, perpendicular $\mathbf{p}_\perp$ and parallel $\mathbf{p}_{//}$ to its tilt plane. As shown in diagrams (d) and (e) of figure C.IV.31, the average orientations of dipoles $\mathbf{p}_{//}$ are different in the surface and in the bulk layers and lead to a zero global contribution for odd N and a non-zero one for even N. As to the dipoles perpendicular to the tilt plane, they have the opposite effect, being exactly compensated only for even N. Consequently, the global polarization of a film is switching from parallel to perpendicular to the tilt plane when N goes from even to odd values.

To conclude this account on the identification of anticlinic structures by optical methods, we would like to cite the ellipsometry experiment of Bahr and Fliegner [23], described in detail in chapter C.VIII dedicated to smectic films. In this experiment, the ellipsometric phase shifts Δ_+ and Δ_- are measured as a function of temperature in thin suspended films of MHPOBC (Fig. C.IV.32). In the SmC* phase, the films consisting of a very small number of layers (N = 3 and 4) have a spontaneous polarization $\mathbf{P}$ oriented parallel to the static electric field $\mathbf{E}$ applied in the plane of the film. When the electric field is reversed, the director $\mathbf{c}$ changes sign, leading to a difference between the phase shifts Δ_+ and Δ_-, irrespective of the film thickness N. In the SmC_A^* phase, on the other hand, the phase shifts depend on N. Thus, a difference between Δ_+ and Δ_- exists for odd N, while $\Delta_+ = \Delta_-$ for even N.

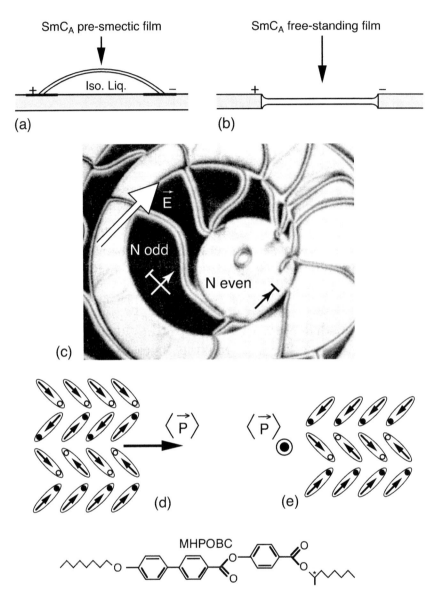

Fig. C.IV.31 Proof of the anticlinic structure in thin films of an SmC$_A$ phase: a) the geometry of the setup used by Galerne and Liébert [22] with presmectic MHTAC films; b) geometry of the experiment by Link et al. on suspended MHPOBC films; c) image of a suspended MHPOBC film between slightly uncrossed polarizers (from Link et al. [16]); d) orientation of the electric dipoles in a four-layer film; e) orientation of the electric dipoles in a three-layer film. To better understand the optical aspects of these experiments, see section C.IV.8.d.

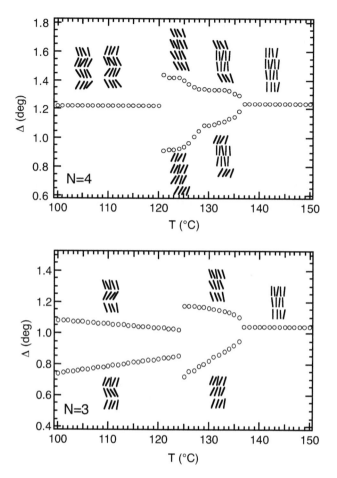

Fig. C.IV.32 Ellipsometry measurements for suspended MHPOBC films. In the SmC$_A$* phase, $\Delta_+ = \Delta_-$ for N = 4, while $\Delta_+ \neq \Delta_-$ for N = 3. On the other hand, $\Delta_+ \neq \Delta_-$ in the SmC* phase and $\Delta_+ = \Delta_-$ in the SmA phase for both N = 3 and 4 (from ref. [23]).

IV.8.d) Dispirations in the SmC$_A$ phase

A very elegant argument in favor of the anticlinic structure of the SmO phase was provided by Takanishi et al. [24]. These authors show that the conjectured $\pm\pi$ disclinations, visible in the optical textures of this phase, are in fact **dispirations** - topological defects typical of the anticlinic structure.

To comprehend the arguments of Takanishi et al., one must remember that to any order parameter resulting from a broken symmetry are associated certain topological defects that can be considered as a signature of that particular type of order. Thus, dislocations are topological defects of the **translational** order parameter. In the SmA phase, this order parameter is complex $|\Psi|e^{i\phi}$, and the dislocations correspond to a $\pm 2\pi$ jump in the phase ϕ on the Burgers circuit enclosing the dislocation. In the SmC phase, to the dislocations, defined in the same way, one must add disclinations, which are topological defects of the **orientational** order parameter $c = |c|(\cos\phi, \sin\phi)$ and characterized by a $\pm 2\pi$ jump of the azimuthal angle ϕ on a Burgers circuit enclosing the disclination. In smectics C, the dislocations and disclinations can be defined independently of one another; the symmetry group (space group) of this phase is **symmorphic** (i.e., the direct product of the translation group and the point symmetry group). On the other hand, the symmetry group of the anticlinic phase SmC$_A$ is **non-symmorphic** because, besides the pure translations (such that $t = 2Na_o$; $N = 1,2,3...$) and the pure rotations (such as the 180° rotation around the axis perpendicular to the plane of the image in figure C.IV.25), it contains the second-order helical axis perpendicular to the plane of the layers (see Fig. C.IV.25). This symmetry operation is not symmorphic, being composed of a 180° rotation followed by a translation of a_o, operations that, taken separately, are not symmetry operations of the anticlinic phase. Let us now use this symmetry operation to construct a topological defect by the Volterra method (Fig. C.IV.33).

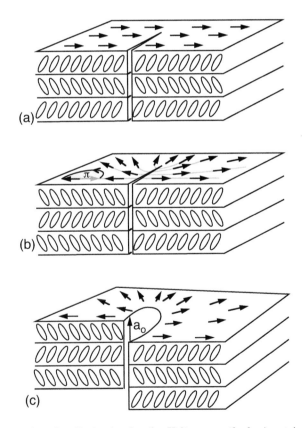

Fig. C.IV.33 Construction of a dispiration by the Volterra method: a) cut in the perfect SmC$_A$ structure; b) azimuthal rotation of the molecules on one side of the cut; c) vertical translation.

The successive operations are the following:

1. Make a cut in the perfect SmC_A phase (Fig. C.IV.33a).

2. Turn the molecules by 180° on one side of the cut (Fig. C.IV.33b).

3. Shift vertically by $\mathbf{a_o}$ (Fig. C.IV.33c).

4. "Reseal the cut" and let the system relax.

The defect thus obtained is termed a **dispiration**.

IV.8.e) Optical properties of the SmC_A phase

What does a dispiration look like under the microscope? To answer this question, let us first examine the optical properties of the SmC_A phase. On the length scale of the optical wavelength (much larger than the layer thickness), the point symmetry of the SmC_A phase is D_{2v}. It ensues that the average dielectric tensor ε_{ij} must be symmetric with respect to three binary axes (indicated in figure C.IV.25), perpendicular to each other. Consequently, ε_{ij} is diagonal in the reference system $(\mathbf{x}, \mathbf{y}, \mathbf{z})$ formed by these symmetry axes. The three eigenvalues of this diagonalized tensor are a priori different, $\varepsilon_{xx} \neq \varepsilon_{yy} \neq \varepsilon_{zz}$, as no symmetry element exists that would allow swapping of the $\mathbf{x}$, $\mathbf{y}$ and $\mathbf{z}$ axes. Using the notations in section C.IV.3.c, one can express them as a function of three locally defined eigenvalues ε_i:

$$\varepsilon_{xx} = \varepsilon_2$$

$$\varepsilon_{yy} = \varepsilon_1 \cos^2\theta + \varepsilon_3 \sin^2\theta \qquad\qquad\qquad\qquad (C.IV.69)$$

$$\varepsilon_{zz} = \varepsilon_3 \cos^2\theta + \varepsilon_1 \sin^2\theta$$

In conclusion, the anticlinic phase should be optically biaxial, a feature actually evidenced by Isozaki et al. [25] by conoscopy (for a description of the method, see figure C.IV.7). In the conoscopic figure of figure C.IV.34 (which, one should remember, is an interference pattern at infinity), one can distinguish the two optical axes A and B that are, by definition, the directions for which the refractive index does not depend on the polarization direction. These optical axes are tilted with respect to the layer normal and are contained in the $(\mathbf{x}, \mathbf{z})$ plane orthogonal to the $(\mathbf{y}, \mathbf{z})$ plan of molecular tilt.

In incidence normal to the layers (along the $\mathbf{z}$ direction), two eigenmodes exist for light propagation, polarized along $\mathbf{x}$ and $\mathbf{y}$. Thus, extinction occurs between crossed polarizers when the molecular tilt plane is parallel or perpendicular to one of the polarizers.

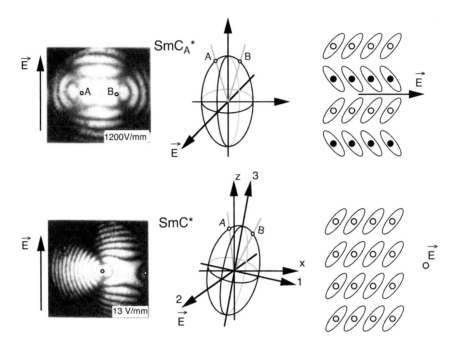

Fig. C.IV.34 Conoscopic figures, index ellipsoids and structures, for a) the anticlinic phase oriented by an electric field and, b) the SmC* phase unwound by the electric field (from ref. [24]).

It follows that, in the photograph of figure C.IV.30c, the plane of molecular tilt is parallel or perpendicular to the polarizers wherever extinction occurs. As this plan turns by π around the z axis of a dispiration, each such defect has two extinction brushes. Thus, the dispirations have the same optical aspect as the $\pm\pi$ disclinations in a planar nematic sample (as noticed by Sigaud in ref. [21]).

IV.8.f) Optical properties of the SmC$_A^*$ phase

A very convincing argument in favor of the anticlinic chiral structure was provided by optical experiments. Chandani et al. [26] observed that, in oblique incidence (with respect to the helical axis), the transmission spectra of the SmC* and SmC$_A^*$ phases are different.

It can be seen in figure C.IV.35 that, in the SmC* phase of MHPOBC, the spectrum contains two Bragg peaks, with the ratio of the wavelengths being equal to two. In the reciprocal space, one has then two wave vectors $q = 2\pi/p$ and $2q = 2\pi/(p/2)$. This is in agreement with the Fourier series expansion of the dielectric tensor $\underline{\varepsilon}(z)$ (see eq. C.IV.3) that, for symmetry reasons (with respect to the helical axis $\mathbf{t}$), is composed of only three terms in the SmC* phase:

$$\underline{\varepsilon}(z) = \begin{pmatrix} \varepsilon_\perp & 0 & 0 \\ 0 & \varepsilon_\perp & 0 \\ 0 & 0 & \varepsilon_{//} \end{pmatrix} + 2\varepsilon_q \begin{pmatrix} 0 & 0 & \cos(qz) \\ 0 & 0 & \sin(qz) \\ \cos(qz) & \sin(qz) & 0 \end{pmatrix}$$

$$+ 2\varepsilon_{2q} \begin{pmatrix} \cos(2qz) & \sin(2qz) & 0 \\ \sin(2qz) & -\cos(2qz) & 0 \\ 0 & 0 & 0 \end{pmatrix} \qquad\qquad (C.IV.70)$$

The first term characterizes, as in the cholesteric phase, the global uniaxial anisotropy of the SmC* phase. The last term is also present in the cholesteric phase (eq. B.VII.32, volume 1). As to the second term, however, it is only present in the SmC* phase (it vanishes in the cholesteric phase as it changes sign under symmetry with respect to one of the two binary axes orthogonal to **t**). This second term is at the origin of the second Bragg reflection observed in the SmC* phase, when the direction of light propagation is tilted with respect to **t**. In the particular case when **k** // **t**, the intensity of this reflection vanishes (see, for instance, the theory of light diffraction by the Blue Phases in Hornreich and Shtrikman [28]).

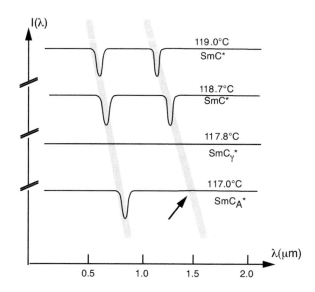

Fig. C.IV.35 Transmission spectra recorded in oblique incidence for the SmC* and SmC$_A$* phases of MHPOBC (from ref. [26]). In the SmC* phase, one notices two peaks, the ratio of the corresponding wavelengths being equal to two. The arrow indicates the absence of the second peak in the SmC$_A$* phase.

Chandani et al. found that in the SmC$_A$* phase, on the other hand, the spectrum only contains one peak, irrespective of the incidence corresponding to the wave vector $q = 2\pi/(p/2)$. The optical behavior of SmC$_A$* is thus similar to that of the cholesteric phase. This result is very easily understood. Indeed, consider the double helix of the SmC$_A$* phase represented in figure C.IV.36b.

By regrouping the smectic layers in pairs, one obtains the simplified diagram given in figure C.IV.36c. From the point of view of symmetry, this simplified structure is almost identical to the cholesteric phase since, if the small angle shift $\delta\phi$ is neglected, a pair of neighboring layers is symmetric with respect to a new binary axis C_2 perpendicular to the binary axis already present in the SmC* phase. As a consequence, the second term in eq. C.IV.70 must go to zero in the $SmC_A{}^*$ phase.

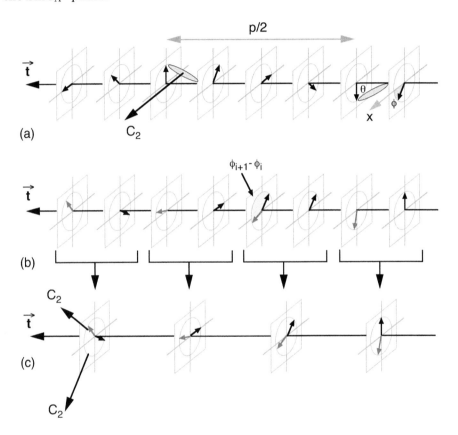

Fig. C.IV.36 Comparison of the helical structures in the SmC* (a) and $SmC_A{}^*$ (b) phases. The structure represented in (b) can be simplified by regrouping the smectic layers in pairs. One obtains then structure (c), which, from the point of view of symmetry, is almost identical to the cholesteric phase. It is noteworthy [17] that the rotation direction of the double helix typical of the $SmC_A{}^*$ phase (c) is contrary to that of the simple helix of the SmC* phase.

Diagram (b) of figure C.IV.36 also shows that the structure of the $SmC_A{}^*$ phase can be seen in two different manners:

– either as a very "tightly wound" simple helix, in which the variation of the azimuthal angle between two neighboring layers $\Delta\phi = \phi_{i+1} - \phi_i$ is close to π. In diagram (b), this variation is smaller than π, such that the helix preserves the same chirality as the helix of the SmC* phase in diagram (a). This can be expressed by:

$$\phi_i = (\pi - \delta\phi) \; l \; ; i = 0, \; \pm 1, \; \pm 2,... \tag{C.IV.71}$$

where $\delta\phi > 0$, so that $\Delta\phi = \pi - \delta\phi > 0$;

      – or as a superposition of two "loose" helices, formed by the layers with even and odd indices, respectively. For layers of even index i:

$$\phi_i = -\delta\phi \, i \; ; i = 0, \; \pm 2, \; \pm 4,... \tag{C.IV.72}$$

while for these where i is odd:

$$\phi_i = \pi - \delta\phi \, i \; ; i = \pm 1, \pm \; 3, \; ... \tag{C.IV.73}$$

A noteworthy feature is that these two "super-helices," shifted by π, are of opposite chiralities to that of the simple helix [17].

IV.8.g) Ferrielectric phases

These optical investigations of Chandani et al. reveal the presence of phases with structures different from the SmC* and SmC$_A$* phases. Indeed, it can be seen in figure C.IV.35 that at $T = 177°C$ the spectrum becomes flat. This anomaly is attributed to the presence of a novel phase, named SmC$_\gamma$*, which appears between the ferroelectric and the antiferroelectric phases, in a range indicated in the phase diagram of figure C.IV.37.

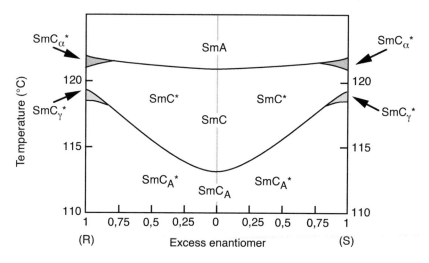

Fig. C.IV.37 Ferrielectric phases in MHPOBC (from ref. [27]).

      Another phase, named SmC$_\alpha$*, was discovered between the SmA and SmC* phases, at high chirality.

      Subsequent studies [29, 30] have shown that the domain indicated as

SmC_γ^* must in fact be subdivided into a multitude of phases. One explanation, held as true for a long time (but that has recently been questioned again, as we shall see in section C.IV.8.g) is that the phases are distinguished by sequences of "synclinic" and "anticlinic" stackings, such as those drawn in figure C.IV.38.

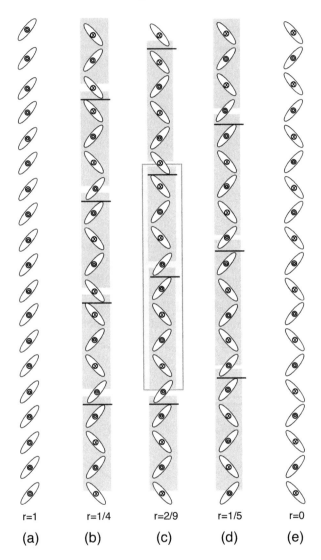

r=1	r=1/4	r=2/9	r=1/5	r=0
(a)	(b)	(c)	(d)	(e)

Fig. C.IV.38 Some stacking sequences corresponding to different values of the parameter $r = q/p$. The synclinic stacking is marked by a thick line. One can see that the $r = 2/9$ sequence is obtained by alternating slices of the sequences $r = 1/4$ and $r = 1/5$.

In this way, the system can go from the SmC* phase (Fig. C.IV.38a), where all the stackings are synclinic, to the SmC_A^* phase (Fig. C.IV.38e) where they are all anticlinic. Isozaki et al. [29] equally noticed that the sequences of synclinic and anticlinic stackings are similar to the Ising spin chains, provided that the $S_i = +1$ spin is associated with the synclinic stacking and the $S_i = -1$ spin with the anticlinic stacking. Based on this analogy, the energy of the stacking system can be expressed using the Hamiltonian proposed by Bak and Bruinsma [31]:

$$E = -\sum_i HS_i + \frac{1}{2} \sum_{ij} J(i-j)(S_i+1)(S_j+1) \qquad\qquad \text{(C.IV.74)}$$

The first term of this Hamiltonian, with $H > 0$, favors the sequences with many synclinic stackings $S_i = 1$. The second term introduces a repulsive interaction between synclinic stackings, as the $J(i-j)$, a parameter that is assumed to be positive, decreases with the distance $(i-j)$, for instance, according to

$$J(i-j) = |i-j|^{-2} \qquad\qquad \text{(C.IV.75)}$$

Let us now assume that the ratio $r = q/p$ (with p and q integers) represents the proportion of synclinic stackings present in a sequence. The $q/p = 1$ and $q/p = 0$ cases being trivial (they correspond to the SmC* and SmC$_A$* phases, respectively), let us consider sequences of the type $r = 1/p$ with $p \neq 1$.

For a given $r = 1/p$, one must find the distribution of synclinic stackings in the sequence. Due to their repulsion, the synclinic stackings must be regularly spaced. It follows that a synclinic stacking is followed by $1 - p$ anticlinic stackings, etc. For 1/5, this results in the repetition of the following pattern:

$$...(+1,-1,-1,-1,-1)... \qquad\qquad \text{(C.IV.76)}$$

represented graphically in figure C.IV.38d. The $r = 1/4$ case represented in figure C.IV.38b is completely similar. The $r = 2/9$ case, intermediate between 1/4 and 1/5 (as 2/10 < 2/9 < 2/8), is interesting insofar as the repulsion term is minimized when the two patterns corresponding to sequences 1/4 and 1/5 alternate, resulting in:

$$...[(+1,-1,-1,-1),(+1,-1,-1,-1,-1)]... \qquad\qquad \text{(C.IV.77)}$$

To find the sequence S_i corresponding to a fraction $r = q/p$, one can use the following construction, which consists of tracing the Diophantine function defined by:

$$\Sigma_i = \text{Int}(ri) \qquad\qquad \text{(C.IV.78)}$$

where Int designates the integer part of ri. In the diagram of figure C.IV.39, this function is traced for three values of parameter r. The steps of this function indicate the positions of synclinic stackings in the sequence. The Σ_i function thus gives the number of synclinic stackings encountered in the sequence between its beginning ($i = 0$) and site i.

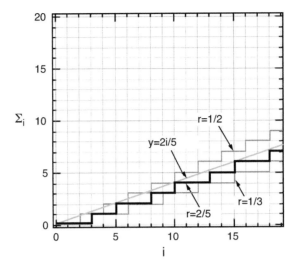

Fig. C.IV.39 Geometrical construction of sequences corresponding to fractions r = 1/2, 1/3 and 2/5.

Note that the shape of function Σ_i brings to mind the surface of a crystal, composed of steps. This resemblance is not accidental, as in the problem of crystal faceting the facets are composed of steps elastically repelling each other.

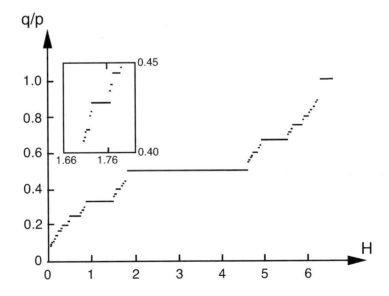

Fig. C.IV.40 "Devil's staircase." Ferrielectric phase diagram computed from the eq. C.IV.74. Each step of this fractal function corresponds to a phase characterized by a rational number q/p indicating the ratio of synclinic stackings in the stacking sequence (from ref. [31]).

The choice of the ratio $r = q/p$ depends on the value of parameter H. Obviously, in the H = 0 limit the Hamiltonian C.IV.74 is minimized by taking $r = 0$, which corresponds to the SmC_A^* phase (absence of synclinic stackings). In the "very large H" limit, the SmC^* phase, composed only of synclinic stackings, is favored. Bak and Bruinsma [31] showed that, by varying H, one obtains the "devil's staircase" phase diagram shown in figure C.IV.40. This fractal curve exhibits an infinite number of steps, each one corresponding to a rational number $r = q/p$. Note that the width ΔH of the steps depends on r and decreases with increasing p and q.

IV.8.h) Resonant X-ray scattering

The description of ferrielectric phases by periodic sequences of synclinic and anticlinic stackings was recently questioned by the results of beautiful experiments using synchrotron radiation diffraction [32, 33].

The starting idea of these experiments is very astute, as it goes against the free electron approximation discussed in appendix 2 of chapter C.I. According to this approximation, in order to calculate the amplitude of X-ray scattering by matter, it is enough to consider that all electrons of indices j placed at points r_j are free, and to apply for each of them eqs. 2.12 and 2.13 of appendix 2 of chapter C.I. In the following, we shall combine these two equations in one and write:

$$E_{si} = E_o \, (\mathbf{s}.\underline{f}(r_j)\mathbf{i}) \, e^{-i\mathbf{q}\cdot r_j} \qquad \text{(C.IV.79)}$$

where $\mathbf{i}$ and $\mathbf{s}$ are the unit vectors describing the polarizations of the incident and scattered waves, while $\underline{f}$ is an isotropic second-rank tensor:

$$\underline{f} = b \begin{pmatrix} 1 & 0 & 0 \\ 0 & 1 & 0 \\ 0 & 0 & 1 \end{pmatrix} \quad \text{with} \quad b = \frac{e^2}{mc^2}\frac{1}{R} \qquad \text{(C.IV.80)}$$

It is easily checked that for σ polarization perpendicular to the scattering vector $\mathbf{q}$, one has:

$$\mathbf{s}.\mathbf{i} = 1 \qquad \text{(C.IV.81)}$$

yielding

$$E_{\sigma i} = E_{o\sigma} \, \frac{e^2}{mc^2}\frac{1}{R} \, e^{-i\mathbf{q}\cdot r_j} \qquad \text{(C.IV.82)}$$

while for π polarization,

$$\mathbf{s}.\mathbf{i} = \cos(2\Psi) \qquad \text{(C.IV.83)}$$

where 2ψ is the scattering angle, so that:

$$E_{\pi i} = E_{o\pi} \frac{e^2}{mc^2} \frac{1}{R} \cos(2\Psi)\, e^{-i\mathbf{q}\cdot\mathbf{r}_j} \tag{C.IV.84}$$

where $E_{o\sigma}$ and $E_{o\pi}$ are the electric field components of the incident plane wave and R the distance from sample to detector.

In this free-electron approximation, the polarization of the scattered beam is the same as that of the incident beam and, for a given polarization, the total amplitude of the scattered beam is simply the sum over the contribution of all electrons:

$$E(\mathbf{q}) = E_o b_{\sigma,\pi} \sum_{j=1}^{N} e^{-i\mathbf{q}\cdot\mathbf{r}_j} \tag{C.IV.85}$$

or

$$E(\mathbf{q}) = E_o b_{\sigma,\pi} \int_V \rho(\mathbf{r})\, e^{-i\mathbf{q}\cdot\mathbf{r}}\, d^3\mathbf{r} \tag{C.IV.86}$$

where $\rho(\mathbf{r})$ is the free electron density. In conclusion, X-ray scattering can only provide information on the Fourier components of the scalar parameter $\rho(\mathbf{r})$.

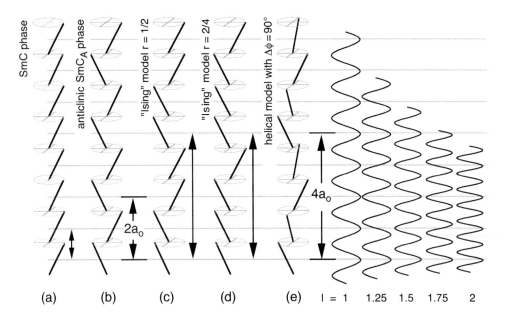

Fig. C.IV.41 Structure of liquid layer phases with tilted molecules: a) synclinic SmC phase with period a_o; b) anticlinic SmC$_A$ phase with period $2a_o$; c,d,e) competing structures with the same period $4a_o$: (c) and (d) are "Ising-type" models, while (e) is the helix model.

In conclusion, it should be impossible, using X-ray scattering, to distinguish between the SmC phase, with period a_o (Fig. C.IV.41a), and the anticlinic phase of period $2a_o$ (Fig. C.IV.41.b) or any other sequence of synclinic or anticlinic stackings such as, for instance, the sequences $r = 1/2$ and $r = 2/4$ in figures C.IV.41c and d. It should equally be impossible to discriminate between structures c and d of the "Ising type" and the helical structure in figure C.IV.41.e, all with the same period $4a_o$. Indeed, in all these phases, the Fourier components of the electron density $\rho(z)$ have the same wave vectors $q_n = nq_o$, integral multiples of $q_o = 2\pi/a_o$.

A trick, proposed by A.-M. Levelut with the goal of sidestepping this indetermination in the choice of the structure, consists of abandoning the free-electron approximation by introducing inside the molecules a sulphur atom and by selecting the wavelength of **polarized X-rays** such that it is on the edge of the K absorption line of sulphur ($hc/\lambda = 2474.8$ eV).

The summary diagram of the experimental setup used in the experiments in refs. [32, 33] is shown in figure C.IV.42a. Here, the X-ray source is a beam of synchrotron radiation, polarized de facto along the horizontal direction y.

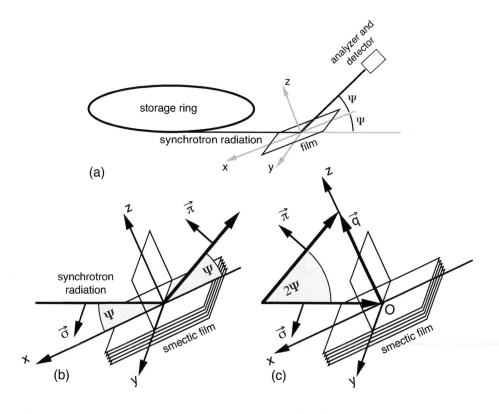

Fig. C.IV.42 Structural study by resonant X-ray scattering of ferro-, antiferro- and ferri-electric phases. a) General diagram; b) geometry of the experiment; c) scattering geometry in the reciprocal space.

It turns out indeed that in these particular conditions, known as **resonant scattering** and resulting from the choice of the wavelength, the amplitude E_d of X-rays scattered by the electrons on sulphur atom j become sensitive to the orientation $n_j(\theta, \phi)$ of molecule j to which the atom belongs. This can be expressed by writing that the tensor $\underline{f}$ in formula

$$E_{si} = E_o \, [\mathbf{s} \cdot \underline{f}(\mathbf{r}_j)\mathbf{i}] \, e^{-i\mathbf{q} \cdot \mathbf{r}_j} \qquad \text{(C.IV.87)}$$

is now a symmetric second-rank tensor of proper axes attached to the molecule. By analogy with the dielectric permittivity tensor, this tensorial structure factor can be split into two parts, the first one represented by the isotropic tensor δ_{ij} and the second by traceless anisotropic tensors. In the $(1, 2, n)$ reference system of the eigenaxes of the tensor, attached to the molecule, this tensor structure factor has the very simple form:

$$\underline{f} = b \begin{pmatrix} 1 & 0 & 0 \\ 0 & 1 & 0 \\ 0 & 0 & 1 \end{pmatrix} + f \begin{pmatrix} 1 & 0 & 0 \\ 0 & 1 & 0 \\ 0 & 0 & -2 \end{pmatrix} + \Delta \begin{pmatrix} +1 & 0 & 0 \\ 0 & -1 & 0 \\ 0 & 0 & 0 \end{pmatrix} \qquad \text{(C.IV.88)}$$

In order to go to the reference system (x, y, z) defined in figure C.IV.42b, with axis z orthogonal to the layers, two rotations must be applied to this tensor: first, a tilt by θ obtained by rotation around the y axis, then rotation by ϕ around the z axis.

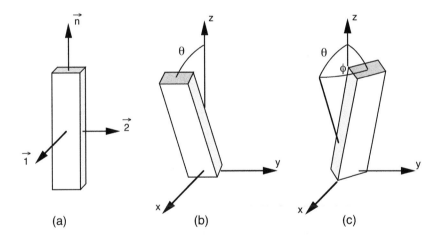

(a) (b) (c)

Fig.C.IV.43 Rotations of tensor $\underline{f}$.

Finally, one obtains an expression similar to the one given in eq. C.IV.70:

$$\underline{f} = b \begin{pmatrix} 1 & 0 & 0 \\ 0 & 1 & 0 \\ 0 & 0 & 1 \end{pmatrix} + f_o \begin{pmatrix} 1 & 0 & 0 \\ 0 & 1 & 0 \\ 0 & 0 & -2 \end{pmatrix} + f_1 \begin{pmatrix} 0 & 0 & \cos\phi \\ 0 & 0 & \sin\phi \\ \cos\phi & \sin\phi & 0 \end{pmatrix}$$

$$+ f_2 \begin{pmatrix} \cos 2\phi & \sin 2\phi & 0 \\ \sin 2\phi & -\cos 2\phi & 0 \\ 0 & 0 & 0 \end{pmatrix} \qquad \text{(C.IV.89)}$$

with

$$f_o = f \frac{2\cos^2\theta - \sin^2\theta}{2} - \Delta \frac{\sin^2\theta}{2} \qquad \text{(C.IV.90a)}$$

$$f_1 = -(3f + \Delta) \cos\theta \sin\theta \qquad \text{(C.IV.90b)}$$

$$f_2 = -f \frac{3\sin^2\theta}{2} + \Delta \frac{\cos^2\theta + 1}{2} \qquad \text{(C.IV.90c)}$$

It can be seen that the last two terms of this tensor are explicitly dependent on the azimuthal angle ϕ, so that its variation with the layer index k will now allow discrimination between the different phases.

In order to detect the contribution of these two terms in their X-ray scattering experiments, the authors of references [32, 33] used the scattering geometry shown in figure C.IV.42b. Here, the scattering plane (x,z) is perpendicular to the smectic layers (x,y) while the natural polarization of the incident synchrotron beam is σ, corresponding to $\mathbf{i} = [0,1,0]$. The calculation yields:

$$\underline{f}(\mathbf{r}_j) \, \mathbf{i} = [f_2\sin(2\phi_j), \, -f_2\cos(2\phi_j), \, f_1\sin(\phi_j)] \qquad \text{(C.IV.91)}$$

Say the polarization $\mathbf{s}$ of the scattered beam, selected using the analyzer (a crystal of pyrolitic graphite), is π. In this case, $\mathbf{s} = [\sin\Psi, 0, \cos\Psi]$ and we have

$$E_{sj} = E_o[\mathbf{s} \cdot \underline{f}(\mathbf{r}_j)\mathbf{i}]e^{-i\mathbf{q} \cdot \mathbf{r}_j} = E_o[f_2\sin(2\phi_j)\sin\Psi + f_1\sin(\phi_j)\cos\Psi]e^{-i\mathbf{q} \cdot \mathbf{r}_j} \quad \text{(C.IV.92)}$$

Finally, the total amplitude of the beam reaching the detector is given by the sum:

$$E(\mathbf{q}) = E_o \sum_j [f_2\sin(2\phi_j)\sin\Psi + f_1\sin(\phi_j)\cos\Psi]e^{-i\mathbf{q} \cdot \mathbf{r}_j} \qquad \text{(C.IV.93)}$$

The vector $\mathbf{r}_j$, indicating the position of the sulphur atoms in the smectic film, can be written as a sum of two components — one in the plane of the layers and the other perpendicular to the layers:

$$\mathbf{r}_j = \mathbf{r}_{j//} + k_j \, [0, 0, a_o] \qquad \text{(C.IV.94)}$$

Here, $\mathbf{r}_{//} = (x, y, 0)$ is an arbitrary vector (the layers are liquid), while k_j stands for the index of the layers in the film.

To simplify the analysis, we shall first consider scattering by a domain inside the film, of size dxdy small enough so that, inside it, the angle ϕ_j only depends on the index k of the layers. Within this approximation, the sum

C.IV.93 can be replaced by

$$dE(\mathbf{q}) = E_o \, \rho dxdy \sum_{k=1}^{N} [f_2 \sin(2\phi_k)\sin\Psi + f_1 \sin(\phi_k)\cos\Psi] e^{-i\mathbf{q}\cdot\mathbf{r}_k} \qquad (\text{C.IV.95})$$

where $\rho dxdy$ is the number of molecules in one layer of the domain $dxdy$.

In the SmC phase (Fig. C.IV.41a), the angle ϕ is identical throughout the layers, so that

$$dE(\mathbf{q}) = E_o \, \rho dxdy \, [f_2 \sin(2\phi)\sin\Psi + f_1 \sin(\phi)\cos\Psi] \sum_{k=1}^{N} e^{-i\mathbf{q}\cdot\mathbf{r}_k} \qquad (\text{C.IV.96})$$

The amplitudes of the radiation scattered by the layers in the domain are added coherently and give a Bragg reflection for

$$\mathbf{q} = l\,(0,0,2\pi/a_o) \quad \text{with} \quad l \text{ an integer} \qquad (\text{C.IV.97})$$

Indeed, in these conditions, all the terms of the sum are identical, as $e^{-i\mathbf{q}\cdot\mathbf{r}_k} = e^{-i(2\pi kl)} = 1$, resulting in:

$$dE(l) = E_o \, \rho dxdy \, N \, [f_2 \sin(2\phi) \sin\Psi + f_1 \sin(\phi) \cos\Psi] \qquad (\text{C.IV.98})$$

In a film of large size, the texture of the director field **c** can be quite complex and the angle ϕ can vary as a function of the coordinates (x,y). If the angle ϕ is randomly distributed, the scattering becomes incoherent and the total intensity of the signal scattered by the film becomes:

$$I(l) = E_o^2 \, \rho S N \, [f_2^2 <\sin^2(2\phi)> \sin^2\Psi + f_1^2 <\sin^2(\phi)> \cos^2\Psi] \qquad (\text{C.IV.99})$$

In the SmC$_A$ phase, the azimuthal angle ϕ_k is, for instance, ϕ for even k and $\phi + \pi$ for odd k (Fig. C.IV.41b). One then has:

$$dE(l) = E_o \, \rho dxdy \, f_1 \sin(\phi)\cos\Psi \sum_{k=1}^{N} (-1)^k e^{-i(2\pi kl)}$$

$$+ \, E_o \, \rho dxdy \, f_2 \sin(2\phi) \, \sin\Psi \sum_{k=1}^{N} e^{-i(2\pi kl)} \qquad (\text{C.IV.100})$$

The second term of this expression results in the same selection rule as in the SmC phase: l must be an integer. As to the first term, it generates constructive interference if the sign of the $e^{-i(2\pi kl)}$ factor changes sign at the same time as the sign of the $(-1)^k$ factor. This occurs for

$$l = \frac{2n-1}{2} \quad \text{with} \quad n \text{ an integer} \qquad (\text{C.IV.101})$$

for in this case, $e^{-i(2n-1)k\pi} = 1$ for even k and $e^{-i(2n-1)k\pi} = -1$ for odd k. In agreement with this analysis, the spectrum (e) in figure C.IV.44 exhibits peaks at $l = 2$ and $l = 1.5$.

To analyze the diffraction pattern given by chiral phases with helical structure, Levelut and Pansu [34] used an alternative expression of the $\underline{f}$ tensor

$$\underline{f} = b\,\underline{\delta} + g_0\,\underline{M}_0 + g_1\,e^{i\phi}\,\underline{M}_1 + g_{-1}e^{-i\phi}\,\underline{M}_{-1} + g_2\,e^{i2\phi}\,\underline{M}_2 + g_{-2}\,e^{-i2\phi}\,\underline{M}_{-2} \qquad \text{(C.IV.102a)}$$

where

$$g_0 = f_0 \;,\quad g_1 = g_{-1} = \frac{f_1}{2}\;,\quad g_2 = g_{-2} = \frac{f_2}{2} \qquad \text{(C.IV.102b)}$$

The five matrices

$$\underline{M}_0 = \begin{pmatrix} 1 & 0 & 0 \\ 0 & 1 & 0 \\ 0 & 0 & -2 \end{pmatrix},\quad \underline{M}_{\pm 1} = \begin{pmatrix} 0 & 0 & 1 \\ 0 & 0 & \pm i \\ 1 & \pm i & 0 \end{pmatrix},\quad \underline{M}_{\pm 2} = \begin{pmatrix} 1 & \pm i & 0 \\ \pm i & -1 & 0 \\ 0 & 0 & 0 \end{pmatrix} \qquad \text{(C.IV.102c)}$$

are identical to those already employed in the expression of the order parameter of Blue Phases (eq. B.VIII.12, volume 1 or in eqs. C.IV.3a–c of the dielectric tensor for the SmC* phase). The amplitude of the scattered field becomes in this case:

$$dE(l) = E_0\,\rho dxdy \sum_{m=-2}^{2} \left[g_m\,(\mathbf{s}\cdot\underline{M}_m\,\mathbf{i})\,\sum_{k=1}^{N} e^{i(m\phi_k - 2\pi lk)}\right. \qquad \text{(C.IV.103)}$$

If

$$\phi_k = k\,\Delta\phi \qquad \text{with} \qquad \Delta\phi = 2\pi v \qquad \text{(C.IV.104)}$$

then

$$m\phi_k - 2\pi lk = 2\pi(mv - l)k \qquad \text{(C.IV.105)}$$

To obtain constructive interference, the factor $(mv - l)$ must be an integer. This leads to the selection of the scaled scattering vector

$$l_{n,m} = n + mv \qquad \text{with } n \text{ an integer and } m = -2, -1, 0, 1, 2 \qquad \text{(C.IV.106)}$$

In the SmC* phase, the v factor is equal to the ratio

$$\varepsilon = \frac{a_0}{P} \qquad \text{(C.IV.107)}$$

between the layer thickness a_0 and the helix pitch P. Of course, the ratio ε depends on the chirality. Of the order of 10^{-2} in pure enantiomers, it tends toward 0 in a racemic mixture. Consequently, in the conditions of resonant scattering, each Bragg reflection of the SmC* phase, of integer index n, splits into five peaks, very close in wavelength, defined by eq. C.IV.106.

In the SmC$_A$* phase, the v factor is close to 1/2

$$\nu = \frac{1}{2} - \varepsilon \qquad\qquad\qquad\qquad\qquad\qquad\qquad \text{(C.IV.108)}$$

so that the Bragg reflection of the achiral SmC_A phase, corresponding to $l = 3/2$ in eq. C.IV.101, splits, in the SmC_A^* phase, into two very close peaks

$$l_{1,+1} = \frac{3}{2} - \varepsilon \qquad\qquad \text{and} \qquad\qquad l_{2,-1} = \frac{3}{2} + \varepsilon \qquad\qquad \text{(C.IV.109)}$$

This effect due to the chirality is clearly illustrated by spectra (e) and (d) in figure C.IV.44.

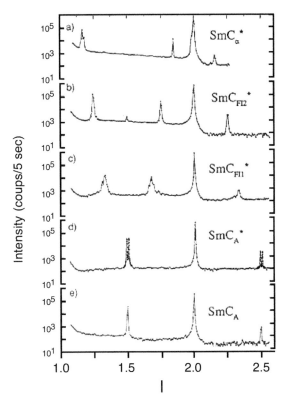

Fig. C.IV.44 Scattered intensity as a function of the scaled wave vector (from ref. [32]). The compounds employed are the R enantiomer and the racemic mixture of 10OTBBB1M7.

The same reasoning leads to the conclusion that spectrum (c) in figure C.IV.44 is due to a phase where the ν factor is close to 1/3. The successive peaks of this spectrum correspond to

$$l_{1,+1} = \frac{4}{3}; \quad l_{2,-1} = \frac{5}{3}; \quad l_{2,0} = 2 \quad \text{and} \quad l_{2,+1} = \frac{7}{3} \qquad \text{(C.IV.110)}$$

In this ferrielectric phase, named SmC_{FI1}, the azimuthal angle increases from layer to layer in steps of $\Delta\phi \approx 120°$. As to the spectrum (b), characteristic of another ferrielectric phase, called SmC_{FI2}, the disposition of the peaks suggests a structure of period $4a_0$ where the azimuthal angle varies in increments of $\Delta\phi \approx 90°$ (Fig. C.IV.41e). The peak indices are then, in increasing order:

$$l_{1,+1} = \frac{5}{4}; \quad l_{1,2} = l_{2,-2} = \frac{6}{4}; \quad l_{2,-1} = \frac{7}{4}; \quad l_{2,0} = \frac{8}{4}; \quad l_{2,+1} = \frac{9}{4} \qquad \text{(C.IV.111)}$$

The experimental analysis of the polarization of the scattered signal shows that the peaks $l_{1,+1}$ and $l_{1,-1}$ are of polarization π, while peak $l_{1,+2}$ is of polarization σ. To verify that this experimental fact is in agreement with the selection rules given by eq. C.IV.103, one must determine the $(\mathbf{s}.\underline{\mathbf{M}}_m\mathbf{i})$ product. With $\mathbf{i} = \sigma_{in} = (0,1,0)$, $\mathbf{s} = \sigma_{sc} = (0,1,0)$ for the σ polarization and $\mathbf{s} = \pi_{sc} = (\sin\Psi, 0, \cos\Psi)$ for the π polarization one gets

$$(\sigma_{sc}.\underline{\mathbf{M}}_{\pm 1}\sigma_{in}) = 0 \; ; \qquad (\pi_{sc}.\underline{\mathbf{M}}_{\pm 1}\sigma_{in}) = \pm i \cos\Psi \qquad \text{(C.IV.112)}$$

$$(\sigma_{sc}.\underline{\mathbf{M}}_{\pm 2}\sigma_{in}) = -1 \; ; \qquad (\pi_{sc}.\underline{\mathbf{M}}_{\pm 2}\sigma_{in}) = \pm i \sin\Psi \qquad \text{(C.IV.113)}$$

For $m = \pm 1$, the polarization of the scattered signal is thus π. For $m = \pm 2$, the scattered signal contains, in principle, the two components σ and π but, in practice, the σ polarization dominates, since the scattering angle Ψ is small ($< 10°$).

In conclusion, let us point out that the "Ising type" models for the ferrielectric phases, introduced in the previous section, give for $r = 1/2$ and $r = 2/4$ structures where the period is also $4a_0$ (Fig. C.IV.41c and d). The scattering spectrum given by these models could then also contain peaks with the same scaled wavelengths 1, 1.25, 1.5, 1.75, 2... as the one given by the helical model in figure C.IV.44c. It is, however, easily checked from figure C.IV.44, that, in an Ising model, the intensity of the peak $l_{2,2} = 1.5$ must be strictly zero, in disagreement with experimental results.

IV.9 Mesophases formed by banana-shaped molecules

In 1992, Brand, Cladis and Pleiner [35] put forward the idea that the spontaneous polarization could appear in a smectic phase formed by achiral molecules shaped as bananas. Such molecules can be stacked in a smectic layer, as shown in diagram (a) of figure C.IV.45.

This stacking, symmetric with respect to the binary axis $\mathbf{C}_2$, can also be symmetric under reflection in the $(\mathbf{z}, \mathbf{p})$ plane. If this is the case, the symmetry of the stacking is C_{2v} and the molecules are arranged in the plane $(\mathbf{z}, \mathbf{p})$ of the figure. On the other hand, the mirror symmetry is broken if, as in the diagram of figure C.IV.45b, the molecules are tilted with respect to the (**z**,**p**) plane; the symmetry of the stacking is then only C_2. In both cases, however, and for the symmetry reasons already invoked in the beginning of this chapter, the stackings can possess a spontaneous polarization **P** along the binary axis C_2.

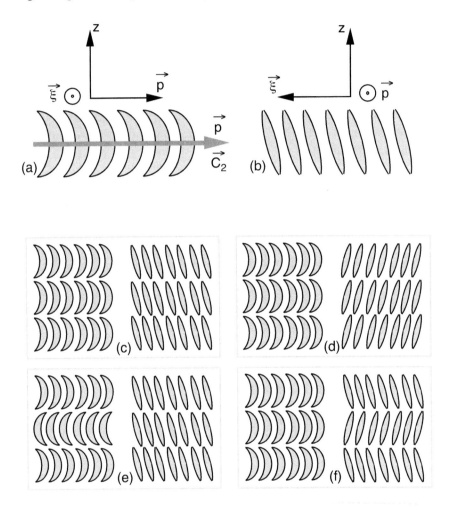

Fig. C.IV.45 Polarized smectic phases formed by banana-shaped molecules.

    The case of the C_2 symmetry is very interesting because, if the ξ vector gives the direction of molecular tilt, the three vectors **z**, **p** and ξ form a chiral trihedron. One can see that, even for achiral molecules, the stacking can exist in two versions of different chirality (Fig. C.IV.45c and d). This chirality can alternate from one layer to the next (Fig. C.IV.45e and f) and form, in this case, an antiferroelectric phase.

    Experimental research on these remarkable phases is in progress [36–40].

BIBLIOGRAPHY

[1] Meyer R.B., Liébert L., Strzelecki L., Keller P., "Ferroelectric liquid crystals," *J. Physique (France)*, **36** (1975) L-69. (This paper describes the first ferroelectric mesophase.)

[2] Clark N.A., Lagerwall S.T., *Appl. Phys. Lett.*, **36** (1980) 899.

[3] Lagerwall S.T., *Ferroelectric and Antiferroelectric Liquid Crystals*, Wiley-VCH, Weinheim, 1999.

[4] Bahr Ch., "Smectic liquid crystals: ferroelectric properties and electroclinic effect" in *Chirality in Liquid Crystals*, Eds. Kitzerow H.-S. and Bahr Ch., Springer, New York, 2001.

[5] Beresnev L.A., Blinov L.M., Osipov M.A., Pikin S.A., "Ferroelectric Liquid Crystals", *Mol. Cryst. Liq. Cryst.*, **158A** (1988) 1-150.

[6] a) Martinot-Lagarde Ph., *J. Physique Lett. (France)*, **38** (1977) L-17.
b) Dumrongrattana S., Huang C.C., Nounesis G., Lien S.C., Viner J.M., *Phys. Rev. A*, **34** (1986) 5010.

[7] a) Pieranski P., Guyon E., Keller P., *J. Physique (France)*, **36** (1975) 1005.
b) Pieranski Pawel, Guyon E., Keller P., Liébert L., Kuczynski W., Pieranski Piotr, *Mol. Cryst. Liq. Cryst.*, **38** (1977) 275.

[8] a) Jakli A., Bata L., Buka A., Eber N., Janossy I., *J. Physique Lett. (France)*, **46** (1985) L-759.
b) Jakli A., Bata L., Buka A., Eber N., *Ferroelectrics*, **69** (1986) 153.

[9] Garoff S., Meyer R.B., *Phys. Rev. Lett.*, **38** (1977) 848.

[10] Bahr Ch., Heppke G., Sabaschus B., *Ferroelectrics*, **84** (1988) 103.

[11] Bahr Ch., Heppke G., Sabaschus B., *Liq. Cryst.*, **9** (1991) 31.

[12] Bahr Ch., Heppke G., *Liq. Cryst.*, **2** (1987) 825.

[13] Stegemeyer H., Meister R., Hoffmann U., Kuczynski W., *Liq. Cryst.*, **10** (1991) 295.

[14] Takezoe H., Takanishi Y., "Smectic Liquid Crystals: Antiferroelectric and Ferrielectric Phases" in *Chirality in Liquid Crystals*, Eds. Kitzerow H.-S. and Bahr Ch., Springer, New York, 2001.

[15] Michelson A., Cabib D., Benguigui L., *J. Physique (France)*, **38** (1977) 961.

[16] Link D.R., Maclennan J.E., Clark N.A., *Phys. Rev. Lett.*, **77** (1996) 2237.

[17] Marceroux J.P., Private communication.

[18] Beresnev L.A., Blinov L.M., Baikalov V.A., Pozhidayev E.P., Pourvanetskas G.V., Pavluchenko A.I., *Mol. Cryst. Liq. Cryst.*, **89** (1982) 327.

[19] Chandani A.D.L., Hagiwara T., Suzuki Y., Ouchi Y., Takezoe H., Fukuda A., *Jap. J. Appl. Phys.*, **27** (1988) L729.

[20] Keller P., Liébert L., Strzelecki L., *J. Physique Coll. (France)*, **37** (1976) C3-27.

[21] Levelut A.-M., Germain C., Keller P., Billard J., *J. Physique (France)*, **44** (1983) 623.

[22] Galerne Y., Liébert L., *Phys. Rev. Lett.*, **64** (1990) 906.

[23] Bahr Ch. and Fliegner D., *Phys. Rev. Lett.*, **70** (1993) 1842.

[24] Takanishi Y., Takezoe H., Fukuda A., Watanabe J., *Phys. Rev. B*, **45** (1992) 7684.

[25] Isozaki T., Fujikawa T., Takezoe H., Fukuda A., Hagiwara T., Suzuki Y., Kawamura I., *Phys. Rev. B.*, **48** (1993) 13439.

[26] Chandani A.D.L., Gorecka E., Ouchi Y., Takezoe H., Fukuda A., *Jap. J. Appl. Phys.*, **28** (1989) L1265.

[27] Fukuda A., Yakanishi Y., Isozaki T., Ishikawa K., Takezoe H., *J. Mat. Chem.*, **4** (1994) 997.

[28] Hornreich R.M., Shtrikman S., *Phys. Rev. A*, **28** (1983) 1791.

[29] Isozaki T., Hiraoka K., Takanishi Y., Takezoe H. Fukuda A., Suzuki Y., Kawamura I., *Liq. Cryst.*, **12** (1992) 59.

[30] Bahr Ch., Fliegner D., Booth C.J., Goodby J.W., *Phys. Rev. E*, **53** (1995) R3823.

[31] Bak P., Bruinsma R., *Phys. Rev. Lett.*, **49** (1982) 249.

[32] Mach P., Pindak R., Levelut A.-M., Barois P., Nguyen H.T., Huang C.C., Furenlid L., *Phys. Rev. Lett.*, **81** (1998) 1015.

[33] Mach P., Pindak R., Levelut A.-M., Barois P.,Nguyen H.T., Baltes H., Hird M., Toyne K., Seed A., Goodby J.W., Huang C.C., Furenlid L., *Phys. Rev. E*, **60** (1999) 6793.

[34] Levelut A.-M., Pansu B., *Phys. Rev. E*, **60** (1999) 6803.

[35] Brand H.R., Cladis P.E., Pleiner H., *Macromolecules*, **25** (1992) 7223.

[36] Akutagawa T., Matsanuga Y., Yasuhara K., *Liq. Cryst.*, **17** (1994) 659.

[37] Niori T., Sekine T., Watanabe J., Furukawa T., Takezoe H., *J. Mat. Chem.*, **6** (1996) 1231.

[38] Link D.R., Natale G., Shao R., MaclennanJ.E., Clark N.A., Körblova E., Walba D.M., *Science*, **278** (1997) 1924.

[39] Heppke G., Moro D., *Science*, **279** (1998) 1872.

[40] Coleman D.A., Fernsler J., Chattham N. et al., *Science*, **301** (2003) 1204.

Chapter C.V

The twist-grain boundary smectics

By analogy with a type II superconductor placed in a magnetic field (see section C.I.1), de Gennes, in 1972, put forward the idea that a smectic composed of chiral molecules can twist by developing screw dislocations [1]. In this model (see appendix 1 to chapter C.I), the twist of the director field is localized in the core of each dislocation and in its immediate vicinity, over a distance λ termed penetration length: in this region, the molecules are tilted with respect to the layer normal, as in a smectic C.

In the same way that the chiral smectic C is denoted by SmC*, one could denote by SmA* the twisted version of the smectic A.

The hypothetical SmA* phase that, like the Blue Phases, is a phase consisting of defects, was for more than 20 years only a theoretical speculation, before being discovered in 1989 and prompting extensive investigations, which are summarized in this chapter.

The plan of the chapter is the following. First (section V.1), we shall present the theoretical model of Renn and Lubensky, who revisited de Gennes' idea and specified, among other things, the distribution of screw dislocations (all of the same sign) in a SmA* phase. We shall see that this model leads to the concept of twist-grain boundary phases (TGB$_A$), which can be commensurate or incommensurate. We shall also recall the main features of their structure factors, before discussing the physical phenomena that can lead to their proliferation. We shall continue by presenting the conditions in which the incommensurate TGB$_A$ phase was discovered by the Bell group, and we shall describe their experimental results (section V.2). It will then be shown that other TGB phases can exist, such as the TGB$_C$ phase first predicted theoretically by Lubensky et al. in 1991, and discovered two years later by the liquid crystal group in Bordeaux (section V.3). Finally, we discuss the Smectic Blue Phases, which exhibit double twist and a locally smectic structure and which are presently being studied (section V.4).

V.1   The Renn-Lubensky model

V.1.a)   Structure of the dislocation lattice

In 1988, possibly inspired by the Blue Phases, which are distinguished by a different architecture of the disclination lattice, Renn and Lubensky asked themselves the following question [2]:

If the SmA* phase exists, how are the dislocations arranged?

This question is very relevant, especially because answering it provided the experimentalists with the first unequivocal criterion allowing to this phase to be distinguished from all other mesophases.

Paradoxically, among the possible configurations of the dislocations, one must start by ruling out the one adopted by vortices in superconductors, namely a lattice of parallel and regularly spaced lines (forming, for instance, a hexagonal lattice). In superconductors, this vortex configuration is imposed by an external source, viz. the applied magnetic field. In smectics, on the other hand, the origin of the defects is intrinsic, so that a uniform density of screw dislocations, all parallel, is simply impossible.

To prove it, consider the diagram in figure C.V.1, showing a smectic cylinder containing a uniform density n (number of dislocations per unit surface in the (x, y) plane) of screw dislocations parallel to the **z** axis, all of the same Burgers vector $\mathbf{a_o}$ (with a_o the layer thickness).

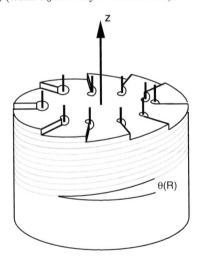

Fig. C.V.1 A uniform density of screw dislocations with Burgers vector $\mathbf{a_o}$ // **z** generates a tilt of the layers with respect to the (x, y) plane. This tilt increases with the radius R. The vertical distance between the layers (measured along z) being preserved (and equal to a_o), the real layer thickness decreases with the tilt, which is forbidden. Note that the dislocations acquire a wedge component that increases with the tilt angle.

Let u be the layer displacement. The tilt $\boldsymbol{\delta m}$ of the smectic layers with respect to the (x, y) plane is given by the gradient of vertical displacement $\mathbf{grad}(u)$. To determine this tilt as a function of the distance R, let us consider the contour integral

$$\int \mathbf{grad}(u).\mathbf{dl} \tag{C.V.1}$$

representing the vertical displacement accumulated over the integration contour. For a circular contour of radius R, this displacement is determined by the number $N = \pi R^2 n$ of screw dislocations contained within the contour. Each dislocation having a Burgers vector a_o, we obtain:

$$\int \mathbf{grad}(u).\mathbf{dl} = \pi R^2 n a_o \tag{C.V.2}$$

The total length of the integration contour being $2\pi R$, the gradient along the circuit, which is also the tangent of angle θ, increases linearly with R:

$$\text{tg } \theta = \frac{a_o n}{2} R \tag{C.V.3a}$$

As the smectic layers are more and more tilted, their thickness, measured along the normal to the layers, and of expression:

$$a = a_o \cos \theta \tag{C.V.3b}$$

decreases.

This layer compression is unphysical. To avoid this problem, Renn and Lubensky [2] put forward another arrangement of the dislocations, wherein their axis follows the rotation of the smectic layers (Fig. C.V.2). In this model, the dislocations are grouped in flat and parallel boundaries, regularly spaced (the distance between two boundaries is l_b). Inside each boundary (Fig. C.V.3), the dislocations (all of the same sign) are equidistant (separated by a distance l_d) and thus parallel to each other. Their collective orientation turns regularly from one boundary to the next.

The angle $\Delta\theta$ between dislocations belonging to two consecutive boundaries is equal to the angle of rotation of the smectic layers across a boundary of screw dislocations. This angle also depends on the distance l_d between the dislocations within a boundary and is given by (see the diagram in Fig. C.V.3b):

$$\Delta\theta = 2 \text{ Arcsin}\left(\frac{a_o/2}{l_d}\right) \approx \frac{a_o}{l_d} \tag{C.V.4}$$

Since the smectic layers turn by $\Delta\theta$ across each boundary, the average twist $\mathbf{n.rot\ n}$ along the $\mathbf{q}$ axis is:

$$\langle \mathbf{n.curl\ n} \rangle = \frac{\Delta\theta}{l_b} \approx \frac{a_o}{l_d l_b} \qquad\qquad (C.V.5)$$

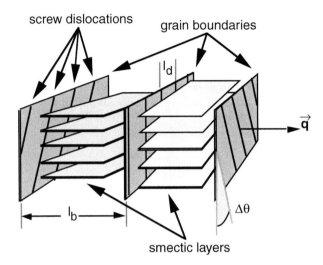

Fig. C.V.2 Model of the SmA* phase by Renn and Lubensky. The screw dislocations are grouped in boundaries. Within each boundary, the dislocations are parallel and equidistant. The orientation of the dislocations turns by $\Delta\theta$ between two consecutive boundaries, such that the dislocations remain symmetrically placed (along the bisector) with respect to the layers in the slabs. Thus, the layer thickness can be preserved within each slab.

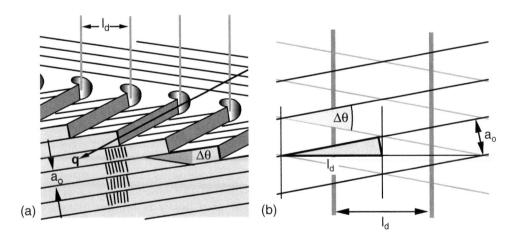

Fig. C.V.3 Twist wall consisting of a row of regularly spaced dislocations.

In summary, in the model of Renn and Lubensky, the smectic phase is carved in slices of thickness l_b, separated by boundaries of screw dislocations.

Inspired by the terminology of solid state physics, Renn and Lubensky named this SmA* phase **T**wist-**G**rain **B**oundary state or **TGB**$_A$ for short.

V.I.b) Commensurate or incommensurate TGB$_A$ phases

The model of the TGB$_A$ phase devised by Renn and Lubensky brings into play at least four characteristic length scales:

1. The "cholesteric" pitch P, over which the smectic layers turn by 2π.
2. The thickness l_b of a smectic slab.
3. The distance l_d between dislocations within the grain boundaries.
4. The thickness a_o of the smectic layers.

Since the cholesteric pitch P and the thickness of the smectic slabs l_b define two repeat distances along the same space direction, Renn and Lubensky considered two possible versions for the TGB$_A$ phase:

1. The **commensurate** TGB$_A$ phase, where the ratio α between the two steps is rational:

$$\alpha = \frac{l_b}{P} = \frac{p}{q} \qquad\qquad (C.V.6)$$

where p and q are integers.
2. The **incommensurate** TGB$_A$ phase, where α is an irrational number.

To provide the experimentalists with the clearest criterion possible for identifying the TGB$_A$ phases, Renn and Lubensky also determined their structure factors S(**q**). Before discussing these results, however, let us describe the structural changes occurring in the cholesteric phase close to the transition toward a smectic or TGB phase.

V.1.c) Structure factor of the cholesteric phase
close to a TGB$_A$ phase

The calculations of Renn and Lubensky lead to the following expression of the structure factor of the cholesteric phase when approaching the transition to the TGB$_A$ phase:

$$S(\mathbf{q}) = S_o \frac{e^{-q_x^2 \xi_{//}^2}}{(k_o - k_{c2})/k_{c2} + \xi^2(q_\perp - q_o)^2} \qquad\qquad (C.V.7)$$

In this expression, S_o is a constant, ξ the coherence length of the smectic order parameter (equal to $1/\sqrt{2|\alpha|M_\perp}$, using the customary notations to be reviewed in section V.1.e), $q_o = 2\pi/a_o$ a wave vector corresponding to the local layer thickness and $k_o = 2\pi/P$ the cholesteric twist. One also has:

$$\xi_{//}^{-2} = \sqrt{q_o q_\perp} \; k_o \tag{C.V.8}$$

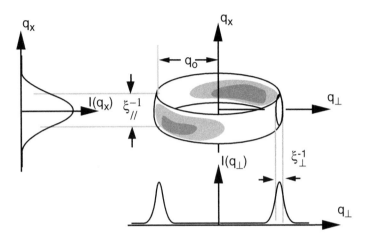

Fig. C.V.4 Structure factor of the cholesteric phase close to the transition to the TGB$_A$ phase.

This structure factor is shaped like a torus of radius $q_o = 2\pi/a_o$, of thickness $\xi_\perp^{-1}$ and of height $\xi_{//}^{-1}$ (Fig. C.V.4). Its presence signals the existence of a short-range order of the smectic type, appearing within the cholesteric phase close to the transition to the TGB$_A$ phase. This structure factor exhibits revolution symmetry along the q_x axis, since the direction of the smectic planes, perpendicular to the director **n**, turns with **n** along the cholesteric helix. The thickness $\xi_\perp^{-1}$ of the torus along a direction $q_\perp$ perpendicular to the helical axis varies as:

$$\xi_\perp^{-1} = \xi^{-1} \left[(k_o - k_{c2})/k_{c2} \right]^{1/2} \tag{C.V.9a}$$

and tends to 0 when the wave vector k_o of the cholesteric helix reaches the critical value k_{c2}. This signifies that the range of the smectic correlations, along a direction perpendicular to the helix, diverges at the transition. On the other hand, the height $\xi_{//}^{-1}$ of the torus remains finite at the transition as k_o tends to k_{c2}, since:

$$\xi_{//}^{-2} = q_o k_o = q_o k_{c2} \tag{C.V.9b}$$

for $q_\perp = q_o$.

In conclusion, on approaching the Chol → TGB$_A$ transition, the structure factor of the cholesteric phase is a torus, the height of which remains finite and of vanishing width.

It is worthwhile to compare this behavior of the structure factor with that of the nematic phase studied by McMillan close to the second-order N → SmA transition in achiral materials (Fig. C.I.11).

Instead of a torus, one now has two spherical caps, a distance $2\pi/a_o$ away from the origin. This distance, connected to the layer thickness a_o, is analogous to the radius of the torus in the cholesteric phase. One should also note that the thickness of the cap ($\approx \xi_{//}^{-1}$, defined as the width of the Lorentzian profile of S(**q**) along the director) corresponds to the thickness of the torus ($\approx \xi_{\perp}^{-1}$, defined as the width of the Lorentzian profile of S(**q**) along the normal of the smectic layers). These two length scales diverge at the transition. Finally, the radius of the cap ($\approx \xi_{\perp}^{-1}$, defined as the width of the Lorentzian profile of S(**q**) along a direction perpendicular to the director) is similar to the height of the torus ($\approx \xi_{//}^{-1}$, defined as the width of the Gaussian profile of S(**q**) along the cholesteric helix). In the case of nematics, $\xi_{\perp}$ diverges. On the other hand, the range of the smectic correlations along the helix must remain finite at the transition due to the twist of the director field (incompatible with a long-range lamellar order).

V.1.d) Structure factor of the TGB$_A$ phases

The shape of the structure factor of a TGB$_A$ phase is different for commensurate and incommensurate phases.

In the incommensurate case, all azimuthal directions of the smectic layers are possible. Consequently, the structure factor (Fig. C.V.5) consists of a sequence of continuous cylindrical strips, of radii

$$q_j = jq_o = j\frac{2\pi}{a_o} \qquad \text{(j integer)} \tag{C.V.10}$$

and of height

$$\xi_{//}^{-1} = \sqrt{q_o k_o} \tag{C.V.11}$$

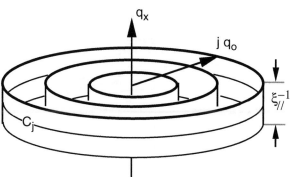

Fig. C.V.5 Structure factor of the incommensurate TGB$_A$ phase. Close to a second-order Chol → TGB$_A$ transition, the intensity of the first ring of radius q_o is stronger than for the others.

In the case of the commensurate TGB_A phase, the possible azimuthal directions of the smectic layers are given by the expression:

$$\phi_i = i\,\Delta\phi = i\,2\pi\,\alpha = 2\pi\,i\,\frac{p}{q} \qquad \text{(i, p and q integers)} \qquad \text{(C.V.12)}$$

This means that the maxima of $S(\mathbf{q})$ are arranged in vertical planes Σ_i (containing the helical axis q_x) of azimuthal directions $\phi_I + \pi/2$.

Taking into account the periodicity of the smectic layers, one finds that, in each vertical plane, the maxima must be placed on vertical rods D_{ij}, at a distance $q_{\perp j}$ from the origin which is a multiple of q_o:

$$q_{\perp j} = j\,q_o \qquad \text{(C.V.13)}$$

Since along the helix (of pitch P), the repeat distance of each direction ϕ_i is equal to

$$pP = ql_b \qquad \text{(C.V.14)}$$

the maxima of $S(\mathbf{q})$ must also be arranged in planes Ξ_k orthogonal to the helical axis and of height

$$q_{xk} = k\,\frac{2\pi}{pP} = k\,\frac{k_o}{p} \qquad \text{(C.V.15)}$$

with integer k.

This analysis of the structure factor is summarized in the diagram of figure C.V.6. Each maximum of the structure factor is identified by three indices (ijk).

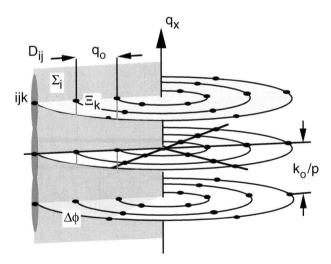

Fig. C.V.6 Structure factor of the commensurate TGB_A phase. In order to simplify the drawing, only one D_{ij} rod is shown. The diagram is strongly expanded in the x direction.

In this diagram, the vertical distances are exaggerated. In fact, the wave vector k_o/p being much smaller than q_o, the Ξ_k planes are very close, so that the maxima (ijk) with equal i and j indices fuse and form the vertical rods D_{ij}. The number of rods D_{ij} on a circle of radius $q_\perp = jq_o$ is q if q is an even number. However, this number is 2q for odd q because each family of smectic layers of azimuthal direction ϕ_i (i = 1, 2,... q) generates two diametrically opposed maxima. Similarly, the number of horizontal planes Ξ_k in the case where q is odd is twice that for q even, as the repeat unit of the TGB_A phase along the vertical axis in the first case is that of corresponding to the second case.

V.1.e)   Where do we search for TGB phases?

Besides the criterion for the TGB_A phases provided by Renn and Lubensky, the experimentalists had at their disposal some information about the regions of the phase diagrams where they could look for TGB phases. This information is mostly provided in the seminal paper by de Gennes, who assumed that the molecular chirality must play, in smectics A, a similar part to the role of the magnetic field **B** created by the solenoid surrounding a superconductor.

At this point, let us recall that, when a material is placed inside a solenoid sustaining a current j_{ext}, the thermodynamical potential to minimize is no longer the Landau-Ginzburg free energy:

$$F = \int f_{SC}\, dV \tag{C.V.16}$$

with f_{SC} given by eq. C.I.34, but rather the Gibbs free energy:

$$G = F - \int \mathbf{B}.\mathbf{H}\, dV \tag{C.V.17}$$

including the coupling term between the magnetic field $\mathbf{B} = \mathbf{curl\,A}$ and the field **H** created by the current j_{ext} going through a solenoid around the superconducting sample:

$$\mathbf{curl\,H} = \mathbf{j}_{ext} \tag{C.V.18}$$

In the case of chiral smectics, one proceeds in a similar way, adding to the Landau-Ginzburg-de Gennes free energy a term **linear** in the twist:

$$G = \int g\, dV = \int (f_L + f_G + f_F)\, dV - K_2 \int (\mathbf{n}.\mathbf{curl\,n})\, k_o\, dV \tag{C.V.19}$$

where

$$f_L = \frac{1}{2}\alpha(T)\,|\Psi|^2 + \frac{1}{2}\beta\,|\Psi|^4 \qquad \text{(Landau energy)} \qquad \text{(C.V.20)}$$

$$f_G = \frac{1}{2M_\perp}\left|\left(\nabla_\perp - iq_o\,\delta\mathbf{n}_\perp\right)\Psi\right|^2 + \frac{1}{2M_{//}}\left|\nabla_{//}\Psi\right|^2 \quad \text{(Ginzburg term)} \qquad \text{(C.V.21)}$$

$$f_F = \frac{1}{2}K_2\,(\mathbf{n}.\mathbf{curl}\,\mathbf{n})^2 \qquad \text{(Frank energy)} \qquad \text{(C.V.22)}$$

The wave vector k_o plays here the role of a field coupled to the twist $\mathbf{n}.\mathbf{curl}\,\mathbf{n}$, while the K_2 coefficient is similar to $\mu_o^{-1} = \varepsilon_o c^2$.

In the cholesteric phase, the smectic order parameter Ψ goes to zero and minimizing the potential G with respect to the twist yields:

$$\mathbf{n}.\mathbf{curl}\,\mathbf{n} = k_o \qquad \text{(C.V.23)}$$

and

$$g_{chol} = -\frac{1}{2}K_2\,k_o^2 \qquad \text{(C.V.24)}$$

This cholesteric phase is competing with the smectic phase where $\Psi \neq 0$ (constant in space), and where the Gibbs free energy G depends on the chirality k_o and on the temperature, via $\alpha(T)$.

Taking advantage of the superconducting analogy, and without going through the detailed calculations, de Gennes proposes two phase diagrams, depending on the value of the ratio $\kappa = \lambda/\xi$ between the penetration length of the twist in a direction perpendicular to the director $(\lambda = \lambda_{2\perp})$ and the smectic correlation length ξ.

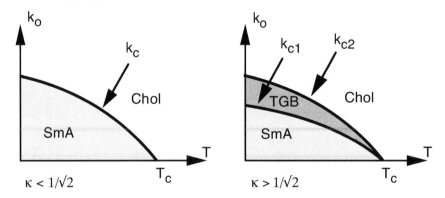

Fig. C.V.7 Phase diagrams of smectics A^* of the first and second type.

In the case of **type I smectics A**, the emergence of the smectic order parameter comes at the expense of complete twist expulsion. If $\mathbf{n}.\mathbf{curl}\,\mathbf{n} = 0$, minimizing G with respect to Ψ results in:

$$\Psi^2 = -\frac{\alpha}{\beta} \tag{C.V.25}$$

and

$$g = -\frac{1}{4}\frac{\alpha^2}{\beta} \tag{C.V.26}$$

The cholesteric-smectic A phase transition takes place when the two phases have the same energy:

$$\frac{\alpha^2}{\beta} = 2\,k_o^2\,K_2 \tag{C.V.27}$$

yielding the equation of the transition line close to T_c (Fig. C.V.7a):

$$k_o = k_c = \sqrt{\frac{\alpha^2}{2K_2\beta}} = Cst\,(T_c - T) \tag{C.V.28}$$

In the case of **type II smectics A**, the cholesteric and smectic A phases are not only competing between them, but also against the TGB phase containing screw dislocations. As shown in the diagram of figure C.V.7b, the SmA phase does not go directly into the cholesteric phase with increasing chirality k_o, but rather starts by accommodating the twist through the formation of screw dislocations. Note that the appearance of screw dislocations in the smectic phase starts when the chirality exceeds the critical value $k_{c1}(T)$. This value is necessarily smaller than k_c, and k_{c1} can be estimated by writing the energy balance in the limit of low dislocation density, when the dislocations do not interact with each other. One can show that, in the $\kappa = \lambda/\xi \gg 1$ limit, the energy of a dislocation is given by:

$$F_{dis} \approx \frac{k_c K_2 a_o}{\sqrt{2}\kappa}\ln(\kappa) \tag{C.V.29}$$

The variation $\delta g = \delta G/V$ in the average density of the Gibbs free energy of the smectic phase due to the presence of screw dislocations is then:

$$\delta g = \frac{F_{dis}}{l_b l_d} - K_2 k_o <\mathbf{n.curl\ n}> \tag{C.V.30}$$

Knowing that

$$<\mathbf{n.curl\ n}> = \frac{\Delta\theta}{l_b} \approx \frac{a_o}{l_d\,l_b} \tag{C.V.31}$$

we obtain:

$$\delta g = \frac{F_{dis} - K_2 k_o a_o}{l_d\,l_b} \tag{C.V.32}$$

Creating dislocations becomes favorable for

$$\delta g < 0 \qquad\qquad\qquad\qquad (C.V.33a)$$

meaning that

$$k_o > k_{c1} \qquad\qquad\qquad\qquad (C.V.33b)$$

with

$$k_{c1} = \frac{k_c}{\sqrt{2}\kappa} \ln(\kappa) \qquad\qquad\qquad\qquad (C.V.34)$$

In conclusion, when κ is large enough, the critical chirality k_{c1} becomes lower than k_c. Under these conditions, the TGB phase must appear before the cholesteric phase.

V.1.f)   How do we increase the penetration length λ?

Consider the situation shown in the diagram of figure C.V.8, where the smectic layers are everywhere flat and perpendicular to the z axis. We intend to determine the penetration length of a twist distortion.

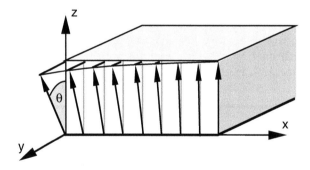

Fig. C.V.8 Definition of the twist penetration length λ.

Imagine that the director is tilted in the (y, z) plane by an angle θ_o with respect to the layer normal:

$$\theta(x = 0) = \theta_o \qquad\qquad\qquad\qquad (C.V.35)$$

To see how the director returns to its normal position, let us minimize the free energy (per unit surface in the y,z plane) $F = F_G + F_F$
    The penetration of the twist deformation inside the smectic layers brings into play the tilt

$$\mathbf{i} = \delta\mathbf{m} - \delta\mathbf{n} \qquad\qquad\qquad\qquad (C.V.36)$$

of the director $\mathbf{n}$ with respect to the layer normal $\mathbf{m}$. In de Gennes' expression C.V.19 of the free energy density, this tilt appears in the first Ginzburg term, since

$$|\text{grad}_\perp \Psi - iq_o\delta\mathbf{n}\, \Psi|^2 = q_o^2\, |\Psi|^2\, |(\delta\mathbf{m} - \delta\mathbf{n})|^2 \tag{C.V.37}$$

The director tilt $\delta\mathbf{n}$ also appears in the Frank term:

$$f_F = (1/2)\, K_2\, (\mathbf{n}.\text{curl}\, \delta\mathbf{n})^2 \tag{C.V.38}$$

Setting

$$\mathbf{n} \approx [\, 0, \theta, 1\,] \tag{C.V.39}$$

yields:

$$F = \int dx \left[\frac{q_o^2 |\psi|^2}{2M_\perp}\, \theta^2 + \frac{K_2}{2}\left(\frac{\partial\theta}{\partial x}\right)^2 \right] \tag{C.V.40}$$

This equation can be rewritten as:

$$F = \frac{q_o^2|\Psi|^2}{2M_\perp} \int dx \left[\theta^2 + \lambda^2 \left(\frac{\partial\theta}{\partial x}\right)^2 \right] \tag{C.V.41}$$

where we have set:

$$\lambda^2 = \; = \frac{K_2 M_\perp \beta}{|\alpha| q_o^2} \tag{C.V.42}$$

Minimizing this functional leads to the Euler equation

$$\lambda^2 \frac{d^2\theta}{dx^2} = \theta \tag{C.V.43}$$

of solution:

$$\theta = \theta_o\, e^{-x/\lambda} \tag{C.V.44}$$

The expression C.V.42 for the **penetration length** λ shows that **it varies as** $\mathbf{M_\perp^{1/2}}$. On the other hand, we know that, close to a second-order SmA $\to$ SmC transition, $M_\perp$ diverges as a function of temperature or composition.

Combining this conclusion with that of the previous section leads to the conclusion that one must look for the TGB phase close to the SmA $\to$ Chol transition in materials of strong twist exhibiting a SmC phase close to the SmA $\to$ Chol transition.

According to this rule, the neighborhood of the NAC multicritical (Lifshitz) point in chiral materials is an ideal area to search for the TGB phase, as the three phases – cholesteric, smectic A, and smectic C – meet there.

V.2 Discovery of the TGB$_A$ phase

V.2.a) Announcement of the discovery by the Bell group

The discovery of the TGB phase was announced in 1989 in a short paper published in *Nature*. The article, titled "Characterization of a new helical smectic liquid crystal" [4], is signed by several authors, all researchers at the Bell Laboratories, with the physicist R. Pindak and the chemist J.W. Goodby among them.

The chemical structure and the phase sequence of the products studied by the Bell group are shown in figure C.V.9.

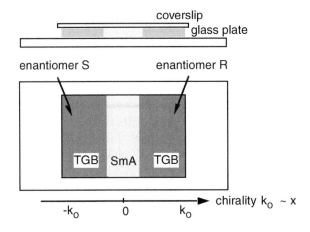

$$C_nH_{2n+1}O - \hspace{-2pt}\bigcirc\hspace{-2pt} - C \equiv CCOO - \hspace{-2pt}\bigcirc\hspace{-2pt}\bigcirc\hspace{-2pt} - COOC^*H(CH_3)C_6H_{13}$$

$$S4 \longleftrightarrow S3 \longleftrightarrow C^* \longleftrightarrow A^* \longleftrightarrow I$$

Fig. C.V.9 Chemical structure and phase sequence of the materials (nP1M7) studied by the Bell group (n = 12, 14 and 16). Sm3 and Sm4 are crystalline smectic phases, and their structures have not yet been identified.

One notices immediately that this phase sequence does not correspond to the theoretical diagram in figure C.V.7 because both the cholesteric and the ordinary smectic A phases are absent. However, the theory suggests that, starting from the TGB phase at a given temperature, one retrieves the SmA phase as the chirality decreases.

Fig. C.V.10 Contact preparation of the two enantiomers S and R induces the SmA phase between two TGB phases of opposite chirality.

To vary the chirality, Goodby et al. bring into contact the two enantiomers of the same compound (Fig. C.V.10) and let them interdiffuse [4]. They noticed then that the SmA phase indeed appears in the middle of the sample, sandwiched between two TGB phases of opposite chirality, at the point where the chirality goes through 0. They also noticed that the boundaries between the SmA phase and its neighbors are not sharp, as if the three phases were of the same nature. By increasing the temperature, they also observed that this SmA phase goes directly to the isotropic liquid, with the phase sequence of the racemic mixture being this time:

$$Sm4 \rightarrow Sm3 \rightarrow SmC \rightarrow SmA \rightarrow Iso\ Liq.$$

Note that, since the nematic phase does not appear in the racemic mixture, there are slim chances of obtaining the cholesteric phase in the chiral mixtures.

V.2.b) Grandjean textures of the TGB phase

If the identification of the SmA phase by its very distinctive focal conics texture is straightforward, the situation is completely different for the adjacent phases.

Goodby et al. indeed observed that the textures developing on either side of the SmA phase in the contact preparation are very similar to those of the cholesteric phase. More specifically, they noticed that, in planar anchoring, the textures are of the Grandjean type, with the helix axis perpendicular to the glass plates. These textures, that we will subsequently discuss in detail (see Fig. C.V.29), give rise to coloured Bragg reflections and contain disclination lines due to the change of thickness between slide and coverslip. **This observation is surprising, as the cholesteric phase does not appear in the phase sequence.**

V.2.c) X-ray diffraction spectra

In order to elucidate the puzzle, Goodby et al. studied this "strange" phase using X-rays.

The powder diagram (Fig. C.V.11), reveals the presence of a Bragg peak, of width and amplitude comparable to those of the SmC* phase. It is then obvious that the investigated phase consists of smectic layers, with a long-range lamellar order. From the wave vector of the Bragg peak, they infer that the smectic layer thickness $a_0(T)$, constant in the TGB phase, strongly decreases in the SmC* phase.

Later, Srajer et al. [5] obtained the first diffraction diagrams of aligned samples of the original Bell materials (nP1M7) (Fig. C.V.12). The structure factor $S(\mathbf{q})$ obtained from these diagrams has the characteristic cylindrical shape of the Lubensky model for the incommensurate TGB$_A$ phase, the radius of the cylinder yielding the thickness of the smectic layers $a_o \approx 42$ Å. For completeness, Srajer et al. performed optical measurements and found that the cholesteric pitch P varies from 0.38 to 0.63 μm (Fig. C.V.13).

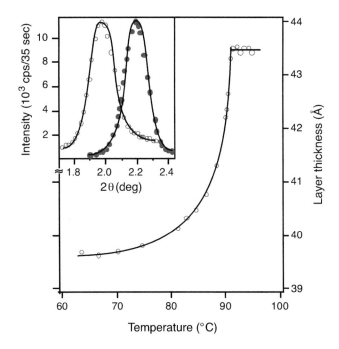

Fig. C.V.11 Results of X-ray studies (powder diagrams). Layer thickness as a function of the temperature for the R enantiomer, n = 14. Two scattering profiles are shown in the inset for this compound in the SmA* phase (open dots, T = 93.3°C) and in the SmC* phase (solid dots, T = 82.8°C) (from ref. [4]).

These results are not sufficient for determining l_b and l_d separately. To estimate the thickness of the smectic slabs, let us assume that $l_b = l_d = l$. With P = 0.5 μm and $a_o = 42$ Å, one obtains:

$$l = \sqrt{a_o P / 2\pi} = 183 \text{ Å} \tag{C.V.45}$$

as well as the twist angle per grain boundary:

$$\Delta\theta = 13° \tag{C.V.46}$$

After its discovery by the Bell group, the TGB$_A$ phase was also found by other teams, in different materials. In particular, Lavrentovich and Nastishin demonstrated that, in certain mixtures, the TGB$_A$ phase is sandwiched between the cholesteric and the smectic A phases, as initially predicted by de Gennes. This important contribution to the knowledge on TGB$_A$ phases is the more interesting for the fact that Lavrentovich and Nastishin were very close to discovering this phase six years before the Bell group.

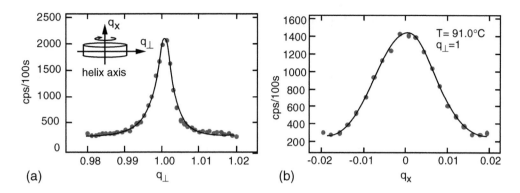

(a)   (b)

Fig. C.V.12 Results of X-ray studies on aligned samples of the TGB_A phase. a) Scattered intensity as a function of $q_\perp$. The solid curve is obtained by convolving the instrument resolution function and the theoretical structure factor of the TGB phase, which is sketched in the inset. $q_\perp$ is given in units of $2\pi/a_o = 0.146$ Å^{-1}; b) scattered intensity as a function of q_x. The solid curve is the best Gaussian fit (q_x is in units of 1.676 Å^{-1}) (from ref. [5]).

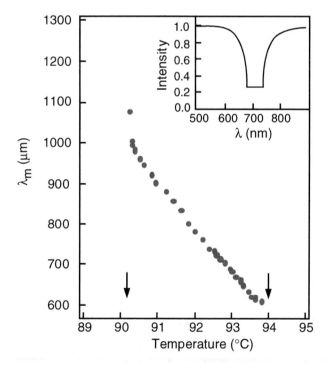

Fig. C.V.13 Since the SmA* phase of (+)-14P1M7 reflects light of right-handed circular polarization, a gap appears in the transmission spectrum, centered on the wavelength λ_m (inset graph). This wavelength varies with the temperature and gives access to the value of the cholesteric pitch. The two arrows give the transition temperatures toward the SmC* phase (at low temperature) and toward the isotropic phase (at high temperature) (from ref. [5]).

V.2.d) Shapes of TGB droplets

In 1984, Lavrentovich and Nastishin [3] noticed a strange behavior of droplets of a mixture of cholesteryl pelargonate and p-nonyloxybenzoic acid in suspension in glycerol. In the phase sequence of this mixture, they first identify the cholesteric phase above 84°C, then the smectic A phase below 82°C. In those two phases, the droplets of the mixture have the usual spherical shape, but they elongate and become cylindrical between 82°C and 84°C, taking the bizarre shape of insect larvae.

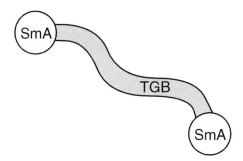

Fig. C.V.14 Typical shape of a droplet at the Chol-TGB-SmA transition.

The explanation given by these authors invokes the **pretransitional** appearance of smectic layers at the surface of the cholesteric droplets. More precisely, they estimate that the interfacial energy of the SmA/Chol interface thus created is given by:

$$\sigma_{ns} \approx (K_2 + K_3) \, q^2 \, (\xi - \lambda) \tag{C.V.47}$$

thus positing that it becomes negative for $\lambda > \xi$ (a more detailed energy balance would lead to $\kappa = \lambda/\xi > 1/2^{1/2}$). Under these conditions, the droplets assume a cylindrical shape in order to increase the area of the SmA/Chol interface, and decrease the surface energy. Though they emphasize the analogy with the mechanism leading to the formation of vortices in type II superconductors, Lavrentovich and Nastishin do not venture the statement that their elongated droplets are in fact composed of the SmA* phase.

They did take this step six years later, after submitting a similar mixture to more detailed structural investigations using X-rays and visible light. These experiments confirm that the mixture possesses the classical phases – cholesteric below 82°C and smectic A above 90°C – and, more important, a hybrid structure between these two temperatures. Since this latter phase is optically similar to the cholesteric phase (its optical axis is wound as a helix) and since the X-ray diagrams show at the same time the presence of smectic layers, Lavrentovich and Nastishin conclude that a TGB$_A$ phase is present between 82°C and 90°C. This result was confirmed by subsequent measurements of ultrasound propagation velocity and attenuation, as well as by calorimetric measurements, revealing anomalies when crossing the limits of this interval.

This study, titled "**Helical smectic A**" was published in October 1990 [6], almost two years after the announcement of the Bell group. It is, however, important to say that the phase sequence:

$$\text{SmA} \rightarrow \text{TGB}_\text{A} \rightarrow \text{Chol}$$

observed by Lavrentovich et al. corresponds to de Gennes' initial prediction.

V.2.e) TGB_A phases in other materials

The same "classical" phase sequence was found in homochiral substances synthesized by Slaney and Goodby [7] (Fig. C.V.15). Furthermore, as they had obtained the two optical antipodes for one of these products, Slaney and Goodby were able to determine the phase diagram of the binary mixture of these optical isomers (Fig. C.V.16). The structure of the TGB_A was determined by Inh et al. [8] using freeze fracture and electron microscopy.

Fig. C.V.15 New molecule synthesized by Slaney and Goodby.

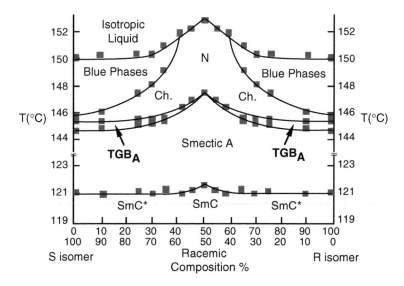

Fig. C.V.16 Phase diagram of the R-S binary mixture of the compound shown in figure C.V.15 (from ref. [7]).

The TGB$_A$ phase forms here a "wedge," inserted between the cholesteric and smectic A phases. This result is predicted by the theory, since close to the racemic mixture ($x_{rac} = x_R = x_S = 0.5$), the wave vector k_o of the spontaneous twist must go to zero, being proportional to the concentration shift with respect to its racemic value, $\Delta x = x - x_{rac}$. Consequently, in such a diagram the concentration plays the same role as the k_o parameter in the theoretical diagrams (see Fig. C.V.7) [9, 10]. Other examples are given in refs. [11–13a,b].

V.2.f) Commensurate TGB$_A$ phases

If the pioneer work we have just presented yielded the first examples of TGB$_A$ phases, and thus confirmed the theoretical predictions of Renn and Lubensky, the structure of the TGB$_A$ phase still needed refining and, in particular, the distinction between its the two variants, commensurate and incommensurate, was still to be made by precise measurements of the twist angle $\Delta\theta$. Indeed, the 13° value (eq. C.V.46) obtained by Srajer et al. [5] is only an approximation and does not establish the type of the TGB phase.

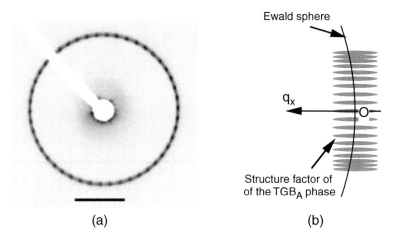

(a)   (b)

Fig. C.V.17 Experimental evidence for the commensurate TGB$_A$ phase by X-ray diffraction. a) Diffraction pattern obtained in incidence parallel to the helix axis; b) Ewald construction in the reciprocal space (from ref. [12a]).

In the face of this new challenge, conclusive progress was made by Navailles et al. [12a] who used X-rays to study a new substance, $10F_2BTFO_1M_7$; its phase diagram, to be discussed further (see Fig. C.V.20), contains the TGB$_A$ phase. The diffraction patterns of this phase, in incidence parallel to the $\mathbf{q}_x$ axis of the helix, are indeed remarkable, as a **finite** number of Bragg spots can be distinguished (Fig. C.V.17a) - a signature of the commensurability between the helix pitch P and the thickness l_b of the slabs (eq. C.V.14). In the example of figure C.V.17a, 46 Bragg spots can be counted.

We can thus conclude that if

$$q \, \Delta\theta = p \, 2\pi \qquad \text{and} \qquad p = 1 \qquad\qquad\qquad \text{(C.V.48)}$$

then $q = 46$, or $q = 46/2 = 23$. This uncertainty in the choice of the integer q stems from the fact that each smectic slab generates **two** diametrically opposed spots. In order to remove this uncertainty, Navailles et al. followed the evolution of the Bragg spots as a function of temperature. The results are summarized in the diagram of figure C.V.18. It turns out that the number of Bragg spots varies as a function of the temperature, and that by heating the sample by about 0.2°C, one goes from 46 to 44 spots. For 44 Bragg spots, the only possible choice for the number q is $q = 44$, because if q were 22, only 22 Bragg spots would appear.

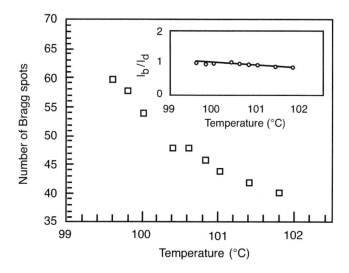

Fig. C.V.18 Variation of the number of Bragg spots as a function of temperature. Inset: variation of the l_b/l_d ratio as a function of temperature (from ref. [12a]).

In conclusion, in the case of the pattern in figure C.V.17a, one has

$$\Delta\theta = \frac{2\pi}{46} \approx 0.136 \, \text{rad} \approx 7.8° \qquad\qquad\qquad \text{(C.V.49)}$$

It should be noted that the radius of the ring in the pattern of figure C.V.17a yields the thickness of the smectic layers

$$a_o = 3.8 \, \text{nm} \qquad\qquad\qquad \text{(C.V.50)}$$

Finally, the optical measurements of the pitch of the cholesteric helix published in Isaert et al. [13a] give a value

$$P = 950 \, \text{nm} \qquad\qquad\qquad \text{(C.V.51)}$$

Knowing $\Delta\theta$ and a_o, eq. C.V.4 gives access to the distance between two dislocations in a boundary. Its value is:

$$l_d = \frac{a_o}{\Delta\theta} \approx 27.8\,\text{nm} \qquad\qquad (\text{C.V.52})$$

Furthermore, once the pitch P and the ratio $\alpha = 1/q$ are known, equation C.V.6 yields the thickness l_b of the smectic slabs:

$$l_b = \frac{P}{q} \approx 28\,\text{nm} \qquad\qquad (\text{C.V.53})$$

The distance between dislocations is thus almost identical to the distance between the boundaries. The inset of the diagram in figure C.V.18 shows that the l_b/l_d ratio remains almost equal to unity over the entire existence range of the TGB$_A$ phase.

V.3   Other TGB phases

V.3.a)   Theory of the TGB phases close to the NAC multicritical Lifshitz point

From the theoretical point of view, the vicinity of the NAC multicritical Lifshitz point in chiral materials turns out to be the ideal spot for finding TGB phases [9, 10]. Indeed, the three phases, cholesteric, smectic A, and smectic C, meet at this point, with two favorable consequences:

1. The amplitude of the smectic order parameter Ψ vanishes when the Chol $\rightarrow$ SmA transition is second order. For this reason, the critical chirality k_c also goes to zero.

2. The elastic coefficient $C_\perp = 1/M_\perp$ goes to zero when the SmA $\rightarrow$ SmC transition is second order. Thus, the penetration length diverges.

To treat the vicinity of the NAC point, where the $C_\perp$ coefficient (multiplying the quadratic term in the tilt $\mathbf{i} = \delta\mathbf{n} - \delta\mathbf{m}$) changes sign, Lubensky and Renn added to the expansion of the free energy f_G a quartic term of the following form:

$$D_\perp\,\delta_{ij}{}^T\delta_{kl}{}^T(D_iD_j\,\Psi)^*(D_kD_l\Psi) \qquad\qquad (\text{C.V.54})$$

where

$$\delta_{ij}{}^T = \delta_{ij} - n_i\,n_j \qquad\qquad (\text{C.V.55})$$

and

$$D_i = V_i - iq_o \delta n_i \qquad\qquad (C.V.56)$$

The results of this model are represented in the diagrams of figure C.V.19.

Diagram (a): $k_o = 0$ (no chirality)

When the chirality is zero (for instance in the racemic mixture), the phase diagram in the parameter plane $(C_\perp, A)$ exhibits the three phases N, SmA and SmC. These phases are separated by second-order transition lines joining at the Lifshitz point LP. The $N \rightarrow SmA$ and $SmA \rightarrow SmC$ transitions correspond to a change in sign of the quadratic terms $\alpha |\Psi|^2$ and $C_\perp |\delta n|^2$. As to the $N \rightarrow SmC$ transition, its temperature ($\sim \alpha$) increases as $C_\perp^2$, since the negative value of the $C_\perp$ coefficient favors the presence of SmC order.

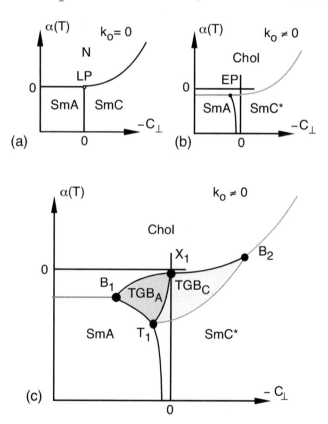

Fig. C.V.19 Theoretical phase diagrams. a) Without chirality; b) in the presence of chirality but without considering the TGB phases; c) in the presence of chirality and taking into account the TGB phases.

Diagrams (b) and (c): $k_o \neq 0$ (with chirality)

When the chirality is not zero, the NAC phase diagram undergoes several modifications. If one neglects, in a first approximation, the TGB phases, these changes are as follows (diagram b):

1. The Chol → SmA and Chol → SmC* phase transition lines become first order and get shifted toward lower temperatures. The cholesteric phase (Chol or chiral nematic N*) gains in territory, as it favors the twist of the director field.

2. The SmA → SmC* transition preserves its "second-order" character, but is shifted toward the region of positive $C_\perp$. Again, the SmC* phase gains over the SmA phase, as it accommodates the twist through molecular tilt, something the A phase cannot do.

3. The Lifshitz point becomes a critical terminal point, indicated by EP (*critical end point*).

When the TGB phases are brought into competition with the three classical phases, they win in a region close to the NAC point (diagram c). For $C_\perp > 0$, the TGB_A phase appears in the $B_1T_1X_1$ triangle. For $C_\perp < 0$, on the other hand, Lubensky and Renn imagine a new kind of TGB phase, denominated TGB_C, where the molecules are tilted with respect to the smectic layers.

At the end of the paper presenting this new phase diagram, Lubensky and Renn present experimentalists with the following challenge:

"In view of the above discussion, we hope that our experimental colleagues rise to the challenge of finding the TGB_C phase."

V.3.b) Discovery of the TGB_C phase

This challenge was taken up by the Bordeaux group, among others. Nguyen et al. [11] synthesized, and then studied, a multitude of materials exhibiting the SmC phase instead of the SmA phase, and where $C_\perp$ (negative in the SmC phase) goes to zero at the SmC → N transition. In particular, they examined some tolane derivatives (designated $nF_2BTFO_1M_7$), with long aliphatic chains (n = 10, 11 and 12). As shown in the phase diagram of the n = 10 compound in figure C.V.20, the TGB_A phase is sandwiched between the cholesteric and SmC* phases (see section C.V.2f). For n = 11, another twist grain boundary smectic phase appears at the transition between the TGB_A and SmC* phases. By a combination of optical and X-ray techniques, Nguyen et al. determine that, in this new phase, the molecules are tilted with respect to the plane of the smectic layers. The TGB_C phase had been discovered. For n = 12, the TGB_A phase disappears, and only the TGB_C phase persists.

By studying the X-ray diffraction diagrams of these novel phases, Navailles et al. revealed some very interesting structural features:

1. In incidence parallel to the helical axis q_x, the diffraction diagram exhibits 18 Bragg spots regularly spaced on a circle (Fig. C.V.21b), showing that the TGB_C phase is **commensurate**, with q = 9 (for p = 1) [12b].

2. The intensity of certain Bragg spots **increases** when the incidence of

the incoming X-ray beam (synchrotron radiation) becomes oblique with respect to the helical axis $\mathbf{q}_x$. In terms of structure factor, this means that $S(\mathbf{q})$ consists of **two systems of rods**, parallel to the $\mathbf{q}_x$ axis and arranged on two circles, as depicted in the diagrams (a) and (c) of figure C.V.22 [16].

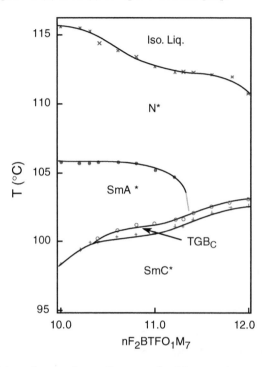

Fig. C.V.20 Superposition of two phase diagrams for binary mixtures of the $nF_2BTFO_1M_7$ materials, with n = 10 and 11 and n = 11 and 12 (from ref. [11]).

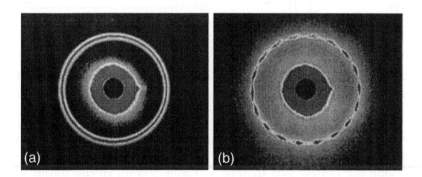

Fig. C.V.21 X-ray diffraction diagrams obtained in the plane perpendicular to the helical axis; a) for the $12F_2BTFO_1M_7$ compound in the cholesteric phase at T = 103°C. The intensity scale spans the interval between white (< 500 cps) and black (> 800 cps). The ring position is 0.168 Å^{-1}, and its width is equal to the resolution: 0.016Å^{-1}; b) same compound in the commensurate TGB$_C$ phase at T = 102.15°C. The intensity scale spans the interval between white (< 500 cps) and black (> 630 cps). The ring, at 0.168 Å^{-1}, is strongly modulated and contains 18 perfectly visible spots (from the Ph.D. thesis of L. Navailles [14]).

This last result is very important, as it allows us to discriminate among several competing models for the TGB_C phase (Fig. C.V.23).

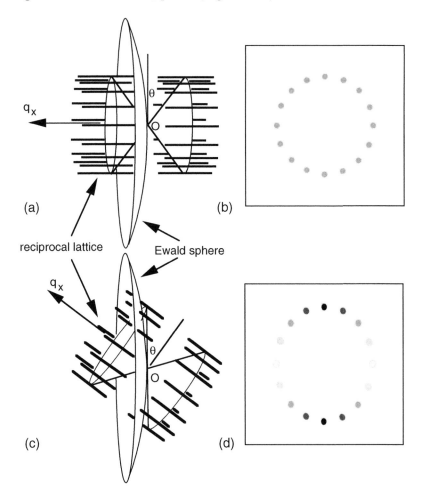

(a)

(b)

reciprocal lattice

Ewald sphere

(c)

(d)

Fig. C.V.22 Structure factor of the TGB_C phase and the corresponding X-ray diffraction diagrams. a) and b) Ewald construction in incidence parallel to the helical axis; c) and d) Ewald construction in oblique incidence (from ref. [16]).

Indeed, one can suppose that the TGB_C phase is obtained from the TGB_A phase by tilting the molecules with respect to the plane of the layers, which remain oriented along the helical axis $\mathbf{q}_x$ (Fig. C.V.23a and b). In the initial model of Renn and Lubensky (Fig. C.V.23a), the molecules are tilted in the direction orthogonal to the helix axis $\mathbf{q}_x$. For symmetry reasons already discussed in the previous chapter, all the slabs have the same spontaneous polarisation $\mathbf{P}$, parallel to $\mathbf{q}_x$. Such a structure is then ferroelectric. On the other hand, in the model presented in figure C.V.23b, the molecules are tilted in the direction of the helical axis. This second structure is obviously helielectric. These two models "a" and "b" have structure factors $S(\mathbf{q})$ identical to that of the TGB_A phase, which has a single system of rods parallel to the q_x axis and arranged on a circle of radius $q = 2\pi/a_o$.

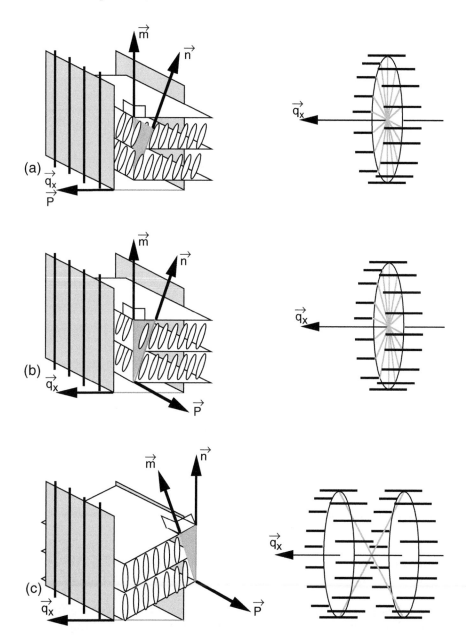

Fig. C.V.23 Three possible models for the TGB_C phase: a) ferroelectric model of Renn and Lubensky; b) helielectric model with the smectic layers parallel to the helical axis; c) helielectric model with tilted layers.

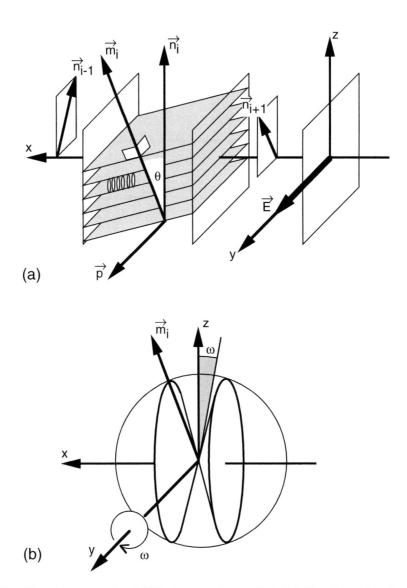

Fig. C.V.24 Experimental study of TGB phases under an electric field orthogonal to the helix. a) Setup geometry; b) definition of the sweep angle ω in X-ray diffraction.

One can also imagine that the TGB_C phase is obtained from the TGB_A phase by keeping the orientation of the **n** director (of the molecules) perpendicular to the helical axis; the smectic layers must then be tilted with respect to the director. If this tilt takes place in the direction orthogonal to the helix, one finds again the "a" model already discussed. If, on the other hand, one assumes, following Navailles et al., that the smectic layers tilt in the direction of the helix, one obtains a new model (model "c" in figure C.V.23), which is compatible with the diffraction diagrams in figure C.V.21.

Additional arguments in favor of the model with layers oblique with respect to the helix were provided by X-ray diffraction experiments on highly aligned samples. The scattering geometry is specified in the diagram of figure C.V.24b. It consists of exploring, at a fixed wavelength λ, a circle of radius $2\pi/\lambda$ in reciprocal space, by an appropriate rotation of the sample and of the detector around direction y perpendicular to the helical axis. This circle is thus contained in the (x, z) plane, orthogonal to the y axis. Two "ω sweep" diffraction spectra $I(q_x)$, measured on a $12F_2BTFO_1M_7$ sample, are shown in figure C.V.25.

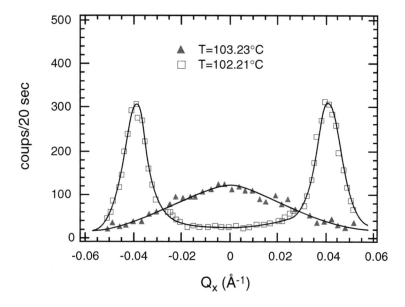

Fig. C.V.25 "ω sweep" diffraction spectra obtained under zero electric field on a $12F_2BTFO_1M_7$ sample (from ref. [15]).

The one obtained at 103.23°C is typical of the cholesteric phase (to be compared with that in Fig. C.V.12b). The other, measured at 102.21°C, contains two symmetrical maxima corresponding to the intersection with the two cones of the structure factor in model "c" for the TGB_C phase.

V.3.c) Behavior of the TGB phases in an electric field

Having understood the arrangement of the smectic layers (or of the normal **m** to the layers) and of the director **n** with respect to the helical axis of the TGB_C phase, Petit et al. [17] then considered the action of an electric field on this helielectric phase, applied along the y direction perpendicular to the helix (see Fig.C.V.24a). To detect the effect of the field, these authors recorded X-ray diffraction spectra in the "ω sweep" configuration (defined in Fig. C.V.24b). The results obtained on an $11F_2BTFO_1M_7$ sample are shown in figure C.V.26. In the absence of the field, the spectrum, which exhibits two maxima symmetrically placed on either side of $\omega = 0$, is similar to the one of $12F_2BTFO_1M_7$ in the TGB_C phase (Fig. C.V.25). On the other hand, the amplitudes of these maxima appear to be weak since, close to the $TGB_C \rightarrow TGB_A$ transition, the tilt angle of the layers with respect to the helical axis is low, such that the maxima are close together and superpose to a large extent.

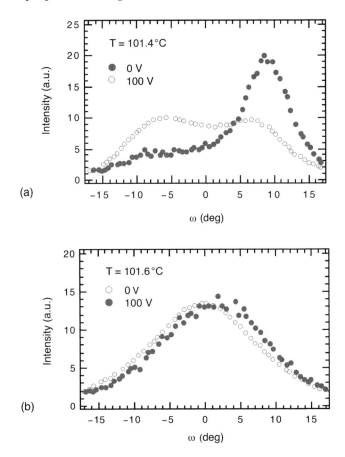

Fig. C.V.26 Effect of an electric field on the TGB_A and TGB_C phases of an oriented $11F_2BTFO_1M_7$ sample. a) "ω sweep" diffraction spectra obtained for the TGB_C phase; b) similar spectra for the TGB_A phase (from ref. [17]).

The action of the electric field is spectacular: the spectrum becomes very asymmetric, to the advantage of the right-hand-side maximum. This behavior validates the structural model for the TGB_C phase. Indeed, the two maxima in the spectrum are given by smectic slabs with opposite polarizations, as shown in the diagram of figure C.V.27a. To decrease the electrostatic energy

$$f_{él} = -\mathbf{P}.\mathbf{E} \tag{C.V.57}$$

the width l_b of the slabs having the polarization $\mathbf{P}$ oriented "the right way" (parallel to **E**) increases, while the width of the slabs where the polarization P points "the wrong way" (antiparallel to **E**) decreases (Fig. C.V.27b).

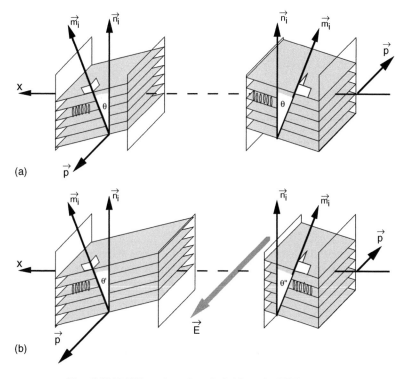

Fig. C.V.27 Effect of an electric field on the TGB$_C$ phase.

A second, more subtle, feature adds to this dominant effect: the right-hand-side maximum shifts toward $\omega > 0$ values, signalling an increase in the tilt angle of the layers with respect to the helical axis. This is an **electroclinic** effect, similar to the one encountered in the SmC* phase (see section C.IV.7.c). The diagram in figure C.V.26b shows that the electroclinic effect is also present in the TGB$_A$ phase.

V.4   Elasticity of the TGB phases

A TGB phase is characterized by four repeat distances:

1. The smectic layer thickness a_o.
2. The distance l_d between dislocations in a twist boundary.
3. The thickness l_b of the smectic slabs.

4. The cholesteric pitch $P = \dfrac{2\pi}{k}$.

These length scales are related as:

$$k = \frac{2\pi}{P} = <\mathbf{n}.\mathbf{curl\ n}> = \frac{\Delta\theta}{l_b} \approx \frac{a_o}{l_d l_b} \qquad\text{(C.V.58)}$$

In an infinite medium, they have values $(a_o, l_{do}, l_{bo}, P_o)$ minimizing the free energy of the system:

$$g_\infty = g(a_o, l_d, l_b, P_o) \qquad\text{(C.V.59)}$$

Any deviation with respect to these equilibrium values leads to an increase δg in the free energy.

V.4.a) TGB phases confined in a Cano wedge

Such deviations occur in a Cano wedge (Fig. C.V.28) where the liquid crystal is confined between two planar surfaces making a small angle and treated to ensure strong planar anchoring along the x direction of the dihedral edge [13a,b, 14].

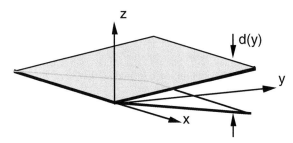

Fig. C.V.28 Cano wedge.

Due to the strong anchoring conditions, the angle θ by which the director turns between the two plates must be a multiple of π:

$$\theta = N\pi \qquad\text{(C.V.60)}$$

At the same point in the wedge, M smectic slabs separated by M−1 twist boundaries are contained between the two surfaces. The angle by which the director turns across each of these walls is:

$$\Delta\theta = \frac{N\pi}{M-1} \qquad\text{(C.V.61)}$$

The conclusion is that the confined TGB phase can choose between different states, indexed by the integers N and M.

In the case of an incommensurate TGB phase, none of the accessible states (N,M) corresponds to the free energy minimum g_{min}, by definition. However, certain (N,M) states minimize the energy difference:

$$\delta g_{N,M} = g_{N,M} - g_\infty \qquad\qquad (C.V.62)$$

V.4.b) An approximation of the deformation energy

To calculate the energy difference $\delta g_{N,M}$, one must first of all write down δg as a function of three of the four variables a, l_d, l_b and k (or P). How do we choose the independent variables? The problem can be simplified by assuming that the thickness a of the smectic layers is constant and equal to a_o, thus reducing the problem to the choice of two variables among the remaining three, l_d, l_b and k. In a TGB phase, the most important variable is the average twist k = $2\pi/P$, which is actually the reason for the existence of the TGB phase. We shall consequently take as the first variable of $\delta g_{N,M}$ the difference:

$$\delta k = k - k_o \qquad\qquad (C.V.63)$$

It is noteworthy that the same twist k can be obtained by changing the arrangement of the dislocations without changing their global density, since:

$$(l_d l_b)^{-1} = \frac{k}{a_o} \qquad\qquad (C.V.64)$$

For a given twist k, only one configuration of the dislocations among all those allowed minimizes the interaction energy of the dislocations. We assume that this state is described by the same thickness l_{bo} of the smectic slabs as in the equilibrium configuration k = k_o. We then choose:

$$\delta\beta = l_b^{-1} - l_{bo}^{-1} \qquad\qquad (C.V.65)$$

as the second relevant variable.

The Taylor series expansion of g (k,β) around (k_o,β_o) yields:

$$\delta g_{N,M} = A\,\delta k^2 + B\,\delta\beta^2 + C\,\delta k\delta\beta \qquad\qquad (C.V.66)$$

Note that the C coefficient must be zero, since we assumed that, for all δk, the energy has its minimum for δβ = 0.

V.4.c) Quantitative analysis of the texture of a TGB phase in a Cano wedge

At a given thickness H, the deviations δk and δβ depend on M and N. To calculate δβ, suppose that the M smectic slabs, contained between the two surfaces, all have the same thickness l_b. In these conditions:

$$l_{bM} = \frac{d}{M} \tag{C.V.67}$$

whence

$$\delta\beta = \frac{M}{d} - l_{bo}{}^{-1} \tag{C.V.68}$$

Obtaining δk is trivial:

$$\delta k = \frac{N\pi}{d} - k_o \tag{C.V.69}$$

Note that these variables depend on M and N separately. Thus, in the C = 0 approximation, M and N can be independently chosen.

In the Cano wedge, the transition from one (N,M) state to the neighboring one (N + 1,M) occurs when the energy densities $\delta g_{N,M}$ and $\delta g_{N+1,M}$ are identical. This leads to the relation:

$$\left[\frac{(N+1)\pi}{d} - k_o\right]^2 = \left[k_o - \frac{N\pi}{d}\right]^2 \tag{C.V.70}$$

yielding the transition thickness $d_{N,N+1}$:

$$d_{N,N+1} = \frac{(N + 1/2)\pi}{k_o} \tag{C.V.71}$$

In the same way, the transition from the (N,M) state to the (N,M+1) state occurs for

$$d_{M,M+1} = (M + 1/2)\, l_{bo} \tag{C.V.72}$$

In conclusion, in a TGB phase where the k and β variables are decoupled, the two transition sequences are independent and do not superpose.

V.4.d) An alternative expression for the deformation energy

Another possible choice for the second relevant variable of the deformation energy could be:

$$\delta\gamma = \frac{\pi}{a_o} \frac{l_d}{l_b} - \gamma_o \tag{C.V.73}$$

where

$$\gamma_o = \frac{\pi}{a_o} \frac{l_{do}}{l_{bo}} \tag{C.V.74}$$

Taking the deformation energy of the form:

$$\delta g = A\delta k^2 + B\delta\gamma^2 \qquad\qquad (C.V.75)$$

without the cross-term, the affine configurations (l_d/l_b = Cst) of the dislocations are favored.

Let us now show that, in a Cano wedge, this new expression of the deformation induces a coupling between the transitions involving N and M.

With l_b = d/M and $d/N\pi = (l_d l_b)/a_o$, one gets:

$$\frac{\pi}{a_o}\frac{l_d}{l_b} = \frac{\pi}{a_o}\frac{da_o/N\pi}{(d/M)^2} = \frac{M^2}{Nd} \qquad\qquad (C.V.76)$$

and the expression of the free energy δg becomes:

$$\delta g_{N,M} = A\left[(N\pi/d) - k_o\right]^2 + B\left[(M^2/Nd) - \gamma_o\right]^2 \qquad\qquad (C.V.77)$$

The second term of this expression depends both on M and N, leading to a coupling between the two transitions sequences.

Thus, the transition between the (N,M) and (N,M+1) states occurs for:

$$\gamma_o - \frac{M^2}{Nd} = \frac{(M+1)^2}{Nd} - \gamma_o \qquad\qquad (C.V.78)$$

or, equivalently:

$$\frac{(M+1)^2 + M^2}{Nd} = 2\gamma_o \qquad\qquad (C.V.79)$$

and

$$\gamma_o\, d_{M,M+1} = \frac{M^2 + M + 1/2}{N} \qquad\qquad (C.V.80)$$

So far, the systematic observations of the two transition types are not sufficiently advanced to convincingly discriminate between the two models. However, it will be shown that, in the TGB$_C$ case, a coupling seems to exist between the two transition types.

V.4.e) Texture of the TGB$_C$ phase confined in a Cano wedge

Recently, Isaert et al. [13a] and Navailles [14] studied the two transition sequences in a TGB$_C$ phase, by polarized microscopy observations. In the photograph of figure C.V.29a), one notices first of all a system of strongly contrasted lines, very similar to those observed in the cholesteric phase. These lines mark the changes in the number N of cholesteric half-pitches contained in the wedge. In between these widely spaced lines, one can discern (with some difficulty) another system of lines, of much weaker contrast. These lines correspond to changes in the number M of smectic slabs. One can observe that the distance between the weakly contrasted M-type lines vary from one N stripe to the next. This seems to point to the existence of a coupling between the transitions M and N-type transitions. Stoenescu et al. very recently complemented these observations by using samples with a sphere-plane geometry and with "sliding" planar anchoring on one of the surfaces [13b]. As shown in the photographs of figure C.V.29b, taken while approaching the sphere to the plane, the Grandjean-Cano lines have disappeared, unlike the dislocation lines associated to the stacking of the smectic slabs, which remain perfectly visible. It is noteworthy that these lines are not circular, but rather anisotropic in shape, when they nucleate during the approach of the sphere to the plane. These discrete orientation changes clearly show that the smectic layers turn discontinuously from one slab to the next.

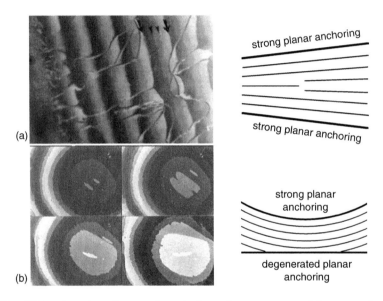

Fig. C.V.29 TGB$_C$ phase in a Cano wedge: a) plane-plane geometry with strong planar anchoring on the two limiting surfaces, $11F_2BTFO_1M_7$ at 102.2°C. The two larger arrows indicate the Grandjean-Cano lines; the smaller arrows point to the dislocations related to the stacking of the smectic slabs (from ref. [13a]); b) sphere-plane geometry. This time, the planar anchoring is strong on the sphere (the orientation is fixed) and degenerate planar, or "sliding" on the plane, $12F_2BTFO_1M_7$. Sequence of photos taken for decreasing distances between the sphere and the plane. The Grandjean-Cano lines have disappeared, unlike the dislocation lines due to the stacking of the smectic slabs (courtesy Y. Dozov, 2001).

V.5   Smectic Blue Phases

We have just seen that, in the TGB$_A$ and TGB$_C$ phases, the **large-scale** cholesteric structure of the director field **n(r)** – **twisted along a unique direction q$_x$** – coexists with a **local smectic order** of the A or C type, owing to the presence of a system of screw dislocation walls. By extension, it is tempting to wonder whether the **three-dimensional twist**, typical of the Blue Phases, could also coexist with the smectic order, provided the dislocations are arranged in a different configuration.

From the theoretical point of view, the smectic Blue Phases are very attractive, as they solve a problem involving a double frustration:

– the first one results from the topological impossibility of constructing a director field $\mathbf{n(r)}$ such that the local tendency of the chiral molecules to form twisted configurations in more than one direction be satisfied at each point $\mathbf{r}$ (see chap. B.VIII of volume 1);

– the second frustration is a result of the topological incompatibility between the smectic order and a twisted director field (see chapter C.I).

V.5.a) Experimental evidence

After the discovery of TGB phases, the possibility of encountering smectic Blue Phases appeared on the horizon when Slaney and Goodby (Fig. C.V.16) found materials containing both Blue Phases and the TGB_A phase, in the sequence: $SmC^* \rightarrow TGB_A \rightarrow Chol \rightarrow$ Blue Phases $\rightarrow$ Iso. Liq.

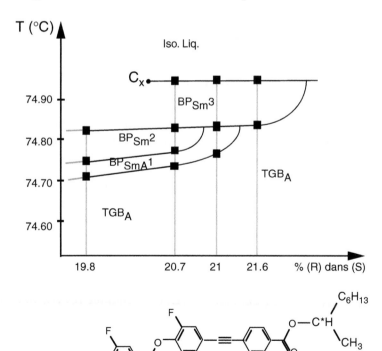

Fig. C.V.30 Phase diagram of 16BTMHC, according to E. Grelet [19].

This hope became even stronger as Li et al. [18a,b], Pansu et al. [18c,d], Young et al [18e], and Grelet [19] demonstrated, by calorimetry, X-ray diffraction, and optical microscopy, that in compounds from the nBTMHC series, three Blue Phases appear **directly** between the isotropic liquid and the TGB_A phase (Fig. C.V.30).

Since, this time, the TGB_A phase crosses directly into the Blue Phases, one could hope that the smectic order typical of the TGB_A phase persists in the Blue Phases. To verify this hypothesis, Li et al. [18b] and Pansu et al [18c,d] studied the Blue Phases in 18BTMHC using X-rays. A typical example of a diffraction diagram in one of these phases is shown in figure C.V.31.

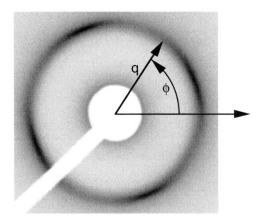

Fig. C.V.31 X-ray diffraction diagram of a monocrystal of the $BP_{Sm}2$ phase (courtesy B. Pansu, 1998). The geometry of the setup is shown in figure C.V.32.

In this diagram, the clearly visible ring has a radius q almost identical to the one recorded in the TGB_A phase, showing the presence of a smectic order with repeat distance $a_o = 2\pi/q$. On the other hand, the I(q) and I(ϕ) profiles of the ring are different. The I(q) profile is Gaussian: its width Δq (measured at half-maximum), characterizing the range of smectic correlations, strongly depends on the temperature. Pansu et al. measured it and found that, in Blue Phases, the correlation length of the smectic order does not exceed 1000 Å. As to the angular intensity distribution I(ϕ) it exhibits, **four maxima** at 90° angles (Fig. C.V.31), suggesting a possible fourth-order symmetry axis 4.

These results demonstrate the presence of smectic Blue Phases. They were denoted by $BP_{Sm}1$, $BP_{Sm}2$ and $BP_{Sm}3$, to distinguish them from the "classical" Blue Phases (BPI, BPII and BPIII) (see, chapter B.VIII, volume 1).

V.5.b) Properties of smectic Blue Phases

To determine the structure factor S($\mathbf{q}$) of one of these phases, Grelet [19] and Pansu et al. [20a–c] performed X-ray diffraction experiments on monocrystals of the BP$_{Sm}$2 phase. These are obtained from the amorphous phase BP$_{Sm}$3, by extremely slow growth in a capillary tube kept in an axial temperature gradient of 0.1°C/mm. The diagram in figure C.V.32 shows that the sequence of diffraction patterns obtained by rotating the sample around the axis of the tube allow reconstruction of the structure factor S($\mathbf{q}$).

In the case of the BP$_{Sm}$2 phase, eight maxima were detected, arranged as shown in figure C.V.33.

Due to the inversion symmetry of the structure factor S($\mathbf{q}$), these eight maxima form, for obvious reasons, four pairs AA', BB', CC' and DD'. The angular distance between A and B is about 90° (see the pattern in figure C.V.31). Between B and C, as well as between C and D, the angular distance is about 60°. In conclusion, X-ray studies suggest that the BP$_{Sm}$2 phase has hexagonal symmetry [20a,b]. This hypothesis is strengthened by polarizing microscopy observations, conclusively showing that the BP$_{Sm}$2 phase is birefringent.

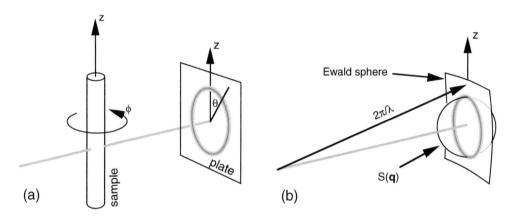

Fig. C.V.32 Diffraction geometry allowing reconstruction of the structure factor S($\mathbf{q}$). a) In real space; b) in reciprocal space.

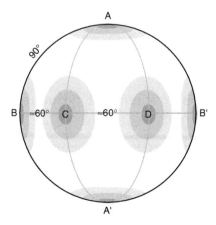

Fig. C.V.33 Structure factor S(q) of the BP$_{Sm}$2 phase determined by Pansu [18b] and Grelet [19].

However, recent observations by Grelet [19] on the shape of facetted crystallites of the $BP_{Sm}2$ phase, call these conclusions into question. Indeed, as shown in photo a) of figure C.V.34, the facets form a dodecahedron fairly similar to that of the classical Blue Phase II, where 12 facets, all identical, belonging to the {110} family, form a regular dodecahedron. The resemblance is limited to the relative orientation of the facets, since they have different sizes and shapes. Indeed, one can see in the photo of figure C.V.34 that facets 1 and 2 are hexagonal, while the others are rhombic. This is a result of the fact that, at a fixed orientation, the size and shape of the facets depend on their distance to the center of mass of the crystal. To illustrate this, we drew a crystal habit, derived from the regular dodecahedron, on which two opposing facets (1 and 1', only facet 1 being visible) were moved by 25% toward the center of the crystal (Fig. C.V.34b). This habit is completely similar to the one observed in experiments. Since this transformation of the dodecahedron preserves the Bravais lattice, we are faced with a paradox, since the X-ray experiments and the optical birefringence suggest a hexagonal symmetry, while the shape of the crystallites would rather indicate cubic symmetry.

This paradox has not yet been solved.

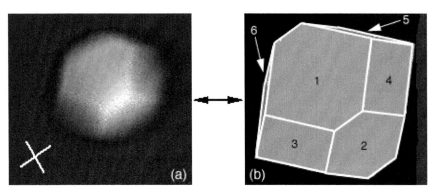

Fig. C.V.34 Facetting of the $BP_{Sm}2$ phase. a) Transmission view of a monocrystal between crossed polarizers (parallel to the edges of the image). The white cross indicates the polarizer orientation for which extinction is obtained. b) Crystal habit obtained from a regular dodecahedron by moving the 1 and 1' facets 25% closer to the center of the crystal (1', situated on the opposite side, is not visible) (from ref. [19]).

V.5.c)   Smectic order in double-twist cylinders

A plausible structure for the smectic Blue Phases can be obtained on the basis of the uniaxial model for the classical Blue Phases, presented in chapter B.VIII (volume 1). In this model, of cubic symmetry O_2, the director field twisted in three dimensions is obtained by meshing together three sets of double-twist cylinders, parallel to the x, y, and z axes. We recall that, in a double-twist cylinder, the tilt angle of the director with respect to the main axis is linear in the radius r:

$$\theta(r) = q_o r \qquad \text{(C.V.81)}$$

To create a smectic Blue Phase, one must additionally endow these cylinders with smectic order, as Kamien did [21]. We emphasize that this task is much more difficult than creating the TGB_A phase by introducing smectic slabs in a cholesteric structure twisted along only one direction. Indeed, in the TGB_A phase all the slabs, identical in structure and of equal thickness l_b, simply turn following the helical axis. In a double-twist cylinder, on the other hand, the slabs form concentric tubes of average curvature radii r_{oi}, all different, due to the revolution symmetry, and their thicknesses l_{bi} can be variable.

Let us now take a closer look at the structure of one such slab (of index i). One can see it as being created from an ordinary "flat" slab of thickness l_{bi} and width $2\pi r_{oi}$, by first bending, and then "gluing" it, as schematically shown in figure C.V.36. To allow gluing together of the two sides of the slab, the tilt of the smectic layers in the initial flat slab must be such that the level change Δz_i along one layer, between the two edges to be glued, is an integer multiple of the layer separation, measured along the vertical z. One must then have:

$$\Delta z_i = 2\pi r \, \mathrm{tg} \, \theta_i = n_i \frac{a_o}{\cos \theta_i} \qquad (n_i \text{ integer}) \qquad \text{(C.V.82)}$$

Clearly, the level change Δz_i must also be preserved when the smectic slab is bent and then glued. Consequently, the quantity Δz_i is constant in the final tube, where relation C.V.82 must also hold, irrespective of the radius r.

This construction has several consequences. First of all, the tilt angle $\theta_i(r)$ of the layers in the tube decreases with r, since

$$\tan \theta_i(r) = \frac{\Delta z_i}{2\pi r} \qquad \text{(C.V.83)}$$

On the other hand, the cylindrical tube is made up of n_i continuous and identical layers, shaped as helicoids and fitted together. Finally, the thickness of the layers, measured along their normal, depends on r according to the following formula:

$$a_i(r) = \frac{\Delta z_i \cos \theta_i(r)}{n_i} \qquad \text{(C.V.84)}$$

The level change Δz_i being constant, **the layer thickness increases with r**, as the angle θ_i decreases with r, according to equation C.V.83.

A smectic tube can also be cut out of a bulk smectic sample containing one screw dislocation of Burgers vector $\Delta z_i = n_i a_i(\infty)$, where $a_i(\infty)$ is the layer thickness in the $r \to \infty$ limit. In this case, the previous equation can be rewritten as:

$$a_i(r) = \frac{\Delta z_i \cos\theta_i(r)}{n_i} = a_i(\infty) \cos\theta_i(r) \tag{C.V.85}$$

Finally, the structure of each smectic tube is entirely specified by two parameters:

1. The number n_i of continuous smectic layers it contains.
2. The layer thickness $a_i(r)$, which need only be given for one value of the radius r.

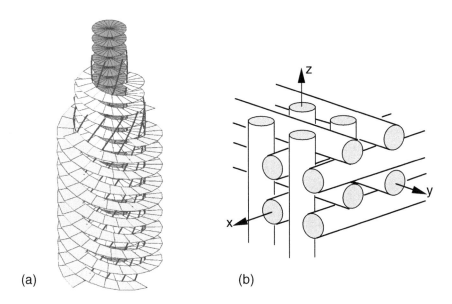

(a)   (b)

Fig. C.V.35 Kamien model of smectic Blue Phase. a) Structure of a double-twist cylinder (from ref. [21]). b) Model built by meshing together double-twist cylinders.

The next step is an appropriate choice of these two parameters so that:

1. The tilt angle $\theta_i(r)$ of the layers in a tube takes values as close as possible to $k_o r$ (in the interval $r_{min} < r < r_{max}$), in order to satisfy as far as possible the natural tendency of the system to become twisted (condition A).

2. The layer thickness $a_i(r)$ remains as close as possible to its equilibrium value a_o (condition B).

Let us now see how these conditions can be fulfilled. Let r_{oi} be the inner radius of tube i. Suppose that, for this radius:

$$a_i(r_{oi}) = \frac{\Delta z_i \cos\theta_i(r_{oi})}{n_i} = a_o \tag{C.V.86}$$

and that the tilt angle of the molecules with respect to the z axis is that for the double-twist cylinder:

$$\theta_i(r_{oi}) = k_o r_{oi} \tag{C.V.87}$$

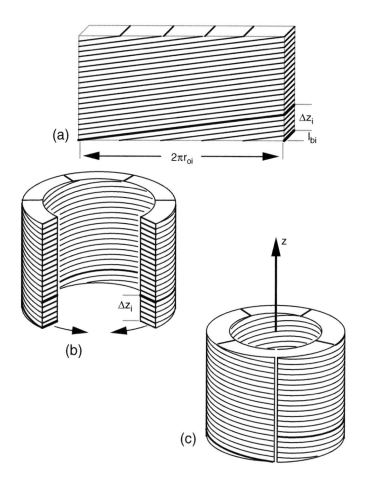

Fig. C.V.36 Formation of a smectic tube by bending and gluing a smectic slab of the TGB_A type. To allow gluing together the two sides of the slab, the level change Δz_i must be an integer multiple of the vertical distance between the layers. It is preserved throughout this process, from (a) until (c).

Equation C.V.83 then imposes the value of the level change Δz_i as:

$$\Delta z_i = 2\pi r_{oi}\, tg\,(k_o r_{oi}) \tag{C.V.88}$$

Finally,

$$a_i(r_{oi}) = \frac{2\pi r_{oi}\, tg\,(k_o r_{oi})\, \cos\,(k_o r_{oi})}{n_i} = a_o \tag{C.V.89}$$

This equation dictates the choice of the number n_i as a function of the radius r_{oi}:

$$n_i = \frac{2\pi r_{oi}\, tg\,(k_o r_{oi})\, \cos\,(k_o r_{oi})}{a_o} \tag{C.V.90}$$

The graph in figure C.V.37 shows the shape of this function, calculated for a double-twist cylinder of radius 100 times larger than the equilibrium layer thickness.

One more thing to consider is the structure of the boundaries between two adjacent tubes, i and i+1. To this end, we turn our attention to the schematic in figure C.V.38, representing the layers in two adjacent tubes, projected onto the plane of the page. The radius R of the boundary is limited by

$$r_{oi} < R < r_{oi+1} \tag{C.V.91}$$

The moiré pattern appearing in this figure indicates the position and orientation of the dislocations. The main purpose of a boundary is to allow a change in the number of helicoids fitted together. In the case of figure C.V.38, $n_i = 5$ and $n_{i+1} = 8$. For this reason, $\Delta n = n_{i+1} - n_i = 3$ dislocations are encountered along the perimeter of the wall.

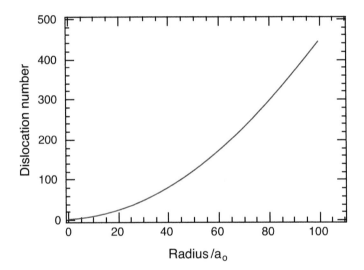

Fig. C.V.37 Variation in the number n_i of helical screw dislocations as a function of the radius. This number is determined from eq. C.V.90 using $k_o = \pi / 400a_o$, such that the tilt angle of the layers at the surface of the double-twist cylinder is equal to 45°.

Having determined the number of dislocations in the boundary, we must now determine their orientation. This orientation depends on the layer thickness at infinity (or measured along the z direction, which amounts to the same):

$$a_i(\infty) = \frac{a_o}{\cos \theta_i(r_{oi})} \tag{C.V.92}$$

Since this thickness is a constant in a smectic tube, we find the same value at the boundary, where

$$a_i(\infty) = \frac{a_i(R)}{\cos\theta_i(R)} \qquad\qquad (C.V.93)$$

The same relations exist in the neighboring tube:

$$a_{i+1}(\infty) = \frac{a_o}{\cos\theta_{i+1}(r_{i+1})} = \frac{a_{i+1}(R)}{\cos\theta_{i+1}(R)} \qquad\qquad (C.V.94)$$

If $a_{i+1}(\infty)$ were equal to $a_i(\infty)$, the dislocations would be purely of the "screw" type and, as a consequence, they would be perpendicular to the layers at infinity (i.e., vertical), as in diagram (a) of figure C.V.38. In our model, this condition is not satisfied, as

$$a_{i+1}(\infty) = \frac{a_o}{\cos\theta_{oi+1}(r_{oi+1})} > \frac{a_o}{\cos\theta_i(r_{oi})} = a_i(\infty) \qquad\qquad (C.V.95)$$

because

$$\theta_{i+1}(r_{oi+1}) = q_o r_{oi+1} > q_o r_{oi} = \theta_i(r_{oi}) \qquad\qquad (C.V.96)$$

According to the inequality C.V.95, the dislocations in the boundary must also allow a change in the layer thickness at infinity. They must therefore acquire a wedge component by tilting with respect to the vertical direction. They wrap around the circular wall, assuming a helical shape. Suppose that, over the vertical distance between two dislocations, m layers of tubes i are crossed, and m – 1 (thicker) layers of tubes i+1 (m is an integer in the simplest commensurability case). Consequently,

$$m\, a_i(\infty) = (m - 1)\, a_{i+1}(\infty) \qquad\qquad (C.V.97)$$

leading to:

$$m/\cos(q_o r_{oi}) = (m - 1)/\cos(q_o r_{oi+1}) \qquad\qquad (C.V.98)$$

or, equivalently

$$1 - \cos(q_o r_{oi+1})/\cos(q_o r_{oi}) = 1/m \qquad\qquad (C.V.99)$$

This equation defines the number m, provided the r_i and r_{i+1} radii are known, and then, knowing $a_i(\infty)$, one can obtain the angle ψ between the dislocations and the z axis:

$$tg\,\psi = \frac{2\pi R}{\Delta n\, m\, a_i(\infty)} \qquad\qquad (C.V.100)$$

Within this model, the orientation of the dislocations is a result of the somewhat arbitrary choice of satisfying at the same time both conditions A and B previously imposed on the orientation and thickness of the smectic layers. A different point of view was adopted by Kamien [21], who assumes that the dislocations are symmetrically arranged with respect to the layers in adjacent tubes (along the bisector of the angle formed by the layers). This hypothesis implies that the thickness of the layers in two adjacent cylinders is identical at the boundary:

$$a_i(R) = a_{i+1}(R) \qquad\qquad\qquad (C.V.101)$$

Or, from eq. C.V.85, one knows that in each tube the layer thickness increases with the radius r. It follows that the condition

$$a_i(r) = a_o \qquad\qquad\qquad (C.V.102)$$

can only be fulfilled for one value of r, in only one of the tubes of the double-twist cylinder, which does not strike us as the ideal solution.

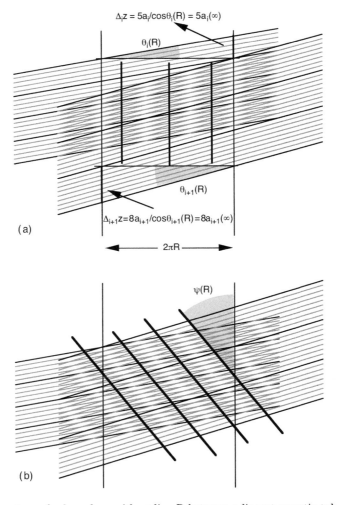

Fig. C.V.38 Structure of a boundary with radius R between adjacent smectic tubes when a) the vertical thickness of the smectic layers is identical in the tubes, and b) the vertical thickness of the layers in tube i+1 is larger than in tube i. In the latter case, the dislocations are tilted with respect to the z axis of the double-twist cylinder.

V.5.d Structure factor of a double-twist cylinder.

The model of smectic Blue Phases obtained by meshing together double-twist cylinders (Fig. C.V.35b) gives an idea of the general features of their X-ray diffraction diagrams. First of all, one must take into account the fact that smectic Blue Phases bring into play two very different length scales: the lattice size, of the order of thousands of Å, and the thickness of the smectic layers, of the order of 30 Å. This scale difference has a notable influence on the structure factor $S(\mathbf{q})$.

Clearly, the crystalline order of Blue Phases, which is periodical, three-dimensional and long-range, is difficult (if not impossible) to resolve using X-rays, due to the lattice size and to the weak electron density modulation.

On the other hand, X-ray investigations will be very sensitive to the appearance of the local smectic order, the structure factor $S(\mathbf{q})$ increasing strongly for all wave vectors $\mathbf{q}$ of absolute value close to $2\pi/a_o$ and orientation parallel to the layer normal $\mathbf{m}$:

$$\mathbf{q} \approx \frac{2\pi}{a_o}\mathbf{m} \qquad\qquad (C.V.103)$$

More precisely, the structure factor $S_{dt}(q;\theta,\phi)$ of a double-twist cylinder depends, at a given $q = |\mathbf{q}|$, of the angles (θ,ϕ). This dependence reflects the probability $p(\theta,\phi)\,d\Omega$ of finding the layers oriented perpendicular to the direction (θ,ϕ) within a small solid angle $d\Omega = \sin\theta\,d\theta\,d\phi$.

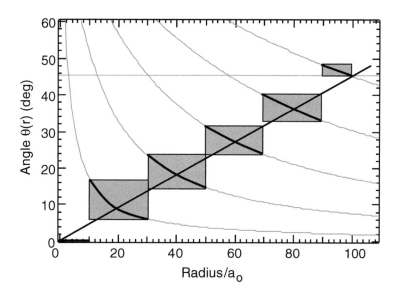

Fig. C.V.39 Variation of the layer tilt with respect to the axis of the double-twist cylinder. The smectic tubes (corresponding to the gray rectangles) were arbitrarily chosen.

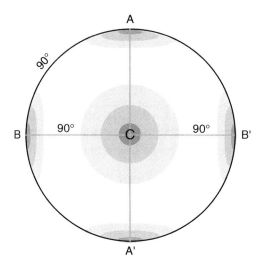

Fig. C.V.40 Structure factor for the case of the Kamien model with O_2 symmetry. Note that this function does not necessarily have maxima along the x, y and z axes, as suggested by the diagram.

In a double-twist cylinder of axis z, the $p(\theta,\phi)$ distribution exhibits azimuthal symmetry about this axis, so that $p(\theta,\phi) = p(\theta)$. As to its dependence on the angle θ, it is determined by the function $\theta(r)$, which exhibits jumps when going from one tube to the next but varies continuously inside each tube. Consequently, as shown in the diagram of figure C.V.39, the angle θ takes almost all values between $0°$ and $45°$. The structure factor $S(\mathbf{q})$ of a Blue Phase built as a mesh of double twist cylinders is then obtained from $S_{dt}(q;\theta,\phi)$ by reproducing this function under the point symmetry operations of the phase. In the case of a model with O_2 cubic symmetry, the structure factor $S(\mathbf{q})$ must then exhibit six enhancements (which are not necessarily maxima) centered on the x, y and z axes (Fig. C.V.40).

BIBLIOGRAPHY

[1]   de Gennes P.-G., *Sol. State Comm.*, **10** (1972) 753.

[2]   Renn S.R., Lubensky T.C., *Phys. Rev. A*, **38** (1988) 2132.

[3]   Lavrentovich O.D., Nastishin Yu.A., Pis'ma Zh., *Eksp. Teor. Fiz.*, **40** (1984) 1015.

[4]   Goodby J.W., Waugh M.A., Stein S.M., Chin E., Pindak R., Patel J.S., *Nature*, **337** (1989) 449.

[5]   Srajer G., Pindak R., Waugh M.A., Goodby J.W., Patel J.S., *Phys. Rev. Lett.*, **64** (1990) 1545.

[6]   Lavrentovich O.D., Nastishin Yu.A., Kulishov V.I., Narkevich Yu.S., Tolochko A.S. , Shiyanovskii S.V., *Europhys. Lett.*, **13** (1990) 313.

[7]   Slaney A.J., Goodby J.W., *Liq. Cryst.*, **9** (1991) 849.

[8]   Ihn K.J., Zasadzinski J.A.N., Pindak R., Slaney A.J., Goodby J.W., *Science*, **258** (1992) 275.

[9]   Lubensky T.C., Renn S.R., *Phys. Rev. A*, **41** (1990) 4392.

[10] Renn S.R., Lubensky T.C., *Mol. Cryst. Liq. Cryst.*, **209** (1991) 349.

[11] Nguyen H.T., Bouchta A., Navailles L., Barois P., Isaert N., Twieg R.J., Maaroufi A., Destrade C., *J. Phys. II (France)*, **2** (1992) 1889.

[12] a) Navailles L., Pansu B., Gorre-Talini L., Nguyen H.T., *Phys. Rev. Lett.*, **81** (1998) 4168.
b) Navailles L., Barois P., Nguyen H.T., *Phys. Rev. Lett.*, **71** (1993) 545.

[13] a) Isaert N., Navailles L., Barois P., Nguyen H.T., *J. Phys. II (France)*, **4** (1994) 1501.
b) Stoenescu D.N., Nguyen H.T., Barois P., Navailles L., Nobili M., Martinot-Lagarde Ph., Dozov I., *Mol. Cryst. Liq. Cryst.*, **358** (2001) 275.

[14] Navailles L., "Synthesis and structural studies of novel helical smectic phases: TGB$_A$ and TGB$_C$" (in French), Ph.D. Thesis, University of Bordeaux I, Bordeaux, 1994.

[15] Navailles L., Pindak R., Barois P. Nguyen H.T., *Phys. Rev. Lett.*, **74** (1995) 5224.

[16] Barois P., Heidelbach F., Navailles L., Nguyen H.T., Nobili M., Petit M., Pindak R., Riekel C., *Eur. Phys. J. B*, **11** (1999) 455.

[17] Petit M., Nobili M., Barois P., *Eur. Phys. J. B*, **6** (1998) 341.

[18] a) Li M.H., Nguyen H.T., Sigaud G., *Liq. Cryst.*, **20** (1996) 361.
b) Li M.H., Lause V., Nguyen H.T., Sigaud G., Barois P., Isaert N., *Liq. Cryst.*, **23** (1997) 389.
c) Pansu B., Li M.H., Nguyen H.T., *J. Phys. II (France)*, **7** (1997) 751.
d) Pansu B., Li M.H., Nguyen H.T., *Eur. Phys. J. B*, **2**, (1998) 143.
e) Young M., Pitsi G., Li M.H., Nguyen H.T., Jamée P., Sigaud G., Thoen J., *Liq. Cryst.*, **25** (1998) 387.

[19] Grelet E., "Structural study of smectic Blue Phases" (in French), Ph.D. Thesis, University of Paris XI, Orsay, 2001.

[20] a) Pansu B., Grelet E., Li M.H., Nguyen H.T., *Phys. Rev. E*, **62** (2000) 658.
b) Grelet E., Pansu B., Li M.H., Nguyen H.T., *Phys. Rev. Lett.*, **86** (2001) 3791.
c) Grelet E., Pansu B., Nguyen H.T., *Phys. Rev. E*, **64** (2001) 10703.

[21] Kamien R., *J. Phys. II (France)*, **7** (1997) 743.

Chapter C.VI

Hexatic smectics

We already know that some **molecular crystals** composed of anisotropic molecules sometimes exhibit liquid crystalline phases (smectics or nematics) before melting into the isotropic phase (Fig. C.VI.1). These **mesophases** are optically anisotropic and they are all characterized by a **long-range order of the molecular orientation**.

What happens, on the other hand, when the crystal is formed by spherical atoms or molecules? Does it melt directly to the isotropic liquid or can intermediate phases still appear?

Before answering this question, let us first return to the fundamental concept of long-range order and to the correlation functions of the crystalline order parameter. We shall then see that the thermal phonons destroy long-range order in a one-dimensional space, such that one cannot speak of melting in this case [1a,b]. On the other hand, the long-range order is not completely destroyed by the phonons in a two-dimensional space. In this case, the crystal sometimes melts to an intermediate – **hexatic** – phase, which no longer exhibits long-range positional order, but still preserves the long-range orientational order. This phase can itself melt into the isotropic liquid. We shall show that these two phase transitions are **defect-induced** (by dislocations in the first case and by disclinations in the second case), along a process discovered by Kosterlitz and Thouless [2]. After this theoretical reminder (section VI.1), we shall return to liquid crystals (section VI.2) in order to show that hexatic phases can exist in such materials. We shall describe successively the smectic I phase and its defects, then the hexatic smectic B phase. In conclusion, we shall present some results on the dynamics of these phases, obtained by light scattering (section VI.3) and by rheological measurements (section VI.4).

VI.1 Theory of two-dimensional melting: the hexatic phase

VI.1.a) The concept of crystalline order

Before describing the way a crystal melts, one must first define the notion of crystalline order. Let us recall that this concept was invented by Haüy and Bravais in order to describe the arrangements of molecules (or atoms) in **three-dimensional** crystalline minerals [3]. In their description, the molecules are placed on the sites **t** of a **lattice**, named after Bravais, and generated from three basis vectors **a**, **b**, **c**:

$$\mathbf{t} = m\mathbf{a} + n\mathbf{b} + p\mathbf{c} \qquad \text{with} \qquad m, n, p = 0, \pm 1, \pm 2 \qquad \text{(C.VI.1)}$$

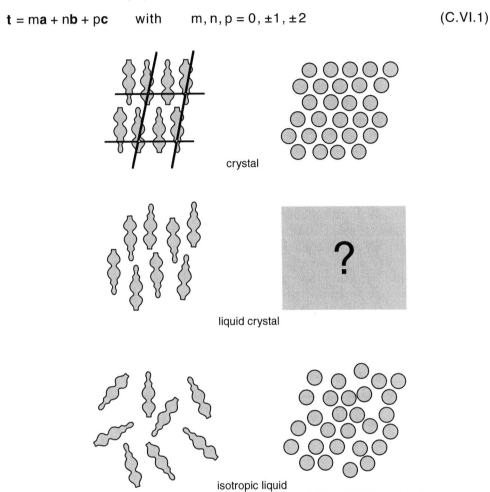

crystal

liquid crystal

?

isotropic liquid

Fig. C.VI.1 A molecular crystal can melt without completely losing the orientational order of the molecules; this can lead, for instance, to the nematic phase. What happens, however, when the crystal is composed of atoms? Does it melt directly into the isotropic liquid?

The presence of this lattice determines many of the physical properties of crystals (elasticity, optical anisotropy, etc.)

A more rigorous way of describing crystalline order is in terms of the symmetry elements of the mass (or electron) density $\rho(\mathbf{r})$. This function can be written as the sum of two terms:

$$\rho(\mathbf{r}) = \rho_o(\mathbf{r}) + \rho_{cr}(\mathbf{r}) \qquad\qquad\qquad (C.VI.2)$$

representing the average density $\rho_o(\mathbf{r}) = \text{cst}$ and the **crystalline order parameter** $\rho_{cr}(\mathbf{r})$, respectively. This latter is invariant with respect to:

1. The discrete set of translations $\mathbf{t}$.
2. Certain discrete rotations and/or reflections.
3. Certain superpositions of type 1 and 2 operations (for instance, the helical axes 4_1 and 4_2 of Blue Phases I and II).

We remind that the spatial periodicity of the density modulation $\rho_{cr}(\mathbf{r})$ appears explicitly in its three-dimensional Fourier series expansion:

$$\rho(\mathbf{r}) = \sum_{\mathbf{G}} \rho_{\mathbf{G}} e^{i\mathbf{G}.\mathbf{r}} ; \qquad \rho_{\mathbf{G}}^* = \rho_{-\mathbf{G}} \qquad\qquad (C.VI.3)$$

The wave vectors $\mathbf{G}$ form the **reciprocal lattice** and are generated from three basis vectors $\mathbf{a}^*$, $\mathbf{b}^*$ and $\mathbf{c}^*$:

$$\mathbf{G} = h\,\mathbf{a}^* + k\,\mathbf{b}^* + l\,\mathbf{c}^* ; \qquad h, k, l = 0, \pm 1, \pm 2... \qquad (C.VI.4)$$

themselves related to the basis vectors of the Bravais lattice by the relations:

$$\mathbf{a}^* = 2\pi(\mathbf{b}\times\mathbf{c})/V ; \; \mathbf{b}^* = 2\pi(\mathbf{c}\times\mathbf{a})/V ; \; \mathbf{c}^* = 2\pi(\mathbf{a}\times\mathbf{b})/V \qquad (C.VI.5)$$

where $V = \mathbf{a}.(\mathbf{b}\times\mathbf{c})$ is the volume of the unit cell.

An example of a Bravais lattice and of its reciprocal lattice is given in figure C.VI.2.

Owing to this particular choice of the wave vectors $\mathbf{G}$, the function $\rho_{cr}(\mathbf{r})$ is invariant with respect to all translations $\mathbf{t}$ of the Bravais lattice:

$$\sum_{\mathbf{G}} \rho_{\mathbf{G}} e^{i\mathbf{G}.(\mathbf{r}-\mathbf{t})} = \sum_{\mathbf{G}} \rho_{\mathbf{G}} e^{i\mathbf{G}.\mathbf{r}} \qquad\qquad (C.VI.6)$$

as

$$\mathbf{G}.\mathbf{t} = 2\pi(mh + nk + pl) \qquad\qquad\qquad (C.VI.7)$$

is a multiple of 2π. On the other hand, any other translation $\mathbf{u}(\mathbf{r},t)$ induces a phase change

$$\phi_{\mathbf{G}}(\mathbf{r}, t) = -\mathbf{G}.\mathbf{u}(\mathbf{r},t) \qquad\qquad\qquad (C.VI.8)$$

of the Fourier components:

$$\rho_{\mathbf{G}} \rightarrow \rho_{\mathbf{G}} e^{i\phi_{\mathbf{G}}(\mathbf{r},t)} \qquad\qquad\qquad (C.VI.9)$$

Thus, every Fourier component in the expansion of the density $\rho_{cr}(\mathbf{r})$ is defined by an **amplitude** $|\rho_{\mathbf{G}}|$ and a **phase** $\phi_{\mathbf{G}}(\mathbf{r},t)$, in the manner of the smectic, superfluid and superconducting order parameters, or of any other order parameter of a broken symmetry phase.

To quantify **the positional order** in terms of the order parameters $\rho_{\mathbf{G}}(\mathbf{r}) = |\rho_{\mathbf{G}}| e^{i\phi_{\mathbf{G}}(\mathbf{r},t)}$, let us introduce the **correlation functions**:

$$g_{\mathbf{G}}(\mathbf{r}) = <\rho_{\mathbf{G}}(\mathbf{r})\,\rho_{\mathbf{G}}(\mathbf{0})>$$ (C.VI.10)

We shall say that **the positional order is long-range** if the correlation function tends toward a finite value as $\mathbf{r} \to \infty$:

$$\lim_{r \to \infty} <\rho_{\mathbf{G}}(\mathbf{r})\,\rho_{\mathbf{G}}(\mathbf{0})> = \text{const.}$$ (C.VI.11)

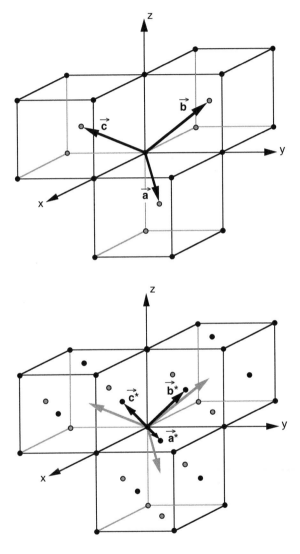

Fig. C.VI.2 Body-centered cubic Bravais lattice generated by three base vectors **a**, **b** and **c**. Its reciprocal lattice, generated by the wave vectors **a***, **b*** and **c***, is face-centered.

We shall speak of **quasi-long range order** when it decays **algebraically** with the distance. The order is considered **short-range** when the decay is exponential.

Clearly, this definition can be extended to any broken symmetry phase described by a complex order parameter. The order, be it positional or orientational, short- or long-range, is then defined in terms of correlation functions of the order parameter. In the following section, we shall examine the influence of the thermal phonons on the order in a broken symmetry phase. This problem, first considered by Landau and Peierls, highlights the crucial role of the space dimension.

VI.1.b) Peierls disorder: the generic case of a broken symmetry phase described by a complex order parameter

Landau and Peierls were the first to study the effect of **phonons** (i.e., of the lattice vibrations) on the crystalline order.

Rather than deal with this in the particular case of solids (see the appendix), let us calculate the correlation function in the generic case of a broken symmetry phase described by a complex order parameter $\Psi(r) = \Psi_0 e^{i\phi(r)}$. In this context, a **phonon** is a **continuous** (hence uniquely defined at each point in space and time) and thermally activated distortion of the phase of the order parameter. **Consequently, it does not change the topology of the order parameter**. In crystals, the phonons are **continuous deformations of the u(r,t) lattice** or of the phases $\phi_G = - G.u(r,t)$. In the case of superconductors or superfluids, they are fluctuations of the phase $\phi(r,t)$ of the wave function Ψ. In nematics, they are fluctuations of the director orientation, etc.

Obviously, these fluctuations represent a source of uncertainty in the phase and tend to decrease the correlations of the order parameter. **Peierls posited the idea that the effects of the fluctuations depend on the dimensionality of the space where they take place.**

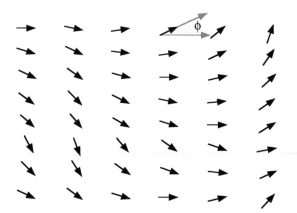

Fig. C.VI.3 The continuous fluctuations of the phase ϕ of the complex order parameter reduce the $<\psi^*(r)\psi(0)>$ correlations at large distances.

This can be shown by calculating the correlation function of the order parameter:

$$<\psi^*(\mathbf{r})\psi(\mathbf{0})> = |\psi_0|^2 <e^{-i\phi(\mathbf{r})} e^{i\phi(\mathbf{0})}>$$

$$= |\psi_0|^2 e^{-<[\phi(\mathbf{r}) - \phi(\mathbf{0})]^2>/2} \qquad (C.VI.12)$$

The first step is to expand the phase $\phi(\mathbf{r})$ in a Fourier series:

$$\phi(\mathbf{r}) = \sum_{\mathbf{q}} \phi(\mathbf{q}) e^{i\mathbf{q}\cdot\mathbf{r}} \qquad (C.VI.13)$$

In a system of size L along each of its n dimensions, the wave vectors $\mathbf{q}$ are quantized, taking values:

$$q_i(N_i) = (2\pi/L) N_i \; ; \qquad N_i = 0, \pm 1, \pm 2, ..., \pm N_{max} \; ; \qquad i = 1,...,n \qquad (C.VI.14)$$

with

$$N_{max} = L/c_0 \qquad (C.VI.15)$$

where c_0 is the intermolecular distance.

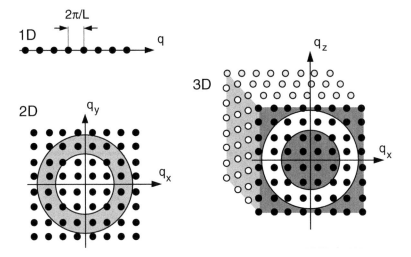

Fig. C.VI.4 The number of modes $g_n(q)\,dq$ in the interval $[q, q + dq]$ depends on the space dimension n. In 1D, $g_1(q) = L/2\pi$; in 2D, $g_2(q) \sim q$; in 3D, $g_3(q) \sim q^2$.

The second step consists of determining the mean square deviation of the phase at point $\mathbf{r}$ with respect to the origin $\mathbf{0}$:

$$<[\phi(\mathbf{r}) - \phi(\mathbf{0})]^2> = \sum_{\mathbf{q}} <|\phi(\mathbf{q})|^2> |e^{i\mathbf{q}\cdot\mathbf{r}} - 1|^2 \qquad (C.VI.16)$$

In a macroscopic system, exhibiting a very large number of modes, it is convenient to replace this sum by an integral over q:

$$<[\phi(r) - \phi(0)]^2> = \int dq\{g_n(q)<|\phi(q)|^2> |e^{iq \cdot r} - 1|^2 \qquad \text{(C.VI.17)}$$

The $g_n(q)dq$ factor counts the number of modes with an absolute value of the wave vector q in the interval [q, q + dq]. It depends on the space dimension, being given by:

$$g_1(q)dq = \frac{dq}{2\pi/L} \qquad \text{in 1 D} \qquad \text{(C.VI.18a)}$$

$$g_2(q)dq = \frac{2\pi q dq}{(2\pi/L)^2} \qquad \text{in 2 D} \qquad \text{(C.VI.18b)}$$

$$g_3(q)dq = \frac{4\pi q^2 dq}{(2\pi/L)^3} \qquad \text{in 3 D} \qquad \text{(C.VI.18c)}$$

In the next step, we determine the mean square amplitude $<|\phi(q)|^2>$ of each mode. To this end, one must calculate the total energy of the fluctuations using the "elasticity" relation:

$$F = \frac{K}{2} \int |\text{grad} \, \phi|^2 d^n r \qquad \text{(C.VI.19)}$$

In this expression, K is an elasticity coefficient with units of erg.cm in 1D, of erg in 2D and of erg/cm in 3D. From eq. C.VI.13, one obtains:

$$F = \sum_q (K/2) L^n q^2 |\phi(q)|^2 \qquad \text{(C.VI.20)}$$

yielding, after use of the equipartition theorem:

$$<(K/2) L^n q^2 |\phi(q)|^2> = k_B T/2 \qquad \text{(C.VI.21a)}$$

or, equivalently:

$$<|\phi(q)|^2> = \frac{k_B T}{L^n K q^2} \qquad \text{(C.VI.21b)}$$

We notice that the mean square amplitude diverges in the limit of small wave vectors q, since the long wavelength fluctuations have a low energy cost.
Finally, one must calculate the integral C.VI.17. It contains the factor

$$|e^{iq \cdot r} - 1|^2 = 2 [1 - \cos(q \cdot r)] \qquad \text{(C.VI.22)}$$

which, for large values of the distance r, oscillates rapidly between 0 and 4. It can thus be replaced by its average value 2 in the integral C.VI.17. In these conditions, at a distance L from the origin one has:

$$<[\phi(L) - \phi(0)]^2> = \int dq\, g_n(q) \frac{k_B T}{L^n K q^2} 2 \qquad \text{(C.VI.23)}$$

yielding:

in 1D: $<[\phi(L) - \phi(0)]^2> = \int \frac{dq}{2\pi/L} \frac{k_B T}{L K q^2} 2 = \frac{k_B T}{\pi K}\left(\frac{1}{q_{min}} - \frac{1}{q_{max}}\right)$

or $\qquad <[\phi(L) - \phi(0)]^2> \approx \frac{k_B T}{2\pi^2 K} L \qquad\qquad\qquad$ (C.VI.24a)

in 2D: $<[\phi(L) - \phi(0)]^2> = \int \frac{2\pi q\, dq}{(2\pi/L)^2} \frac{k_B T}{L^2 K q^2} 2 = \frac{k_B T}{\pi K}[\ln q_{max} - \ln q_{min}]$

or $\qquad <[\phi(L) - \phi(0)]^2> = \frac{k_B T}{\pi K} \ln\left(\frac{L}{c_0}\right) \qquad\qquad$ (C.VI.24b)

in 3D: $<[\phi(L) - \phi(0)]^2> = \int \frac{4\pi q^2 dq}{(2\pi/L)^3} \frac{k_B T}{L^3 K q^2} 2 = \frac{k_B T}{\pi^2 K}[q_{max} - q_{min}]$

or $\qquad <[\phi(L) - \phi(0)]^2> = 2\frac{k_B T}{\pi K c_0} \qquad\qquad\qquad$ (C.VI.24c)

Starting from these three expressions, we can determine the corresponding correlation functions:

in 1D:

$$<\psi^*(L)\psi(0)> = |\psi_0|^2\, e^{-L/\xi} ; \qquad \xi = 4\pi^2 K/(k_B T) \qquad \text{(C.VI.25a)}$$

in 2D:

$$<\psi^*(L)\psi(0)> = |\psi_0|^2\, (a/L)^\eta ; \qquad \eta = k_B T/(2\pi K) \qquad \text{(C.VI.25b)}$$

in 3D:

$$<\psi^*(L)\psi(0)> = |\psi_0|^2\, e^{-W(T)} ; \qquad W(T) = k_B T/(\pi K c_0) \qquad \text{(C.VI.25c)}$$

These results lead to the following conclusions:

     1. The "true" long-range order is only possible in three dimensions where, in the $L \to \infty$ limit, the correlation function $<\psi^*(L)\psi(0)>$ tends toward a

finite value. This value decreases exponentially with the temperature (in crystallography, $e^{-W(T)}$ is known as the Debye-Waller factor).

2. In one dimension, the order can only be short-range. The range ξ of the order decreases with the temperature.

3. In two dimensions, the correlation function decreases algebraically as $(1/L)^{\eta}$. The exponent η increases with the temperature.

In the following, we shall consider more closely the two-dimensional case, which turns out to be the most interesting, and restrict our discussion to the case of crystals.

VI.1.c) Effect of phonons on the positional order in two-dimensional crystals

In 2D crystals, the fluctuations of the phase $\phi_G(\mathbf{r})$ are due to the $\mathbf{u}(\mathbf{r})$ displacements of the lattice [4, 5]. It is noteworthy that a phase gradient, **grad** ϕ_G, can be induced either by longitudinal or by transverse phonons (Fig. C.VI.5).

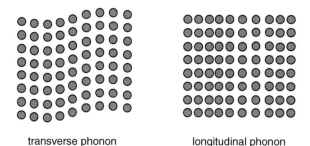

transverse phonon   longitudinal phonon

Fig. C.VI.5 Transverse and longitudinal phonons in a two-dimensional crystal.

To take them into account explicitly, one must use a more accurate expression of the elastic energy C.VI.19. We should recall that the energy of an elastic deformation of the crystal has, in linear elasticity, the generic expression:

$$F = \frac{1}{2} \int c_{ijkl} \varepsilon_{ij} \varepsilon_{kl} d^2\mathbf{r} \tag{C.VI.26}$$

with c_{ijkl} the tensor of the elastic coefficients and

$$\varepsilon_{ij} = \frac{1}{2} \left(\frac{\partial u_i}{\partial x_j} + \frac{\partial u_j}{\partial x_i} \right) \tag{C.VI.27}$$

the strain tensor (symmetric by definition). In two dimensions, the tensors ε_{ij} and c_{ijkl} have three and six independent coefficients, respectively (since $c_{ijkl} = c_{jikl} = c_{ijlk} = c_{klij}$). The elastic energy must also be invariant under the symmetry operations of the lattice, resulting in additional relations between the c_{ijkl} coefficients, and in a further decrease of the number of independent coefficients. In the case of a triangular lattice, exhibiting sixfold symmetry, only two independent coefficients (known as Lamé coefficients) remain, and they can be chosen in the same way as for an isotropic medium:

$$F = \frac{1}{2} \int d^2r \, [\, 2\mu \, \varepsilon_{ij}\varepsilon_{ji} + \lambda \, \varepsilon_{ii}^2] \tag{C.VI.28}$$

One can easily check that the **transverse** modes only involve the shear constant μ, as opposed to **longitudinal** modes, which bring into play the combination $2\mu + \lambda$ of the two elasticity coefficients. It ensues that the exponent η_G describing the algebraic decrease of the positional correlation functions:

$$<\rho_G(L) \, \rho_G(0)> \sim (1/L)^{\eta_G} \tag{C.VI.29}$$

depends on a combination of the Lamé coefficients:

$$\eta_G = \frac{|G|^2 k_B T}{4\pi K_{eff}} \qquad \text{with} \qquad K_{eff} = \frac{\mu \, (2\mu + \lambda)}{\mu + (2\mu + \lambda)} \tag{C.VI.30}$$

We emphasize that the positional quasi-order in two-dimensional crystals is reflected in the structure factor

$$S(\mathbf{q}) = <\rho(\mathbf{q})\rho(-\mathbf{q})> \tag{C.VI.31}$$

describing the intensity distribution close to a Bragg peak. Due to the phonons, the shape of the peaks is no longer that of a delta function, but rather a Lorentzian of expression :

$$S(\mathbf{q}) \sim \sum_{G} |\mathbf{q} - \mathbf{G}|^{-2 + \eta_G} \tag{C.VI.32}$$

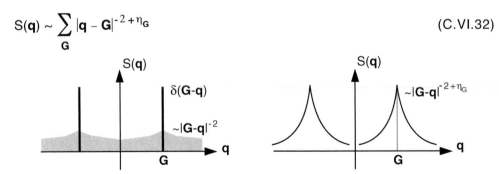

Fig. C.VI.6 The structure factor in the case of a three-dimensional crystal is a delta function; the Bragg peak, very narrow, is superposed over the continuous background created by the phonons. In a two-dimensional crystal, the long-wavelength phonons widen the peak and give it an algebraically decaying profile.

Let us now try to define an orientational order in a 2D crystal.

VI.1.d) Orientational order of near-neighbor molecular bonds in 2D crystals

The algebraic decay of positional order in two-dimensional crystals is due to the long-wavelength phonons. As shown in the diagram of figure C.VI.7, these high-amplitude phonons induce considerable relative displacements between two distant points of the crystal. However, Mermin [6] noticed that these displacements take place without changing the relative orientation of the two parts of the crystal.

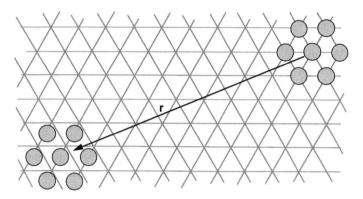

Fig. C.VI.7 The long-wavelength phonons induce relative displacements between two points of the crystal, but do not change their orientation with respect to the ideal lattice.

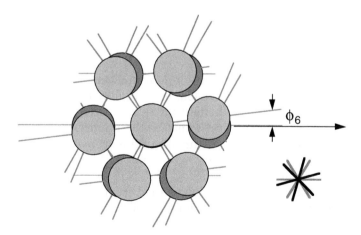

Fig. C.VI.8 Definition of the orientational order parameter in a crystal of hexagonal symmetry based on the anisotropic mass distribution in the vicinity of an atom. The phase ϕ_6 of the hexatic order parameter indicates the rotation of a group of atoms with respect to an axis. The hexatic order parameter is usually figured by a six-arm star.

This leads to the following question: how does one define mathematically the orientation of one part of the crystal with respect to another?

To search for an answer, let us introduce a function describing the anisotropy of mass distribution in the vicinity of an atom of the 2D crystal. This distribution is a function $\rho_a(r,\theta)$ of the distance r with respect to the central atom and of the angle θ with respect to a given direction in the plane. The function can be written as a Fourier series:

$$\rho_a(r,\theta) = \sum_n \rho_n(r) e^{i(n\theta + \phi_n)} \qquad\qquad (C.VI.33)$$

In the case of the hexagonal symmetry crystal shown in the diagram of figure C.VI.8, the first anisotropic term ($n \neq 0$) is the sixth-order one:

$$\rho_a(r,\theta) = \rho_o + \rho_6\, e^{i[6(\theta + \phi_6)]} + \dots \qquad\qquad (C.VI.34)$$

The amplitude ρ_6 and the phase $6\phi_6$ of this term play the role of an orientational order parameter:

$$\psi_6 = \rho_6\, e^{i 6\phi_6} \qquad\qquad (C.VI.35)$$

As the crystal is "shaken" by phonons, the amplitude and phase of this order parameter fluctuate. Mermin showed, however, that the phonons do not destroy the long-range orientational order in a two-dimensional crystal, the correlation function $<\psi_6^*(L)\psi_6(0)>$ reaching a finite value in the $L \to \infty$ limit. This behavior reveals the existence of a **long-range orientational order of the bonds between nearest-neighbor molecules.**
Let us now describe the melting of a 2D crystal.

VI.1.e) Theory of melting in two dimensions: the hexatic phase

Having refined the concept of two-dimensional crystalline order, we are now better prepared to tackle the problem of crystal melting. The scenario we shall describe in the following was thought up by Kosterlitz and Thouless in 1973 [2]. These two authors were the first to notice the importance of the role played by topological defects of the order parameter in this transition.
    Their reasoning starts with the simple, but crucial observation that, notwithstanding the algebraic decay of the positional correlations, the two-dimensional crystalline order is long-range from the topological point of view (as the orientational order still remains). Thus, the deformations due to the phonons preserve the crystalline lattice which, although stretched or compressed, maintains its identity.
    These authors then noticed that one of the ways of destroying the positional correlations is to introduce topological defects of the dislocation type [7].
    At this point, one must distinguish between a pair of dislocations of opposite signs, as the one shown in figure C.VI.9, and an isolated dislocation (Fig. C.VI.10).

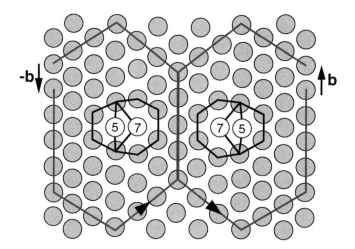

Fig. C.VI.9 Pair of dislocations of opposite Burgers vectors. The deformation of the crystalline lattice is localized in the vicinity of the pair.

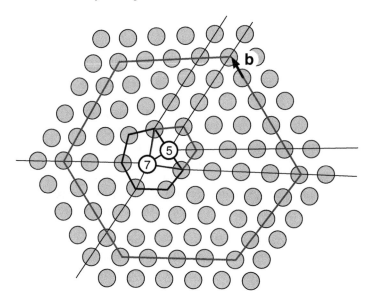

Fig. C.VI.10 Elementary dislocation in a hexagonal lattice. The core of the dislocation presents two molecules with anomalous environments: one of them has seven near neighbors, while the other has five. The closure defect of the Burgers circuit surrounding the dislocation defines the **Burgers vector b**.

Indeed, a pair of dislocations does not perturb the lattice over large distances, the deformations produced by the two dislocations compensating exactly far away from the pair. It ensues that the energy of a pair is finite, a classical elasticity calculation showing that:

$$E_2(\mathbf{r}) = 2K^*b^2 \left[\ln\left(\frac{r}{r_c}\right) - \frac{(\mathbf{b}.\mathbf{r})^2}{b^2 r^2} \right] + 2E_c \qquad \text{(C.VI.36)}$$

where $\mathbf{b}$ (respectively $-\mathbf{b}$) are the Burgers vectors of the two dislocations, $\mathbf{r}$ the position vector separating them, E_c their core energy and K^* an effective elasticity constant:

$$K^* = \frac{\mu\,(\mu + \lambda)}{2\pi\,(2\mu + \lambda)} \qquad \text{(C.VI.37)}$$

These dislocation pairs do not destroy the long-range positional order and must be present in any crystal (they are thermally activated).

    If the paired dislocations do not change the positional order over long distances, the situation is completely different for **free dislocations**. Indeed, an isolated dislocation perturbs the crystal over large distances. This is obvious from the expression of its elastic energy:

$$E_1 = K^* b^2 \ln\left(\frac{A}{A_c}\right) + E_c \qquad \text{(C.VI.38)}$$

diverging with the surface area A of the crystal, A_c being the surface of the dislocation core ($\approx r_c^2$).

    Suppose now that these free dislocations start to proliferate in the crystal. In practice, they are generated by the dissociation of the dislocation pairs already present in the crystal. To preserve the global topological charge of the sample, there must be an equal number of dislocations of each sign. Before determining the critical temperature T_c above which these dislocations appear spontaneously, let us describe their effect on the crystalline order.

    Clearly, these **dislocations destroy the positional order**, each of them introducing a 2π defect in the phase of the order parameter ρ_G. Let ξ be the average distance between two dislocations. Due to the phase jumps generated by the dislocations, it will be impossible to predict the correlation of the phases ϕ_G between two points separated by a distance larger than ξ. The position correlations must therefore decrease exponentially over the distance ξ:

$$<\rho_G(\mathbf{r})\,\rho_G(0)> \sim e^{-r/\xi} \qquad \text{(C.VI.39)}$$

The positional order only extends over a short range as soon as the free dislocations appear.

    Let us now consider the **orientational order defined by the order parameter ψ_6**. Halperin and Nelson realized that this parameter is only perturbed in the immediate vicinity of a dislocation. Everywhere else, the dislocation has no effect on ψ_6. We can see this by noticing that the integral of the variation in the phase $\phi_6(\mathbf{r})$ vanishes on a circuit enclosing a dislocation (Fig. C.VI.11):

$$\int \delta\phi_6(\mathbf{r}) = \int \mathbf{dl}.\mathbf{grad}\,\phi_6(\mathbf{r}) = 0 \qquad\qquad\qquad (\text{C.VI.40})$$

Thus, everything happens as if the dislocation was not there. From this, Nelson and Halperin concluded that above T_c the orientational order ψ_6 is preserved. This phase, in which the free dislocations proliferate, is then topologically different from the crystal and from the isotropic liquid. Nelson and Halperin proposed the term **hexatic phase**, as a reference to the hexagonal symmetry of the order parameter ψ_6.

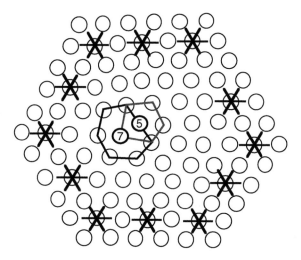

Fig. C.VI.11 The variation in the phase ϕ_6 of the hexatic order parameter on a circuit enclosing a dislocation is zero; the dislocations do not destroy the orientational order of the bonds between nearest-neighbor molecules.

Before describing the process leading from the hexatic phase to the isotropic liquid, let us see how to determine the critical temperature for the transition between the crystal and the hexatic phase. The following calculation is rather crude, but nevertheless contains the essential physics of this transition.

The calculation consists of striking a balance between the elastic energy lost by introducing the dislocations, and the gain in configurational entropy. Let ξ be the average distance between two dislocations of opposite signs. Each dislocation of Burgers vector $b = c_0$ (c_0 is the spacing of the hexagonal lattice) has an energy

$$E = K^* c_0^2 \ln\left(\frac{\xi}{r_c}\right) + E_c \qquad\qquad\qquad (\text{C.VI.41})$$

and a configurational entropy

$$S = k_B \ln \left(\frac{\xi^2}{c_0^2} \right) \tag{C.VI.42}$$

Consequently, the variation in free energy of the system upon introducing $1/\xi^2$ dislocations per unit surface reads:

$$\Delta F = \frac{1}{\xi^2} \left[K^* c_0^2 \left(\ln \left(\frac{\xi}{c_0} \right) + x \right) - 2 k_B T \ln \left(\frac{\xi}{c_0} \right) \right] \tag{C.VI.43}$$

Here, we took $r_c \approx c_0$ and posed $E_c = K^* c_0^2 x$ (with $x \approx 1$). Setting:

$$T_c = \frac{K^* c_0^2}{2 k_B} \tag{C.VI.44}$$

the free energy can be rewritten under the form:

$$\Delta F = \frac{2 k_B}{\xi^2} \left[(T_c - T) \ln \left(\frac{\xi}{c_0} \right) + x T_c \right] \tag{C.VI.45}$$

This expression reveals the presence of two different situations:
  1. Either $T < T_c$; in this case $\Delta F > 0$, and nothing is gained by creating dislocations.
  2. Or $T > T_c$, in which case one must minimize the free energy with respect to ξ, leading to:

$$\xi(T) = c_0 \exp \left(\frac{x T_c}{T - T_c} \right) \tag{C.VI.46}$$

and

$$\Delta F = \frac{k_B}{\xi^2} (T_c - T) < 0 \tag{C.VI.47}$$

This calculation shows that above T_c, the free energy decreases when creating dislocations. The temperature T_c is thus the transition temperature toward the hexatic phase. One can see that the correlation length ξ diverges "exponentially" on approaching T_c. A more detailed theory, using renormalization group techniques, confirms that the divergence is of the "stretched exponential" type:

$$\xi(T) \approx \exp \left[\frac{Cst}{(T - T_c)^{\upsilon}} \right] \quad \text{with} \quad \upsilon \approx 0.37 \tag{C.VI.48}$$

At this point, it becomes interesting to look for materials where the hexatic phase might appear. The transition temperature T_c given by formula C.IV.44 must not be too high with respect to the room temperature, of about 300 K. With $c_0 = 5$ Å, this requires a material where $K^* \approx 2 k_B T_c / c_0^2 \approx 30 \ erg/cm^2$, corresponding to a bulk elastic modulus $B = K^*/c_0 \approx 6\times10^8$ erg/cm^3. It turns out that the elastic moduli of the SmB phases of thermotropic liquid crystals are of the right order of magnitude, so it should come as no surprise that the hexatic phase was found in these systems. Note that in a metal, $B \approx 10^{12}$ erg/cm^3, yielding $T_c \approx 10^5$ K. Clearly, the metal melts well below this temperature.

To conclude this section, let us describe the scenario along which the hexatic phase melts into the isotropic liquid. We have just seen that the dislocations destroy the positional order. To break the orientational order, one must consider the thermal fluctuations of the order parameter ψ_6 and distinguish again between the continuous (topology-preserving) fluctuations, and the rotation defects (or disclinations), which perturb the topology.

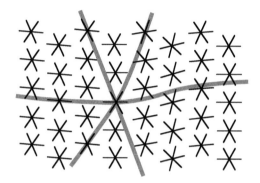

Fig. C.VI.12 Continuous fluctuations of the hexatic order parameter.

One already knows that the continuous fluctuations (see section VI.1.b and figure C.VI.12) reduce the correlations of the order parameter ψ_6 at large distances, the correlation function decreasing as a power law:

$$<\psi_6^*(L)\psi_6(0)> = |\rho_6|^2 \, (c_0/L)^{\eta_6}; \qquad \eta_6 = 36k_BT/(2\pi K_6) \qquad (C.VI.49)$$

In this formula, K_6 is the elasticity coefficient associated with the phase ϕ_6 of the order parameter ψ_6 of the hexatic phase:

$$F_6 = \int \frac{K_6}{2} |\mathbf{grad}\,\phi_6|^2 d^2\mathbf{r} \qquad (C.VI.50)$$

To destroy the orientational order completely, one must introduce the elementary disclinations of the hexatic phase, namely defects of angles $2\pi/.6$ and $-2\pi/6$ of the phase ϕ_6 of the hexatic order parameter (figures C.VI.13 and 14). These defects already exist in the hexatic phase, since the core of each free dislocation can be considered as the association of a $2\pi/6$ defect with a $-2\pi/6$ defect (Fig. C.VI.10). In the same way a pair of opposite-sign dislocations does not perturb the positional order, a pair of opposite-sign disclinations only perturb the orientational order in the immediate vicinity of the cores, as illustrated in figure C.VI.15.

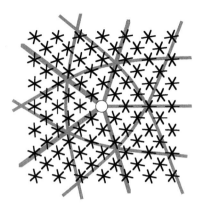

Fig. C.VI.13 $+2\pi/6$ disclination of the hexatic phase. A $+2\pi/6$ variation of the phase ϕ_6 is accumulated on a circuit enclosing the disclination.

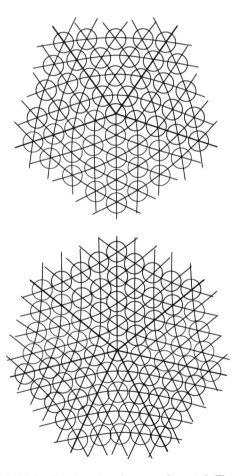

Fig. C.VI.14 $+2\pi/6$ and $-2\pi/6$ disclinations in a hexagonal crystal. The disclination cores consist of molecules with an anomalous number of neighbors: five for the $+2\pi/6$ disclination and seven for the $-2\pi/6$ disclination.

Imagine now that these pairs dissociate, like the crystal dislocation pairs do at T_c. This phenomenon occurs at a higher temperature, T_f, which can be determined in the same way as for T_c. This is the temperature at which the hexatic phase melts into the isotropic liquid. Indeed, the free disclinations "kill" the long-range orientational order, leading to an exponential decrease of the correlation function of the orientational order parameter, the hallmark of an isotropic liquid.

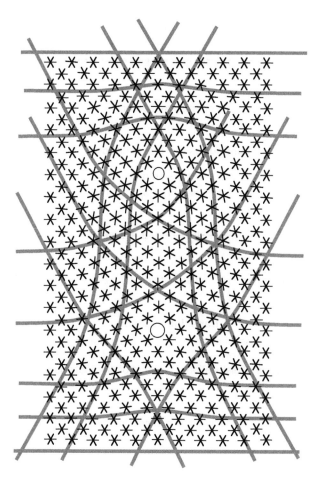

Fig. C.VI.15 Pair of $+2\pi/6$ and $-2\pi/6$ disclinations. The distortion of the hexatic order parameter is localized close to the defect pair.

To summarize, the melting of two-dimensional crystals takes place in two steps (Fig. C.VI.16):

Step 1: The crystalline phase contains bound dislocations pairs. These pairs are thermally excited, just like phonons. Above a critical temperature T_c, the dislocation pairs dissociate. The proliferation of free dislocations destroys the crystalline positional order, but not the orientational order of near-neighbor molecular bonds. The hexatic phase appears.

Step 2: Same scenario, but this time concerning free disclinations, instead of the dislocations. Their proliferation above T_f destroys the hexatic order and creates the isotropic phase.

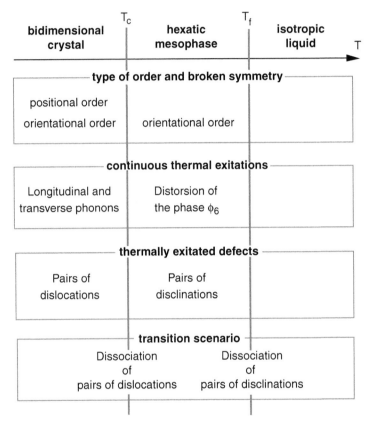

Fig. C.VI.16 Two-step melting scenario for two-dimensional crystals proposed by Halperin and Nelson.

VI.2   Hexatic smectics

VI.2.a) The search for hexatic mesophases

The "invention" of the hexatic mesophase by Halperin and Nelson prompted the interest of researchers working on two-dimensional systems. It would not be an overstatement to say that the publication of their theoretical work spurred a true "hunt for the hexatic phase." Where should one look for such a phase?

Many systems held the promise of possessing this extraordinary phase. Some examples are:

1. Atomic or molecular monolayers adsorbed on solid surfaces.

2. Electrons or ions confined in a two-dimensional layer at the surface of liquid helium.

3. Monolayers of colloidal particles confined between two solid walls.

4. Smectic mesophases, which are 3D stacks of monomolecular layers and can be stretched to form free-standing films with a small number of layers.

The research started concurrently along all these directions, the crucial question being how to recognize a hexatic mesophase.

Since this phase exhibits long-range orientational order of the bonds between near-neighbor molecules, but no positional order, the most straightforward technique to detect it would be X-ray or electron diffraction. Another method, more elegant from our point of view, is to search for the topological defects characteristic of the hexatic order parameter.

In this section, we discuss these two methods. We shall mostly consider experiments performed on free-standing smectic films, as they have multiple advantages with respect to all the other systems [8, 9] (see also chapter C.VIII). Indeed:

1. They are suspended in the air, eliminating the effects related to the anisotropy of a solid support.

2. They exist at temperatures close to the room temperature, so that the experiments are easier to perform.

3. They are perfectly flat, with layers parallel to the free surfaces, which is an advantage with respect to bulk samples, often difficult to align.

4. The number N of molecular layers can be easily varied between $N = 2$ and $N = \infty$.

VI.2.b) Observation of $2\pi/6$ disclinations in SmI films

In 1986, Dierker, Pindak and Meyer [10] observed under the polarizing microscope the elementary disclinations of the smectic I hexatic phase in a film of (S)-4-(2-methylbutyl)phenyl 4'-n-octyloxybiphenyl-4-carboxylate. We recall that the sequence of bulk phases in this material is the following:

$$\text{Crystal} \rightarrow \text{SmG} \rightarrow \text{SmF} \rightarrow \text{SmI} \rightarrow \text{SmC} \rightarrow \text{SmA} \rightarrow \text{Isotropic liquid}$$

The experiment begins by drawing a film in the smectic C phase. Due to the molecular anchoring on the edge of the frame holding the film, a 2π disclination appears in the center of the film, visible between slightly uncrossed polarizers as a singular point setting off two dark branches. At equilibrium, the disclination is in the center of the film, this position corresponding to the minimum elastic deformation energy of the director field.

By lowering the temperature below the SmC → SmI transition temperature, the aspect of the disclination changes, the singular point in the middle being replaced by a five-arm star (Fig. C.VI.17). A very important detail is that the arms of the star appear as discontinuity lines of the light intensity, and their length increases on moving away from the SmC → SmI transition temperature. Dierker, Pindak and Meyer proposed that this transformation of the point defect in a five-arm star is due to the appearance of the hexatic order within the smectic layers below the SmC → SmI transition. As shown in the diagram of figure C.VI.18, this texture consists of six $2\pi/6$ disclinations, one of them in the center of the star and the five others at the end of each arm. According to this model, a 2π disclination splits into six $2\pi/6$ disclinations at the SmC → SmI transition.

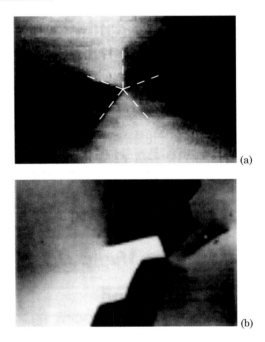

(a)

(b)

Fig. C.VI.17 Five-arm star: a) typical texture in a smectic I film obtained from a 2π disclination of the smectic C phase; b) the same texture observed in the crystalline smectic G phase at lower temperature (from ref. [10]).

The explanation of the observed phenomenon cannot stop here, as one still needs to understand the nature of the optical discontinuity lines forming the five arms of the star. Indeed, these arms are not disclinations of the hexatic phase, these latter being singular **points** surrounded by a continuous deformation field. Moreover, the true hexatic phase must be optically isotropic, owing to its sixfold symmetry. Consequently, **the disclinations of the true hexatic phase must be invisible under the polarizing microscope**.

To understand the nature of the lines connecting the disclinations of the five-arm star, one must take into account the tilt of the molecules with respect to the plane of the smectic layers. We shall therefore discuss the structure of the smectic phase I.

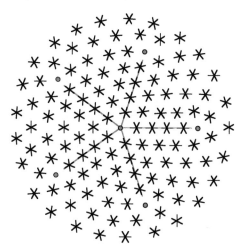

Fig. C.VI.18 Dissociation of a 2π disclination into six $2\pi/6$ disclinations.

VI.2.c) The two order parameters of the SmI phase

The smectic I phase is characterized by two coupled orientational order parameters:

1. The first one, inherited from the smectic C phase:

$$\psi_1 = |\psi_1| e^{i\phi_1(r)} \tag{C.VI.51}$$

describes the orientational order of the elongated molecules within the layers. Its amplitude $|\psi_1|$ is the molecular tilt angle with respect to the layers; its phase ϕ_1 is the angle between the projection of the molecule onto the plane of the layers and a fixed direction. This order parameter gives the optical anisotropy of the smectic I phase.

2. The second order parameter is specific to the hexatic phase:

$$\psi_6 = |\psi_6| e^{i\phi_6(r)} \tag{C.VI.52}$$

and only appears below the SmC $\rightarrow$ SmI transition temperature: its amplitude $|\psi_6|$ and phase ϕ_6 correspond to the amplitude and the phase of the sixth-order Fourier component of the mass distribution in the vicinity of each molecule.

Clearly, due to the molecules being tilted with respect to the layers, the smectic I phase has the same point symmetry C_{2h} as the smectic C phase. Consequently, the Fourier series expansion of the anisotropic mass distribution should contain all even terms, of order 2, 4, 6 .... The quadratic and quartic orders should thus be present below the SmC $\rightarrow$ SmI transition, at the same time as the hexatic order. However, X-ray analysis shows that the six neighbors of a molecule in the smectic I phase are placed on an almost perfectly regular hexagon: the sixth-order term is thus larger than all the others (Fig. C.VI.19). For this reason, in the following we shall only be concerned with the hexatic order parameter.

The optical anisotropy of the SmI phase being due to the molecular tilt in the plane of the smectic layers, the photo in figure C.VI.17 only gives access to the spatial variation of the phase ϕ_1 of the order parameter ψ_1. This yields the diagram in figure C.VI.20, clearly showing that the spatial variation of the phase ϕ_1 of order parameter ψ_1 follows the one of the phase ϕ_6. The two order parameters are hence coupled. This brings us to discussing the elasticity of the smectic I phase.

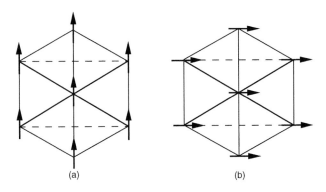

(a)   (b)

Fig. C.VI.19 Structure of the smectic I (a) and F (b) phases: the arrows represent the projection of the molecules onto the plane of the layers. In both cases, the hexagonal lattice is only slightly distorted.

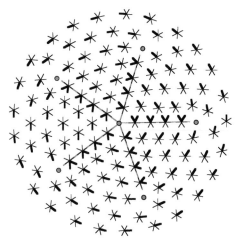

Fig. C.VI.20 Distribution of the phase of the order parameter ψ_1 inferred from the photo in figure C.VI.17. The phase undergoes a 60° discontinuity across the lines connecting the five peripheral disclinations to the central disclination.

VI.2.d) Pindak scar and elasticity of the SmI phase

On the lines connecting the five peripheral disclinations to the central one, the phase ϕ_1 undergoes a $2\pi/6$ rotation. One can see these lines as scars left by the Volterra process on creating the disclinations of the hexatic phase (Fig. C.VI.21). Indeed, in a true hexatic phase, the order parameter ψ_6 heals perfectly where the cut surfaces meet, the only trace left by the Volterra process being a singular point.

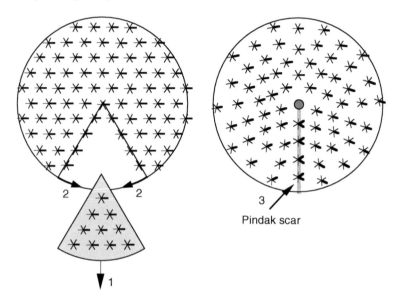

Fig. C.VI.21 Pindak scar created by the Volterra process: 1. 60° circular sector is cut out of the SmI phase. 2. The lips of the cut are brought together. 3. The SmI phase is glued shut: the hexatic order parameter is perfectly healed, but the ψ_1 order parameter remains scarred.

There still remains, however, in the smectic I phase a discontinuity of the order parameter ψ_1 along the joining line, that we shall call a "Pindak scar."

More generally, we call "Pindak wall" any wall across which the phase difference $\phi_1 - \phi_6$ undergoes a 60° jump.

The internal structure of a Pindak wall thus involves spatial variations of the phases $\phi_1(x)$ and $\phi_6(x)$ along the x axis taken as perpendicular to the wall (Fig. C.VI.22). At equilibrium, the functions $\phi_1(x)$ and $\phi_6(x)$ must minimize the energy F per unit length of wall:

$$F = \int dx\, f(\phi_1, \phi_6, \nabla\phi_1, \nabla\phi_6) \qquad\qquad (C.VI.53)$$

where f is the elastic deformation energy of the smectic I phase:

$$f = \frac{1}{2}\{K_1(\mathbf{grad}\,\phi_1)^2 + K_6(\mathbf{grad}\,\phi_6)^2$$

$$+ 2g\,(\mathbf{grad}\,\phi_1)\cdot(\mathbf{grad}\,\phi_6) - 2h\cos[6\,(\phi_1 - \phi_6)]\} \tag{C.VI.54}$$

This expression, put forward by Nelson and Halperin [11], contains four terms. The first three correspond to the elastic energy associated with the phase gradients of the two order parameters. The last term expresses the coupling between the two order parameters. Owing to this coupling, the difference $\delta\phi = \phi_1 - \phi_6$ between the two phases tends at each point toward a multiple of $2\pi/6$ (for $h > 0$), in agreement with the structure proposed in figure C.VI.22.

At the wall shown in figure C.VI.22, the angle $\delta\phi$ goes from 120° for $x < -\zeta/2$ to 60° for $x > \zeta/2$. The phases ϕ_1 and ϕ_6 vary in opposite ways across the wall, so that the elastic torques associated with the phase gradients of the two order parameters are balanced. As usual, the equilibrium equations for the torques are obtained by minimizing the total energy C.VI.53 with respect to the angle variable ϕ_1 and ϕ_6:

$$+ 6h\sin[6(\phi_1 - \phi_6)] - K_1\frac{d^2\phi_1}{dx^2} - g\frac{d^2\phi_6}{dx^2} = 0 \tag{C.VI.55a}$$

and

$$- 6h\sin[6(\phi_1 - \phi_6)] - K_6\frac{d^2\phi_6}{dx^2} - g\frac{d^2\phi_1}{dx^2} = 0 \tag{C.VI.55b}$$

Taking the sum and the difference of these two equations yields:

$$(K_1 + g)\frac{d^2\phi_1}{dx^2} = -(K_6 + g)\frac{d^2\phi_6}{dx^2} \tag{C.VI.56a}$$

and

$$12h\sin[6(\phi_1 - \phi_6)] - (K_1 - g)\frac{d^2\phi_1}{dx^2} + (K_6 - g)\frac{d^2\phi_6}{dx^2} = 0 \tag{C.VI.56b}$$

Let us look for solutions of the type (see figure C.VI.22):

$$\phi_1(x) = \phi_0 + \phi_{10}f(x)\;; \qquad \phi_6(x) = -\phi_{60}f(x) \tag{C.VI.57}$$

By replacing in eq. C.VI.56a, one obtains a first relation between the amplitudes ϕ_{10} and ϕ_{60} of the deformations:

$$\phi_{60} = \frac{K_1 + g}{K_6 + g}\phi_{10} = \kappa\,\phi_{10} \tag{C.VI.58}$$

Experimentally, the variation of the ϕ_1 angle is close to 60°, so that the ϕ_6 angle changes very little. Consequently,

$$\kappa = \frac{K_1 + g}{K_6 + g} << 1 \qquad\qquad (C.VI.59)$$

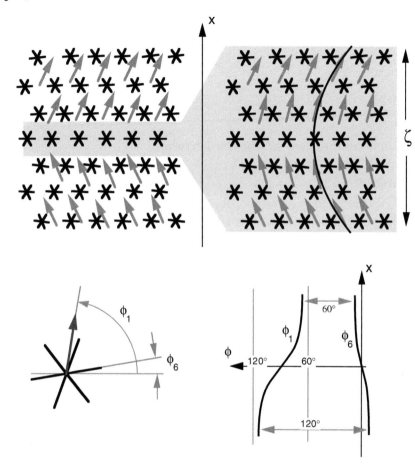

Fig. C.VI.22 Internal structure of an isolated Pindak wall: the angle $\delta\phi = \phi_1 - \phi_6$ goes from 120° for $x < 0$, to 60° for $x > 0$. For $K_6 >> K_1$, this is essentially obtained by a rotation of the ϕ_1 phase. The ϕ_6 phase also turns a little in the opposite direction, so that the elastic torques are balanced. As a result, the rotation angle of the ϕ_1 phase across the wall is below 60°.

This inequality is only fulfilled if

$$g << K_6 \qquad \text{and} \qquad K_1 << K_6 \qquad\qquad (C.VI.60)$$

In the following, we shall assume that $g = 0$, leading simply to:

$$\phi_{60} = (K_1 / K_6)\, \phi_{10} \qquad\qquad (C.VI.61)$$

Let us now replace ϕ_6 and ϕ_1 by their expressions C.VI.57 in equation C.VI.56b. We obtain a differential equation for the function $f(x)$:

$$12h \sin\{6\,[\phi_o + (1 + \kappa)\phi_{10}f(x)]\} - [(K_1 - g) + (K_6 - g)\kappa]\phi_{10}\frac{d^2f}{dx^2} = 0 \qquad (C.VI.62)$$

Introducing the new variable :

$$\psi = 6\,[\phi_o + (1 + \kappa)\phi_{10}f(x)] \qquad (C.VI.63)$$

and the typical length:

$$\zeta^2 = \frac{K_1}{36h(1 + \kappa)} \qquad (C.VI.64)$$

this equation takes the classical sine-Gordon form:

$$\zeta^2 \frac{d^2\psi}{dx^2} - \sin\psi = 0 \qquad (C.VI.65)$$

Its particular solution is:

$$\psi(x/\zeta) = 4\left[\pi - arctg\left(e^{x/\zeta}\right)\right] \qquad (C.VI.66)$$

Taking $\phi_6(0) = 0$, we finally find $\phi_o = \pi/2$, with:

$$\phi_1 = \phi_o + \phi_{10}\,f(x) = \frac{\pi\kappa}{2(1 + \kappa)} + \frac{\psi(x/\zeta)}{6(1 + \kappa)} \qquad (C.VI.67a)$$

and

$$\phi_6 = -\,\phi_{60}\,f(x) = \frac{\pi\kappa}{2(1 + \kappa)} - \frac{\kappa\psi(x/\zeta)}{6(1 + \kappa)} \qquad (C.VI.67b)$$

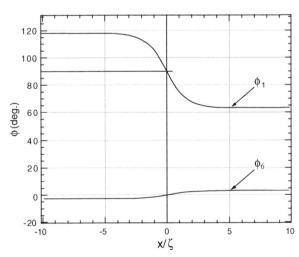

Fig. C.VI.23 Variation of the phases ϕ_1 and ϕ_6 across a Pindak wall; the difference $\delta\phi = \phi_1 - \phi_6$ goes from 120° on the left to 60° on the right.

The two phases ϕ_1 and ϕ_6 are plotted as a function of x/ζ in the diagram of figure C.VI.23. Across the wall, the difference $\delta\phi = \phi_1 - \phi_6$ between the two phases goes from 120° on the left to 60° on the right. However, each of the two phases ϕ_1 and ϕ_6 taken individually varies by less than 60°. For the ϕ_1 phase, measurable under the polarizing microscope, this variation is:

$$\phi_1(-\infty) - \phi_1(\infty) = 60°/(1 + \kappa) \tag{C.VI.68}$$

thus depending on the ratio κ between the elastic coefficients. Specifically, for $\kappa \ll 1$, this variation is 60°, the hexatic order being too "stiff" to be deformed. If, on the other hand, $\kappa = 1$, the two order parameters undergo deformations of similar amplitude (30°), but of opposite sign.

In conclusion, the variation of the angle ϕ_1 across a Pindak wall yields information about the ratio between the elastic coefficients of the two order parameters.

Finally, let us determine the energy of a Pindak wall per unit length. From equations C.VI.54, 66 and 67, we get:

$$F \approx (1/2) K_1 \left(\frac{2\pi/6}{\zeta}\right)^2 \zeta \approx (1/2) \frac{K_1}{\zeta} \approx 3\sqrt{K_1 h} \tag{C.VI.69}$$

It is now time to go back to the experiment of Dierker, Pindak and Meyer, where five Pindak walls are created in the smectic I phase instead of the 2π disclination of the smectic C phase. Clearly, the elastic deformation can be relaxed by the creation of these walls. Let L be the length of these walls. From figure C.VI.20, one can see that the energy gain ΔF_1 is essentially due to the lack of deformation of the hexatic order parameter ϕ_6 inside a disk of radius L centered on the five-arm star. The order of magnitude of this energy gain is:

$$\Delta F_1 \approx -\pi K_6 \ln\left(\frac{L}{c_0}\right) \tag{C.VI.70a}$$

where c_0 is a molecular distance. At the same time, the creation of five walls of length L increases the energy by:

$$\Delta F_2 \approx 15 \sqrt{K_1 h} \, L \tag{C.VI.70b}$$

Finally, the total energy variation in the system is:

$$\Delta F \approx 15 \sqrt{K_1 h} \, L - \pi K_6 \ln\left(\frac{L}{c_0}\right) \tag{C.VI.70c}$$

when the 2π disclination is replaced by a five-arm star of radius L. This quantity has its minimum at:

$$L = L^* \approx \frac{\pi}{15} \frac{K_6}{\sqrt{K_1 h}} \tag{C.VI.71}$$

It is also necessary that the energy variation ΔF be negative for $L = L^*$, requiring that:

$$\frac{K_6}{\sqrt{K_1 h}} > 13 \, c_0 \qquad\qquad\qquad (\text{C.VI.72})$$

a condition comfortably satisfied in practice.

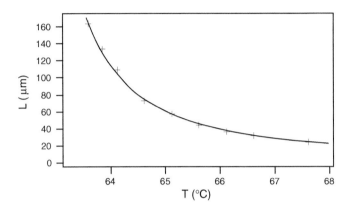

Fig. C.VI.24 Arm length as a function of temperature. The solid curve is the best fit with the theoretical law (from ref. [10]).

In conclusion, we recall that the theory predicts for the elastic constant $K_6 \propto \xi^2$ behavior, result confirmed by the light scattering experiments of Sprunt and Litster [12]. Consequently, the length of the arms L must be proportional to the positional correlation length ξ. This parameter is shown as a function of temperature in figure C.VI.24. The best fit with a function of the type $\exp[\mathrm{const}/(T - T_c)^\upsilon]$ (solid curve in the figure) yields $\upsilon \approx 0.38 \pm 0.15$, in good agreement with the theoretical prediction ($\upsilon = 0.37$).

VI.2.e) Experimental demonstration of the hexatic order by X-ray and electron diffraction

Optical microscopic observations of $2\pi/6$ disclinations represent a simple method for detecting the hexatic order in the SmI phase. This method is unfortunately inapplicable to the real hexatic phase, which is optically isotropic due to the molecules being normal to the layers. In this case, one must resort to X-ray or electron diffraction.

Equipped with powerful X-ray sources (rotating anode devices or synchrotrons), the hunters of hexatic phases believed they had detected the hexatic order in the smectic B phase of 4O.8 (see chapter C.VIII). It was however a false alert, as Pindak and his collaborators later showed that this phase was actually crystalline. However, at the same time as this invalidation, these same authors found proof that the smectic B phase present in the phase sequence of another material, the 65OBC, was indeed hexatic [13]:

$$\text{SmE (Crystal)} \xrightarrow{60°C} \text{SmB} \xrightarrow{68°C} \text{SmA} \xrightarrow{85°C} \text{Isotropic liquid}$$

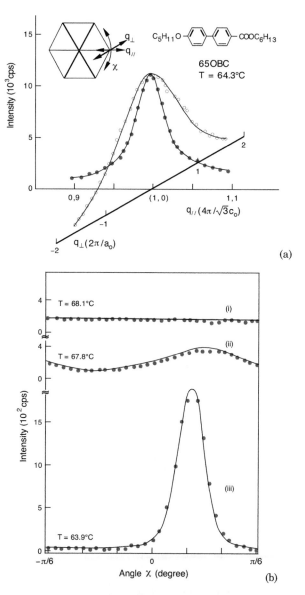

(a)

(b)

Fig. C.VI.25 a) Scattering intensity for the hexatic smectic B phase of 65OBC as a function of the $q_{//}$ (open dots) and $q_\perp$ (solid dots) components of the scattering vector. b) Scattered intensity as a function of the χ angle in the smectic A phase (i), just below the A-B transition (ii) and in the B phase (iii). From these measurements, one can obtain the structure factor $S(\mathbf{q})$ sketched in figure C.VI.26 (from ref. [13]).

They reached this conclusion by studying the X-ray diffraction diagrams of 65OBC smectic films. They measured the scattered intensities as a function of the scattering vectors $q_{//}$, $q_{\perp}$ and the azimuthal angle χ (Fig. C.VI.25), and determined that the structure factor of the SmB phase in 65OBC had the shape sketched in figure C.VI.26.

In terms of real-space structure, the torus surrounding the z axis normal to the smectic layers indicates the existence of short-range positional order within the smectic layers. In the smectic A phase, this torus has azimuthal symmetry around the z axis. In the hexatic smectic B phase, however, the cross section of the torus is modulated as a function of the scattering angle χ, revealing the existence of orientational order in the layers. Mathematically, the structure factor can be expanded in a Fourier series:

$$S(q_{\perp}, q_{//}, \chi) = \sum_{p} S_{6p}(q_{\perp}, q_{//}) \cos(6p\chi) \qquad (C.VI.73)$$

The experiment shows that, just below the temperature of transition toward the A phase, only the $S_6(q_{\perp}, q_{//})$ harmonic is detectable (Fig. C.VI.25ii).

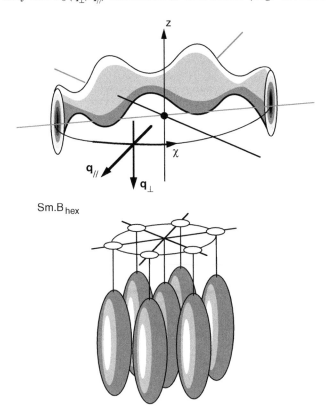

Fig. C.VI.26 Structure factor $S(\mathbf{q})$ of the hexatic smectic B phase. Compare with the structure factor of the smectic A phase.

With decreasing temperature, the other harmonics show up progressively and the angular distribution of the peaks decreases (Fig. C.VI.25 iii). We emphasize that angular broadening of the peaks is an effect of sample "mosaicity," stemming in principle from the algebraic decrease of long-range orientational order (ideal case of an "infinite" monodomain). In practice, it is often very difficult to avoid an accidental disorientation of the lattice (artefact) when preparing the sample and quantitative analysis of the way the orientational order evolves with temperature becomes delicate. In the case of negligible mosaicity (relevant close to the transition temperature toward the crystalline phase), the structure factor has a Lorentzian shape, the positional order in the plane of the layers being short-range:

$$S(\mathbf{q}, q_\perp = 0) = \frac{1}{[\xi^2 (\mathbf{q} - \mathbf{q_o})^2 + 1]} \qquad \text{(C.VI.74)}$$

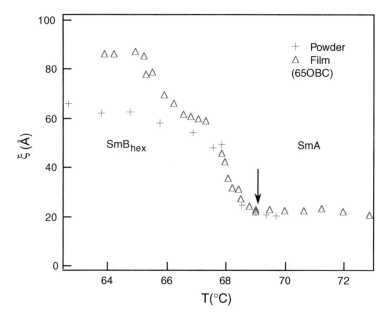

Fig. C.VI.27 Correlation length ξ as a function of temperature in 65OBC. It does not appear to diverge at the transition toward the crystalline E phase (from ref. [14]).

In this expression, $\mathbf{q_o}$ is one of the six basis vectors of the reciprocal lattice and ξ the positional correlation length in the plane of the layers. Generally, the intensity profile no longer has a simple expression (see refs. [15–17] for a detailed analysis). In practice, the intensity distribution along the $q_{//}$ direction is fitted with a Lorentzian, providing an estimate for the correlation length ξ (Fig. C.VI.27).

We would also point out that the films studied by Pindak using X-rays are thick (100 layers) and behave as bulk samples. The fact that six individual peaks were observed shows that **the orientational order of the bonds**

extends in three dimensions and over long distances in the hexatic smectic B. On the other hand, analysis of the diffuse rods along the $q_\perp$ direction (Fig. C.VI.25a) proves that there is **no positional correlation between the layers.**

Cheng et al. also obtained **electron** diffraction diagrams of smectic films [18]. This is quite an achievement, as the film must be placed inside an electron microscope, kept at the correct temperature in a gas atmosphere, all this without breaking the vacuum inside the microscope, but allowing passage of the electron beam. The diagrams in figure C.VI.28 [18, 19] were obtained using smectic films perpendicular to the incident electron beam.

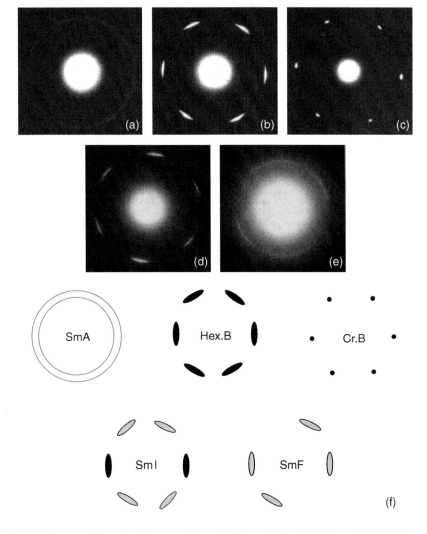

Fig. C.VI.28 Electron diffraction diagram of smectic films. (a) SmA; (b) hexatic SmB; (c) crystalline SmB; (d) SmI; (e) SmF. From (a) to (c), nine layer films (from ref. [19]); (d)–(e) four layer films (from ref. [18]); (f) theoretical diffraction diagrams.

These diagrams can then be seen as cuts through the S($\mathbf{q}$) function by the Ewald sphere tangent to the plane of the film. In diagram (a), one can identify the diffuse ring characteristic for the smectic A phase. In diagram (b), six "wide" peaks appear: it is the hexatic B phase. In (c), the six Bragg peaks are very sharp, the hallmark of a crystalline B phase. In diagrams (d) and (e), the spots are no longer equivalent: the sixfold symmetry is broken, and only twofold symmetry is preserved. These diagrams are typical for the tilted hexatic phases. In (d), two of the six maxima are stronger than the others: it is a smectic I. In (e), two spots are extinct and the four others are widely spread: this time, one is dealing with a smectic F. In this phase, the correlation length ξ is often much smaller than in the smectic I phase.

VI.3 Quasi-elastic light scattering

In this section, we aim to study the dynamics of the orientational order parameter quantified by the angle ϕ_6 that gives the orientation of the hexagonal lattice with respect to a fixed direction. We recall that in the tilted hexatic phases I or F, this variable couples to the angle ϕ_1 describing the direction of the molecular projection with respect to the reference direction. In the I phase, $\phi_6 - \phi_1 = 0$ (mod. $2\pi/6$) at equilibrium, while in the F phase, $\phi_6 - \phi_1 = \pi/6$ (mod. $2\pi/6$) (Fig. C.VI.19).

Quasi-elastic light scattering is a very powerful method for studying the dynamics of hexatic phases. Dierker and Pindak [20] were the first to use it successfully for the study of thin films of the homochiral material (S)-4-(2-methylbutyl)phenyl 4'-n-octylbiphenyl-4-carboxylate. In a two-layer film, the phase sequence is the following:

$$G^* \text{ crystal } \rightarrow J^* \text{ crystal } \rightarrow SmI^* \text{ (hexatic) } \rightarrow SmC^*$$

The star indicates chiral phases exhibiting a spontaneous polarisation P_o perpendicular to the director (see chapter C.IV on chiral smectics C). This property is very useful for calibrating the scattered intensity and gives access to the absolute values of the elastic constants and the viscosities of SmC* and SmI* phases.

In practice, the film is exposed to a laser beam with a power around 20 mW. The average intensity of the scattered signal, as well as its time autocorrelation function, are measured. For this experiment, the scattering vector is chosen to lie within the plane of the film, allowing the analysis of the way light is scattered by thermal orientation fluctuations of the director $\mathbf{n}$ in the plane of the film. We denote by $q_{//}$ and $q_\perp$ the components of $\mathbf{q}$ parallel and perpendicular to the director in the plane of the film. The accessible wave vector ranges from 1000 to 6000 cm^{-1}. Finally, the director field is aligned by a magnetic field.

In chapter B.II, we showed that the intensity I($\mathbf{q}$) scattered along a direction $\mathbf{k}' = \mathbf{k} + \mathbf{q}$ is proportional to the mean square amplitude of director orientation fluctuations $<\delta\phi_1^2(q)>$. To calculate this quantity, let us rewrite the elastic energy density C.VI.54, taking $g = 0$:

$$f = \frac{1}{2}\{K_1(\mathbf{grad}\,\phi_1)^2 + K_6(\mathbf{grad}\,\phi_6)^2 - 2h\cos[6(\phi_1 - \phi_6)\,]\} \qquad \text{(C.VI.75a)}$$

A rigorous treatment should distinguish between the "bend" modes (of wave vector $q_{//}$ parallel to the director) and the "splay" modes (of wave vector $q_\perp$ perpendicular to the director) and assign them distinct elastic constants: $K_{1S} \neq K_{1B}$ and $K_{6S} \neq K_{6B}$. In the following, we use the simple form C.VI.75a, with the convention that $K_1 = K_{1B}$ and $K_6 = K_{6B}$ for measurements performed at $q_\perp = 0$, while $K_1 = K_{1S}$ and $K_6 = K_{6S}$ for $q_{//} = 0$ (for these limiting cases the modes are decoupled, simplifying the calculations). Finally, we assume a strong "h" coupling, so that the difference $\phi_1 - \phi_6$ remains small. Under these conditions, the cosine term can be expanded to second order, yielding:

$$f = \frac{1}{2}\{K_1(\mathbf{grad}\,\phi_1)^2 + K_6(\mathbf{grad}\,\phi_6)^2 + 36h(\phi_1 - \phi_6)^2\} \qquad \text{(C.VI.75b)}$$

Let us now see how to determine $<\delta\phi_1^2(q)>$. This calculation is not trivial, due to the "h" coupling between angles ϕ_1 and ϕ_6. The solution consists of diagonalizing the eq. C.VI.75b by writing it in terms of the normal modes $\phi_- = \phi_6 - \phi_1$ and $\phi_+ = \alpha\phi_6 + \beta\phi_1$, such that:

$$f = \frac{1}{2}\{K_-(\mathbf{grad}\,\phi_-)^2 + K_+(\mathbf{grad}\,\phi_+)^2 - 36h\phi_-^2\} \qquad \text{(C.VI.75c)}$$

It is easily checked that the appropriate values are:

$$K_+ = K_1 + K_6, \quad K_- = \frac{K_1 K_6}{K_1 + K_6}, \quad \alpha = \frac{K_6}{K_1 + K_6} \quad \text{and} \quad \beta = \frac{K_1}{K_1 + K_6} \qquad \text{(C.VI.76)}$$

Using the equipartition theorem, one then obtains:

$$<\delta\phi_-^2(q)> \approx \frac{k_B T}{K_- q^2 + 36h} \quad \text{and} \quad <\delta\phi_+^2(q)> \approx \frac{k_B T}{K_+ q^2} \qquad \text{(C.VI.77)}$$

whence, knowing that $<\delta\phi_1^2> = <\delta\phi_+^2(q)> + \alpha^2 <\delta\phi_-^2(q)>$:

$$<\delta\phi_1^2> \approx \frac{k_B T}{(K_1 + K_6)q^2} + \frac{k_B T}{K_1(1 + K_1/K_6)q^2 + 36h(1 + K_1/K_6)^2} \qquad \text{(C.VI.78)}$$

Experimentally, the "h" coupling is strong, such that $h \gg K_6 q^2$ over the accessible range of wave vectors. Under these conditions, the second term in the preceding equation can be neglected, and the scattered intensity can be written as:

$$I^{-1} \propto (K_1 + K_6)q^2 = K_+ q^2 \qquad \text{(C.VI.79)}$$

This prediction, fully supported by the experimental results, gives access to the value of K_+ as a function of temperature in the C^* and I^* phases (Fig. C.VI.29). One can notice that K_+ exhibits discontinuous behavior at the transition, this jump being obviously related to the appearance of hexatic order. This feature is in agreement with the theoretical predictions.

Even more interesting are the results concerning the autocorrelation function of the scattered intensity $g(t) = <I(q,0)I(q,t)>$, shown in figure C.VI.30a for a five-layer film of SmI*. This dependence, a superposition of two mono-exponentials, reveals the existence of two separate relaxation modes, with dispersion relations shown in figure C.VI.30b.

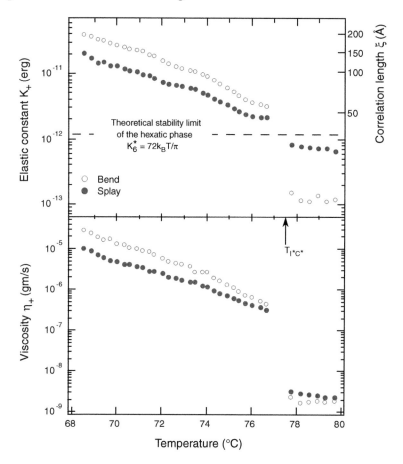

Fig. C.VI.29 Elastic constant K_+ and viscosity η_+ as a function of temperature in a two-layer film of (S)-4-(2methylbutyl)phenyl 4'-n-octylbiphenyl-4-carboxylate. In the I^* phase, $K_+ = K_6 + K_1$, while in the C^* phase, $K_+ = K_1$, as K_6 vanishes. One can see that in this material the "splay" and "bend" elastic constants exhibit a jump at the transition, which is first-order (note the absence of experimental points in the coexistence range of the two phases, typically extending over 1°C) (from ref. [20]).

The first mode, absent in the SmC* phase, is a **"non-hydrodynamic mode"** since its relaxation rate Γ_- does not vanish as $q \rightarrow 0$. Corresponding to the variable $\phi_- = \phi_6 - \phi_1$, the relaxation rate is obtained by solving in Fourier space the following phenomenological equation:

$$\frac{\delta f}{\delta \phi_-} = -\eta_- \frac{\partial \phi_-}{\partial t} \qquad \text{(C.VI.80)}$$

where η_- is a rotational viscosity. The calculation yields:

$$\Gamma_- = \frac{K_- q^2 + 36h}{\eta_-} \approx \frac{36h}{\eta_-} \qquad \text{(C.VI.81)}$$

This mode is specific to the hexatic I and F phases.

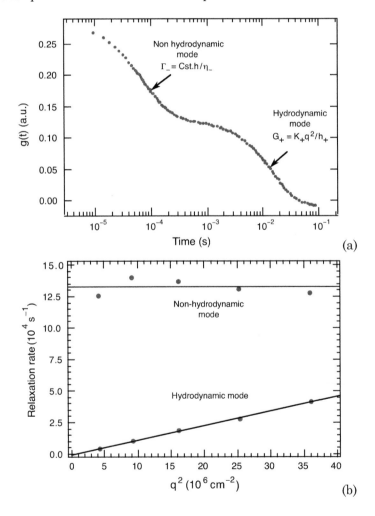

(a)

(b)

Fig. C.VI.30 a) Autocorrelation function of the scattered intensity as a function of time in a five-layer SmI* film. b) Relaxation rate of the two modes as a function of the wave vector (from ref. [20]).

The second mode, also present in the SmC* phase, is a **hydrodynamic mode**, as $\lim_{q \to 0} \Gamma_+ = 0$.

For this mode, associated to the variable ϕ_+, one obtains, using an equation similar to eq. C.VI.80:

$$\Gamma_+ = \frac{K_+ q^2}{\eta_+} \qquad\qquad (C.VI.82)$$

These two dispersion relations are in agreement with the experimental data (Fig. C.VI.30b). The temperature evolution of the "bend" and "splay" rotational viscosities η^+ is also displayed on the lower graph in figure C.VI.29. They differ by a factor 3, at most.

Similar measurements were performed by Sprunt and Litster using thick films (several μm) of a different material exhibiting a SmC → SmI transition. In this case, the twist modes along the normal of the film come into play. The results are, however, qualitatively identical, except for the presence of an additional, non-hydrodynamic mode in the SmI phase. These measurements are important insofar as they show that the three-dimensional stacking of the layers does not fundamentally change the properties of hexatic phases.

VI.4   Rheology

We have seen that the hexatic phase is in some way a crystal loaded with a very high density of free dislocations of opposite signs. At least, this is the image one would get if the system was frozen at a given moment. However, contrary to the case of ordinary crystals, these dislocations are at thermal equilibrium and decrease the free energy of the phase. They are also very mobile and move around, with a diffusion coefficient D. For these reasons, the theorists like to speak of a "dislocation plasma" [21]. Under stress, these dislocations move collectively, resulting in a deformation. Under these conditions, the phase should behave as a Newtonian liquid with a viscosity depending on the density of dislocations, hence on ξ. This is what we shall try to show experimentally in this section.

Two rheological experiments will be considered. The first one was performed on free-standing films, while the second concerns bulk samples. In this second case, 3D effects are expected.

VI.4.a) Viscoelasticity of a free-standing film

This experiment, fundamental since it reveals the true nature of the hexatic phase, is again due to Pindak et al. [22]. It consists of shearing a free-standing film in the plane of the layers and detecting its viscoelastic response. Obviously, the question is whether the phase behaves as a solid or as a liquid.

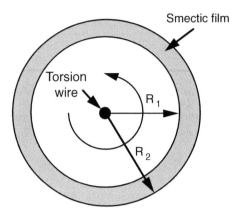

Fig. C.VI.31 Experimental setup.

In this experiment (described in detail in section C.VIII.5c), an annular film is stretched between two disks of different diameters (Fig. C.VI.31). The outer disk is fixed, while the inner one, attached to a torsion wire, can oscillate. In practice, this torsion oscillator is excited and its resonance frequency f, as well as its quality factor Q, are measured with a relative precision of 10^{-5} for the former and of 10^{-3} for the latter. We recall that the quality factor is a dimensionless quantity, defined as the product of the resonance frequency of the oscillator and its viscous damping time. It is also the ratio between the resonance frequency and the half-width of the resonance peak in the power spectrum of the oscillator ($Q = f/\Delta f_{1/2}$). The quantity Q^{-1} quantifies the dissipation, being proportional to the viscosity of the medium.

The experiment consists of first determining the resonance frequency and the quality factor of the oscillator in the empty state (typically, $f_o \approx 500$ Hz, $Q_o \approx 1000$), and then comparing these values with those (f and Q) obtained in the presence of a smectic film. In the following, we will rather use the period difference:

$$\Delta P = \frac{1}{f_o} - \frac{1}{f} \qquad \text{(C.VI.83)}$$

First of all, we notice that a jump in the period ΔP would point to the presence of a finite shear elastic modulus μ in the film. Indeed, the period P of the oscillator behaves as:

$$P = 2\pi \sqrt{\frac{I}{K_o + K_F}} \qquad \text{(C.VI.84)}$$

where I is the moment of inertia of the oscillator, K_o the twist modulus of the torsion wire, and K_F the correction due to the elasticity of the film:

$$K_F = 4\pi\mu \, \frac{R_1^2 R_2^2}{R_2^2 - R_1^2}$$

(C.VI.85)

Let us now see (Fig. C.VI.31) how ΔP and Q^{-1} behave as a function of temperature in the presence of a 65OBC film.

We first notice that ΔP remains zero at the SmA → SmB (hexatic) transition. This means that **the shear modulus is zero** in the hexatic phase, as in the smectic A phase. **The hexatic phase behaves like a liquid, a result consistent with the absence of long-range positional correlation.** The dissipation also seems to remain constant at this transition. This result is unfortunately not very relevant: it simply shows that dissipation within the film is negligible in comparison with the intrinsic dissipation (without the film) of the oscillator.

The situation changes 3°C lower in temperature. At this temperature, ΔP and Q^{-1} both exhibit a jump. This behavior is observed in all films and in the existence domain of the (bulk) hexatic phase. It is attributed to the crystallisation of the two surface layers of the film. This transition is hysteretic upon heating or cooling, proving that it is first-order.

Below 61°C, the film crystallises in the bulk (smectic E phase), and ΔP and Q^{-1} strongly increase.

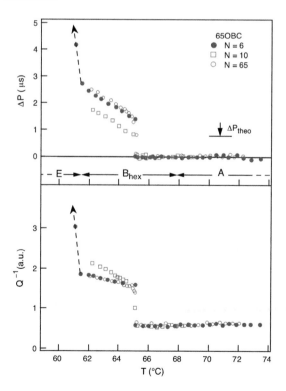

Fig. C.VI.32 Period difference (top) and dissipation (bottom) as a function of temperature for 65OBC films with different numbers of layers N (from ref. [22]).

341

VI.4.b) Rheology of bulk samples

Experiments performed on films are not sensitive enough to give the viscosity of hexatic phases under shear, hence the interest in using bulk samples. In this case, the difficulty lies in orienting the smectic layers; the problem is especially difficult when the material (65OBC, for instance) does not exhibit a nematic phase at higher temperatures. Two different geometries were used.

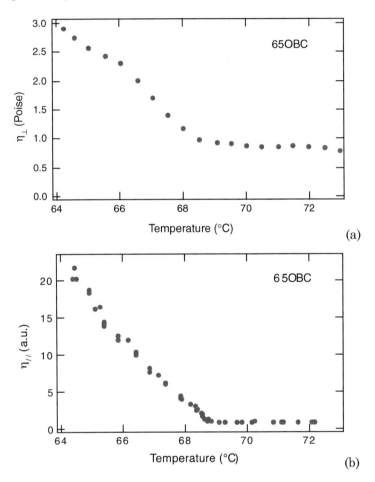

Fig. **C.VI.33** Temperature dependence of the viscosities for shear parallel to the layers (a) and in the plane of the layers (b) in the A and hexatic B phases of 65OBC. Comparison with the measurements of the correlation length ξ in figure C.VI.26 shows that $\eta_\perp$ varies like ξ, while $\eta_{//}$ is closer to $\xi^{2.5}$ (from refs. [23] and [24]).

In the first one, the sample undergoes a shear deformation parallel to the smectic layers [23]. The experiment, performed using the hexatic B phase of 65OBC and the F phase of 9O.4, shows that these two phases have a

Newtonian behavior entirely similar to that observed for the SmA phase. This result is compatible with the absence of positional correlations between the layers evidenced using X-rays. A less trivial result, lacking an immediate explanation, is that the measured viscosity $\eta_\perp$ varies linearly with the positional correlation length ξ in the plane of the layers (Fig. C.VI.33a):

$$\eta_\perp \propto \xi \qquad\qquad (C.VI.86)$$

The second geometry directly brings into play the free dislocations of the hexatic phase [24]. It consists of compressing a planar 65OBC sample (layers perpendicular to the glass plates) inducing a flow in the smectic layers. The experiment shows that the stress relaxes exponentially with time and finally vanishes. Thus, the hexatic phase behaves like a Newtonian fluid of viscosity $\eta_{//}$ in the plane of the layers. The results of the measurements (Fig. C.VI.33b) show that the apparent viscosity of the sample (proportional to $\eta_{//}$, up to a geometrical prefactor) is no longer proportional to the correlation length ξ, varying as:

$$\eta_{//} \propto \xi^{2.5} \qquad\qquad (C.VI.87)$$

Treating the hexatic phase as a dislocated solid (a dislocation plasma), Zippelius, Halperin and Nelson showed that the viscosity of a two-dimensional phase is proportional to ξ^2 [21]:

$$\eta_{//}(\text{theo.}) = \frac{k_B T}{D}\left(\frac{\xi}{a}\right)^2 \qquad\qquad (C.VI.88)$$

This dependence is not rigorously verified here. The discrepancy might be due to the three-dimensional character of the phase or to a temperature variation of the diffusion coefficient D of the dislocations. We recall that the same anomaly affects the rotational viscosity η_+ of the orientational order parameter measured in films. As shown in figure C.VI.29b, this viscosity varies much faster than ξ^2, in disagreement with theory, which predicts for this viscosity a law identical to eq. C.VI.88 [21].

BIBLIOGRAPHY

[1]   a) Peierls R., *Surprises in Theoretical Physics*, section 4: "Melting in one, two and three Dimensions," Princeton Series in Physics, Princeton University Press, Princeton, New Jersey, 1979.
b) Landau L., Lifchitz E., *Statistical Physics* (3rd edition), Pergamon Press, Oxford, 1980.

[2]   Kosterlitz J.M., Thouless D.J., "Ordering, metastability and phase transitions in two-dimensional systems," *J. Phys. C*, **6** (1973) 1181.

[3]   Kittel C., *Introduction to Solid State Physics*, Wiley, New York, 1995.

[4]   Halperin B.I., "Superfluidity, melting and liquid-crystal phases in two dimensions," in Proceedings of Kyoto Summer Institute 1979 - *Physics of Low-Dimensional Systems*, Eds. Nagaoka Y. and Hikami S., Publications Office, Progress of Theoretical Physics, Kyoto, Japan, 1979, p. 53.

[5]   Halperin B.I., Nelson D.R., *Phys. Rev. Lett.*, **41** (1978) 121.

[6]   Mermin N.D., *Phys. Rev.*, **176** (1968) 250.

[7]   Friedel J., *Dislocations*, Pergamon Press, Oxford, 1964.

[8]   Brock J.D., Birgeneau R.J., Litster J.D., Aharony A., *Contemporary Physics*, **30** (1989) 321.

[9]   Pershan P.S., *Structure of Liquid Crystal Phases*, World Scientific, Singapore, 1988.

[10] Dierker S.B., Pindak R., Meyer R.B., *Phys. Rev. Lett.*, **56** (1986) 1819.

[11] Nelson D.R., Halperin B.I., *Phys. Rev. B*, **21** (1980) 5312.

[12] Sprunt S., Litster J.D., *Phys. Rev. Lett.*, **59** (1987) 2682.

[13] Pindak R., Moncton D.E., Davey S.C., Goodby J.W., *Phys. Rev. Lett.*, **46** (1981) 1135.

[14] Moussa F., Benattar J.-J., Williams C., Proceedings of the Ninth International Liquid Crystal Conference, India, 6–12 Dec. 1982, *Mol. Cryst. Liq. Cryst.*, 1983.

[15] Aeppli G., Bruinsma R., *Phys. Rev. Lett.*, **53** (1984) 2133.

[16] Brock J.D., Aharony A., Birgeneau R.J., Evans-Lutterodt K.W., Litster J.D., Horn P.M., Stephenson G.B., Tajbakhsh A.R., *Phys. Rev. Lett.*, **57** (1986) 98.

[17] Aharony A., Birgeneau R.J., Brock J.D., Litster J.D., *Phys. Rev. Lett.*, **57** (1986) 1012.

[18] Cheng M., Ho J.T., Hui S.W., Pindak R., *Phys. Rev. Lett.*, **59** (1987) 1112.

[19] Jin A.J., Veum M., Stoebe T., Chou C.F., Ho J.T., Hui S.W., Surendranath S., Huang C.C., *Phys. Rev. E*, **53** (1996) 3639.

[20] Dierker S.B., Pindak R., *Phys. Rev. Lett.*, **59** (1987) 1002.

[21] Zippelius A., Halperin B.I., Nelson D.R., *Phys. Rev. B*, **22** (1980) 2514.

[22] Pindak R., Sprenger W.O., Bishop D.J., Osheroff D.D., Goodby J.W., *Phys. Rev. Lett.*, **48** (1982) 173.

[23] Oswald P., *Liq. Cryst.*, **1** (1986) 227.

[24] Oswald P., *J. Physique (France)*, **47** (1986) 1279.

Appendix 1   PEIERLS DISORDER

1D atom chain

The simplest case, studied by Peierls in his book *Surprises in Theoretical Physics* [1a], is the one of a chain consisting of N atoms placed, at T = 0, at positions:

$$x_n = n\, c_o \; ; \qquad n = 0, \pm 1, \pm 2... \tag{1.1}$$

and interacting with their nearest neighbors.

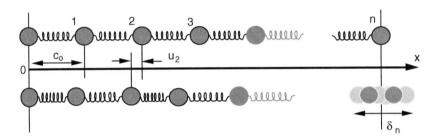

Fig. A.1 Atom chain undergoing thermal fluctuations.

Under the influence of thermal fluctuations, the atoms are displaced by u_n with respect to their ideal position x_n. Thus, the distance between the atom $n = 0$ (placed at the origin 0 and taken as reference) and its first neighbor, $n = 1$, differs from its ideal value c_o by δ. The mean quadratic value of δ depends on the temperature and the interaction potential U(x):

$$< \delta^2 > = k_B T \left(\frac{d^2 U}{dx^2}\right)^{-1} \tag{1.2}$$

The next atom, $n = 2$, is only influenced by its nearest neighbor, $n = 1$, with respect to which it tries to maintain the equilibrium distance, with the same uncertainty δ. The position error δ_n of the n-th atom will then be the superposition of n independent errors, each one of the order of δ:

$$\delta_n = n^{1/2} \delta \tag{1.3}$$

Peierls concludes that, as soon as $\delta_n > c_o$, the positional order of the ideal configuration is lost.

For a more rigorous proof, let us determine

$$\delta_n^2 = <(u_n - u_o)^2> \tag{1.4}$$

where u_o is the displacement of the central atom with respect to the origin and u_n the displacement of the n-th atom with respect to its ideal position $x_n = nc_o$.

This average value can be obtained by finding the vibration eigenmodes of the chain and then determining their mean square values from the energy equipartition theorem.

We start by writing the equation of motion for the s-th atom:

$$m\frac{d^2u_s}{dt^2} = \sum_p C_p(u_{s+p} - u_s)$$ (1.5)

The right-hand side represents the interaction force between this atom and its neighbors, identified by the indices $p = \pm 1, \pm 2 \dots$ (with $C_p = C_{-p}$). We seek for solutions of the form:

$$u_{s+p} = A_q\, e^{i[q(s+p)c_o - \omega t]}$$ (1.6)

After replacing in eq. 1.5, we obtain the dispersion relation for the phonons:

$$\omega^2 m = -\sum_p C_p\left(e^{ipqc_o} - 1\right)$$ (1.7)

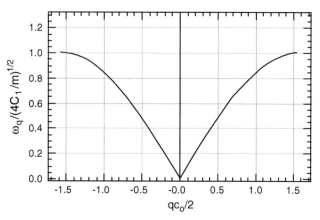

Fig. A.2 Dispersion relation for the phonons of the atom chain.

This relation simplifies drastically when only the nearest-neighbor interactions are taken into account ($C_1 = C_{-1}$). In this case,

$$\omega_q^2 = \frac{2C_1}{m}[1 - \cos(qc_o)]$$ (1.8a)

or, equivalently:

$$\omega_q = \sqrt{\frac{4C_1}{m}}\left|\sin\left(\frac{qc_o}{2}\right)\right|$$ (1.8b)

Note that, in the limit of small wave vectors q, the dispersion relation is linear and reads:

$$\omega_q = cq \qquad \text{with} \qquad c = c_0 \sqrt{\frac{C_1}{m}} \qquad (1.9)$$

The constant c represents the velocity of large-wavelength phonons.

We now apply the energy equipartition theorem. Each mode, of frequency ω_q and mean square amplitude $<A_q^2>$, stores a kinetic energy:

$$(1/2)\, M\omega_q^2 <A_q^2>, \qquad (1.10)$$

which is equal to $(1/2)k_BT$. In this expression, M = Nm is the mass of the chain. We obtain:

$$<A_q^2> = \frac{k_BT}{M\omega_q^2} \qquad (1.11)$$

Replacing in eq. 1.5 the fluctuations u_n and u_o by their expressions 1.6 yields:

$$<\delta_n^2> = \sum_q \{<A_q^2>[1 - \cos(qnc_o)]\} \qquad (1.12)$$

or

$$<\delta_n^2> = \frac{k_BT}{M} \sum_q \frac{1 - \cos(qnc_o)}{\omega_q^2} \qquad (1.13)$$

For sufficiently large values of N, the sum over all the eigenmodes of wave vector q can be replaced by an integral

$$\sum_q [\ ] = \int [\ ]g_1(q)dq \qquad (1.14)$$

where $g_1(q)dq$ accounts for the number of modes in the interval $[q, q+dq]$. The function $g_1(q)$ is the density of eigenmodes. We have already shown that, in 1D, it is given by (see figure C.VI.4):

$$g_1(q) = 1/(2\pi/L) \qquad (1.15)$$

Transforming the sum into an integral in eq. 1.13, we finally obtain:

$$<\delta_n^2> = \frac{k_BT}{M} \int dq \frac{Nc_o}{2\pi} \frac{1 - \cos(qnc_o)}{\omega_q^2} = 2\frac{k_BT}{m} \frac{c_o}{2\pi} \int dq \frac{\sin^2(nqc_o/2)}{\omega_q^2} \qquad (1.16)$$

For large n, the integrand is equal to $n^2q^2c_o^2/(4c^2q^2)$ over a small fraction, of the order of 1/n, of the integration interval $[-\pi/c_o, \pi/c_o]$. Elsewhere in this interval, it oscillates between 0 and $1/(c^2q^2)$, leading to the following estimate:

$$\int dq \, \frac{\sin^2(nqc_o/2)}{\omega_q^2} = \frac{n^2 c_o^2}{4c^2} \frac{2\pi}{nc_o} + \frac{c_o^2}{c^2} \frac{2\pi}{c_o}$$ (1.17)

yielding, in the limit of large n, the final result:

$$<\delta_n^2> = n \frac{k_B T}{2mc^2} c_o^2$$ (1.18)

In conclusion, for a one-dimensional chain of atoms, the displacement with respect to the ideal positions of the atoms increases like $n^{1/2}$. For large enough n, the long-range positional order is destroyed.

2D and 3D crystals

In the case of two- and three-dimensional crystals, the dispersion relations are more complex, as for a given q one has several modes of different polarisations, and hence with different dispersion relations, although these still behave linearly with q in the small-q limit.

Using the same arguments as in the previous case, one obtains the following estimate:

$$<\delta_n^2> = C \frac{k_B T}{M} \int dq \, g(q) \frac{1 - \cos(nqc_o)}{\omega q^2}$$ (1.19)

where C is a numerical factor and g(q) the mode density, given by

$$g_2(q) = \frac{2\pi q dq}{(2\pi/L)^2} = \frac{L^2 q}{2\pi} dq \qquad \text{in} \qquad \text{2D}$$ (1.20a)

and

$$g_3(q) = \frac{4\pi q^2 dq}{(2\pi/L)^3} = \frac{L^3 q^2}{2\pi^2} dq \qquad \text{in} \qquad \text{3D}$$ (1.20b)

These relations are then plugged into eq. 1.19, and calculating the integral yields:

$$<\delta_n^2> = \frac{k_B T}{mc^2} c_o^2 [\ln(n) + C_2] \qquad \text{in} \qquad \text{2D}$$ (1.21a)

and

$$<\delta_n^2> = \frac{k_B T}{mc^2} c_o^2 C_3 \qquad \text{in} \qquad \text{3D}$$ (1.21b)

In conclusion, the true long-range positional order only exists in three-dimensional crystals, where $<\delta_n^2>$ is independent of n and remains finite. In two-dimensional crystals, on the other hand, the crystalline order slowly disappears over large enough distances.

Chapter C.VII

The smectic B plastic crystal

"Ordered" smectics were first observed in 1932 by Herrmann and Krummacher in 4-phenylbenzylideneaminocinnamate [1]. These authors found no less than four distinct phases in this material, between the isotropic liquid and the crystalline phase. They pointed out that the one just above the crystalline phase in temperature is very viscous, and inferred that it must be of the smectic type. The following year, Bernal and Crowfoot confirmed this hypothesis by analyzing its structure using X-rays [2] and concluded that the molecules are placed on the sites of a hexagonal lattice, and that they undergo a rotational motion about their elongation axis. At the time, they did not go further in their analysis and it is only in the 1970s that physicists again started to show an interest in these phases.

The smectic B phases, to be considered in this chapter, belong to the family of ordered smectics. Their common features are that the molecules are oriented along the layer normal and that they exhibit hexagonal symmetry [2–7]. As a consequence, they are optically uniaxial. However, it has long remained undecided whether they were real crystals or just stackings of crystallized layers, uncorrelated one with the other, as proposed by de Gennes and Sarma in 1972 [8]. Today we know that they are real three-dimensional crystals, but highly sophisticated X-ray scattering experiments were needed to prove this result conclusively [9, 10]. Consequently, from a structural point of view, these phases are fundamentally different from the hexatic phases, which lack a long-range positional order. They do, however, exhibit some physical properties similar to hexatic phases, and even to more fluid phases such as the smectic A. We shall see, for instance, that the smectic B exhibits a layer buckling instability, analogous to the one encountered in smectics A. On the other hand, as far as their interfacial and growth properties are concerned, they are much closer to ordinary crystals than to fluid phases.

In this chapter we shall try to describe this very diverse behavior. Most experiments were performed on 4O.8 (of the nO.m series), the most studied material in this area of research. We recall that this material exhibits the following phase sequence (on heating) [11]:

Crystal–33°C–SmB–49.9°C– SmA–63.7°C–Nem.–79°C– Isotropic liquid

The chapter is structured as follows. First of all, we shall consider in detail the structure of SmB phases, the subject of controversy for a long time (section VII.1). The subsequent discussion will be confined to the most common phase, of hexagonal compact structure (hcp) and corresponding to a layer stacking of the ABAB... type. We shall see that this phase is highly anisotropic, with immediate consequences on its elastic and plastic properties. In particular, we demonstrate the existence of a layer buckling instability, somewhat surprising in a 3D crystal (section VII.2). We shall then describe the static properties of the smectic B-smectic A interface and take this opportunity to review the basic notions of facet and of forbidden orientation (section VII.3), and then to describe the Herring instability that appears when such an orientation is forced to develop (section VII.4). We shall show afterward how this interface gets destabilized in directional growth (Mullins-Sekerka instability, section VII.5). We shall see that its morphology above the threshold and its nonlinear evolution are strongly dependent on the orientation of the layers with respect to the interface. Finally, we shall also analyze the facet dynamics and show that the growth process creates a stress distribution that can lead to layer buckling.

VII.1 Structure of smectic B phases

The first systematic X-ray studies of smectic B phases were performed in the 1970s by Levelut et al. [4–6] and Richardson et al. [7]. They quickly established that:

1. The molecules are arranged on a triangular lattice within the layers.

2. The positional order in the plane of the layers is long-range.

3. The molecules are perpendicular to the layers, their orientation fluctuating by a few degrees about their average position ($\pm$ 4° in 4O.8, from Levelut et al. [4]).

4. And that the molecules turn rapidly about their long axis, bestowing upon these phases a hexagonal symmetry. Thus, they are optically uniaxial, the optical axis being perpendicular to the layers.

On the other hand, these experiments gave little information on the stacking of the layers themselves and on the correlations between layers, obviously much weaker than in a classical crystal. Thus, one had to wait for the high-resolution X-ray experiments performed by Moncton and Pindak on free-standing films in 1979 [9], and for those of Pershan and his collaborators on bulk samples in 1981 [10], in order to find out the nature of the stacking and whether the layers exhibit long-range correlation. For the B phase of 4O.8, the measurements show that the positions of the molecules are correlated over a distance $\xi_\perp$ exceeding 1 µm in the plane of the layers and over a distance $\xi_{//}$ of more than 0.2 µm across the layers. Within the resolution limit of X-ray experiments, the B phase of 4O.8 can thus be considered as a true three-dimensional crystal. It is nevertheless a **plastic** crystal, since the orientation of the molecules is not completely fixed. This crystal is also characterized by a considerable intrinsic disorder, revealed by the strong diffuse scattering around the Bragg spots; this signal is partly due to the thermally excited phonons (shear waves propagating across the layers) but also originates in the numerous stacking faults between the smectic layers. Thus, the energetic coupling between the layers is very weak.

Let us point out that the layers are most often stacked according to a ABAB... pattern (hexagonal compact structure), but this pattern can change with temperature [10, 12]. In 4O.8, for instance, the stacking is successively of the ABAB... type between 49.9°C and 22°C, then of the ABCABC... type from 22°C to 12°C, and then again ABAB... below 12°C. Below 7°C, the lattice becomes monoclinic, the molecules becoming tilted with respect to the layer normal [10]. These recurrent phase transitions are due to the very weak energetic coupling between the layers. We note that the AAA... stacking was observed in 7O.7, a material of the same series as 4O.8 [13, 14]. This particular stacking was never encountered in non-mesogenic materials.

Finally, we mention the presence in these phases of linear (but localized) defects, consisting of a molecular vacancy surrounded by an interstitial defect delocalized over a dozen molecular sites along the layer normal [15–17]. These defects are very numerous (typically, 1% of the total number of molecules) and very mobile, their residence time on a given site of the hexagonal lattice being estimated as 10^{-11}s [15]. Other defects, such as dislocations, are also present in the B phase. Their existence was indirectly revealed by the elasticity and plasticity experiments to be described in the next section.

VII.2 Elastic and plastic properties

The experiments we are about to describe were performed in the "high temperature" B phase, of hexagonal compact structure, of 4O.8. One of the main characteristics of this plastic crystal, uncovered by X-ray scattering experiments, is the weak energetic coupling between the layers. This feature can be partly explained by the strong structural anisotropy, defined as the ratio of the layer thickness and the parameter of the hexagonal lattice:

$$A = \frac{a_o}{c_o} \qquad\qquad (C.VII.1)$$

Indeed, it can be shown that the coupling energy between the layers varies as $\exp(-2\pi A)$ [18]. In the B phase of 4O.8, $A \approx 6$, since $a_o \approx 30$ Å and $c_o \approx 5$ Å. It is thus not surprising that the layers exhibit very weak energetic coupling in this material.

This property is directly reflected by the elastic and plastic properties of the B phase. We shall illustrate this by describing two creep experiments in shear parallel to the layers. The first one involves very small deformations, belonging to the realm of micro-plasticity. The second one is a true plasticity experiment, the sample undergoing considerable deformation. It will be shown that these two experiments are complementary and can be interpreted using a unique model for the dislocations. In conclusion, we shall comment upon an old experiment by Ribotta on layer buckling, showing that the layers slide easily on top of each other.

VII.2.a) Microplasticity

The first measurements of the elastic moduli in the crystalline B phase, and in particular of its C_{44} modulus in shear parallel to the layers were performed in the beginning of the 1980s. We recall that, in a hexagonal phase, the elastic deformation energy depends on six elastic constants C_{ij} and has the general expression [19]:

$$\rho f_e = \frac{C_{11}}{2}(u_{xx}^2 + u_{yy}^2) + C_{12}\,u_{xx}u_{yy} + (C_{11} - C_{12})u_{xy}^2 + \frac{C_{33}}{2}\,u_{zz}^2$$

$$+ C_{13}\,u_{zz}\,(u_{xx} + u_{yy}) + 2C_{44}(u_{xz}^2 + u_{yz}^2) \tag{C.VII.2}$$

The z axis is taken along the C_6 axis and u_{ij} is the strain tensor: $u_{ij} = \frac{1}{2}(u_{i,j} + u_{j,i})$.

A classical method for determining the elastic constants consists of measuring the velocity of ultrasonic waves (with a frequency between 1 and 100 MHz) along several directions. This method was employed by different authors [20–22] leading to the conclusion that all elastic constants are of the same order of magnitude, (of the order of 10^9 erg/cm³), except for C_{44} which is much smaller (in 40.8, $C_{44} < 10^7$ erg/cm³). Unfortunately, this technique does not give access to the value of C_{44}. In the following, we shall assume that $C_{11} \approx C_{12} \approx C_{33} \approx C_{13} = C = 10^9$ erg/cm³.

To prove that the C_{44} constant is finite, and that it remains finite at low frequency, Cagnon and Durand performed two micro-plasticity experiments.

In the first one [23, 24], the smectic B is sheared parallel to the layers. The sinusoidal excitation is applied by a piezoelectric ceramic element. The stress, which yields the mechanical impedance of the sample, is measured using a second ceramic glued onto the opposite glass plate (Fig. C.VII.1). This experiment shows that the smectic B has an elastic response between 10 Hz and 7 kHz with a shear modulus $C_{44} \approx 10^6$ erg/cm³.

In the second experiment, performed using the same device [25], the system is sheared at a constant strain rate $\dot{\varepsilon}$ parallel to the layers and the stress σ is measured as a function of time. The deformations ε are still very small, below 10^{-3}. The time evolution $\sigma(t)$ of the stress is shown in figure

C.VII.2. We see that σ increases and then saturates at a constant value σ_s, a signature of plastic deformation.

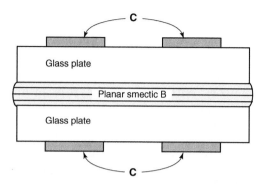

Fig. C.VII.1 Experimental setup of Cagnon and Durand used to measure the mechanical impedance of a planar sample in shear parallel to the layers (from [23]). The ceramics C are used to impose the strain and to measure the stress.

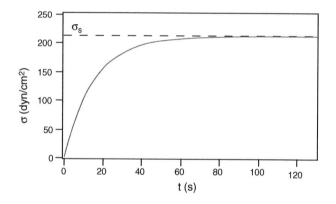

Fig. C.VII.2 Stress as a function of time at fixed shear rate ($\dot{\varepsilon} = 1.8 \times 10^{-5}$ s^{-1}, from ref. [25]).

To explain this behavior, Cagnon and Durand propose a Maxwell model, well-known in viscoelasticity, which amounts to replacing the sample by a sandwich composed of a purely elastic solid (with elastic modulus C_{44} and thickness αd), coupled in series with a Newtonian fluid (of viscosity η and thickness $(1 - \alpha)d$). Under these conditions, the sample responds to a deformation ε according to the following material law:

$$\dot{\varepsilon} = \frac{\alpha\dot{\sigma}}{C_{44}} + \frac{(1-\alpha)\sigma}{\eta} \qquad\qquad (C.VII.3)$$

which (solved at fixed $\dot{\varepsilon}$) yields:

$$\sigma(t) = \sigma_s \left[1 - \exp\left(-\frac{t}{\tau}\right) \right] \qquad\qquad (C.VII.4a)$$

with

$$\sigma_s = \frac{\dot{\varepsilon}\eta}{1 - \alpha} \qquad \text{and} \qquad \tau = \frac{\eta}{C_{44}} \frac{\alpha}{1 - \alpha} \qquad\qquad \text{(C.VII.4b)}$$

This law accounts for the experimental results if we take $\eta_{app} = \eta/(1-\alpha) \approx 10^7$ poise and $C_{44}/\alpha \approx 10^6$ erg/cm^3. Cagnon and Durand were also able to check that the apparent viscosity is independent of the applied stress (Newtonian behavior). Within this model, α is an unknown parameter, probably close to 1 (this result will be justified later), since $1-\alpha$ represents the volume fraction of "bad crystal" occupied by the dislocations responsible for the deformation. Consequently, C_{44} itself would be of the order of 10^6 erg/cm^3, a result in good agreement with the previous results.

This experiment illustrates the high ductility of the smectic B in shear parallel to the layers. However, it gives little information on the microscopic mechanisms of plastic deformation. This is why we shall now describe another creep experiment, involving much larger stresses and strains.

VII.2.b) Plastic behavior under strong deformation

In this new creep experiment in shear parallel to the layers [26], the σ stress is applied (and kept constant), and the strain rate $\dot{\varepsilon}$ is measured using a linear displacement sensor. Here, $\sigma > 1000$ dyn/cm^2 and $\varepsilon \gg 10^{-3}$, while in the experiments of Cagnon and Durand $\sigma < 200$ dyn/cm^2 and $\varepsilon < 10^{-3}$. Under these conditions, the dislocations move over much larger distances, providing additional details on the mechanisms of plastic deformation. This experiment, performed using 4O.8, shows that the creep is stationary at all temperatures ($\dot{\varepsilon}$ is constant in time and only depends on σ and T), and that it is no longer Newtonian, but rather exponential in the stress and thermally activated. As shown in figure C.VII.3, an Arrhenius law of the type:

$$\dot{\varepsilon} = \dot{\varepsilon}_o \exp\left(-\frac{Q}{k_B T}\right) \exp\left(\frac{\sigma\Omega}{k_B T}\right) \qquad\qquad \text{(C.VII.5)}$$

with $\dot{\varepsilon}_o = 3 \times 10^7$ s^{-1}, an activation energy $Q \approx 0.64$ eV and an activation volume Ω close to 5×10^{-17} cm^3 (see figure C.VII.4 for the exact values) accounts for all the experimental results, measured over a wide temperature range. It is noteworthy that the activation volume is very large (10^5 molecular volumes) compared with the values found in ordinary solids (a few hundred atomic volumes, generally).

On the other hand, the experimentally observed Arrhenius law suggests that the movement of dislocations is controlled by thermally activated crossing of localized obstacles. The lattice friction being negligible due to the low energetic coupling between the layers, the only possible obstacles are the other dislocations, especially screw dislocations of Burgers vector **ST** (Fig. C.VII.5).

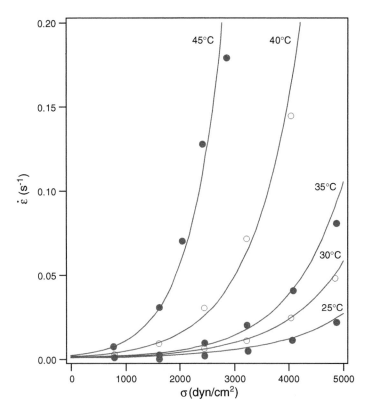

Fig. C.VII.3 Creep rate of a 4O.8 sample, of thickness 50 μm, as a function of the applied stress and temperature. The solid lines were computed with the values of the activation volume given in figure C.VII.4 (from ref. [26]).

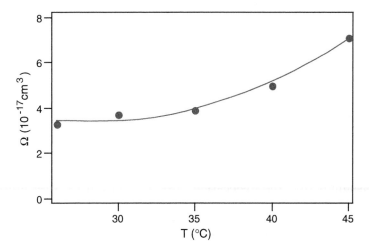

Fig. C.VII.4 Activation volume as a function of temperature. These values were used to fit the experimental curves in figure C.VII.3 (from ref. [26]).

357

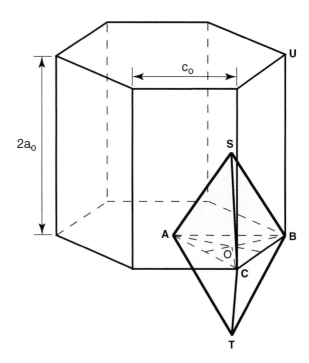

Fig. C.VII.5 Burgers vectors of the principal dislocations in a hexagonal compact structure. The **AB**, **ST** and **CU** dislocations are perfect. The **AO** dislocation is imperfect.

These dislocations are called trees in the very suggestive language of metallurgists. As shown in figure C.VII.6, the crossing of a tree and a basal dislocation (whose the Burger vector is in the plane of the layers) is accompanied by the formation of a jog on each dislocation. This crossing is therefore thermally activated, since an energy at least equal to the one of the two jogs must be provided so that the mobile dislocation can cross the obstacle. This energy is provided by the applied stress, exerting a glide force on the basal dislocations. Without going into all the details of the model, this kind of mechanism leads to the following creep law (see appendix 1 for a simplified demonstration) [18, 26]:

$$\dot{\varepsilon} = \rho_m c_o \sqrt{\frac{\sqrt{CC_{44}}}{\rho}} \exp\left(\frac{\Delta S}{k_B}\right) \exp\left(-\frac{U_k + U_k'}{k_B T}\right) \exp\left(\frac{(\sigma - \sigma_i)\,\Omega}{k_B T}\right) \qquad (C.VII.6)$$

This law assumes that the dislocation does not move back after crossing the obstacle, which holds true if the stress is strong enough. In the opposite case, the possibility of backward jumps must be considered; we shall return to this point later, in order to interpret the micro-plasticity experiments of Cagnon and Durand.

In expression C.VII.6, ρ_m is the density of mobile dislocations, ρ the density of the material (close to 1 g/cm^3) and ΔS an entropic term of the order of $5k_B$, stemming mainly from the temperature dependence of the energy barrier via the elastic moduli [26].

The activation energy Q is the sum of the energy U_k of the jog formed on the tree:

$$U_k \approx 0.4\sqrt{CC_{44}}\, a_o^2 c_o \approx 0.1 \text{ eV} \qquad\qquad \text{(C.VII.7a)}$$

And the energy U_k' of the jog formed on the mobile dislocation:

$$U_k' \approx 0.2\, Ca_o c_o^2 \approx 0.4 \text{ eV} \qquad\qquad \text{(C.VII.7b)}$$

These approximate formulae were obtained using the results of Chou [27], who gave the exact expressions of the stress field around dislocations in anisotropic elasticity. They yield for the creep an activation energy $Q = U_k + U_k' \approx 0.5$ eV, in good agreement with the experiments.

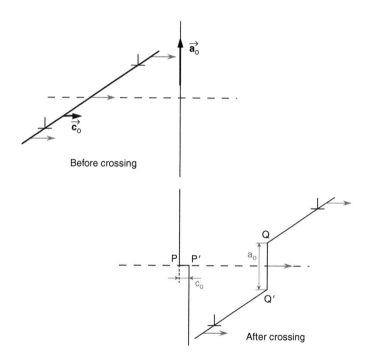

Fig. C.VII.6 Crossing of a mobile edge dislocation and a screw dislocation ("tree"). Two jogs of the "edge" type appear, of length $PP' = c_o$ and $QQ' = a_o$, respectively.

As to the activation volume Ω, it represents the number of molecules involved in an activated jump. It is given by [18]:

$$\Omega = \Lambda a_o L \qquad\qquad \text{(C.VII.8)}$$

where Λ is the average distance between two trees and L the dissociation width of the core of a mobile dislocation in the plane of the layers. It is indeed likely that, due to the weak coupling between the layers, these dislocations dissociate into two partial dislocations separated by a stacking fault (Fig. C.VII.7). To estimate L, we make the – reasonable – assumption that the density of trees is comparable to that of screw dislocations in the smectic A phase found at higher temperature. In this case, $\Lambda \approx 5 - 10 \; \mu m$ (see chapter C.III), yielding $L \approx 20 - 40 \; \text{Å}$ since $\Omega \approx 5 \times 10^{-17} \; cm^3$. This result provides an estimation for the surface energy γ of the stacking fault. Comparing the energy gained by decomposing the dislocation into two partial dislocations (since $\mathbf{AB}^2 > \mathbf{AO}^2 + \mathbf{OB}^2$), with the energy cost of creating the stacking fault (of surface energy γ), yields:

$$L \approx \frac{\sqrt{CC_{44}} \; \mathbf{AO.OB}}{2\pi\gamma} \approx \frac{\sqrt{CC_{44}} \; a_o^2}{20\gamma} \qquad (C.VII.9)$$

With $L = 20 - 40 \; \text{Å}$, one obtains $\gamma \approx 1 - 2 \times 10^{-2} \; erg/cm^2$, which is a very low value (100 times lower than in graphite, for instance). This dissociation also results in a significant decrease of the Peierls-Nabarro stress, thus facilitating the glide motion.

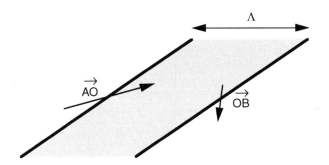

Fig. C.VII.7 Dissociation of a basal dislocation **AB** into two partials **AO** and **OB**. The cross-hatched area corresponds to a stacking fault.

Finally, σ_i is the internal stress related to the elastic deformation around the mobile dislocations. This stress sets the yield threshold for the smectic B phase. It can be calculated in anisotropic elasticity as [26]:

$$\sigma_i \approx \frac{\sqrt{CC_{44}}}{2\pi} a_o c_o \rho_m \qquad (C.VII.10)$$

with ρ_m the density of mobile dislocations. Using this expression, after comparing the theoretical creep law C.VII.6 with the one found experimentally (eq. C.VII.5), one finds $\rho_m \approx 5 \times 10^8 \; cm^{-2}$ and $\sigma_i \approx 30 \; dyn/cm^2$. This particularly weak value of the internal stress (which cannot be detected by the devices employed) is an important consequence of the elastic anisotropy of the material.

To conclude this discussion, let us return to the micro-plasticity experiment of Cagnon and Durand already described. Here, the applied stress is very weak ($\sigma < 200$ dyn/cm^2) and one must consider the possibility of backward jumps once a dislocation crosses an obstacle. In this limit, the exponential dependence on the stress is replaced by a hyperbolic sine ($\dot{\varepsilon} \propto \text{arcsinh}(\sigma\Omega/k_BT)$), which can also be linearized if $\sigma\Omega/k_BT \ll 1$. This condition is fulfilled in micro-plasticity, as $\sigma\Omega/k_BT$ is always below 0.2. One recovers then the Newtonian behavior described by the Maxwell relation C.VII.3 ($\dot{\varepsilon} \propto \sigma$ for $\sigma = $ Cst). We can also estimate the order of magnitude of the fraction of bad crystal $1 - \alpha$, a phenomenological parameter introduced in the Maxwell model:

$$1 - \alpha \approx L\, a\, \rho_m \approx 6 \times 10^{-5} \qquad\qquad \text{(C.VII.11)}$$

In agreement with our assumptions, α is very close to 1, validating a posteriori the estimation of C_{44}.

In summary, the plastic behavior of the B phase can be explained using an Arrhenius model. As a consequence, one is led to consider a deformation mechanism by activated glide motion of basal dislocations with obstacle crossing, and can thus obtain a first estimation of their density. Nevertheless, it would be very useful to visualize these defects (e.g., by electron microscopy) in order to validate the initial assumptions conclusively. Creep experiments were also performed in compression normal to the layers. They can also be interpreted using defect models, but the mechanisms are different since in this case there is flow across the layers. It has thus been shown that the creep is of the Nabarro-Herring type at low stress, then of Harper-Dorn type at higher stress, with a climb motion of the transverse wedge dislocations **ST** [28].

In dilation normal to the layers, the situation is more complicated due to the layer buckling.

VII.2.c) Layer buckling instability

Ribotta [29] was the first to study a layer buckling instability in ordered smectics of the B or H types. He performed his experiments on 7O.7, but we checked that a similar behavior is observed in the B phase of 4O.8. This instability is of the same nature as the one observed in smectics A, but its dynamics are much slower. Thus, it appears several seconds (up to one minute) after dilation, while in a smectic A, a few ms are enough. On the other hand, the undulations remain linear far from the threshold in smectic B phases, while smectics A very quickly develop a square undulation lattice, followed by the nucleation of focal parabolae.

To interpret this instability, one must keep in mind that a tilt of the layers allows the system to relax the imposed dilation stress. If the molecules are strongly anchored on the surface, the layers have no choice but to develop an undulation, of maximal amplitude in the middle of the sample and progressively going to zero toward the plates. The instability threshold is reached when the energy gain of tilting the layers exactly compensates the loss in elastic energy due to bending the layers. In a smectic A, the bending energy of the layers is reduced to the Frank energy, as the layers can easily slide one over the other. This bending energy is associated with the disorientation of the director field and its expression is, in a smectic B as well as in smectics A:

$$\rho f_c = \frac{1}{2} K_B \left(\frac{\partial^2 u_z}{\partial x^2} + \frac{\partial^2 u_z}{\partial y^2} \right) \tag{C.VII.12}$$

where K_B is the bending modulus of the layers. However, in the smectic B phase there is an additional energy cost due the layers sliding with respect to one another, since C_{44} is finite. It is then easy to go through the same calculations as for a smectic A, while taking into account this additional term, and to show that, in a sample of thickness d, the undulation develops above a critical dilation:

$$\alpha_c = \frac{C_{44}}{C} + 2\pi \frac{\lambda_B}{d} \tag{C.VII.13a}$$

where

$$\lambda_B = \sqrt{\frac{K_B}{C}} \tag{C.VII.13b}$$

is a typical penetration length of the material. The wave vector at the threshold of this undulation is:

$$q_c = \sqrt{\frac{\pi}{\lambda_B d}} \tag{C.VII.14}$$

Fig. C.VII.8 Layer undulation in a smectic B, obtained by dilating a homeotropic sample (photo by R. Ribotta).

Thus, everything happens as in the smectics A, except that the critical dilation is shifted by a constant term C_{44}/C, independent of the thickness, of the order of 10^{-3}.

To our knowledge, the laws C.VII.13–14 and their dependence on the thickness were never tested experimentally. Nevertheless, Ribotta gives a few indications concerning the critical wave vector. He finds a value typically twice larger in the smectic B than in the smectic A phase ($q_{cB} \approx 2q_{cA}$). At 60°, in a 20 μm thick 7O.7 sample, he found $q_{cB} \approx 2.6 \times 10^4$ cm^{-1}, yielding $\lambda_B \approx 6$ Å from relation C.VII.14. Considering that in this material $C \approx 10^9$ erg/cm^3, one obtains $K_B \approx 4 \times 10^{-6}$ dyn. This value is of the same order of magnitude as in smectics A.

To conclude this discussion, let us briefly discuss the nonlinear regime. The experiments show that "high" amplitude undulations appear very easily above the instability threshold (Fig. C.VII.8), involving layer sliding. To show that this is possible, let us determine the minimal amplitude u_{om} of the layer undulation above which the shear stress locally exceeds the internal stress σ_i. A simple calculation yields:

$$C_{44}u_{om}q_c = \sigma_i \qquad \text{(C.VII.15)}$$

Leading to $u_{om} = 30/2.6 \times 10^4 \times 10^6$ cm ≈ 0.1 Å, using the previously obtained values. This extremely small displacement shows that the plastic flow threshold is rapidly reached, as soon as the instability appears. In these conditions, the dynamics of the instability are driven by the apparent viscosity η of the sample. Let us call σ_c the critical stress ($\sigma_c = C\alpha_c$) and τ the evolution time of the instability. By analogy with smectics A:

$$\tau \approx \frac{\eta}{\sigma - \sigma_c} \qquad \text{(C.VII.16)}$$

At twice the threshold stress, one gets $\tau \approx 10^7/10^6 \approx 10$ s using the value given by Cagnon and Durand for the viscosity. We shall see that the instability develops much more slowly than in a smectic A (where it only takes a few ms to appear), a feature that we verified qualitatively in the case of 4O.8.

In the following section, we shall describe some static and dynamic properties of the smectic B-smectic A interface. Although this interface is very particular, it is also very interesting as it can be rough, facetted or spontaneously unstable, according to its orientation with respect to the layers. This very strong interfacial anisotropy has important consequences in directional growth, expressed by very different destabilisation mechanisms depending on the chosen orientation. Let us start by describing the equilibrium shape of a smectic B germ.

VII.3 Equilibrium shape of a smectic B germ

The experiment we are about to describe was performed on a planar 4O.8 sample [30]. To obtain this orientation, the glass plates are covered by a polyimide layer brushed along one direction. In these conditions, the smectic layers orient perpendicularly to the glass plates and to the micro-scratches of the polyimide layer. A smectic B germ is obtained by gradually lowering the temperature from the nematic phase (where the molecules spontaneously align along the desired orientation) until the coexistence range of the A and B phases. Several germs usually nucleate at the same time, but it is always possible to select one that is far enough from the others to be considered as isolated. The next step is to equilibrate its shape, which is never easy, especially in the presence of facets. To achieve this goal, one must control the temperature of the oven such that the germ, seen here in phase contrast microscopy, remains at a constant surface area over time. This operation is performed automatically using a computer that digitizes the image and treats it (in order to determine the contour, and then the surface of the germ), and then transmits the needed commands to the temperature regulator setting the temperature of the oven. Experience shows that one day is typically needed to equilibrate a small germ (of the order of a few tens of µm), much faster than the equilibration time of a metal in the same conditions, but much slower than a He^4 crystal, even of millimetric size. One should keep in mind that it is never obvious to know whether a domain is well equilibrated.

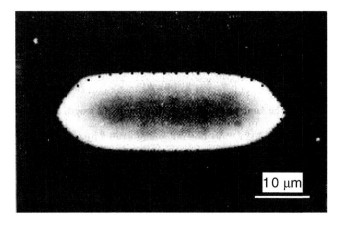

Fig. C.VII.9 Smectic B germ in equilibrium with the smectic A phase. The points correspond to the shape calculated from the Wulff diagram of figure C.VII.11 (15-µm thick 4O.8 sample observed in phase contrast in polarized light) (from ref. [30]).

One way of checking it is to compare several germs of different size and to verify that their shapes are similar to each other, i.e., related by a homothetic transformation.

Figure C.VII.9 shows a smectic B germ in equilibrium with the smectic A phase. Its main characteristics are the following:
1. It is elongated along the plane of the layers, with an aspect ratio length/width of the order of 3.

2. It is facetted parallel to the layers.

3. Its two facets tangentially join the curved adjoining regions.

4. It exhibits two pointy ends, a signal that the facets oriented perpendicular to the smectic layers (or close to that direction) are absent on the equilibrium shape. In this case, one speaks of **missing** or **forbidden** orientations.

Before trying to interpret these observations, let us first recall that all equilibrium shapes must satisfy the Gibbs-Thomson equation. In its complete form, this equation contains elastic terms (see chapter B.VII, volume 1), but we shall neglect them here, as the layers remain perfectly oriented during crystallisation. In these conditions, the Gibbs-Thomson equation takes the following simple form in two dimensions (cylindrical geometry):

$$\frac{\gamma + \gamma''}{R} = \Delta\mu \qquad\qquad\qquad (\text{C.VII.17})$$

In this expression, $\gamma(\theta)$ is the surface energy of a facet of orientation θ with respect to the layers (the facet corresponding to $\theta = 0$) and R represents the local curvature radius of the interface (it is taken as positive for a convex crystal).

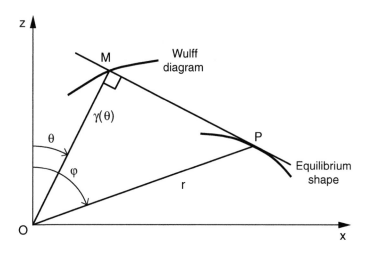

Fig. C.VII.10 Wulff construction.

The quantity $\gamma + \gamma'' = \tilde{\gamma}$ defines the **stiffness** of the interface and $\Delta\mu$ is the imposed supersaturation fixing the size of the germ at equilibrium. Wulff [31] demonstrated that the solutions of this equation can be obtained geometrically in the following manner (Fig. C.VII.10):

1. Trace, with respect to an arbitrary origin, the polar diagram $\gamma(\theta)$ of the surface free energy [Wulff diagram (Γ)].

2. In each point of (Γ), raise the normal to the vector radius. The equilibrium shape is, up to a homothecy of center O, the convex envelope of these normals [the pedal of (Γ)].

It is well worth the effort to prove that the shape obtained in this way fulfils the Gibbs-Thomson relation. Let M be the running point of (Γ) and P the corresponding point of the equilibrium shape (Fig. C.VII.10). Let r = OP and denote by φ the polar angle (Oz, OM) of the equilibrium shape. The line MP is described, in polar coordinates, by the equation:

$$r \cos(\varphi - \theta) = \gamma(\theta) \qquad (C.VII.18)$$

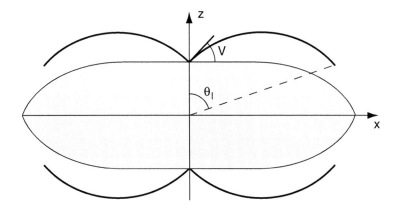

Fig. C.VII.11 Experimental Wulff diagram, constructed from the equilibrium shape of a smectic B germ coexisting with the smectic A phase (4O.8).

This equation defines a family of lines with parameter θ. Their envelope is obtained by taking the derivative of eq. C.VII.18 with respect to θ:

$$r \sin(\varphi - \theta) = \gamma'(\theta) \qquad (C.VII.19)$$

These two equations yield the parametric equation of the equilibrium shape:

$$r = \sqrt{\gamma^2 + \gamma'^2} \qquad (C.VII.20a)$$

$$\varphi = \theta + \arctan\left(\frac{\gamma'}{\gamma}\right) \qquad (C.VII.20b)$$

It is then easy to check that:

$$R = \sqrt{r'^2 + r^2\varphi'^2} = \gamma + \gamma'' \qquad (C.VII.21)$$

completing the proof.

Equations C.VII.18 and 19 also give a parametric representation in Cartesian coordinates of the equilibrium shape:

$$x\,\Delta\mu = \gamma(\theta)\,\sin\theta + \gamma'(\theta)\,\cos\theta \qquad\qquad \text{(C.VII.22a)}$$

$$z\,\Delta\mu = \gamma(\theta)\,\cos\theta - \gamma'(\theta)\,\sin\theta \qquad\qquad \text{(C.VII.22b)}$$

These formulas (written here in dimensional variables) are due to Landau [32].

Obviously, using the Wulff construction one can also determine the polar energy diagram starting from the equilibrium shape. In the case of smectic B phases, one obtains the diagram in figure C.VII.11, which exhibits two essential features:

1. The presence of two cusps in $\theta = 0$ and π, corresponding to the two observed facets.

2. The absence of any facets with orientation $\theta_1 < \theta < \pi - \theta_1$ and $\pi + \theta_1 < \theta < 2\pi - \theta_1$. These missing (or forbidden) facets are contained in the two pointy ends of the germ.

To understand better the origin of the missing facets on an equilibrium shape, imagine that the surface stiffness $\tilde{\gamma}(\theta) = \gamma + \gamma''$ goes to zero and becomes negative in an angle range $[\theta_1^o, \theta_2^o]$ (points 1 and 2 in figure C.VII.12a). Two cusps (denoted by 1 and 2 in figure C.VII.12b) appear in the pedal of the Wulff diagram, separating regions of positive curvature from a region of negative curvature. In this case, the equilibrium shape observed (cross-hatched area in the figure) possesses a double point A, corresponding to two distinct orientations θ_1 and θ_2, different from θ_1^o and θ_2^o. Hence, in this point the equilibrium shape exhibits an angular discontinuity.

Note that these two angles obey the Landau relations C.VII.22:

$$\gamma(\theta_1)\,\sin\theta_1 + \gamma'(\theta_1)\,\cos\theta_1 = \gamma(\theta_2)\,\sin\theta_2 + \gamma'(\theta_2)\,\cos\theta_2 \qquad \text{(C.VII.23a)}$$

$$\gamma(\theta_1)\,\cos\theta_1 - \gamma'(\theta_1)\,\sin\theta_1 = \gamma(\theta_2)\,\cos\theta_2 - \gamma'(\theta_2)\,\sin\theta_2 \qquad \text{(C.VII.23b)}$$

so their values can be determined. These two relations translate the equilibrium of the chemical stresses [33] at point A (Fig. C.VII.12c). We shall prove in the following section that the facets with negative stiffness are absolutely unstable. On the other hand, the other missing facets, with orientations θ between θ_1 and θ_1^o and between θ_2 and θ_2^o, are metastable. Note that, in the particular case of the smectic B, the two previous relations are reduced to only one, by symmetry:

$$\tan\theta_1 = \frac{\gamma(\theta_1)}{\gamma'(\theta_1)} \qquad\qquad \text{(C.VII.24)}$$

with $\theta_1 = \theta_1$ and $\theta_2 = \pi - \theta_1$.

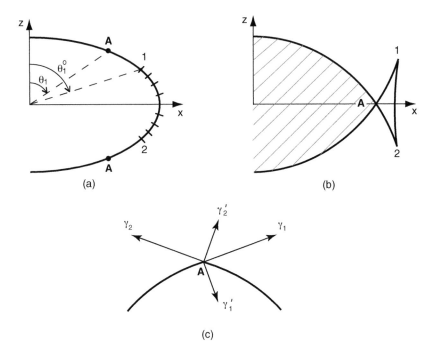

Fig. C.VII.12 Wulff diagram (a) and equilibrium shape (b) in the case when the stiffness goes through zero and becomes negative in an angular range [θ_1^o, θ_2^o]. A cusp A (orientations θ_1 and θ_2) appears on the equilibrium shape. In this point, the "chemical" stresses are balanced (c). In this diagram, we assumed that the crystal has a symmetry plane parallel to the x axis (case of the smectic B in planar orientation).

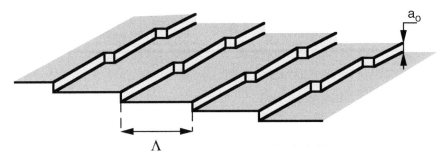

Fig. C.VII.13 Representation of a vicinal surface in the T.L.K model. Note that, at the SmA–SmB interface, the steps are somewhat particular as can be seen in figure C.VII.26.

To conclude this section, let us comment on the facets and the way they are joined to the vicinal surfaces (which have, by definition, orientations close to the one of the facet). In smectic B phases, this junction is tangential. To describe it, one usually employs the T.L.K. model (from "Terraces, Ledges, Kinks"), which we already introduced briefly in chapter B.VIII. We recall that vicinal surfaces are represented by singular planes (terraces) separated by steps (ledges), themselves bearing kinks (Fig. C.VII.13). Clearly, this description is only meaningful within a very narrow angular range, since the distance between steps must be larger than their width. Assume this is the case and let θ be the miscut angle of the vicinal surface with respect to the facet. The number of steps n per unit length along x is:

$$n = \frac{1}{\Lambda} = \frac{\tan\theta}{a_o} \qquad\qquad\qquad \text{(C.VII.25)}$$

where a_o is the height of a step and Λ the distance between steps. The steps being well-separated, the surface energy per unit length along x has the shape:

$$E(n) = \gamma_o + e(n) \, n \qquad\qquad\qquad \text{(C.VII.26)}$$

where e(n) is the free energy of a step. Since the length of the interface increases by a factor $1/\cos\theta$ when it is tilted by an angle θ, we have:

$$\gamma(\theta) = E(n)\cos\theta \qquad\qquad\qquad \text{(C.VII.27)}$$

e(n) is usually developed as a function of n:

$$e(n) = \beta + \Phi n^2 \qquad\qquad\qquad \text{(C.VII.28)}$$

The first term β, only depending on the temperature, represents the free energy of an **isolated** step. To be precise, this term also contains the energy $K\Delta a_o/a_o$ of the epitaxial dislocations (with energy $\approx K$) which one must add in the SmA phase to compensate the difference in the parameter a_o between the two phases (see section VII.5.e). The other term is related to the interactions between steps, which can be entropic or elastic. In a usual solid, the entropic and elastic interactions are repulsive and both of them vary as $1/\Lambda^2$; this dependence leads to the Φn^2 term in the series expansion C.VII.28 with $\Phi > 0$.

In the following, we shall need the energy of a step β. To determine it from the experimental Wulff diagram, let us expand $\gamma(\theta)$ around $\theta = 0$ to lowest order:

$$\gamma(\theta) = \gamma_o + \beta\frac{\theta}{a_o} + o(\theta^2) \qquad\qquad\qquad \text{(C.VII.29)}$$

and calculate the angle V of the Wulff diagram at the cusp (Fig. C.VII.11). We find immediately:

$$\tan V = \frac{a_o\gamma_o}{\beta} \qquad\qquad\qquad \text{(C.VII.30)}$$

This angle is directly measurable on the experimental Wulff diagram with a reasonable precision. The height of a step a_o being known, this relation gives access to the ratio β/γ_o, but not to the step energy separately. In the case of 4O.8, this yields $\beta/\gamma_o \approx 35$ Å.

Another method that should in principle give access to β, is to measure the size l_F of the facet. Indeed, this size is given by applying the Landau relations C.VII.22a and b to the edge of the facet with coordinates (x_F, z_F), yielding:

$$l_F = 2x_F = \frac{2\gamma'(0)}{\Delta\mu} = \frac{2\beta}{a_o \Delta\mu} \qquad \text{(C.VII.31)}$$

This relation tells us that the energy is stationary at equilibrium. Indeed, it shows that the bulk energy $a_o l_F \Delta\mu$, gained by adding a layer to the facet compensates the energy 2β lost by creating two new steps at the extremities. Unfortunately, $\Delta\mu$ is very difficult to measure precisely (one would need to know the phase diagram perfectly and the sample composition at the time of the measurement, which is almost impossible). Additional imprecision arises from the determination of the edge of the facet, and hence the value of l_F, as the junction is tangential.

This method is thus unusable for the determination of the step energy β. One is left with the option of measuring the facet energy γ_o, since the ratio β/γ_o is already known.

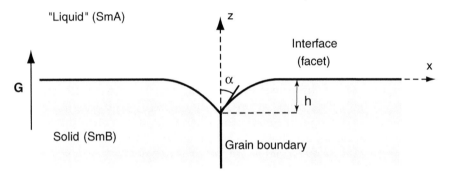

Fig. C.VII.14 Measurement of the surface energy by the grain boundary method. In the 4O.8 experiment, the interface is facetted.

The the grain boundary method, widely used in metallurgy, allows us to determine γ_o. It consists of observing the deformations induced by a grain boundary at the smectic B-smectic A interface. In practice, the sample (in the planar orientation) is placed in a temperature gradient G with the layers parallel to the growth front. The interface is therefore facetted. Since the orientation of the hexagonal lattice can change from one point to another, domain walls (or grain boundaries) appear that intersect the interface and deform it locally (Fig. C.VII.14). Let $z = f(x)$ be the equation of the interface close to a junction. It must obey the Gibbs-Thomson relation:

$$\frac{\tilde{\gamma}f''}{(1+f'^2)^{3/2}} = \frac{LGf}{T_c} \qquad \text{(C.VII.32)}$$

where L is the latent heat of transition per unit volume ($L \approx 5\times10^7$ erg/cm³) and T_c the transition temperature. Here, the surface stiffness is a known function of θ (since the Wulff diagram of 4O.8 is known), and thus of f', as $\tan\theta = f'$. The only unknown being γ_o, one only needs, in practice, to find the numerical solution of the differential equation C.VII.32 with the boundary conditions:

$$f (x = \pm \infty) = 0 \qquad\qquad\qquad\text{(C.VII.33a)}$$

$$f'(0) = \cotan \alpha \qquad\qquad\qquad\text{(C.VII.33b)}$$

where α is the angle of the junction at the bottom of the groove (Fig. C.VII.14), which can be determined experimentally. One then seeks for the value of γ_o that yields the best adjustment of the computed curve with the experimental profile. Note that, from eq. C.VII.32, the depth of the groove must vary as $(\tilde{\gamma}T_c/LG)^{1/2}$. This $1/\sqrt{G}$ dependence is well satisfied experimentally. This method of measurement yields $\gamma_o \approx 0.1$ erg/cm² in 4O.8. Since for this product $\beta/\gamma_o \approx 3.5\times10^{-7}$ cm, one gets $\beta \approx 3.5\times10^{-8}$ dyn.

We shall see later that this quantity is very relevant for the dynamics of the facets. Before that, however, let us return to forbidden orientations and try to find out what happens once such an orientation is forced to appear.

VII.4 Herring instability

We saw in the previous section that one necessarily has, among the missing facets of a crystal, facets of negative stiffness $\tilde{\gamma}$, and we admitted that they were spontaneously unstable. This can be shown by considering the stability of an interface with orientation θ_o with respect to a sinusoidal deformation of equation $z = \xi(x) = \xi_o\cos kx$ and wavelength $\lambda = 2\pi/k$ (Fig. C.VII.15).

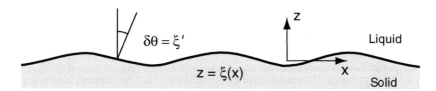

Fig. C.VII.15 Undulation at the solid-liquid interface.

The surface energy E per unit area along x is worth:

$$E = \frac{1}{\lambda}\int_0^\lambda \gamma(\theta_o + \delta\theta)\sqrt{1 + \xi'^2}\, dx \qquad\qquad\text{(C.VII.34)}$$

Expanding to second order in $\delta\theta = \xi'$, one finds:

$$E \approx \frac{1}{\lambda} \int_0^\lambda \left(\gamma(\theta_o) + \gamma'(\theta_o)\, \xi' + \gamma''(\theta_o)\, \frac{\xi'^2}{2} \right) \left(1 + \frac{\xi'^2}{2} \right) dx$$

$$\approx \gamma(\theta_o) + \frac{1}{2\lambda} \int_0^\lambda \tilde{\gamma}(\theta_o)(\mathbf{grad}\ \xi)^2 dx \qquad\qquad (C.VII.35)$$

This formula shows that the surface undulates in order to decrease its energy if its *stiffness is negative*: in this case, the interface is *absolutely unstable*. Conversely, an interface is *stable* (or metastable) if its *stiffness* $\tilde{\gamma}$ is *positive*.

Imagine now that an unstable orientation is forced to appear. This experiment is easily performed by placing a planar 40.8 sample in a temperature gradient parallel to the layers. In this case, the smectic layers are perpendicular to the macroscopic interface between the two phases. Experience shows that a zigzag structure spontaneously develops at the interface (Fig. VII.16a), its average wavelength following a $G^{-1/3}$ law (Fig. VII.16b). This instability was predicted by Herring in 1951 [34]. Its periodicity is the result of a competition between the surface energy gained by creating the zigzag, and the one "locally" lost at the tips and, in the bulk, in the supersaturated areas between the interface and the transition isotherm $T = T_c$ (shown in gray in figure C.VII.17). Rigorously calculating these different contributions is not easy, since one must know the shape of the interface, which requires solving the Gibbs-Thomson relation while preserving the equilibrium of chemical stresses at each tip (eq. C.VII.24). This condition imposes that the tilt angle of the surface with respect to the smectic layers at each tip be equal to the one formed at the tip of a germ in equilibrium (Fig. C.VII.17).

Calculating the wavelength λ of the microstructure is considerably easier if the local radius of curvature of the interface is everywhere much larger than λ [33]. In this case, each portion of the Herring structure can be seen as a straight segment tilted by an angle θ_l with respect to the layers. This condition is fulfilled if:

$$\lambda \ll l_c \sqrt{2\tan\theta_l} \qquad\qquad (C.VII.36)$$

where $l_c = \sqrt{\dfrac{2\gamma T_c}{LG}}$ is a thermal capillary length. In the experiment on 40.8, this relation is well enough respected for all the gradients employed. In this limit, the gain in free surface energy becomes independent of the wavelength λ of the microstructure, which is then only fixed by the competition between the bulk free energy (tending to decrease λ) and the tip energy (tending to increase λ). These two contributions are easily calculated and their value, per unit length along x, is:

$$E = E_{bulk} + E_{tip} = \frac{LG\lambda^3}{96\,T_c(\tan\theta_l)^2} + \frac{2\gamma\zeta}{\lambda} \qquad\qquad (C.VII.37)$$

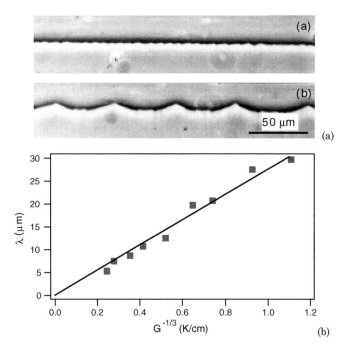

(a)

(b)

Fig. C.VII.16 a) Herring microstructure observed at the smectic A-smectic B interface of 4O.8. In this sample, the smectic layers are parallel to the imposed temperature gradient; (a) G = 22 °C/cm; (b) G = 1.25°C/cm. b) Average wavelength as a function of the temperature gradient. The solid curve is the best fit with the theoretical $G^{-1/3}$ law (from ref. [35]).

Denoting by $\gamma\zeta$ the energy of a tip (ζ is an undefined microscopic length). Minimizing with respect to λ, yields:

$$\lambda = \sqrt[3]{48\, l_c^2\, \zeta\, (\tan\theta_l)^2} \qquad\qquad (C.VII.38)$$

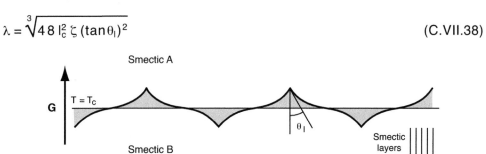

Fig. C.VII.17 Schematic representation of the Herring microstructure. In the gray areas, the liquid is undercooled and the solid is overheated. The angle θ_l is determined by the equilibrium of chemical stresses at each tip.

This law shows that λ varies like $G^{-1/3}$, in agreement with the experiments. Fitting the experimental results with the theoretical law C.VII.38 (Fig. C.VII.16b), one finds $\zeta \approx 50$ nm; this (rather large in this case) distance sets the size of the cusps. At lower length scales, the macroscopic theory is no longer valid.

VII.5   Instabilities of the SmA-SmB front in directional growth

We have just seen that the A–B interface can be facetted, rough or spontaneously unstable, according to its orientation with respect to the smectic layers. This behavior is related to the very strong structural anisotropy of the B phase and to its lamellar structure. In this section, we shall prove that the destabilization mechanisms of a growth front and its nonlinear evolution crucially depend on its equilibrium interfacial properties [36]. We examine successively the rough, the "forbidden" and the facetted case.

VII.5.a) The rough case

The A-B interface of 4O.8 is rough at the atomic scale when the layers make an angle α between typically 10° and 90° with the thermal gradient (in the notations of the preceding section, $\alpha = \pi/2 - \theta$). In these conditions, the interface is straight at rest (image "a" in figure C.VII.18). Its evolution at constant pulling velocity is shown in this figure for $\alpha = 45°$. We see that the front becomes unstable above a critical velocity V_c of the order of 1 μm/s. The cellular bifurcation is slightly subcritical, the cells having a finite amplitude at the instability threshold (images "b" and "c"). They are rounded, and also slightly asymmetric, so that they drift along the macroscopic front. As shown in figure C.VII.19, this drift velocity V_d depends on the angle α and goes through a maximum for an angle close to 45°. Generically, this drift is related to the breaking of the mirror symmetry with respect to the plane perpendicular to the interface, due to the layer tilt with respect to the gradient. The consequence is an asymmetry of the impurity diffusion field in the smectic A phase, since $D_{//}$ and $D_\perp$ have no reason to be equal in this phase (we assume here that diffusion in the B phase is negligible). In these conditions, a finite imaginary component appears in the dispersion relation $\omega(k)$ of the front and its value at the threshold sets the drift velocity. A linear stability calculation gives the following approximate formulae (see chapter B.IX):

$$V_c = \frac{(D_\perp \cos^2\alpha + D_{//}\sin^2\alpha)GK}{\Delta T} \tag{C.VII.39a}$$

$$\frac{V_d}{V} = \frac{(D_\perp - D_{//})\sin\alpha\cos\alpha}{D_\perp\cos^2\alpha + D_{//}\sin^2\alpha} \tag{C.VII.39b}$$

where $\Delta T = T_{\text{liquidus}} - T_{\text{solidus}}$ (0.2 to 0.5°C, typically) and K the partition coefficient of the impurities ($K \approx 0.5$). The best fit of the experimental curve in figure C.VII.19 with the theoretical law C.VII.39b leads to $D_{\perp}/D_{//} \approx 1.6$, in qualitative agreement with other determinations in smectics A [37]. We emphasize that one can prove experimentally [36] that the drift of the cells at the threshold is not due to the anisotropy of molecular attachment kinetics at the interface [38]. In the nonlinear regime, the cell amplitude increases. The drift is stronger, partly due to the lateral facetting of the cells (Fig. C.VII.18d).

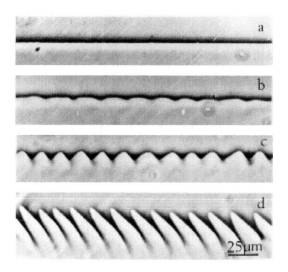

Fig. C.VII.18 Destabilization of the A-B interface in the rough case when the layers make an angle of 45° with the temperature gradient. (G = 76°C/cm, ΔT = 0.4°C). a) Plane interface at rest; b) and c) V = 0.8 mm / s; d) V = 0.95 mm / s (from ref. [36]).

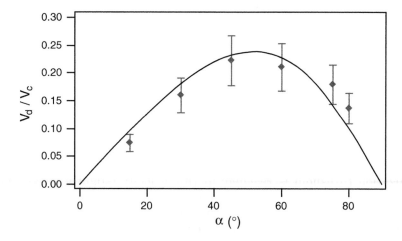

Fig. C.VII.19 Drift velocity of the cells at the threshold as a function of the tilt angle of the layers with respect to the thermal gradient (from ref. [36]).

VII.5.b) The "forbidden" case

This is a pathological case, as the interface is unstable, even at rest. One can nevertheless wonder about the evolution of the Herring microstructure in directional growth [39]. The answer to this question is given in figure C.VII.20.

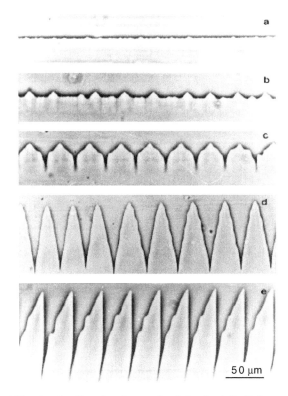

Fig. C.VII.20 Destabilization in directional growth of the SmA-SmB front, when the layers are parallel to the temperature gradient (unstable orientation, $\Delta T = 0.7°C$, $G = 76°C/cm$). a) $V = 0.2 \,\mu m/s$, the Herring microstructure is unchanged; b) and c) $V = 0.38 \,\mu m/s$, destabilization of the Herring microstructure and formation of pointy symmetric cells; d) $V = 0.43 \,\mu m/s$, symmetric cells of large amplitude; e) $V = 0.45 \,\mu m/s$, asymmetric cell domain drifting to the right due to facet growth (from ref. [36]).

Experience shows that the Herring microstructure (image "a") is almost unchanged at low velocity. However, a spectacular change occurs when the destabilization threshold is reached: at this velocity (slightly lower than the value predicted by the Mullins-Sekerka calculation), isolated triangles start to grow (image "b"); they quickly form symmetric cells of finite amplitude and well-defined wavelength. Note that the bifurcation is here subcritical and the hysteresis is stronger than in the rough case. This result is hardly surprising, the front being unstable from the start. On the other hand, the cells have an unusual shape, exhibiting a cusp at the tip. This discontinuity does not disappear when the velocity is increased, proving that the orientations remain forbidden during the growth process, as at equilibrium (for low velocities, at least). One can equally notice, in image "c", that the sides of the symmetric cells rapidly become modulated. These modulations are here correlated over several cells and may be the result of a collective oscillation involving the entire front. This instability, if it exists, could not be revealed experimentally. Finally, we see in image "d" that the cells soon become asymmetric with increasing velocity. In this case, a large facet appears on one side of each cell. This spontaneous breaking of the mirror symmetry leads to the formation of domains where the facets appear alternatively on the right and on the left side of the cells. These domains are separated from each other by cell sources and sinks, the dynamics of which is described in detail in Melo and Oswald [39]. In this article one can also find a complete description of the successive bifurcations leading to a "chaotic" cell regime (no well-defined periodicity), then to dendrites (which have, on the contrary, a clearly defined wavelength).

VII.5.c) The facetted case

The interface is facetted when the layers are perpendicular to the temperature gradient ($\alpha = 90°$). In this case, there still exists a critical destabilisation velocity (experimentally very close to the value given by the constitutional undercooling criterion), below which the front is perfectly flat and above which macro-steps appear and rapidly occupy the entire sample (Fig. C.VII.21).

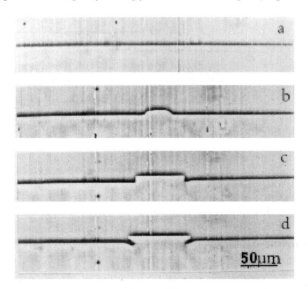

Fig. C.VII.21 Destabilization of a facet. The layers are here perpendicular to the temperature gradient ($\Delta T = 0.9°C$, $G = 76°C/cm$, $V = 0.15\ \mu m/s$). a) Facet immediately before destabilization; b) (t = 0): nucleation of a bump that rapidly becomes facetted, forming a pair of macro-steps; c) (t = 6 mn) and d) (t = 7 mn): drift and shape change of the two macro-steps (from ref. [36]).

The aspect and dynamics of the front are here completely different from the rough case, as new steps keep appearing, without leading to a stationary periodic structure [36, 40]. Note that, immediately following their nucleation, the steps are wide and smoothly joined to the facets. They propagate from the warm facets toward the cold facets, favoring the growth of the B phase. Their drift velocity, of the order of 0.2 V at the start, increases with time. This acceleration is accompanied by a change in the shape of the steps, which start by becoming steeper and steeper (until they resemble the steps of a staircase), and then develop a sharp tip that propagates along the front at a velocity close to the pulling velocity V. This instability is due to the local bunching of the "isoconcentration" lines of the impurity concentration field at the top of the step, and to their fanning out at its bottom (Fig. C.VII.22). When two steps meet, they merge completely if they are of the same height (which is rather uncommon), or partially, with formation of a new, shallower, step (Fig. C.VII.23). This new step starts drifting again, then meets another step, merges with it, etc.

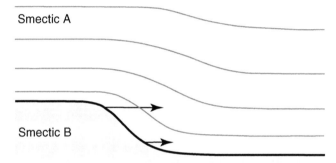

Smectic A

Smectic B

Fig. C.VII.22 "Isoconcentration" lines of the impurity field close to the macro-step. These lines being closer together at the top than at the bottom, the top of the step propagates faster than its bottom, leading to a shape change. This instability is identical to the Mullins-Sekerka instability and stems from the fact that the velocity of the front is proportional to the local gradient of impurity concentration. In this sketch, we assumed that the interface is an isoconcentration line, which amounts to neglecting the stabilizing effects of the temperature gradient and of the surface tension.

At this point, one can ask why the system does not select a stationary periodic structure similar to the crenellated front imagined by the theorists [40]. This solution is formed by a sequence of warm and cold facets separated by rough regions to which they are smoothly joined (Fig. C.VII.24). Such a solution could in principle exist in the smectic B under study. However, it is not observed. Why is that?

To answer this question, the theorists have looked for the conditions in which such a solution can exist. This problem being somewhat complicated, we shall only point out here its difficulties, and then we shall give some results on the existence conditions for a crenellated front.

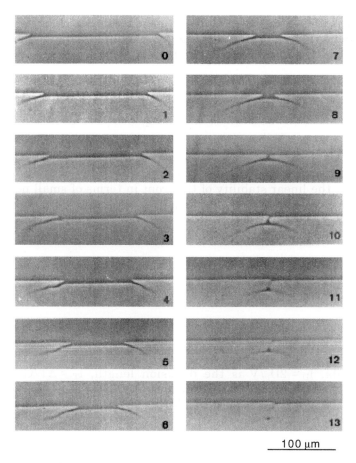

100 µm

Fig. C.VII.23 Encounter of two macro-steps. Since they are not of the same height, a step remains at the end at the process. The time interval between two images is 15 s, except between the two first, where it is 30 s (4O.8, $\Delta T = 0.7°C$, $V = 0.35$ µm/s, $G = 50°C/cm$) (from ref. [40]).

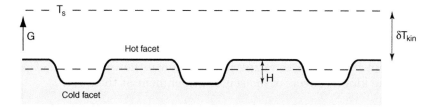

Fig. C.VII.24 Crenellated solution imagined by the theorists [41]. Due to the kinetic effects of molecular attachment at the facet, the entire front recedes in the temperature gradient toward colder regions.

The main difficulty in studying the destabilisation of a facet, consists of the fact that the surface stiffness diverges on the facet. The Gibbs-Thomson relation can thus no longer be applied in its usual local form. A way of avoiding this difficulty is to integrate it formally over the facet. Let 2L be its size, and $\Delta\mu$ the local supersaturation, not necessarily constant along the facet during growth. Keeping in mind that the curvature is given by $d\theta/ds$, one finds from eq. C.VII.29:

$$\int_{-L}^{+L} \Delta\mu \; ds = \int_{\theta^-}^{\theta^+} (\gamma + \gamma'') \; d\theta = \gamma'(\theta_+) - \gamma'(\theta_-) = \frac{2\beta}{a_o} \qquad \text{(C.VII.40)}$$

This relation connects the size of the facet L to the energy of the step β and to the **average** supersaturation $\Delta\mu$ seen by the facet. Since it is **global**, one can no longer treat the linear stability of the front in terms of small perturbations. One possible approach is to look for a crenellated solution of finite amplitude H. Such a solution exists if:

$$H > L_{kin} \qquad \text{(C.VII.41)}$$

with L_{kin} the kinetic length, expressed as:

$$L_{kin} = \frac{2}{\Delta T} \frac{\delta T_{kin}}{\dfrac{2}{l_D} - \dfrac{1}{l_T}} \qquad \text{(C.VII.42)}$$

In this formula, $l_D = 2D_{//}/V$ is the diffusion length, $l_T = \Delta T/G$ the thermal length and $\delta T_{kin} = T_s - T$ the kinetic undercooling (see section VII.5.d). Since L_{kin} diverges for $2/l_D = 1/l_T$, such a solution can only exist above a critical velocity given by:

$$V_c = \frac{D_{//} G}{DT} \qquad \text{(C.VII.43)}$$

This velocity is none other than the critical destabilisation velocity given by the criterion of constitutional undercooling. Note the existence of a continuum of solutions parameterized by their amplitude H. For each solution, analytical expressions exist for the size and temperature of the warm and cold facets as a function of H [41]. We do not consider that it is useful to cite them here, as these solutions were never observed. This result suggests that condition C.VII.42 is not fulfilled in the experiments, hence the importance of precisely measuring the kinetic effects of molecular attachment at the facet.

VII.5.d) On the kinetics of facets

For a facet to grow, it must be supersaturated (i.e., away from its temperature of thermodynamic equilibrium). The larger this deviation, the faster the facet will grow. The kinetic law connecting the growth velocity V to the deviation δT_{kin} from the equilibrium temperature:

$$\delta T_{kin} = F(V) \qquad\qquad\qquad\qquad\qquad\qquad\qquad\qquad\text{(C.VII.44)}$$

In directional growth, $\delta T_{kin} = T_s - T$, since the temperature of the front in the absence of kinetic effects must be the one of the solidus.

Let us first see how to measure the kinetic law experimentally. At this point, several precautions must be taken. First of all, the material must be purified as much as possible, in order to avoid the appearance of the Mullins-Sekerka instability. One must then measure the temperature of the interface as a function of its velocity, for instance, by superposing to the sample under study a reference sample that has a phase transition at the same temperature. In practice, this sample is a 9CB-10CB mixture exhibiting a smectic A-nematic transition at about 50°C. The attachment kinetics being very fast at this front [42], its temperature changes very little with the pulling velocity. It is then sufficient to measure under the microscope the shift of the facet position with respect to this reference front to find out its temperature, with the temperature gradient being known. This method can detect temperature shifts as small as 1/100°C.

The results obtained in 4O.8 using this technique are shown in figure C.VII.25. Two separate growth regimes are apparent in this graph:

1. A "low velocity" regime, well described by a square-root law:

$$\delta T_{kin}(°C) \approx 0.051 \sqrt{V} \ (\mu m/s) \qquad (V < 5 \ \mu m/s) \qquad\qquad \text{(C.VII.45)}$$

2. A regime where the kinetic shift seems to saturate as the velocity increases.

Two distinct models, both of them widely used in metallurgy, allow us to interpret this experimental data.

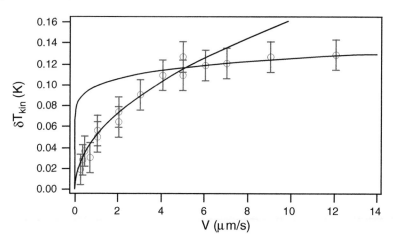

Fig. C.VII.25 Kinetic shift of the facet measured in directional growth in 4O.8 (from ref. [42]).

The first one is a model of growth by screw dislocations, which is well known to give a law of the observed type. This model is based on the probable existence of a large number of screw dislocations perpendicular to the layers and with Burgers vector **ST** (see section VII.5.c and figure C.VII.5). These dislocations, where they emerge to the surface of the crystal, form steps of a thickness $b = 2a_0$. When the interface is supersaturated, a force acts on the steps, which start to move, favoring crystal growth. Each step takes a spiral shape having as center the emergence point of the dislocation that created it. Writing the force balance on each step, one can determine the evolution of an isolated spiral, and then of the collection of spirals. This famous model, invented by Burton, Cabrera and Frank [43], gives the kinetic law as a function of the mobility of the steps m, their line tension β^* and the latent heat L:

$$V \approx \frac{1}{20}\left(\frac{2bL}{T}\right)^2 \frac{m}{\beta^*}(\delta T_{kin})^2 \qquad (C.VII.46)$$

We recall that the mobility is usually defined by the relation:

$$V_{step} = m\,\Delta\mu = m\,\frac{L\delta T_{kin}}{T} \qquad (C.VII.47)$$

Comparison with the experimental law C.VII.45, yields $m/\beta^* \approx 86$ cms³g⁻², giving $m \approx 6 \times 10^{-6}$ cm²sg⁻¹ assuming that $\beta^* \approx 2\beta \approx 7 \times 10^{-8}$ dyn (since the height of the steps created by the screw dislocations is twice that of an elementary step). In the following section, we explain how to calculate the mobility of the steps.

This dislocation model does not describe correctly the "high velocity" measurements (V > 5 μm/s), predicting larger kinetic shifts than those measured experimentally in this regime. This means that another growth mechanism replaces (or, more exactly, complements) the first one at high velocity. One possible mechanism is homogeneous island nucleation. This process is thermally activated, limiting its effect at weak supersaturation. At strong supersaturation, on the other hand, it is very effective due to its exponential behavior. Within this model, the growth velocity of the facet depends on the rate of island nucleation on the considered facet and on the propagation velocity of the steps. Skipping again the details, one can show that the growth rate of a facet in the polynuclear regime [44] is given by a formula of the type:

$$V = mbL\left(\frac{k_B T}{\beta c_0^4}\right)^{1/3}\left[\exp\left(-\frac{\pi\beta^2}{3k_B bL\delta T_{kin}}\right)\right]\frac{\delta T_{kin}}{T} \qquad (C.VII.48)$$

This law for the velocity is very sensitive to the energy of a step β, since β^2 appears in the exponent. We traced it in dotted line in figure C.VII.24 for the value of β that gives the best fit to the experimental points measured at V > 5 μm/s. This yields $\beta \approx 4 \times 10^{-8}$ dyn and

$$V(\mu m/s) = 1.53 \times 10^5\,\delta T_{kin}\exp\left(-\frac{0.943}{\delta T_{kin}(°C)}\right) \qquad (C.VII.49)$$

with $b = 3 \times 10^{-7}$ cm, $c_0 = 5 \times 10^{-8}$ cm, $L = 5 \times 10^7$ erg/cm^3 and for the mobility m, the value previously obtained for a step with height $2a_0$, namely $m = 6 \times 10^{-6}$ cm^{2}sg^{-1} (at the end of this section, we also provide a rationale for this choice). One sees that the value of β we obtain is in good agreement with the one measured in static experiments ($\beta \approx 3.6 \times 10^{-8}$ dyn).

We conclude that the two mechanisms we have seen give a reasonable account of the observed kinetics. The measurements also show that the inequality C.VII.42 concerning crenellated fronts is never satisfied when the macro-steps appear above the Mullins-Sekerka threshold. It is then not surprising that this solution was never observed in experiments, not even as a transient. An additional reason, discovered by the theorists a little later, is that the crenellated front is intrinsically unstable with respect to breather modes [45].

To conclude this section, let us return to the mobility of the steps. If we compare it to that of dislocations in the smectic A phase, we find that it is typically 1000 times larger in the present case. To explain this result, let us consider the origin of the dissipation. Two mechanisms can be invoked: the first one is related to the variation in layer thickness $\Delta a_0 / a_0$ and the second one to the molecular density variation in the layers $2\Delta c_0 / c_0$ between the two phases. Each mechanism leads to local flow of a particular type in the smectic phase A (the more fluid one), and hence to energy dissipation (Fig. C.VII.26). Let v_m be the velocity of the step, η the viscosity of the smectic A phase in shear parallel to the layers and λ_p its permeation coefficient. The first mechanism, associated with a change in layer thickness, entails a displacement and a gliding motion of the layers one over the other. The associated dissipation Φ_c is typically [42]:

$$\Phi_c \approx \frac{\eta}{3\, 2\sqrt{\pi}} \left(\frac{\Delta a_0}{a_0}\right)^2 v_m^2 \qquad \text{(C.VII.50)}$$

The second mechanism brings into play the density change in the layers between the two phases, leading to permeation flow across the layers. The same calculation as for the climb of an edge dislocation in the plane of the layers gives, for a step of height b [42]:

$$\Phi_m \approx 8 \sqrt{\frac{\eta}{\lambda_p}} \left(\frac{\Delta c_0}{c_0}\right)^2 b\, v_m^2 \qquad \text{(C.VII.51)}$$

It is easily seen that the second mechanism is largely dominant. To determine the mobility, let F_s be the supersaturation force acting upon the step. According to the general dissipation theorem (in the stationary regime), $\Phi = F_s v_m$. Since $F_s = b\Delta\mu$ and, by definition, $v_m = m\Delta\mu$, we obtain:

$$m \approx \frac{1}{8} \sqrt{\frac{\lambda_p}{\eta}} \left(\frac{c_0}{\Delta c_0}\right)^2 \qquad \text{(C.VII.52)}$$

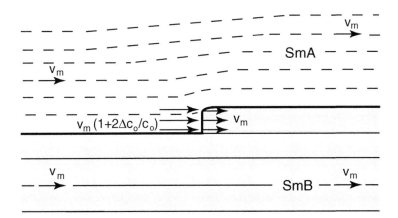

Fig. C.VII.26 Origin of the dissipation for a step propagating at the SmB-SmA interface. The velocity field is given in the reference frame of the step. This model assumes that the flow is negligible in the solid phase (from ref. [42]).

We see that the mobility of the step is independent of its height. On the other hand, it is much larger than that of an edge dislocation in the smectic A phase $(m_{disl} \approx (\lambda_p/\eta)^{1/2})$, by a factor:

$$\frac{m}{m_{disl}} \approx \frac{1}{8}\left(\frac{c_0}{\Delta c_0}\right)^2 \approx 1000 \qquad\qquad (C.VII.53)$$

since $\Delta c_0/c_0 \approx 10^{-2}$ in 4O.8. This order of magnitude is the expected one.

To end this chapter, let us show that the growth process engenders a stress distribution that can lead to layer buckling in the smectic A phase.

V.II.5.e) Stress relaxation and layer buckling

It is well-known in metallurgy that growth generates stress. This latter stems from the migration of impurities in the intercellular areas, or more directly from the hydrodynamic flow produced by the density difference between the two phases. In solids, this can lead to the appearance of defects (dislocations, vacancies, etc.). Their number and their relative positioning (microstructure) have a direct influence on the mechanical, electrical and chemical properties (corrosion resistance) of the material. Thus, understanding these phenomena is of outstanding practical importance.

To illustrate this point, let us show that similar phenomena occur at the smectic A-smectic B interface [46]. In the experiment we discussed, the front is facetted, the layers being parallel to the interface. Careful observation in the polarizing microscope shows that the microparticles contained in the sample suddenly start to move as the front approaches them (to a distance below a few tens of μm). Their motion is parallel to the layer, and only occurs in the smectic A phase. It is faster for particles closer to the interface and its direction is random. It sometimes happens that two particles very close to each other start moving simultaneously in opposite directions. This observation excludes the possibility of a hydrodynamic flow that might carry the particles along. On the other hand, it is very likely (see chapter C.III) that these particles are trapped by edge dislocations, and carried by them as they move. This observation shows that the layers are compressed or dilated ahead of the front. The stress decreases when moving away from the front, being relaxed by the movement of edge dislocations.

To explain these observations, let us assume the presence of a uniform density of edge dislocations ρ_m in the smectic A phase. In the reference frame of the B phase (considered at rest), the front propagates with a velocity V. The choice of the coordinates is shown in figure C.VII.27. Say that, at the considered moment t, the front is at z = 0 and let u be the layer displacement in the smectic A phase and v the molecular velocity. The transition being first-order, it leads to a variation in layer thickness $\Delta a_o = a_o(SmA) - a_o(SmB)$ and to a density variation $\Delta\rho = \rho(SmA) - \rho(SmB)$. To determine the effect of these parameter changes, let us write the conservation equations for the density and the number of layers. As $v = u = 0$ in the B phase (assumed to be much stiffer than the A phase), one has:

$$V_{z0} = \frac{\Delta\rho}{\rho} V \qquad\qquad (C.VII.54a)$$

$$\dot{u}_0 = -\frac{\Delta a_o}{a_o} V \qquad\qquad (C.VII.54b)$$

The index $_0$ indicates that the quantities are taken on the front in z = 0, on the smectic A side. The parameter $\dot{u}$ designates the derivative of u with respect to time.

Let us now write the equation of mass conservation: div $v = 0$. Due to invariance in the (x,y) plane, this condition yields $\partial v_z/\partial z = 0$, leading to:

$$V_z = V_{z0} \qquad\qquad (C.VII.55)$$

in the entire smectic A phase. Since v_{z0} and $\dot{u}_0$ are not identical, there is permeation across the layers in the smectic A phase. This flow is described by the equation:

$$\dot{u} - v_z = \lambda_p G \qquad\qquad (C.VII.56)$$

where λ_p is the permeation coefficient, G the elastic force acting on the layers:

$$G = \frac{\partial \sigma}{\partial z} = B \frac{\partial^2 u}{\partial z^2} \qquad\qquad (C.VII.57)$$

and B, the compressibility modulus of the layers.

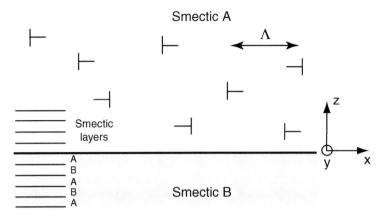

Fig. C.VII.27 The SmA-SmB interface and the edge dislocations allowing the relaxation of the elastic stress generated by the growth process.

Let us now see how to determine the stress field σ(z). To this end, the motion of the edge dislocations must be taken into account. Indeed, their movement under the action of the stress σ(z) leads to a layer displacement at a distance z from the front that is different from the one measured at the interface. A simple, purely geometrical, calculation leads to the Orowan relation (appendix 1) [47]:

$$\dot{u}\,(z) = \dot{u}_0 + \int_0^z \rho_m b v_m(z) dz \qquad\qquad (C.VII.58)$$

where $v_m(z)$ is the climb velocity of a dislocation with Burgers vector b. This velocity is related to the local stress by the relation:

$$v_m(z) = m\sigma(z) \qquad\qquad (C.VII.59)$$

with m the mobility of the dislocation (close to 10^{-6} cm^2/poise, see chapter C.III).

Replacing $\dot{u}$, v_z and G by their expressions in eq. C.VII.56 and using the boundary conditions C.VII.54, we obtain the equation of the stress field:

$$\int_0^z \rho_m b m \sigma_m(z) dz - \left(\frac{\Delta a_o}{a_o} + \frac{\Delta \rho}{\rho}\right) V = \lambda_p \frac{d\sigma}{dz} \qquad\qquad (C.VII.60)$$

Solving it (with boundary conditions σ = dσ/dz = 0 far away from the front) yields:

$$\sigma(z) = \sigma_o \exp\left(-\frac{z}{\xi}\right) \qquad\qquad (C.VII.61)$$

with

$$\sigma_0 = \left(\frac{\Delta a_0}{a_0} + \frac{\Delta\rho}{\rho}\right) V \frac{\xi}{\lambda_p} = -2 \frac{\Delta c_0}{c_0} V \frac{\xi}{\lambda_p} \qquad \text{and} \qquad \xi = \Lambda \sqrt{\frac{\lambda_p}{mb}} \qquad \text{(C.VII.62)}$$

where $\Lambda = (\rho_m)^{-1/2}$ is the average distance between two dislocations.

Thus, the stress goes to zero exponentially over a certain distance ξ ahead of the front. Recalling that, in a smectic A of viscosity μ (see chapter C.III):

$$m \approx \sqrt{\frac{\lambda_p}{\mu}} \qquad \text{(C.VII.63)}$$

we obtain:

$$\xi \approx \Lambda \sqrt{\frac{l_p}{b}} \qquad \text{(C.VII.64)}$$

Since the permeation length $l_p = \sqrt{\lambda_p \eta}$ is comparable to b, we finally find $\xi \approx \Lambda$.

In summary, the stress propagates ahead of the front over a distance comparable with the average distance between dislocations (experimentally, a few tens of μm). The stress has its maximum (absolute value) on the front, and is proportional to the density variation in the layers $-2\Delta c_0/c_0$ and to the velocity. Experimentally, $\Delta c_0/c_0$ is not known, but a positive value $(c_0(\text{SmA}) > c_0(\text{SmB}))$ of the order of 1% seems reasonable. In these conditions, the stress has the same sign as the growth velocity; namely, the layers are dilated during melting $(V > 0)$ and compressed during growth $(V < 0)$.

This latter conclusion is important, since we know that a smectic A can quickly become unstable under layer dilation. This model predicts that the smectic A remains stable during growth, irrespective of the front velocity. This behavior is confirmed by the experiments. On the other hand, a layer buckling instability (rapidly followed by the nucleation of focal parabolae) can occur during melting, if the stress exceeds a certain threshold $\sigma_c \approx 2\pi(KB/\xi)^{1/2}$ (see chapter C.III), and hence if the melting velocity (negative by definition) is lower than a critical velocity V^* of equation:

$$V^* = -\frac{\pi c_0 \lambda_p \sqrt{KB}}{\Delta c_0 \Lambda^2} \qquad \text{(C.VII.65)}$$

With $\lambda_p = 10^{-13}$ cm^2/poise, KB = 100 dyn^{2}cm^{-2} and the values already given for $\Delta c_0/c_0$ and Λ, one obtains $V^* \approx -3$ μm/s. This instability upon melting was observed in experiments for a melting velocity below -10 μm/s. We can conclude that the agreement with the theory is satisfactory. One can equally check that the velocity of the dislocations (estimated by measuring the velocity of the microparticles) is comparable to the theoretical prediction.

In conclusion, we would like to address a final important point concerning dislocation glide. As it happens, we made the implicit assumption that the dislocations could not glide across the layers. This question is relevant, with the interface exerting a repelling force on the dislocations that approach it. In principle, they could glide and gather ahead of the front, or even recombine, in which case our model would need to be revised. It seems this is not the case, experimentally (and this can be justified theoretically [46]). In this case, the dislocations are trapped in the smectic B phase during crystallization. Note that this process generates stacking faults, since each elementary dislocation in the smectic A becomes a partial dislocation in the smectic B, with which is associated a stacking fault.

In the future, it would be interesting to determine the density of defects and of stacking faults as a function of the conditions of growth.

BIBLIOGRAPHY

[1] Herrmann K., Krummacher A.H., *Z. Krist.*, **81** (1932) 317.

[2] Bernal J.D., Crowfoot D., *Trans. Faraday Soc.*, **29** (1933) 1032.

[3] Levelut A.-M. , Lambert M., *C. R. Acad. Sci. (Paris)*, **272B** (1971) 1018.

[4] Levelut A.-M., Doucet J., Lambert M., *J. Physique (France)*, **35** (1974) 773.

[5] Levelut A.-M., *J. Physique Coll. (France)*, **37** (1976) C3-51.

[6] Doucet J., Levelut A.-M., *J. Physique (France)*, **38** (1977) 1163.

[7] Richardson R.M., Leadbetter A.J., Frost J.C., *Ann. Phys.*, **3** (1978) 177.

[8] de Gennes P.-G., Sarma G., *Phys. Lett. A*, **38** (1972) 219.

[9] Moncton D.E., Pindak R., *Phys. Rev. Lett.*, **43** (1979) 701.

[10] Pershan P.S., Aeppli G., Litster J.D., Birgeneau R.J., *Mol. Cryst. Liq. Cryst.*, **67** (1981) 861.

[11] Smith G.W., Gardlund G., Curtis R.J., *Mol. Cryst. Liq. Cryst.*, **19** (1973) 327.

[12] Leadbetter A.J., Frost J.C., Mazid M.A., *J. Physique Lett. (France)*, **4 0** (1979) L325.

[13] Collett J., Sorensen L.B., Pershan P.S., Litster J.D., Birgeneau R.J., Als-Nielsen J., *Phys. Rev. Lett.*, **49** (1982) 553.

[14] Collett J., Sorensen L.B., Pershan P.S., Als-Nielsen J., *Phys. Rev. A*, **3 2** (1985) 1036.

[15] Levelut A.-M., *J. Physique (France)*, **51** (1990) 1517.

[16] Davidson P., Dubois-Violette E., Levelut A.-M., Pansu B., *J. Phys. II (France)*, **3** (1993) 395.

[17] Dubois-Violette E., Pansu B., Davidson P., Levelut A.-M., *J. Phys. II (France)*, **3** (1993) 395.

[18] Friedel J., *Dislocations*, Pergamon Press, Oxford, 1964.

[19] Nye J.F., *Physical Properties of Crystals*, Oxford University Press, Oxford, 1957.

[20] Battacharya S., Shen S.Y., Ketterson J.B., *Phys. Rev. A*, **19** (1979) 1211.

[21] Thiriet Y., Martinoty Ph., *J. Physique Lett. (France)*, **43** (1982) L. 137.

[22] Ali A.H., Benguigui L., *Liq. Cryst.*, **9** (1991) 741.

[23] Cagnon M., Durand G., *Phys. Rev. Lett.*, **45** (1980) 1418.

[24] Cagnon M., Palierne J.-F., Durand G., *Mol. Cryst. Liq. Cryst. Lett.*, **82** (1982) 185.Cagnon M., Durand G., *J. Physique Lett. (France)*, **42** (1981) L451.

[25] Cagnon M., Durand G., *J. Physique Lett. (France)*, **42** (1981) L451.

[26] Oswald P., *J. Physique (France)*, **46** (1985)1255.

[27] Chou Y.T., *J. Appl. Phys.*, **33** (1962) 2747.

[28] Oswald P., *J. Physique Lett. (France)*, **45** (1984) L1037.

[29] Ribotta R., *Phys. Lett. A*, **56** (1976) 130.

[30] Oswald P., Melo F., Germain C., *J. Physique (France)*, **50** (1989) 3527.

[31] Wulff G.Z., *Kristallogr. Mineral.*, **34** (1901) 449.

[32] Landau L., Lifchitz E., *Physique Statistique*, Mir, Moscow, 1967, p. 552.

[33] Nozières Ph., "Shape and growth of crystals" in *Solids Far From Equilibrium*, Ed. C. Godrèche, Aléa-Saclay collection, Cambridge University Press, Cambridge, 1992.

[34] Herring C., *Phys. Rev.*, **82** (1951) 87.

[35] Melo F., Oswald P., *Ann. Chim. (France)*, **16** (1991) 237.

[36] Melo F., Oswald P., *Phys. Rev. Lett.*, **64** (1990) 1381.

[37] Krüger G.J., *Phys. Rep.*, **82** (1982) 229.

[38] Coriell S.R., Sekerka R.F., *J. Cryst. Growth*, **34** (1976) 157.

[39] Melo F., Oswald P., *Phys. Rev. E*, **47** (1993) 2654.

[40] Melo F., Oswald P., *J. Phys. II (France)*, **1** (1991) 353.

[41] Bowley R., Caroli B., Caroli C., Graner F., Nozières Ph., *J. Physique (France)*, **50** (1989) 1377.

[42] Oswald P., Melo F., *J. Phys. II (France)*, **2** (1992) 1345.

[43] Burton W.K., Cabrera N., Frank F.C., *Philos. Trans. Roy. Soc.*, **243** (1951) 299.

[44] Obretenov W., Kashchiev D., Bostanov V., *J. Cryst. Growth*, **96** (1989) 843.

[45] Caroli B., Caroli C., Roulet B., *J. Physique (France)*, **50** (1989) 3075.

[46] Oswald P., Lejcek L., *J. Phys. II (France)*, **1** (1991) 1067.

[47] Adda Y., Dupouy J.M., Philibert J., Quéré Y., *Elements of Physical Metallurgy*, Vol. 5, *Plastic Deformation*, (in French), La Documentation Française, Paris, 1979.

Appendix 1

CREEP BY CROSSING OF LOCALIZED OBSTACLES

It is well-known in metallurgy that a solid does not deform by cooperative glide of the atomic planes one over the other, which would require exceeding the theoretical elastic stress [1, 2]:

$$\sigma_e = \frac{\mu}{2\pi} \tag{A.1}$$

but rather by the propagation of dislocations, which are linear defects breaking locally the translational order of the crystal.

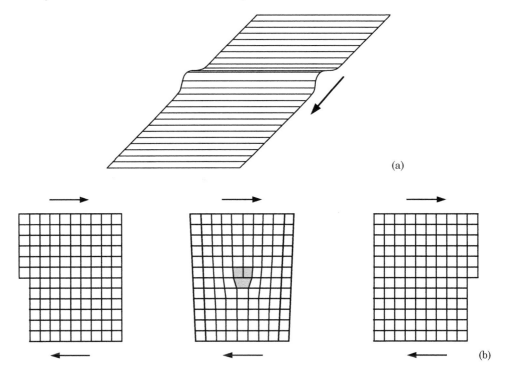

(a)

(b)

Fig. A.1 a) Analogy of the Mott carpet: the best way of moving a carpet is to make a fold and then propagate it. b) Propagation of a wedge dislocation (cross-hatched area) in a crystal under shear. By propagating from left to right, the dislocation generates a plastic deformation $\varepsilon = b/d$, where b is the Burgers vector of the dislocation and d the sample thickness.

N.F. Mott (1905–1996), a famous English physicist who received the Nobel prize in physics in 1977, compared this problem to the one of moving a carpet over a rough floor [1]. There are indeed two ways of doing it: either by pulling it all together by one of its ends, which is difficult, or by making a fold and propagating it along the carpet, which, as everyone knows, is much easier

(Fig. A.1a). In this analogy, the fold plays the same role as the dislocation propagating from one side of the sample to the other (Fig. A.1b).

Referring to the schematic in Figure A.1b, it is immediately apparent that the movement of a dislocation with Burgers vector b from one end of the sample to the other results in a displacement b of the upper face of the crystal with respect to the lower one. Clearly, if the crystal contains N dislocations, the displacements induced by the N dislocations add up. As a result, the strain rate $\dot{\varepsilon}$ of the crystal is proportional to the density of mobile dislocations ρ_m, to their Burgers vector b and to the velocity v they acquire under the action of the applied stress σ. This geometrical formula, very simple but also very general, bears the name of the Orowan relation [1–3]:

$$\dot{\varepsilon} = \rho_m\, b\, v \tag{A.2}$$

Purely geometrical, the relation is fundamental in most of the models of the theory of plasticity [3]. In each of them, the question is to know how the velocity v of the mobile dislocations and their density ρ_m vary as a function of the applied stress (as a matter of fact, the density can change if dislocation sources are present, such as the spiral Frank-Read sources).

In the following, we assume that the **density ρ_m is fixed** and that dislocation motion is hindered by **crossing of localized obstacles**. This is, for instance, the case of the creep experiment in shear parallel to the layers, described in section VII.2.b. Here, the mobile dislocations are basal dislocations of Burgers vector $b = c_o$, and the obstacles are screw dislocations piercing the layers (which form the glide plane of basal dislocations). This is just an example. In practice, other obstacles can exist, such as solute atoms (in the case of an alloy), precipitates or impurity agglomerates. The important thing is that the obstacle can be crossed by thermal activation (requiring that it is localized and of small size).

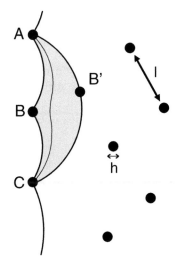

Fig. A.2 Movement of a dislocation in a forest of non-attractive obstacles.

To calculate the average velocity v of a mobile dislocation in its glide plane, consider the sketch in figure A.2 [1, 2]. The obstacles are shown as solid dots of diameter h (the typical size of the obstacle). Let l be the average distance between two obstacles. Assume that, at a given moment, the line, which is "pinned" on obstacles A, B and C, is pushed towards the right by an effective stress σ^*, to be defined later. Under the action of this stress, the line gets bent. At the same time, each dislocation segment of length l starts to vibrate at a characteristic frequency v given by

$$v \approx \frac{c_s}{l} \qquad (A.3)$$

where $c_s = \sqrt{\mu/\rho}$ is the speed of sound (shear wave). This result arises because a dislocation in a solid has a line tension of the order of μb^2 and an effective mass per unit length of the order of ρb^2.

If there is an energy barrier U for crossing obstacle B, the probability that the dislocation crosses this obstacle is given by the Arrhenius factor

$$\exp\left(-\frac{U}{k_B T}\right) \qquad (A.4)$$

such that the crossing frequency is

$$v_c = v \exp\left(-\frac{U}{k_B T}\right) \qquad (A.5)$$

If we now assume that the time needed by the dislocation to reach the next obstacle B' is negligible compared to the time $1/v_c$ it needs to cross an obstacle, we have

$$v = v\, l \exp\left(-\frac{U}{k_B T}\right) \approx c_s \exp\left(-\frac{U}{k_B T}\right) \qquad (A.6)$$

where l is the typical distance covered by a dislocation after each jump. As to the activation energy U, it is equal to the difference between the energy of the two jogs that appear (see figure C.VII.6) and the work of the effective force $\sigma^* b$ acting upon the segment ABC, as it glides over the distance h needed to cross the obstacle and form the two jogs:

$$U = 2U_c - \sigma^* b \times l \times h \qquad (A.7)$$

In the case of crossing between a basal dislocation and a screw dislocation perpendicular to the glide plane, the distance h is of the order of b if the mobile dislocation is not dissociated, and of the order of the dissociation width if it is dissociated into two partial dislocations separated by a stacking fault [2].

Finally, one finds the following creep law by putting together the results of eqs. A.2, 6 and 7:

$$\dot{\varepsilon} = \rho_m \, b \, c_s \exp\left(-\frac{2U_c}{k_B T}\right) \exp\left(\frac{\sigma^* \Omega}{k_B T}\right) \qquad (A.8)$$

In this expression, $\Omega = b \times A$ is an activation volume and $A = l \times h$ an activation area. Ω and A depend on the nature of the obstacle quantified by its width h) and on their average distance l.

In conclusion, let us remember that the effective stress is equal to the applied stress σ minus the internal stress σ_i. This latter is a result of the stress field produced by the other mobile dislocations (the "trees" do not come into play, as they do not interact with the mobile dislocations). This field oscillates around zero with an amplitude σ_M over a typical distance L equal to the average distance between two mobile dislocations (by definition, $\rho_m = 1/L^2$). One then has $\sigma_i = |\sigma_M|$. In isotropic elasticity [2]:

$$\sigma_i = |\sigma_M| \approx \frac{\mu b}{2\pi L} \qquad (A.9)$$

where μ is the shear modulus. It is noteworthy that the internal stress is the minimal shear stress that must be applied to the system in order that the mobile dislocations can move over large distances (larger than L). Thus, it defines a **threshold creep stress**, or **elasticity limit**, since the system does not undergo plastic deformation as long as $\sigma < \sigma_i$. Also note that σ_i slowly decreases with temperature (along with the elastic modulus) and is not thermally activated. On the other hand, σ_i increases with the density of mobile dislocations, which might seem surprising, since the dislocations themselves are responsible for the plastic deformation. One must however keep in mind that, as their density increases, the dislocations hinder each other, blocking their movement. This phenomenon is at the origin of **work hardening**.

BIBLIOGRAPHY

[1] Martin J.-L., *Dislocations et Plasticité des Cristaux*, Presses Polytechniques et Universitaires Romandes, Lausanne, 2000.

[2] Friedel J., *Dislocations*, Pergamon Press, Oxford, 1967.

[3] Poirier J.-P., *Creep of Crystals*, Cambridge University Press, Cambridge, 1985.

C h a p t e r C.VIII

Smectic free films

Owing to their lamellar structure, the smectic liquid crystals form films, **similar to soap bubbles**, that can be stretched on frames of arbitrary shape. This property of smectic mesophases was already known to Georges Friedel at the beginning of the 20th century. In his treatise on mesophases [1], he speaks about it in these terms:

"A similarity can be found between the terrace-shaped droplets and the soap films studied by numerous physicists, and which were in particular the object of observations and measurements by J. Perrin and P.V. Wells."

This remark by G. Friedel went unnoticed for more than fifty years, and it is only at the beginning of the 1970s that Young, Pindak, Clark and Meyer [2a] rediscovered these systems and started the systematic study of their structures and physical properties. Their work rekindled the interest in thermotropic smectic films, which has not ceased to grow during the last two decades, to the point that we now have a considerable corpus of knowledge concerning these systems. One of the reasons for this enthusiasm is the fact that almost all smectic phases, but also the lamellar, cubic and hexagonal phases of lyotropic systems [2b], can form films, allowing the study of the role of confinement and of thickness effects in phase transitions. Surprisingly, the films of lamellar lyotropic phases, the closest in nature to soap bubbles [3], were only studied recently, and to a much lesser extent. The main reason for this delay is that lyotropic liquid crystals are obtained by **dissolving** surfactants in water, **requiring the control of their water concentration.** Obviously, in the case of free-standing films, this concentration can change rapidly due to the very small thickness of the medium. The first experimental setup allowing evaporation to be avoided and, better still, control of water concentration in the film, conceived by Smith et al. [4], is fairly complex. We shall describe later (subsection VIII.1.b) another, simpler, experimental setup for regulating the humidity around the film (and thus setting the water concentration within the film). By modifying this parameter, one can induce in the film structural changes or phase transitions, effects achieved in thermotropic films by changing the temperature. Once the problem of humidity control is solved, manipulating lyotropic films becomes as easy as for thermotropic films.

Let us now explain why we have dedicated an entire chapter to the topic of films.

We have already mentioned the role of thickness effects in phase transitions. Clearly, the films represent model systems for the study of these confinement phenomena.

Another reason is based on the non-trivial observation that a smectic film cannot exist as an isolated system, unlike a thin slab of silicon or a sheet of gold. Clearly, **a film must be stretched on a stiff enough frame in order to exist**. Without its support, it will inevitably collapse under the action of its surface tension. Even in the three-dimensional crystalline SmB phase, the elastic modulus is not high enough to oppose the action of surface tension. These observations naturally lead to discussion of the role of the **meniscus** connecting the film to its frame.

An additional reason for studying films is that, seen in reflecting microscopy, they immediately reveal the lamellar structure of smectic phases. This is immediately apparent in the photograph of a SmA film in figure C.VIII.1a, where one can see a complex system of domains with various thicknesses, separated by steps. In this example, each step corresponds to a change in the thickness of the film by one or several molecular layers. We shall see in the following that, at constant temperature, the steps migrate slowly (over a few hours, sometimes a little faster) toward the meniscus connecting the film to the frame and completely merge with it. At the end, the film thickness is that of the thinnest domain. This observation also deserves an explanation. We point out that the optical appearance of a lyotropic film in the lamellar L_α phase is very similar to that of a thermotropic film in the SmA phase, as shown in figure C.VIII.1b. This similarity also covers the behavior of the steps, at constant temperature and humidity.

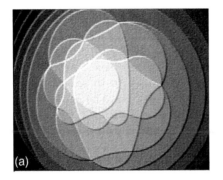 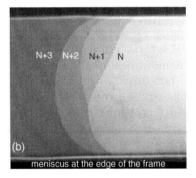

Fig. C.VIII.1 Reflecting microscopy view of a smectic film: a) in the thermotropic SmA phase of the eutectic mixture S2; b) in the lyotropic L_α phase of the binary mixture $C_{12}EO_6$/water (from ref. [10]).

Finally, by manipulating SmA thermotropic or L_α lyotropic films, one quickly realizes that they are surprisingly stable, in contrast with the proverbial fragility of soap bubbles. For instance, they can be pierced with a glass fiber without breaking, made to vibrate like a drumhead, their structure can be changed by varying the temperature or humidity, hydrodynamic flow within the films can be induced, their frame can be deformed, etc. Moreover, thermotropic films, kept under vacuum, can be preserved for months without evaporating.

All these manipulations of smectic films raise a multitude of questions concerning their structure, their creation, and more generally their physical properties, at thermal equilibrium and out of equilibrium.

These are the questions we shall try to answer in this chapter.

However, we shall consider neither the case of soap films (formed by a thin water layer bound by two monolayers), nor that of Newton black films, formed when (almost) all the water is expelled from the film [5].

The plan of the chapter is as follows. In section VIII.1, we first show how to obtain a smectic film, and then give some information on its structure. Section VIII.2 is purely experimental. We show how to determine the film thickness and its surface tension. Section VIII.3 is dedicated to the thermodynamics of smectic films; in particular, the problems concerning the meniscus that connects the film to the frame and the junction between the two will be treated, theoretically and experimentally. In section VIII.4, we approach the topic of steps. We shall see how to control their nucleation using a heating micro-tip, and how to measure their line tension. The growth dynamics of a step loop and the problem of hydrodynamic interactions between steps are also discussed ("step foams"). Section VIII.5 deals with phase transitions and their observation by X-ray scattering, calorimetry and optical or mechanical measurements. In the last section, VIII.6, with the title "Smectic films as model systems," we describe some experiments performed on smectic drums (vibrating films), which provide insight into a variety of topics, such as isospectrality, fractals or quantum billiards. Afterward, we describe an experiment on steps that brings into play nonlinear effects, then a rheology experiment, where the coupling between topology and flow plays an essential role ("topological flow"). We conclude by an experiment on the anomalies of molecular diffusion in films.

VIII.1 Making smectic films

VIII.1.a) How to "stretch" a smectic film?

Any experiment involving smectic films begins, of course, by preparing them. In practice, the films must be stretched over a **frame**. This frame is placed in a temperature-regulated chamber and, if necessary, under controlled humidity in the case of lyotropic films (subsection VIII.1.b). In its simplest original form [6], the frame is a circular hole about 4 mm in diameter, drilled into a duralumin plate of thickness 2 mm, and beveled to give sharp edges (Fig. C.VIII.2a).

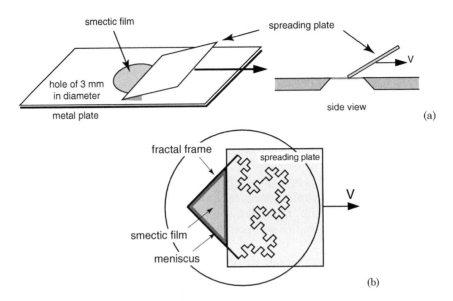

Fig. C.VIII.2 Stretching a thermotropic smectic film by the simplest method. To stretch the film, one uses a small rectangular plate, with one edge coated by a small amount of liquid crystal in the smectic A (for instance, S2 from BDH) or smectic C (SCE4 from BDH) phase. Keeping the plate tilted with respect to the frame, one spreads the material very slowly over the frame. With a little practice, one can stretch films of varying thickness. a) circular frame; b) prefractal frame (from refs. [7a and b]).

To stretch a film over the frame, one uses a second plate (for instance, a flexible metal blade) coated at one edge by a small amount of thermotropic liquid crystal in the smectic phase. Maintaining this plate tilted and in contact with the frame, the film is very slowly stretched over the hole. In his book *Structure of Liquid Crystal Phases*, Pershan [6] presents this spreading technique as follows:

"A small amount of the material, usually in the high temperature region of the SmA phase, is spread around the aperture that is maintained at the necessary temperature, and a wiper is used to drag some of the material across the aperture. If a stable film is successfully stretched, it is detected optically by its finite reflectivity."

This method has the advantage of being applicable to any kind of planar frame. In particular, it was successfully employed to fractal-shaped frames (Fig.C.VIII.2b) [7a,b] as well as to the isospectral polygonal frames of Gordon, Webb and Wolpert [7c] (Fig. C.VIII.89).

However, in order to control film thickness, the manipulation system of the spreading plate must be mechanically stable, ensuring the smoothest possible motion. Indeed, as we shall see further, a peak of the stretching velocity **v** can lead to "spontaneous" thinning of the film and even to its breaking. This requirement of mechanical stability during film stretching is more easily satisfied using the frame of variable surface shown in the photo of figure C.VIII.3. This frame is used to stretch "L"-shaped films.

Fig. C.VIII.3 "L"-shaped mobile frame for controlled film stretching. The fixed parts are beveled at a 45° angle, forming rails over which slide the moving pieces, similarly beveled and equipped with notches fitting the rails. The beveled edges of the four assembled pieces make up the edges of the frame. All parts must be carefully machined, so that the frame is as flat as possible. To stretch a film on this frame, it is first closed and daubed with a small amount of smectic A or C liquid crystal. It is then opened slowly, all the while monitoring the formation of the film.

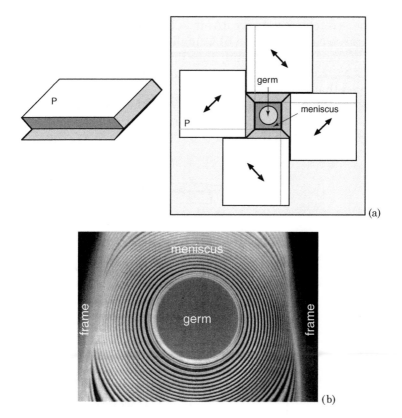

Fig. C.VIII.4 a) Affine square frame made up by four mobile pieces, all identical to P; b) reflecting microscopy photograph of the initial germ.

In another version shown in figure C.VIII.4, the frame, made up by four metal pieces with bevels and fitting notches, is square. The motion of the four pieces is guided and synchronized such that the frame keeps its square shape as the surface changes. This affine square frame is useful for controlling the nucleation of holes or islands in the film (discussed in the next subsection). To control the thickness of the film, one only needs to observe it in reflectivity all through the stretching process. Using the optical setup in figure C.VIII.5, one can observe the steps and measure the reflectivity spectrum of the film as a function of its thickness (see section VIII.2).

Reflecting microscopy observation of films in the SmA phase shows that their final thickness is strongly dependent on the details of the stretching process. Let us consider the example of the **affine square frame**. When it is almost closed, the applied liquid crystal material forms some sort of large and thick drop, where the texture (orientation of the smectic layers) is not well defined. On gently opening it, the initial **"germ"** of the film suddenly appears, as a small circular **domain** of uniform thickness (its initial value N_o can vary from a few layers to a few hundred layers), connected to the frame by a **meniscus** of variable thickness (Fig. C.VIII.4). As the stretching process continues, the surface A of the domain increases. Its thickness remains constant if the stretching velocity dA/dt is low enough (dA/dt < 1–10 mm^2/mn, typically). In these conditions, the film finally has the same thickness N_o as the initial germ. If, on the other hand, the film is stretched too fast, pores with thickness $N < N_o$ can open up within the initial domain. The final film thickness is then that of the thinnest pore.

These important observations show that the films are **metastable** and, whenever possible, become thinner by pore nucleation. We shall explain later the physical reasons for this thinning.

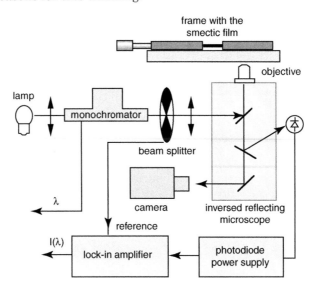

Fig. C.VIII.5 Setup for the observation of smectic films during stretching and for measuring the spectrum of reflected light (from ref. [10]).

In the case of the **fractal frame** in figure C.VIII.2b) the film starts under the shape of a right angled triangle of increasing surface area; at this point, the nucleation process is completely similar to the previous case, viz. only one germ appears, with a certain thickness N_o, and its surface increases (Fig. C.VIII.6a). The stretching process then becomes more complicated with the nucleation of new germs, with thickness N_i. These germs appear each time the edge of the spreading plate starts to form with the edges of the fractal frame a new "small creek" shaped as a right angled triangle. At first, these new films are isolated (and separate from the main film, see Fig. C.VIII.6a and b) up to the moment the spreading plate passes the tip of one of the salient elements of the frame (Fig. C.VIII.6c). At this point, the two neighboring films, of thickness N_i and N_j, come in contact and a step of height $\Delta N = N_i - N_j$ bridges the "straits" formed by the tip P and the edge of the spreading plate (Fig. C.VIII.6c). This step is usually curved toward the thickest film and its curvature radius r_c is of the order of a few tens of micrometers (the physical meaning of r_c will be discussed in subsection VIII.4b). It is noteworthy that this curvature radius is conserved as the straits are widening, while the angular aperture of the step θ (lower than π at the beginning, Fig. C.VIII.6c) increases. However, this process becomes unstable for $\theta = \theta_{crit} = \pi$ (Fig. C.VIII.6d). Beyond this limit, the step advances rapidly, eliminating the thicker domain. At the end of the stretching process, the film thickness is given by the thinnest germ.

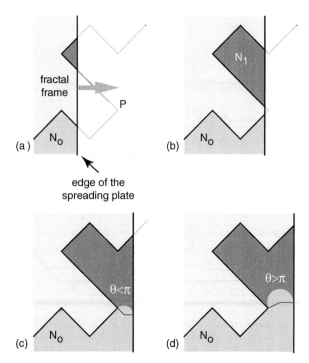

Fig. C.VIII.6 Stretching a film on a fractal frame: a) formation of a new germ with thickness N_1; b) growth of the new germ; c) appearance of a stable step with radius R between the films with thicknesses N_o and N_1; d) elimination of the thicker film by the advancing step.

In the case of the "L"-shaped frame (Fig. C.VIII.3), several germs of different thickness usually nucleate during the stretching process. Once again, competition between these germs selects the thinnest one.

Finally, a smectic film can also be stretched on a vertical frame [8]. In this position, the islands fall and the holes rise, under the action of gravity. This case will be discussed in subsection VIII.3.f.

VIII.1.b) Temperature and humidity control

The frame holding the smectic film is placed inside a chamber equipped with windows for optical observation. When working with thermotropic materials, one only needs to control the temperature of the metal body of the chamber. The gas pressure within the cell can also be controlled using a primary vacuum pump. This is necessary for measuring the frequency of the eigenmodes of a vibrating film, in order to compare them with the solutions of the wave equation for a two-dimensional membrane. These solutions are more easily obtained when neglecting the inertia of the gas set in motion by the vibrations of the film. For very thin films (a few tens of molecular layers), this approximation is only valid at pressures of the order of 1 mbar, or less.

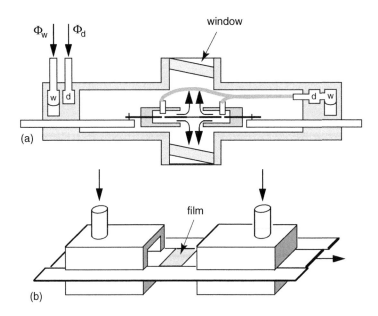

Fig. C.VIII.7 Diagram of the chamber containing the frame holding a thermotropic or lyotropic smectic film. a) For the study of lyotropic films, the metal body of the chamber contains two channels, one of them dry and the other partially filled with water. Two streams of an inert gas (nitrogen) of flux Φ_d and Φ_w are injected in the channels. The Φ_w stream becomes saturated in water vapor during its passage in the water-filled channel. The humidity in the cell is a function of the ratio Φ_w/Φ_d; b) the humid gas is directed onto the surfaces of the film in order to reduce the equilibration time.

When dealing with lyotropic films, on the other hand, control of the gas humidity within the cell is imperative. The principle of the regulation system, shown in figure C.VIII.7, is that used to study the anchoring transitions on mica (Fig. B.V.24, volume 1). It operates by mixing in variable proportions two nitrogen streams, one of them dry and the other at a 100% humidity (Fig. C.VIII.7a). To reduce the time for reaching equilibrium between the gas and the film, the frame is equipped with nozzles that blow the humid gas directly onto the film surfaces (Fig. C.VIII.7b).

VIII.1.c) Basic structural elements of smectic films: menisci, pores, islands, steps and nodes

From the observation of the stretching process of thermotropic films in the SmA phase, one can reach the following conclusions (summarized in the diagram of figure C.VIII.8):

1. If the film maintains a constant thickness (the ideal case) and its surface increases, molecules must be transferred from the meniscus toward the film. Hence, **the meniscus acts as a reservoir of molecules**, supplying the growing film by the necessary flow of matter. If, on the other hand, the frame is slowly closed, the meniscus absorbs the excess molecules; the matter flow is reversed. Finally, **when the surface of the frame remains constant, equilibrium is reached between the film and the meniscus** and the matter flow vanishes.

2. Having established that, as the film is stretched, the molecules are transferred from the meniscus toward the film, it becomes clear that this transfer must be facilitated as much as possible. In practice, it is enough to **heat the meniscus to a temperature where the material used is in the most fluid phase, SmA or SmC**. If these phases do not exist in the phase sequence of the product, stretching a film becomes very difficult, as it is almost impossible to prevent pore opening.

3. If, while stretching a film of thickness N, a pore of thickness N – 1 accidentally appears, it almost always takes over the entire film, even if the stretching process is stopped. We conclude that **films are metastable with respect to any decrease in thickness** and that a pore opening, when stretching is too fast, plays the same role as the growth germ that renders a first-order phase transition possible. However, while the frame maintains a constant surface area, spontaneous nucleation of a pore is so unlikely that **the lifetime of a film, in the absence of impurities, is practically infinite**.

4. Increasing the stretching rate leads to a considerable decrease of the threshold for pore nucleation.

5. If, during film stretching, several pores of different thicknesses appear, the texture of the film can transitorily become very complex, as shown in figure C.VIII.1a. The film is divided into **domains** of different thicknesses,

separated by **steps**. Sometimes, several steps are linked at a **node** (where a microscopic dust particle is usually present).

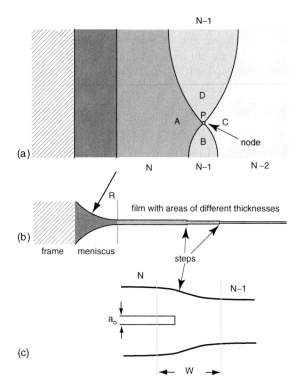

Fig. C.VIII.8 Structural elements of a smectic film. The film is connected to the frame by its meniscus of radius of curvature R. The thickness N of the film can vary from one domain to the next. The domains of different thickness (A, B, C and D) are separated by steps. The steps A/B, B/C, C/D and A/D are linked by the node P. a) Top view of the film; b) side view of the film; c) side view of a step; it is created by an edge dislocation localized in the center of the film due to the repulsion of the free surfaces [9]; the width W of a step (i.e., the extent of the surface deformation) depends on film thickness.

Finally, the behavior of lyotropic films in the L_α phase is, during the stretching process, completely similar to that already described for thermotropic films (at least for films of non-ionic surfactants such as $C_{12}EO_6$). Nevertheless, the microscopic interpretation of this behavior is more complicated, since one must make the distinction between surfactant and water molecules. The surfactant, like the thermotropic liquid crystal, is not volatile and its amount, distributed between the film and the meniscus, must remain constant. As to the water dissolved in the film and the meniscus, one could also assume that the amount remains constant, at constant temperature and vapor pressure, provided that thermal equilibrium between the film and the meniscus is reached. This approximation is, however, restricted to the case of very thick films, where the interactions between the free surfaces are negligible.

III.2   Some crucial experiments

VIII.2.a) Measurement of film thickness

The vocabulary we used to describe SmA or L_α smectic films is based on the fact that the thickness of a smectic film, at constant temperature and vapor pressure, can only vary discontinuously and is therefore given by the number N of layers (or bilayers) it contains. We shall see later that this thickness N plays the role of a thermodynamical parameter, as important as the temperature or the humidity. It is therefore useful, at this point, to emphasize that the thickness N can be precisely determined by film reflectivity measurements. Using the experimental setup shown in figure C.VIII.5, one can measure the intensity $I_r(\lambda)$ of the light reflected by a film as a function of wavelength.

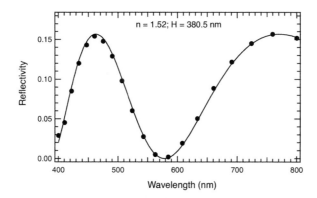

Fig. C.VIII.9 Reflectivity spectrum of a thick film of the liquid crystal 8CB at room temperature (N = 123) (courtesy F. Picano, 1999).

Dividing this spectrum by that of a reference metallic mirror placed at the position of the film, $I_0(\lambda)$, gives the reflectivity:

$$R = \frac{I_r}{I_0} = \frac{f\,\sin^2(2\pi D/\lambda)}{1 + f\,\sin^2(2\pi D/\lambda)} \qquad\qquad \text{(C.VIII.1)}$$

with

$$f = \frac{(n^2 - 1)^2}{4n^2} \qquad\qquad \text{(C.VIII.2a)}$$

and

$$D = nH = nNa_0 \qquad\qquad \text{(C.VIII.2b)}$$

where n is the refractive index, a_o the thickness of one smectic layer (or of a lamellum), H the thickness of the film and D its "optical thickness." This classical formula stems from multiple wave interference within the film, which behaves as a Fabry-Perot interferometer. An example of the spectrum obtained for 8CB is shown in figure C.VIII.9.

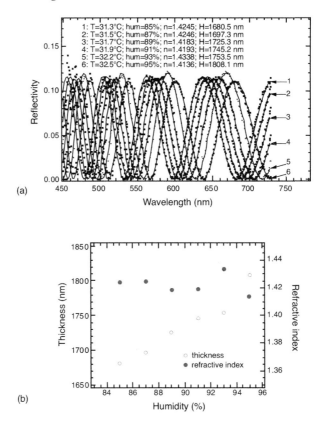

(a)

(b)

Fig. C.VIII.10 a) Reflectivity spectra for a thick film of the L_α phase of the lyotropic mixture $C_{12}EO_6$/water at different humidity rates (N = 363); b) variation of the thickness and of the average refractive index of the film as a function of humidity.

Note that the determination of the refractive index n is much more accurate when the spectrum contains several extremal points. To find n, it is thus convenient to use thick films, as in the example of figure C.VIII.10a) where the spectra are those of a thick $C_{12}EO_6$/water film in the L_α phase. These spectra give the variation, as a function of humidity, of the average refractive index n and of the thickness H at N = const (Fig. C.VIII.10b).

Once the average index n is known, one can determine the thickness of a layer (or lamellum) from the spectra of very thin films. Such spectra, obtained for the $C_{12}EO_6$ / water mixture are shown in figure C.VIII.11a. These spectra were obtained by "peeling" layer by layer a film of unknown initial thickness N_o. Fitting with equation C.VIII.1 yields the thickness $(N_o - \Delta N)a_o$ of the film corresponding to each spectrum. Plotting this thickness as a function of ΔN (Fig. C.VIII.11b), one obtains N_o by extrapolating to 0, and the thickness a_o of a lamellum from the slope.

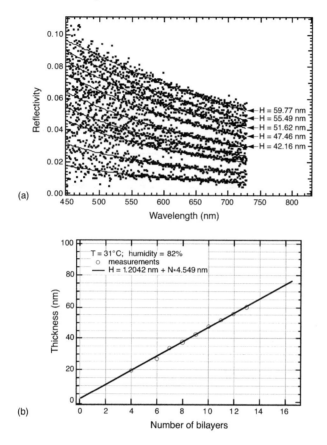

Fig. C.VIII.11 a) Spectra of thin films of the $C_{12}EO_6$/water mixture. From one spectrum to the next, one or two bilayers were removed (at 82% humidity); b) the discontinuous variation in the thickness of thin smectic films gives the thickness of one bilayer ($C_{12}EO_6$/water mixture at 82% humidity).

In Pieranski et al. [10], one can find a similar example for the measurement of the molecular layer thickness for a thermotropic film in the SmA phase.

VIII.2.b) Measurement of film tension

After the thickness N of the film and its surface area A, the surface tension τ is obviously another important parameter for the characterization of the smectic film. It can be measured in several ways.

The most obvious method, based directly on the definition of the surface tension, relies on measuring the force per unit length exerted by the film on its frame. This is done using the setup shown in figure C.VIII.12. It consists of a rectangular frame with two mobile sides, and a trapezoidal plate fixed on a balance pendulum.

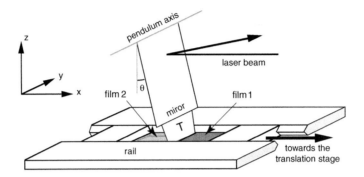

Fig. C.VIII.12 Setup for measuring the surface tension of films (from ref. [10]).

Using the xyz translations of the suspension axis, the trapezoidal plate T can be placed in the center of the frame, leaving between its lateral edges and the rails of the frame a space of the order of 0.01 mm. A film is then stretched on one side of the balance, by moving one of the mobile sides of the frame. The balance is pulled by the film away from the vertical position until the restoring force, due to its own weight, compensates the tension of the film. By detecting (using a photodiode) the deflection of a laser beam reflected by a mirror fixed to the balance, one can determine the static and dynamic tension of the film.

The first result obtained using this setup is to be expected: the static tension τ (at constant surface area A) of an S2 film is, within a few percent, twice the surface tension γ of the liquid crystal:

$$\tau \approx 2\gamma \tag{C.VIII.3}$$

The second – less obvious – result is that the film tension transiently varies as the size of the film is changed. In the diagram of figure C.VIII.13, one can see that the tension $\tau(t)$ of the film increases as the surface area of the film is suddenly increased by an amount $\Delta A > 0$ ("stepwise increase"), then relaxes toward its equilibrium value over a characteristic time t_r:

$$\Delta\tau = \Delta\tau_0\, e^{-t/t_r} \tag{C.VIII.4}$$

The initial tension jump $\Delta\tau_0$ depends on the relative increase in film area $\Delta A/A$ and on the duration Δt of the expansion. More precisely, $\Delta\tau_0$ depends on the expansion rate $(1/A)(\Delta A/\Delta t)$. As this rate increases, so do the tension jump $\Delta\tau_0$ and the relaxation time t_r. Typical orders of magnitude are $\Delta\tau/\tau \approx 1-2\%$ for $(1/A)(\Delta A/\Delta t) \approx 0.02$ s^{-1} and $t_r \approx 1$ s. The effect is reversed (i.e., the tension decreases when the film is compressed) $(\Delta A/A < 0)$.

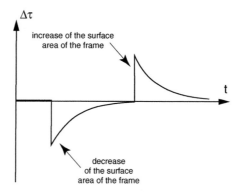

Fig. C.VIII.13 Evolution of the surface tension of a smectic film subsequent to changes in the frame area.

The third non-trivial result, to be explained in the next subsection, is obtained by stretching two films, on either side of the balance. One of them is very thin ($N \approx 20$), while the other film is very thick ($N \approx 10^4$). A small deflection of the balance with respect to the vertical position clearly shows that the pull of the thicker film is stronger than that of the thin film. The diagram in figure C.VIII.14 shows that **the tension of the film increases linearly** with the thickness:

$$\tau = \tau_0 + \alpha N, \qquad \tau_0 = 57.8 \, \text{dyn/cm}, \qquad \alpha = 7.61 \times 10^{-4} \, \text{dyn/cm} \qquad \text{(C.VIII.5)}$$

In a more elaborate – and much more precise – version of this device, the balance carries a magnet subjected to the magnetic field of a solenoid [11]. The signal of the photodiode is used in a feedback loop in order to maintain the balance in the vertical position.

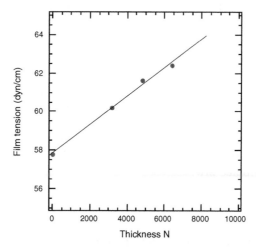

Fig. C.VIII.14 Variation of the surface tension of a smectic film as a function of its thickness (S2 liquid crystal).

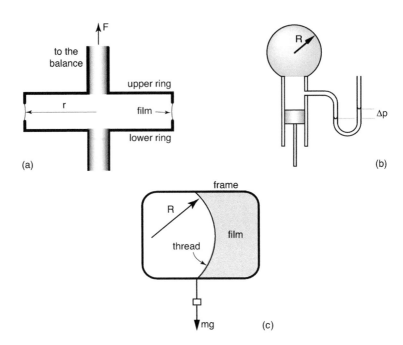

Fig. C.VIII.15 Three other methods for determining the surface tension of smectic films: a) using a balance to measure the force between two thin rings spanned by the film; b) by measuring the pressure inside a smectic bubble; c) from the line tension of a flexible thread making up one of the edges of the film.

The tension of smectic films can also be determined using other methods.

One of the most precise techniques uses a film stretched between two concentric rings [12a]. The force F needed to maintain constant the gap between the two rings is measured using a precise balance. The film tension is given by the simple formula

$$\tau = \frac{F}{2\pi r} \qquad\qquad\text{(C.VIII.6)}$$

where r is the radius of the film at the center (Fig. C.VIII.15a). This technique confirmed the validity of the law C.VIII.5 in thick films with very good accuracy [12b].

Another method consists of measuring the pressure excess ΔP inside a smectic bubble with radius R (Fig. C.VIII.15b) [13, 14a–e]. The tension is obtained from the Laplace law:

$$\tau = \frac{R\,\Delta P}{2} \qquad\qquad\text{(C.VIII.7)}$$

These two methods yield $\tau \approx 60$ dyn/cm at room temperature for 8CB, a value compatible with that obtained for the S2 mixture.

An alternative way of obtaining τ is shown in figure C.VIII.15c. The experiment consists of measuring the tension $\sigma = mg$ and the radius of curvature R of a very flexible (silk) thread making up one of the edges of the film, itself stretched on a vertical frame. In this case, mechanical equilibrium requires:

$$\tau = \frac{\sigma}{R} \qquad\qquad\qquad (\text{C.VIII.8})$$

This original setup was used by Mach et al. [15, 16] to measure the tension of smectic films for several compounds. For instance, these authors find that, for the three "classical" compounds CBOOA, 12CB and 65OBC, **film tension is practically independent of thickness between N = 2 and N = 100**. These results are not really in contradiction with those discussed earlier, as the variations in film tension measured using the balance pendulum or the ring method are too small to be detected using their setup. They also measured the tension of films formed by three other compounds, with the formula

$$C_{11}H_{23}O-\!\!\left\langle\bigcirc\right\rangle\!\!-\overset{\displaystyle O}{\overset{\|}{C}}-O-\!\!\left\langle\bigcirc\right\rangle\!\!-\overset{\displaystyle O}{\overset{\|}{C}}-O-R_n \qquad n = 1,2, \text{ and } 3$$

where $R_1 = -C_5H_{11}$, $R_2 = -CH_2(CF_2)_3CF_2H$ and $R_3 = CH_2C_4F_9$, finding $\gamma_1 = 21.3$ dyn/cm, $\gamma_2 = 18.3$ dyn/cm and $\gamma_3 = 14.6$ dyn/cm, respectively. One can see that in this case the surface tension decreases by almost 25% when going from the R_2 group to the R_3 group, **only by replacing one hydrogen by a fluorine**.

The conclusion to be drawn from these measurements is that **only the atoms very close to the surface contribute to the surface energy**. Thus, the surface tension is practically unchanged when going from 100 to 2 molecular layers; on the other hand, it can decrease by 25% when replacing a single atom at the extremity of the molecule.

To conclude this section, let us point out that the tension of the film can equally be determined from the frequency of its vibration eigenmodes [17] and show that it also depends on the temperature (see subsection C.VIII.5.e).

VIII.3 Thermodynamics of thermotropic films

The experiments we have just described on the creation, the structure and the mechanical properties of smectic films require additional explanations. In particular, we must understand the role of the meniscus and why the films are metastable and generally prone to thinning (unless they are brutally compressed).

In this section, we shall first show how to determine experimentally the profile of the meniscus [18], and then we analyze the thermodynamics of films

[19]. We detail the concepts of **disjoining pressure** and **macroscopic contact angle**, and give a precise definition of **film tension**.

VIII.3.a) Measurement of the meniscus profile and of the contact angle between the film and its meniscus

The photograph in figure C.VIII.16 shows the typical appearance of the meniscus limiting a smectic A film (8CB at room temperature). One can discern the film on the left (uniform light intensity) and the meniscus on the right (variable intensity).

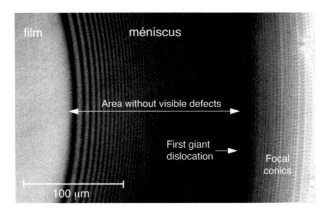

Fig. C.VIII.16 Typical appearance of the meniscus formed between a smectic A film and its frame (here, a circular hole). One can discern the film on the left (thin domain of uniform thickness) and the meniscus to the right. Interference fringes are visible in the thinner region of the meniscus limiting the film. As the thickness increases, the fringes become so narrow as to be no longer discernible. In the thickest region of the meniscus (at the right edge), giant dislocations appear, which split into strings of focal conics. This photograph was taken under combined lighting, both in reflection (to reveal the presence of interference fringes) and in transmission, for a better contrast of the focal conics (photograph by F. Picano, 2000).

Once it reaches equilibrium, two regions must be distinguished within the meniscus [18]: a "smooth" zone, without defects, directly adjacent to the film, and a more "granular" one, filled by a focal conic texture, developing in the thicker part of the meniscus. It should be noted that a giant dislocation with a large Burgers vector (20 layers, typically) delineates the boundary between these two areas. The photograph in figure C.VIII.17a) shows a zoom on the thin region of the meniscus, observed in reflecting microscopy under monochromatic lighting. One can clearly see the interference fringes, and their position suffices (Fig. C.VIII.17b) for reconstructing the profile of the meniscus (Fig. C.VIII.17c). Indeed, from eq. C.VIII.1 we know that the difference in thickness between two fringes is $\lambda/2n$, with n the refractive index. It can be shown experimentally that the profile of the meniscus is perfectly described by a **circular arc with radius R** (Fig. C.VIII.17c). The value of R, the same in any point of the meniscus, depends on the conditions of film preparation (initial amount of matter, spreading velocity, temperature, etc.). Once fixed, however, it remains constant in time, provided the frame does not leak.

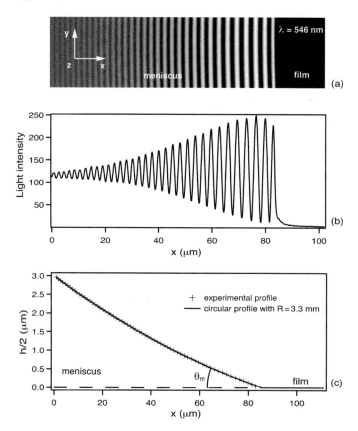

Fig. C.VIII.17 a) Interference fringes observed in reflecting microscopy in the thin region of the meniscus; b) intensity profile of the fringes; c) reconstruction of the height profile of the meniscus h(x) obtained from the position of the fringes. The solid line is a fit of the experimental points with a circle of radius R = 3.3 mm. We point out that the circle crosses the film surface at a macroscopic angle θ_m (here, of the order of 10^{-3} rad). Note the difference in scale between the x and z axes (from ref. [19a]).

Thus, the meniscus has a **circular profile** (rather than exponential, as in the case of a horizontal water layer wetting a vertical wall [19]). The distinctive feature of the smectic is that the curvature of the air interface varies discontinuously when going from the meniscus to the planar film (on the length scale of optical microscopy, at least). We shall see that this discontinuity generates a pressure difference between the planar surface of the film and air, which can only be compensated by a compression of the layers within the film. For a water layer, on the other hand, the curvature of the meniscus must tend to zero continuously, since the planar surface of the liquid cannot bear a pressure difference, according to the Laplace law: this condition is satisfied by an exponential profile. It should also be noted that, in an ordinary liquid, the exponential shape of the meniscus is determined by gravity, while in the smectic case the gravity does not count much, as the pressure difference induced by the curvature of the meniscus is much larger than the gravity-induced variation in pressure over the height of the meniscus.

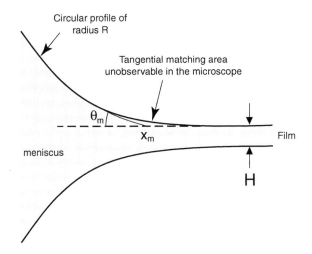

Fig. C.VIII.18 Definition of the macroscopic contact angle θ_m.

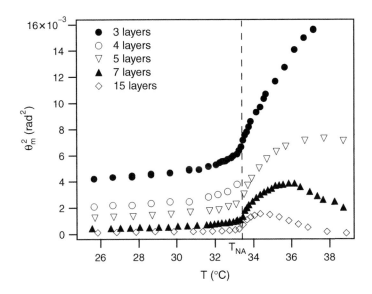

Fig. C.VIII.19 Contact angle θ_m measured as a function of temperature in 8CB films with different thicknesses. The dotted line indicates the temperature of the smectic A-nematic transition in the bulk (from ref. [19a]).

The second result concerns the contact angle θ_m between the meniscus and the surface of the film. If the film is thick (more than 100 layers, typically) the contact occurs at a tangent angle: $\theta_m \approx 0$ (within experimental precision). In films thinner than 100 layers, θ_m is non-zero and increases as the film thickness decreases. This "macroscopic" contact angle (see Fig. C.VIII.18), obtained by extrapolating the circular profile of the meniscus until the surface of the film, is independent of (or very weakly dependent on) the curvature radius of the meniscus, and hence of its volume [19]; it does, however, vary with temperature, strongly increasing close to the temperature T_{NA} of the nematic-smectic A transition (in the bulk) (Fig. C.VIII.19).

This angle is thus an intrinsic feature of the film. As we shall see later, it is a result of the attractive interactions between the free surfaces of the film. Note that, in figure C.VIII.19, we present values of θ_m measured in thin films overheated to temperatures higher than T_{NA}: in these conditions, one speaks of **presmectic films**. Their properties will be studied in detail in subsection VIII.3.e).

VIII.3.b) Meniscus profile and the Laplace law

Let us consider a smectic A (or L_α) sample shaped as a horizontal ribbon stretched between two vertical walls. This sample consists of a thin domain of uniform thickness H (the film) connected to the walls of the frame by menisci (Fig. C.VIII.20a). Denote by h(x) the total sample thickness at the point with abscissa position x: in the film, $h = H$ with $H \approx Na_o$, where N is the number of layers and a_o the thickness of a layer, while in the meniscus h varies from $h = h_o$ at the edge of the frame (in $x = 0$) to $h = H$ at the contact point with the film (in $x = x_m$, Fig. C.VIII.20). Let θ_o be the contact angle between the frame and the meniscus (there is no a priori constraint on the value of this angle, as it occurs at a cusp) and θ_m the "macroscopic" contact angle between the meniscus and the surface of the film (defined in figure C.VIII.18).

To find the profile of the meniscus and the conditions of mechanical equilibrium of the film, we only need to minimize the total energy of the system "film + meniscus":

$$F[h(x)] = \int \left[2\gamma \sqrt{1 + \left(\frac{d(h/2)}{dx} \right)^2} + \Delta Ph(x) + f[h(x)] - E[h(x)] \frac{1}{a_o} \frac{dh}{dx} \right] dx$$

(C.VIII.9)

Several terms can be distinguished in this expression.

The first one, proportional to γ, gives the surface energy. It writes as for an isotropic liquid if the surface of the meniscus is **smooth**. This condition is fulfilled if the deformation fields of the dislocations in the meniscus are substantially superposed, which is always the case in the experiments [18]. We recall that the dislocations accommodate the thickness variation of the meniscus and are localized in the center of the sample, being strongly repelled by the free surfaces (see chapter C.III). A diagram of the meniscus is shown in figure C.VIII.20c.

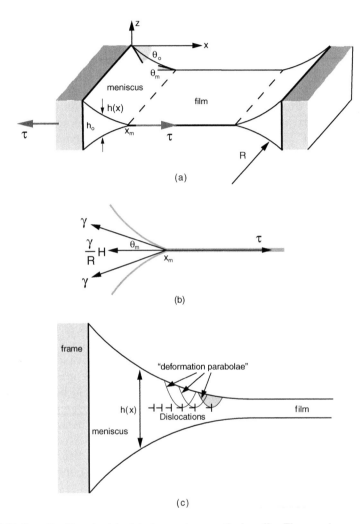

Fig. C.VIII.20 Smectic film stretched between two vertical walls. The meniscus, much bulkier than the film, plays the role of particle reservoir. a) Overall view of the film and the meniscus connecting it to the frame. Note that the force exerted by the meniscus on the frame is, by definition, equal to the film tension τ; b) force equilibrium at the junction between the film and the meniscus. θ_m is the apparent contact angle between the meniscus and the film surface; c) detailed view of the circular profile meniscus: the free surface is practically smooth if the deformation fields of the dislocations (visualized by their "deformation parabolae") are superposed.

The second term corresponds to the work of the pressure. Here, $\Delta P = P_{air} - P_{smectic}$ is the **difference in hydrostatic pressure** between the ambient air and the smectic phase. Note that ΔP can be seen as the Lagrange multiplier ensuring that minimization of the free energy is done at constant total volume of liquid crystal (see eq. C.VIII.17).

The third term depends on the local thickness h(x). **It is conventionally taken as zero in an unstressed infinite system**, but it is non-zero in a sample of finite thickness due to the interactions between surfaces and to layer compression within the film. Thus this term, to be calculated later, corresponds to the **energy excess due to the finite thickness of the system**.

The last term corresponds to the energy of the dislocations in the meniscus (E(h) is the energy of a dislocation, depending in principle on the local thickness h, and $-(1/a_o)(dh/dx)$ the dislocation density). Note that the dislocations can be shown to be elementary (i.e., with a Burgers vector $b = a_o$) in the thinner region of the meniscus, while they bunch together in giant dislocations in the thicker regions (Fig. C.VIII.16) [18].

Minimizing the total energy C.VIII.9 with respect to h immediately yields:

$$\Delta P + \frac{df}{dh} - \gamma \frac{\dfrac{d^2(h/2)}{dx^2}}{\left[1 + \left(\dfrac{d(h/2)}{dx}\right)^2\right]^{3/2}} = 0 \qquad \text{(C.VIII.10)}$$

We emphasize that the energy of the dislocations does not appear in this equation, which can be written under the equivalent form:

$$\Delta P - \Pi_d - \frac{\gamma}{R} = 0 \qquad \text{(C.VIII.11)}$$

with R the local radius of curvature of the surface and Π_d a parameter with pressure dimensions, termed **disjoining pressure** [5, 23]:

$$\Pi_d = -\frac{df}{dh} \qquad \text{(C.VIII.12)}$$

Clearly, eq. C.VIII.11 is satisfied everywhere, in the film and in the meniscus. It is however very instructive to rewrite it for each subsystem, with the appropriate simplifications. Indeed, for a planar film, the radius of curvature $R \to +\infty$, such that

$$\Delta P = \Pi_d \qquad \text{(in the film)} \qquad \text{(C.VIII.13)}$$

Therefore, it is **the disjoining pressure that equilibrates in the film the pressure difference** between the air and the inside of the smectic phase. We shall see later that this disjoining pressure is mainly of elastic origin, being almost equal to the layer compression stress in the film.

In the meniscus, on the other hand, one can assume (and this result will be proven later, see eq. C.VIII.33) that f(h) decreases very rapidly as the thickness h increases. In this case, the disjoining pressure soon becomes negligible and **the pressure difference is equilibrated by the curvature of the interface**:

$$\Delta P = \frac{\gamma}{R} \qquad \text{(far inside the meniscus, where f(h)} \rightarrow 0) \qquad \text{(C.VIII.14)}$$

We recover here the **Laplace law**, well known for ordinary fluids. Note that the presence of "mobile" edge dislocations makes it possible for a smectic A to behave like an ordinary fluid. In particular, these dislocations allow the relaxation of the elastic stress (out of that produced by the dislocations) inside the meniscus (σ(h) = 0).

The Laplace law also gives the profile of the meniscus. Since at equilibrium the pressure is the same in the film and in the meniscus (ΔP = Cst), we deduce that the meniscus must be **circular** in the thicker regions, where the disjoining pressure is negligible:

$$R = \text{const} \qquad \text{(far inside the meniscus)} \qquad \text{(C.VIII.15)}$$

This is indeed observed experimentally. Gravity was neglected; this hypothesis is justified in the case of horizontal film when the thickness of the meniscus is small compared to the gravitational capillary length [24]:

$$h << l_g = \sqrt{\frac{\gamma}{\rho g}} \qquad \text{(C.VIII.16)}$$

In this formula, ρ is the density of the liquid crystal and g the gravitational acceleration. In practice, this condition is always fulfilled for horizontal films. On the other hand, we shall see in subsection C.VIII.3.e, that gravity can play an important role in vertical films.

Let us also point out that **the value of ΔP is not arbitrary, depending on the volume V of matter present in the film,** since we have (neglecting the volume of the film with respect to that of the meniscus):

$$V = \int_{\text{menisci}} h(x, \Delta P) \, dx \qquad \text{(C.VIII.17)}$$

In principle, this equation yields ΔP as a function of V.

On the other hand, **the thickness of the film is arbitrary: it depends on the history** (spreading velocity, presence of dust particles, etc.) and can vary between two and several thousand layers (except for presmectic films, where the number of layers cannot exceed a maximal value; see subsection C.VIII.3.e).

Finally, let us calculate the contact angle between the film and the meniscus. Clearly, the profile h(x) has no cusp and the meniscus is tangentially connected to the planar film, as they consist of the same material, in the same phase. It is however possible that the zone of "tangential junction" is too small to be seen under the microscope. In this case, the experimentalist measures **an apparent contact angle θ_m,** which is nothing else than the angle between the extrapolation of the circular profile (measured at large thickness) with the planar surface of the film. From the mathematical point of view, this angle is perfectly well defined. It is obtained by one integration of eq. C.VIII.10:

$$\Delta P\, h + f[(h(x)] + 2\gamma\, \cos\theta = c \qquad\qquad\qquad (C.VIII.18)$$

where c is an integration constant (equal to the tension of the film; see next subsection) and θ the angle between the free surface and the x axis ($\tan\theta = d(h/2)/dx$). This equation gives:

$$\Delta P\, H + f(H) + 2\gamma = c \qquad \text{(in the film)} \qquad\qquad (C.VIII.19a)$$

and

$$\Delta P\, h + 2\gamma\, \cos\theta = c \qquad \text{(far inside the meniscus, where } f(h) \to 0) \qquad (C.VIII.19b)$$

By extrapolating the profile of the circular meniscus to the film surface (in h = H), we have:

$$\Delta P\, H + 2\gamma\, \cos\theta_m = c \qquad\qquad\qquad (C.VIII.19c)$$

where the constant c is given by eq. C.VIII.19a. It follows that:

$$2\gamma\, (\cos\theta_m - 1) = f(H) \qquad\qquad\qquad (C.VIII.20)$$

Thus, the circular profile makes a finite angle with the surface of the film, in agreement with the observations. In particular, one finds that $\theta_m \to 0$ in thick films, since $f(H) \to 0$ when H increases. On the other hand, measuring θ_m as a function of thickness in thin films provides information on the origin of the disjoining pressure and on the nature of the attractive interactions between free surfaces.

Before discussing this point, let us determine the film **tension**.

VIII.3.c) Film tension

The tension of a film $\tau(H)$ is, by definition, the force (per unit length) that must be exerted on the edge of the frame to maintain the equilibrium. Using the notations in figure C.VIII.20, this force is given by:

$$\tau(H) = 2\gamma\, \cos\theta_o + \Delta P\, h_o \qquad\qquad\qquad (C.VIII.21)$$

We know that $\Delta P = \gamma/R$, while, by geometrical construction, $h_o = H + 2R\,(\cos\theta_m - \cos\theta_o)$. After replacing in the previous equation, we find:

$$\tau(H) = \Delta P\, H + 2\gamma\, \cos\theta_m \qquad\qquad\qquad (C.VIII.22a)$$

a result we could have obtained directly by writing the force balance at the point of macroscopic contact between the film and the meniscus (Fig. C.VIII.20b). This equation can be rewritten as:

$$\tau(H) = 2\,\gamma + \Delta P\, H + f(H) \qquad\qquad\qquad (C.VIII.22b)$$

using eq. C.VIII.20. This formula shows that the film tension is slightly different from 2γ. In particular, it depends on f(H) (and thus on the contact angle θ_m, according to eq. C.VIII.20) and contains a pressure term proportional to film thickness. This latter contribution depends on the meniscus, since it varies as the reciprocal of its curvature radius, but remains small with respect to the surface energy since, in practice, $2\gamma \approx 50$ dyn/cm, while $\Delta PH \approx 0.025$ dyn/cm, with $H = 1\mu m$ and $\Delta P \approx 250$ dyn/cm² (for a radius R of 1 mm). Nevertheless, this pressure term is detected by the balance pendulum experiment presented in subsection VIII.2.b) in thick films (where the f(H) term is negligible). Note that, in this experiment, the menisci on either side of the balance communicate, so their pressures are balanced. Therefore, they have identical radii of curvature, an essential requirement for measuring the dependence of film tension as a function of thickness.

We shall now discuss the origin of disjoining pressure.

VIII.3.d) Disjoining pressure and contact angle θ_m

In this subsection, we shall show that the disjoining pressure (allowing us to equilibrate the pressure difference ΔP between the inside and the outside of the film) is mainly set by the elasticity of the layers, while the contact angle θ_m results from the presence of a different smectic order at the surface and in the bulk [19a]. Rigorously speaking, we should also take into account the van der Waals interactions between the two surfaces, but they are usually very small [19a], so we shall neglect them in the following. The theory we shall present is a generalization of the de Gennes theory of presmectic films [20, 21] (see next subsection). It was also successfully used by Richetti et al. [22] to explain the existence of attractive forces in a lyotropic lamellar phase confined between two solid surfaces close to each other (surface force apparatus).

To describe the profile of the smectic order parameter within a smectic A film, while taking into account the elasticity of the layers, one must use the complex order parameter $\widetilde{\Psi} = \Psi e^{i\phi}$ defined in chapter C.I. Its amplitude $\Psi = |\widetilde{\Psi}|$ describes the density modulation along the direction normal to the layers, while its phase $\phi = 2\pi u/a_0$, related to the layer displacement u, gives their position. Within the general Landau-Ginzburg-de Gennes formalism described in chapter C.I, the energy of a film with thickness H is (up to an additive constant and neglecting the coupling terms with the director **n**):

$$f(H) = \int_{-H/2}^{+H/2} \left[\frac{1}{2}\alpha\,\Psi^2 + \frac{1}{4}\beta\,\Psi^4 + \dots + \frac{1}{2}L\left|\nabla\widetilde{\Psi}^2\right| \right] dz$$

$$= \int_{-H/2}^{+H/2} \left[\frac{1}{2}\alpha\,\Psi^2 + \frac{1}{4}\beta\,\Psi^4 + \dots + \frac{1}{2}L\left(\frac{d\Psi}{dz}\right)^2 + \frac{1}{2}L\,\Psi^2\left(\frac{d\phi}{dz}\right)^2 \right] dz \quad \text{(C.VIII.23)}$$

In this expression, $\alpha = \alpha_o (T - T_{NA})$ and the parameters α_o, β and L are positive constants (with $L = 1/M_{//}$).

Let us denote:

$$\Psi_b = \sqrt{\frac{-\alpha}{\beta}} \qquad \text{and} \qquad \psi = \Psi - \Psi_b \qquad\qquad \text{(C.VIII.24)}$$

The first parameter, Ψ_b, gives the average value of the order parameter at equilibrium in an infinite medium at a temperature $T < T_{NA}$, while ψ describes the shift with respect to this value of the order parameter Ψ in the film. Using ψ as a new order parameter, the film energy can be written under the equivalent form (still up to a constant):

$$f(H) = \int_{-H/2}^{+H/2} \left[-\alpha \psi^2 + \ldots + \frac{1}{2} L \left(\frac{d\psi}{dz}\right)^2 + \frac{1}{2} L \Psi^2 \left(\frac{d\phi}{dz}\right)^2 \right] dz \qquad \text{(C.VIII.25a)}$$

In this expression, the phase is coupled to the order parameter Ψ, complicating the problem. To simplify, we shall assume that the order parameter is not very different from its equilibrium value in the bulk and replace Ψ by Ψ_b.

In these conditions, the film energy writes:

$$f(H) = \int_{-H/2}^{+H/2} \left[-\alpha \psi^2 + \ldots + \frac{1}{2} L \left(\frac{d\psi}{dz}\right)^2 + \frac{1}{2} L \Psi_b^2 \left(\frac{d\phi}{dz}\right)^2 \right] dz \qquad \text{(C.VIII.25b)}$$

Introducing the compressibility modulus of the layers $B = 4\pi^2 L \Psi_b^2/a_o^2$ and the layer displacement $u = a_o \phi/2\pi$, this equation becomes:

$$f(H) = \int_{-H/2}^{+H/2} \left[-\alpha \psi^2 + \ldots + \frac{1}{2} L \left(\frac{d\psi}{dz}\right)^2 + \frac{1}{2} B \left(\frac{du}{dz}\right)^2 \right] dz \qquad \text{(C.VIII.25c)}$$

where the last term describes layer elasticity. Minimizing the energy with respect to the two variables ψ and u is very easy, as they are now decoupled. Let ξ be the correlation length below T_{NA}:

$$\xi = \sqrt{\frac{L}{-\alpha}} = \xi_o \sqrt{\frac{T_{NA}}{T_{NA} - T}} \qquad \text{(C.VIII.26)}$$

Minimizing the energy with respect to ψ yields the following differential equation:

$$\xi^2 \frac{d^2\psi}{dz^2} = 2\psi \qquad \text{(C.VIII.27)}$$

which must be completed by two boundary conditions. The simplest assumption is that the excess order parameter is fixed at the free surface:

$$\psi\,(-H/2) = \psi(H/2) = \psi_s \qquad\qquad\qquad \text{(C.VIII.28)}$$

In these conditions, the solution writes:

$$\psi(z) = \frac{\psi_s}{\cosh\left(\dfrac{H}{\sqrt{2}\xi}\right)}\cosh\left(\frac{z}{\sqrt{2}\xi}\right) \qquad\qquad \text{(C.VIII.29)}$$

Hence, the order parameter decreases exponentially over a distance comparable to the correlation length.

Let us now minimize the energy C.VIII.25c with respect to the layer displacement u. We obtain a well-known elasticity equation (see chapter C.II):

$$B\frac{d^2u}{dz^2} = 0 \qquad\qquad\qquad\qquad \text{(C.VIII.30)}$$

Its first integral gives:

$$B\frac{du}{dz} = B\frac{H - Na_o}{Na_o} = \sigma_N(H) \qquad\qquad \text{(C.VIII.31)}$$

where $\sigma_N(H)$ is the elastic stress along the normal to the layers (independent of z) in an N layer film. This stress can be calculated from the condition of mechanical equilibrium: $\Delta P = \Pi_d = -df/dH$ (eq. C.VIII.13). To perform this calculation, we must first calculate the energy f(H) per unit surface, taking into account the profile of the order parameter already calculated (eq. C.VIII.29). Direct integration yields (keeping in mind that f vanishes in an unstressed sample of infinite thickness):

$$f(H) = \frac{1}{\sqrt{2}}\,\alpha\,\xi\,\psi_s^2\left[1 - \tanh\left(\frac{H}{\sqrt{2}\xi}\right)\right] + \frac{\sigma_N(H)^2}{2B}H \qquad \text{(C.VIII.32)}$$

The first term is given by the space variations of the order parameter within the thickness of the film. It leads to an attractive interaction between the two surfaces, a result easily understandable since the profile of the order parameter becomes "flatter" as the thickness decreases (in these conditions, the gradient term decreases). The second term corresponds to the elastic energy of layer deformation. Note that this term, different from zero in the film (but small, as we shall see later), vanishes in the meniscus where the dislocations allow the relaxation of elastic stresses. The function f(H) is shown in figure C.VIII.21. From this expression, one can obtain the disjoining pressure in the film (knowing that $\sigma_N \ll B$ in linear elasticity):

$$\Pi_d = \frac{1}{2}\alpha\psi_s^2\left[1 - \tanh^2\left(\frac{H}{\sqrt{2}\xi}\right)\right] - \sigma_N \qquad\qquad \text{(C.VIII.33)}$$

Writing that, at equilibrium, $\Delta P = \Pi_d$, gives an equation for the stress σ_N in the film. If the films are thick (typically more than 20 layers) $\alpha\, \psi_s^2$ $[1 - \tanh^2(H/\sqrt{2}\,\xi)] \ll \Delta P \ll B$ and, with a good approximation:

$$\sigma_N = -\Delta P \qquad\qquad\qquad\qquad\qquad \text{(C.VIII.34a)}$$

If the films are thin (less than 10 layers, typically) the Laplace pressure deficit becomes negligible with respect to the attractive interactions between the free surfaces and the stress σ_N is:

$$\sigma_N = \frac{1}{2}\,\alpha\, \psi_s^2 \left[1 - \tanh^2\!\left(\frac{H}{\sqrt{2}\xi}\right)\right] \qquad\qquad \text{(C.VIII.34b)}$$

These two equations show that layer elasticity balances the pressure difference imposed by the meniscus, as well as the attractive interactions between the free surfaces. Note that, since $\Delta P > 0$ and $\alpha < 0$ (below T_{NA}), σ_N is always negative, corresponding to layer compression. In other words, **by compressing, the layers avoid film collapse**. The situation is completely different in a "thick" soap film, which empties by draining and thins continuously until it forms a Newton black film (similar in certain points to a film of the L_α phase consisting of a unique bilayer). In this case, the viscous stress balances the pressure jump during thinning.

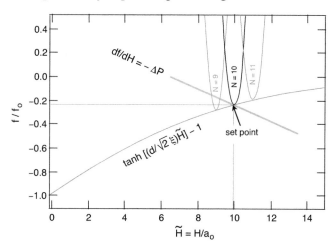

Fig. C.VIII.21 Aspect of the function f(H) in films of 9, 10 and 11 layers. Here, $f_0 = 1/\sqrt{2}\,|\alpha|\,\xi\psi_s^2$ and we took $\sqrt{2}\,\xi/a_0 = 10$. At the "set point" given by equation $\Delta P = \Pi_d$, the film is compressed.

Let us now calculate the contact angle θ_m. In this case, the relevant quantity is the film energy (see eq. C.VIII.20). The film being very stiff in the smectic A phase (B is always above 10^7 erg/cm^2), the elastic energy associated with layer deformation is very small and completely negligible with respect to the energy associated with the profile of the order parameter ψ.

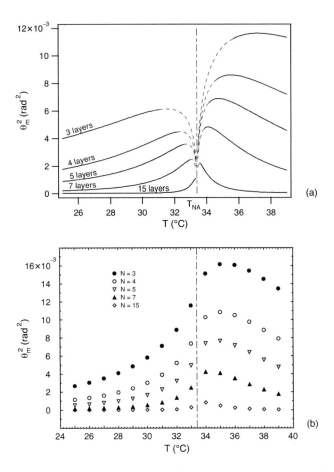

Fig. C.VIII.22 Theoretical contact angle calculated by accounting for an excess of smectic order at the free surface. The curves obtained at $T > T_{NA}$ correspond to the presmectic films discussed in subsection VIII.3.e. a) Theoretical curves obtained analytically by neglecting the quartic term in the expression of the Landau free energy. Only the parts drawn as solid lines are correct ($\xi < Na_o$) [19a]. b) Same curves, numerically calculated taking into account the quartic term. The anomalies observed close to T_{NA} disappear, in good agreement with the experiment [19b].

In this limit, **the contact angle θ_m only depends on the profile of the order parameter**, being given by the following equation:

$$\theta_m^2 = \frac{-\alpha\,\xi\,\psi_s^2}{\sqrt{2}\gamma}\left[1 - \text{th}\left(\frac{Na_o}{\sqrt{2}\xi}\right)\right] \tag{C.VIII.35}$$

This formula provides a prediction for the evolution of the contact angle θ_m as a function of temperature and the number of layers N. In particular, it shows that θ_m must vanish exponentially as the number of layers increases, in agreement with the observations. The theoretical results are shown in figure C.VIII.22a. They are generated for $\xi_o = 8$ Å and $\alpha_o\xi_o\psi_s^2/\gamma = 7.4\times10^{-4}$ K^{-1}, values yielding the best fit to the entire set of experimental data obtained for 8CB.

One can see that these curves are close to the experimental results shown in figure C.VIII.19 as long as $\xi \leq Na_o$.

Beyond this limit (i.e., too close to the T_{NA} transition temperature) the simplified analytical theory predicts a decrease in θ_m^2, which is not observed. In order to retrieve the experimentally observed behavior, one must solve the equations numerically, preserving the quartic term in the expression of the Landau free energy and without decoupling the variables ψ and u. This calculation, which we shall not discuss here, is detailed in Poniewierski et al. [19b]. The numerical results are shown in figure C.VIII.22b.

In the next subsection, we show that this theory is also applicable to the presmectic films observed above T_{NA}.

VIII.3.e) Presmectic films

They are observed in certain materials above the smectic A-nematic or smectic A-isotropic bulk transition temperature. At these temperatures, the meniscus melts but the film can still exist in a metastable state. One then speaks of **presmectic films, since they still have a lamellar structure.**

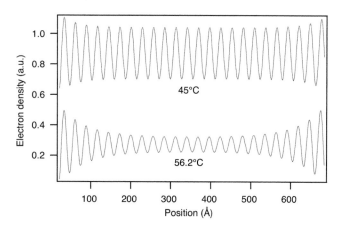

Fig. C.VIII.23 Electron density profile measured by X-rays in a 7AB film 24 layers thick at two different temperatures. The upper profile is measured in the smectic A phase; the lower one corresponds to a presmectic film (from ref. [25]).

The presence of steps proves this conclusively, and so do the structural X-ray studies of Mol et al. [25]. In figure C.VIII.23, we plotted two electron density profiles, measured at two different temperatures in a 24-layer thick 7AB film. This material, with the formula C_7H_{15}-Φ-N_2O-Φ-C_7H_{15}, exhibits a second-order smectic A-nematic phase transition at 52.9°C. The first profile is measured at 45°C in the smectic A phase: it reveals a lamellar structure with, more importantly, a smectic order enhancement at the surface, which penetrates into the film over a distance of the order of three layers. The second profile is measured in the presmectic state at 56.2°C, which is 3.3°C above T_{NA}. In this case, the lamellar structure is preserved, but the amplitude of the electron density modulation is much larger at the surface than in the center of the film. Thus, the layers are much better "defined" at the surface of the film than in the center. This effect becomes stronger on heating and can lead to "melting of the central layer." In this case, the film thins spontaneously by one layer. In the present situation, increasing the temperature by only 0.1°C leads to the film thinning by one layer.

This example shows that, for each temperature T, there exists a maximal layer number N(T) below which the film is metastable and above which it thins systematically. Conversely, for each thickness N there exists a **thinning temperature** T(N) above which the film spontaneously loses one or more layers. To measure the T(N) dependence, it is enough to place the film and its frame in an oven, well regulated in temperature, and to measure its thickness at regular time intervals, while slowly increasing the temperature.

Using this method, it was possible to demonstrate spontaneous thinning transitions in 7AB [25], in 6O.10 [26a] and in 8CB [19a], where the nematic-smectic transition is practically second order, but also in 5O.6 [26b] where the nematic-smectic transition is first order, as well as in fluorinated materials [27a–f] exhibiting a smectic A-isotropic liquid first-order transition.

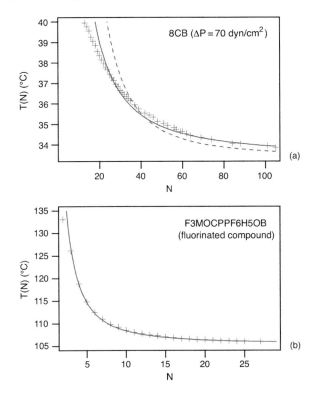

Fig. C.VIII.24 Thinning temperature as a function of the number of layers in 8CB (a) and in a fluorinated material (b). The solid lines are the best fits with a power law. The dotted line in the upper graph corresponds to the best fit with eq. C.VIII.42.

In figure C.VIII.24, we present the curves T(N) for a fluorinated material [27c] and for 8CB [19a]. These curves are qualitatively similar. In both cases, they are well described by a power law of the type:

$$\frac{T-T_{NA}}{T_{NA}} \propto N^{-1/\nu} \qquad\qquad (C.VIII.36)$$

with $\nu \approx 0.68$ in 8CB [19a] and $\nu = 0.61$ in the fluorinated material [27c]. For 7AB, $\nu = 0.68$ [25].

Note that, in the case of 8CB, the thinning temperature was not measured for less than ten layers. The reason is that, in this range of thickness, the thinning temperature exceeds the nematic-isotropic liquid transition temperature. The meniscus melts and droplets of isotropic liquid [28] appear in the film, which finally breaks.

In figure C.VIII.25, we show that thinning of an N layer thick 8CB film is not homogeneous, occurring by heterogeneous nucleation of one (or several) domains of thickness N – 1 at the contact with the meniscus. Once formed, these domains progressively invade the film.

The experiments performed on 8CB also show that the thinning temperatures T(N) are slightly decreased by an increase of excess pressure in the meniscus (i.e., by a decrease of R) [19]. This result shows that the pressure excess in the film depends on the meniscus when the transition is second order, in agreement with the results of the previous subsection.

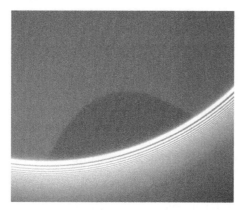

Fig. C.VIII.25 Nucleation of a domain of thickness N – 1 at the contact with the meniscus in a presmectic film of N layers (8CB) (from ref. [19a]).

The situation is different in fluorinated materials, where the transition is first order. In this case, the thinning temperatures are very reproducible and do not seem to depend on the size of the meniscus. This observation suggests that the excess pressure is given by a different mechanism [27d]. This result can be explained by considering the interface formed between the isotropic liquid in the meniscus and the smectic A phase in the film. Clearly, these two phases can only coexist at the temperature $T > T_{AI}$ if the interface between them is curved. This curvature engenders a pressure difference between the film and the meniscus, given by (see chapter B.IX) $L(T - T_{AI})/T_{AI}$ at equilibrium, at temperature T (T_{AI} is the smectic A-isotropic transition temperature in the bulk and L the latent heat per unit volume). This pressure jump adds to the pressure difference due to meniscus curvature. The total pressure difference ΔP between the air and the inside of the film is:

$$\Delta P = (P_{air} - P_{meniscus}) + (P_{meniscus} - P_{film}) = \frac{\gamma}{R} + \frac{L(T - T_{AI})}{T_{AI}} \qquad \text{(C.VIII.37)}$$

This equation reduces to the usual expression for the pressure excess when the transition is second order, since $\Delta H = 0$. When the transition is first order, on the other hand, the second term becomes dominant. Indeed, one has $L(T - T_{AI})/T_{AI} \approx 2 \times 10^5$ dyn/cm^2 for $T - T_{AI} \approx 1$ K, using $L \approx 6 \times 10^7$ erg/cm^3, while $\gamma/R \approx 100$ to 1000 dyn/cm^2 in usual menisci. It is therefore not surprising that in fluorinated materials the size of the meniscus is irrelevant, since the "thermodynamical" excess pressure clearly dominates the capillary excess pressure at $T > T_{AI}$. This property helps us understand not only why the pores, once nucleated in these films, grow extremely fast, at a velocity proportional to $T - T_{AI}$ [27e], but also why the tension of these films strongly increases with temperature, also as $T - T_{AI}$ [27f]. The first observation is easily explained knowing that the velocity of a pore is proportional to ΔP (this result will be demonstrated in subsection VIII.4.d, see eq. C.VIII.64). It ensues that, if the velocity of a pore is measured in µm/s in an 8CB film (see, for instance, figure C.VIII.30), it must be of the order of mm/s one degree above T_{AI} in a film of fluorinated material, since the pressure excess ΔP differs by a factor 100 to 1000 between the two types of films. This is indeed the order of magnitude of the experimentally measured velocities [27e]. As to the increase with $T - T_{AI}$ of the tension τ of presmectic films of fluorinated materials [27f], it merely originates, according to eq. C.VIII.22b, in the ΔPH term in the expression of the tension. In this case too, it is easily checked that the order of magnitude of the $(1/H)(d\tau/dT)$ coefficient as measured experimentally (giving 4×10^5 dyn/cm^2/°C [27f]) is compatible with the theoretical value $L/T_{AI} \approx 2 \times 10^5$ dyn/cm^2/°C.

This being said, let us see how the thinning temperatures T(N) can be calculated theoretically using the Landau-Ginzburg-de Gennes theory already discussed. Clearly, this theory can only be directly applied to materials exhibiting a *second-order* smectic A-nematic transition. Only this case will be considered in the following.

The theoretical approach is the same as in the previous subsection, except that the phase and the amplitude of the order parameter can no longer be decoupled and that the film energy must be calculated by integrating the complete expression of the Landau-Ginzburg-de Gennes free energy. This calculation, presented in Moreau et al. [29a] and Richetti et al. [29b] gives, with the notations defined in the previous subsection:

$$f(H) = \alpha \, \xi \, \psi_s^2 \left[\tanh\left(\frac{H}{2\xi}\right) + \frac{1 - \cos[\phi]_N}{\text{sh}(H/\xi)} - 1 \right] \tag{C.VIII.38}$$

where we used:

$$[\phi]_N = \frac{2\pi}{a_o}(H - Na_o) \tag{C.VIII.39a}$$

and

$$\xi = \sqrt{\frac{L}{\alpha}} \quad \text{with} \quad \alpha = \alpha_o \, (T - T_{NA}) \tag{C.VIII.39b}$$

In this formula, ψ_s is the value of the order parameter at the surface (note that here $\Psi = \psi$, as $\Psi_b = 0$), ξ is the correlation length above T_{NA}, H is the actual film thickness, a_o the thickness of one layer and N the number of layers. As pointed out by Gorodetskiï et al. [21], there are "bands of forbidden thickness" where the compressibility of the film is negative ($\partial^2 f/\partial H^2 < 0$), corresponding to thermodynamically unstable states, and stable (metastable, more precisely) bands where $\partial^2 f/\partial H^2 > 0$. Assuming that $H \gg \xi \gg a_o/2\pi$, it is easily shown that the bands of allowed thickness satisfy the inequality:

$$|H - Na_o| \le \frac{a_o}{4} \tag{C.VIII.40}$$

For these thickness values, the equation of mechanical equilibrium of the film writes:

$$\Delta P = -\frac{4\pi}{a_o} \alpha \, \xi \, \psi_s^2 \exp\left(\frac{-Na_o}{\xi}\right) \sin[\phi]_N \tag{C.VIII.41}$$

In practice, ΔP is the pressure difference imposed by the meniscus, such that this equation gives all possible solutions at a given temperature (Fig. C.VIII.26). In particular, for each value of ΔP there exists a critical thickness H_c above which all films are absolutely unstable. The critical thickness H_c decreases on increasing the temperature, since the disjoining pressure decreases when the temperature increases. In practice, we measure the temperature T(N) at which $H_c = Na_o$. This temperature is obtained by setting $\sin[\phi]_N = -1$ in eq. C.VIII.41. This yields for T(N) an equation of the form:

$$\Delta P = \frac{4\pi}{a_o} \alpha[T(N)] \, \xi[T(N)] \, \psi_s^2 \exp\left(\frac{-Na_o}{\xi[T(N)]}\right) \tag{C.VIII.42}$$

This equation allows the numerical calculation of $T(N) - T_{NA}$ knowing $\alpha_o \xi_o \psi_s^2$ and ξ_o. With the values already given in the previous subsection, one obtains the dotted line in figure C.VIII.24a). This curve is in qualitative agreement with the experimental results obtained for 8CB.

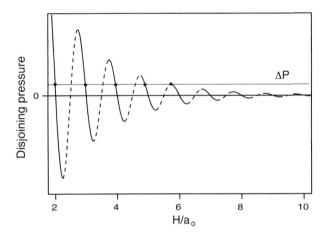

Fig. C.VIII.26 Theoretical disjoining pressure as a function of N. The regions where the curve is plotted as a solid line correspond to the allowed bands, while those in dashed line correspond to thermodynamically unstable states. The horizontal line gives the pressure difference ΔP imposed by the meniscus. The points of intersection with the solid line regions give the spectrum of all possible metastable solutions.

It is noteworthy that eq. C.VIII.42 gives the following scaling law:

$$\frac{Na_o}{\xi[T(N)]} = C \qquad\qquad (C.VIII.43)$$

where C is more or less constant (of the order of 10), up to a logarithmic correction in ΔP and in $T(N) - T_{NA}$. This law predicts that the N dependence of the thinning temperature is of the power law type (eq. C.VIII.36), with a critical exponent v, the same as for the correlation length ξ. In the mean field Landau model we employed $v = 1/2$, while the experiment gives $v = 0.68$ for 8CB. This exponent is in good agreement with that found for the correlation length using X-rays ($v = 0.67$) [30, 31] or by light scattering ($v = 0.72$) [31]. This value is also predicted by the XY model ($v = 0.67$) [32].

We would also like to emphasize that the theoretically predicted temperature T(N) gives the absolute instability limit (spinodal) of a presmectic film of N layers. In practice, the film does not thin in one step, but rather by nucleating domains of thickness N – 1. It so happens that the temperature for spontaneous nucleation of domains N – 1 layers thick is extremely close to the calculated temperature T(N), since the nucleation can only occur if the elastic modulus for layer compression B becomes very small in the center of the film (see next subsection). This is precisely the case as the temperature approaches T(N).

Finally, using the Landau theory one can calculate the macroscopic contact angle θ_m between the film and its meniscus. Numerically, this angle is almost independent of the pressure in the meniscus; it is also larger than below T_{NA} and it goes through a maximum as a function of temperature (Fig. C.VIII.22). These predictions are in agreement with the experiments (Fig. C.VIII.19).

In the next subsection, we analyze the effect of gravitation on the mechanical equilibrium of a film.

VIII.3.f) Gravitational effects on the tension and the shape of the films

Until this point, we have only considered horizontal films and neglected their weight. Obviously, the weight will induce a slight deformation of the films $z(x,y)$ with respect to the horizontal (x,y) plane. To calculate the shape of a film, let us write that the weight per unit surface ρHg, where ρ is the density of the liquid crystal ($\approx 1g/cm^3$), is balanced by the restoring force due to film tension:

$$\tau \Delta z = \rho Hg \qquad \text{(C.VIII.44)}$$

with Δ the two-dimensional Laplacian. If the film is stretched on a circular frame (Fig. C.VIII.27a) of radius R, its equilibrium shape is a paraboloid of equation:

$$z = z_o \frac{x^2+y^2}{R^2} \qquad \text{(C.VIII.45)}$$

The resulting height difference between the center of the film and the plane of the frame

$$z_o = \frac{\rho g R^2 H}{4\tau} \approx H \qquad \text{(C.VIII.46)}$$

is comparable to the film thickness H (i.e., negligible).

Let us now consider the case of a rectangular film, placed in a vertical position (Fig. C.VIII.27b). In practice, this film is stretched from a liquid crystal droplet placed at the bottom of the frame. Now, the weight must be equilibrated by a gradient of film tension:

$$\rho(Na_o)g = \frac{\partial \tau}{\partial z} \qquad \text{(C.VIII.47)}$$

Consequently, if the tension at the bottom of the frame is $\tau(0)$ then it must increase with the height as

$$\tau(z) = \tau(0) + \rho g(Na_o)z \qquad \text{(C.VIII.48)}$$

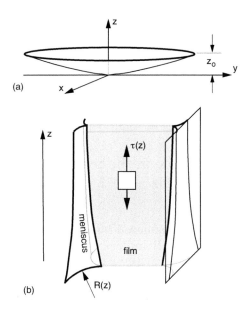

Fig. C.VIII.27 Smectic films under gravity: a) circular horizontal film deformed by its own weight; b) vertical smectic film stretched from a reservoir.

Over a distance L of 1cm, the difference in tension will be $\Delta\tau = L\rho gNa_o = 3\times10^{-4}$ N dyn/cm, or 3×10^{-3} dyn/cm for a film of thickness N = 10. This variation in film tension with the height must be accompanied by a change in the shape of the meniscus, since the equilibrium between the film and the meniscus must be preserved at each level z. As a consequence, the curvature radius R of the meniscus is no longer constant at equilibrium, as in horizontal films, and it varies with z. Let R(0) be the radius of curvature of the meniscus at the bottom of the frame (z = 0). The hydrostatic pressure in the meniscus decreases with height as $-\rho gz$. This decrease in pressure corresponds to an increase in the Laplace pressure deficit:

$$\Delta P(z) = \frac{\gamma}{R(z)} = \frac{\gamma}{R(0)} + \rho gz \qquad (C.VIII.49)$$

itself acting on the film tension given by eq. C.VIII.22. This yields:

$$\tau(z) = 2\gamma\cos\theta_m + \frac{\gamma}{R(z)}Na_o = \tau(0) + \rho gNa_oz \qquad (C.VIII.50)$$

in agreement with eq. C.VIII.48.

As we shall see, this variation in film tension with the height z leads to a step deformation that can be used to measure the line tension (see subsection VIII.4.c).

Finally, let us say that, for a vertical film of N layers, the gravity exerts on a domain of surface area A and thickness N – p (where p is a positive or negative integer, if the domain is thinner or thicker than the rest of the film, respectively) a buoyancy force:

$$F_{buoyancy} = A\rho a_o (\rho - \rho_{air})g \approx A\rho a_o \rho g \qquad (C.VIII.51)$$

equal to the weight of the displaced matter (neglecting the density of air). This gravity effect is clearly visible in the photograph of figure C.VIII.28, where one can see that the thinner domains tend to ascend to the top of a smectic bubble inflated at the tip of a hollow cylinder.

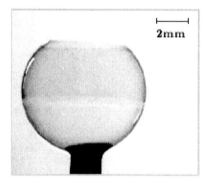

Fig. C.VIII.28 Under the action of gravity, the thin domains ascend to the top of the smectic bubble (from ref. [14a]).

VIII.4 Steps in smectic films

By "brutally" stretching or compressing a smectic film, one can create holes or islands. These domains, thinner or thicker than the rest of the film, are delimited by steps revealing the existence of a dislocation in the center of the film. In general, the steps are not directly visible in reflecting microscopy, their size being well below the resolution limit of the microscope. What we do see is the boundary between two domains of different thicknesses, and hence different reflectivities and contrasts. In this section, we shall study the mechanisms of step nucleation, and then their dynamics of growth or collapse, whether they are isolated or in interaction ("step foam"). We shall also study in detail the properties of the meniscus, which plays an essential role in the physics of films.

VIII.4.a) Nucleation and growth of an isolated loop: experimental results

There are different techniques for thinning an N-layer film. One can heat it (and its meniscus) above T_{NA} until the spontaneous thinning temperature $T(N)$;

in this case, domains of thickness N – 1 appear at the edge of the meniscus (Fig. C.VIII.25). One can also stretch the film suddenly using the deformable frame, inducing the nucleation of circular loops; experiments show that these loops almost always nucleate in the same positions, suggesting the existence of dust particles at these points. Such dust particles are sometimes seen under the microscope. A more elaborate method [33a] consists of using a heating wire folded in the middle with the tip placed a few tens of µm under the film (Fig. C.VIII.29a). By sending a very brief current impulse (of the order of 1 ms) through the wire, the film can locally be overheated up to its spontaneous thinning temperature T(N), inducing the nucleation of a circular domain of thickness N – 1 (Fig. C.VIII.29b).

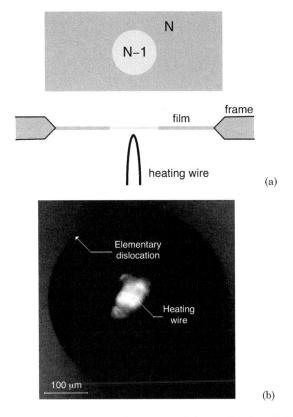

Fig. C.VIII.29 a) Diagram of the experimental setup used for nucleating dislocation loops; b) domain of thickness N − 1 in an N-layer film, nucleated using the heating wire (its tip is visible in transparence under the film) (from ref. [33a]).

Is it enough to nucleate a loop to obtain film thinning? The answer is NO! Indeed, for each loop at the temperature T < T(N) there exists a critical radius r_c below which it collapses and above which it grows. This is demonstrated in figure C.VIII.30, where we plotted the radius of a loop as a function of time for initial radii above and below r_c. In normal temperature conditions, i.e., far from T(N), r_c takes macroscopic values (several tens of μm) and the associated activation energy is considerable (much larger than k_BT). The spontaneous nucleation of a loop (similar to the nucleation of a critical germ at a first-order phase transition) is then highly improbable and film thinning, albeit thermodynamically favorable, becomes impossible.

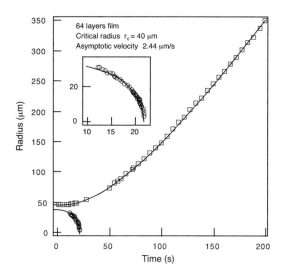

Fig. C.VIII.30 Loop radius as a function of time. Below r_c, the loop collapses; above, it grows without limit, invading the entire film and finally vanishing into the meniscus (from ref. [33a]).

For this reason the lifetime of a smectic film, protected from dust and air currents, is practically unlimited.

Let us now see how the critical radius of a loop can be calculated.

VIII.4.b) Critical radius and associated nucleation energy

The critical nucleation radius is obtained by writing that the free energy of the system is stationary at the critical radius. Let ΔF be the variation in the free energy of the system when a loop with Burgers vector b and radius r is nucleated in an N-layer film (Fig. C.VIII.31a). One obtains:

$$\Delta F = -\pi r^2 b \frac{\gamma}{R} + \pi r^2 [f(Nd - b) - f(Nd)] + 2\pi rE \qquad (C.VIII.52)$$

Three terms can be identified in this expression.

1. The first one gives the global variation in surface energy of the system (for a straight meniscus with a radius of curvature R); in this expression, $\pi r^2 b$ represents the volume of matter transferred from the film toward the

meniscus. If b is positive (the case of a pore), the surface energy decreases and the loop tends to grow. If b is negative (the case of an island), the loop will disappear, its growth being always energetically unfavorable.

2. The second term corresponds to the variation of film energy and takes into account the eventual interactions between the free surfaces.

3. The last term corresponds to the line energy of the loop, which is always positive. Writing that the energy is stationary at the critical radius ($\partial\Delta F/\partial r = 0$ at $r = r_c$), one can calculate:

$$r_c = \frac{E}{b\Delta P + f(Na_o) - f(Na_o - b)} \qquad\qquad (C.VIII.53a)$$

where ΔP is the pressure difference imposed by the meniscus ($\Delta P = \gamma/R$) and $f(Na_o) - f(Na_o - b)$ the difference in energy given by formula C.VIII.32.

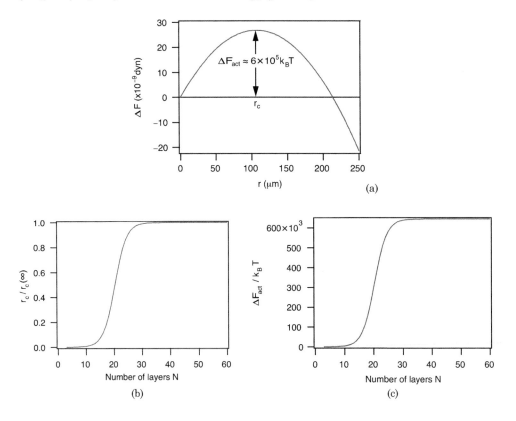

Fig. C.VIII.31 a) Energy variation of the "film + meniscus" system as a function of the loop radius ($b = a_o > 0$). Below r_c, the loop collapses. Above this value, it keeps on growing; b) critical radius as a function of the number of layers. Above 30 layers, the attractive interactions between the free surfaces are negligible. Below 10 layers, these interactions completely dominate the "suction" effect of the meniscus; c) activation energy as a function of the number of layers. The same comments as for the nucleation radius apply ($E = 8\times10^{-7}$ dyn ; $\Delta P = 250$ dyn/cm^2, $\alpha_o\xi_o\psi_s^2 = 0.019$ dyn cm^{-1}K^{-1}, $\xi_o = 8$ Å, $T = 28°$C).

The critical radius is plotted as a function of the number of layers N in figure C.VIII.31b in the case of 8CB at 28°C. It can be seen that r_c is independent of thickness in films thicker than 30 layers, as in this limit the difference $f(Na_o) - f(Na_o - b)$ is negligible in comparison with $b\Delta P$. In this case, eq. C.VIII.53a reduces to [33a]:

$$r_c \approx \frac{E}{b\Delta P} = \frac{E}{\gamma b} R \qquad \text{(in thick films)} \qquad \text{(C.VIII.53b)}$$

This formula shows that, in thick films, the critical nucleation radius must be proportional to the curvature radius of the meniscus. This dependence was verified experimentally in the case of 8CB (Fig. C.VIII.32) [18] and yielded the line tension E of a dislocation in the thick film limit, the surface tension γ and the Burgers vector b being well known.

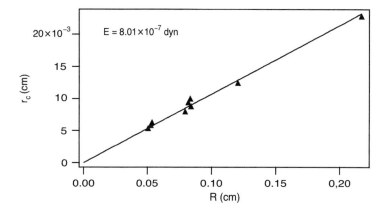

Fig. C.VIII.32 Critical nucleation radius as a function of the curvature radius of the meniscus in 8CB at T = 28°C. The films are all thicker than 100 layers. The slope of this line yields the line tension of an elementary dislocation (from ref. [18]).

Figure C.VIII.31b also shows that the critical nucleation radius strongly decreases below 30 layers. This brutal decrease is due to the $f(Na_o) - f(Na_o - b)$ term, accounting for the attractive interaction between the two free surfaces. This term even becomes overwhelmingly dominant with respect to $b\Delta P$ in thin films of less than 10 layers, where the critical radius becomes less than a micrometer. Note that the critical radius was measured experimentally in thick films using a heating wire by sending successive pulses that progressively increased the initial radius of the germ until the point of unstable equilibrium. This experiment is rather difficult, since the critical radius must no be exceeded (otherwise the germ keeps on growing and the film irreversibly loses one layer). The main difficulty is related to the thermal equilibration of the film, which can require a few seconds after each pulse. During this time, the germ grows anomalously fast, the film being overheated, and it can exceed its critical radius. In this way, one obtains a larger value for the critical radius but one must start over, knowing that the film has already lost one layer. This method makes it possible to bracket the value of the critical radius. In general, several tries are needed to obtain a reasonable bracketing of r_c, which is not a problem in thick films, since r_c only depends very weakly on film thickness in this case. In thin films (less than 30 layers thick), however, the measurement becomes much more delicate since not only the interaction force between the two free surfaces, but also the line tension, and hence r_c, depend on film thickness. It is then imperative to achieve the measurement on the first try, which is very difficult, especially since the film must be heated more to reach its thinning temperature T(N). The measurement then becomes almost impossible and other methods must be used.

Another technique consists in creating an Ω-shaped step, by allowing the circular meniscus created around a fiber piercing the film to approach the straight meniscus. The first steps of thee two menisci coalesce and form a common step shaped as an Ω (Fig. C.VIII.33). At equilibrium, the radius of curvature of the step connecting the two menisci is exactly r_c. This experiment shows that the typical value of the critical radius in a 20-layer thick smectic film of the compound S2 is of the order of 50 μm. However, this method is neither very practical for performing systematic measurements as a function of thickness, nor too precise, since the menisci around the fiber and the frame are rarely at equilibrium (see subsection VIII.4.g). A much more precise method will be presented in the next subsection.

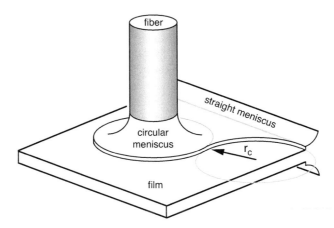

Fig. C.VIII.33 "Mont Saint-Michel" method for measuring the critical radius of a step (from ref. [10]).

Finally, let us give the expression of the activation energy for loop nucleation:

$$\Delta F_{act} = \Delta F(r = r_c) = \pi \frac{E^2}{\Delta P b + f(Na_o) - f(Na_o - b)} = \pi E r_c \qquad \text{(C.VIII.54a)}$$

In thick films, this formula reduces to [33a]:

$$\Delta F_{act} \approx \pi \frac{E^2}{\Delta P b} \qquad\qquad\qquad\qquad (C.VIII.54b)$$

This energy corresponds to the height of the barrier that needs to be crossed in order to nucleate a loop. Note that the attractive interactions between the free surfaces favor the nucleation, since the $f(Na_o) - f(Na_o - b) > 0$ term decreases the activation energy in thin films. This effect is shown in figure C.VIII.31c), where we plotted ΔF_{act} for an elementary dislocation loop as a function of the number of layers N. We can see in this graph that, in the thick film limit, $\Delta F_{act} = 2.6 \times 10^{-8}$ erg $\approx 6 \times 10^5\ k_B T$, which is huge. In thin films, ΔF_{act} is considerably lower, down to "only" 300 $k_B T$ in a film of three layers. This value remains however important, such that even in thin films, spontaneous ex nihilo nucleation of a loop is impossible. This explains the excellent stability of films in usual observation conditions. For the nucleation to occur, the dislocation energy must be considerably lowered. Since the dislocation is in the center of the film and its energy is $E \approx b(KB)^{1/2}$ (see chapter C.III), it is necessary that B vanishes (or almost) in the center of the film. This occurs precisely at the temperature T(N). In practice, the heating wire can locally heat the film to a temperature close to or slightly exceeding T(N). However, the temperature must remain below T(N − 1) to avoid film thinning by two (or even more) layers (as can be seen in figure C.VIII.43, p. 458).

Note that this scenario only applies to materials exhibiting a second-order smectic A-nematic phase transition (this is the case for 8CB, to a very good approximation). The situation becomes more complicated if the nematic-smectic transition is first order, or if the smectic melts directly to the isotropic phase, since, usually, heating the film induces the nucleation of a nematic or isotropic liquid bubble. These bubbles, shaped as flat lenses, are formed due to the finite surface tension between the smectic A and the nematic or isotropic phase (see subsection VIII.4.h). In this case, islands appear instead of holes after the film cools down. These islands subsequently disappear, their line tension and the pressure difference acting in the same direction. It then becomes impossible to thin the film by heating it. This is indeed the case for 10CB and for other materials, such as AMC11, where the smectic A-isotropic transition is first order. Fluorinated materials behave differently, since they can be thinned using a heating wire [27e]. In this case, the surface order is so strong that it "might" induce a nematic phase in the center of the film [27d]. The scenario would then be the same as for a second-order nematic-smectic A transition.

VIII.4.c) Line tension of a step in a vertical film

We have already seen that measuring the critical nucleation radius using the heating wire is a long and difficult procedure, which becomes impracticable in thin films.

A different technique, inspired by the method of the sessile drop used for measuring the surface tension of a liquid (see volume 1, page 374), consists of measuring the deformations of a dislocation loop in a vertical film under the action of gravity [8, 33b]. The experimental setup, shown in figure C.VIII.34, consists of an oven holding the vertical frame, of a system for visualizing the loop and of a system for measuring the thickness of the film by reflectivity. The oven must maintain a very homogeneous temperature to avoid convection in the film and the surrounding air. A heating wire situated behind the film is used to nucleate the loops.

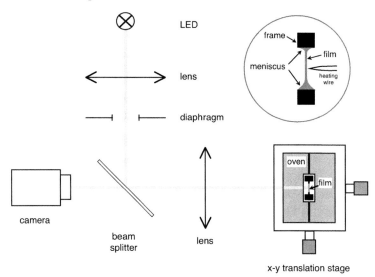

Fig. C.VIII.34 Diagram of the experimental setup used for measuring the line tension of a dislocation loop in a vertical film. The inset shows the film on its vertical frame. The meniscus is thicker and less curved at the bottom of the frame than at the top, due to gravity (from ref. [33b]).

The principle of the measurement is very simple. Once the film is stabilized, a dislocation loop is nucleated using the heating wire. This loop, being lighter, climbs in the film under the action of buoyancy. Its ascension stops as it is blocked by the upper edge of the frame. If the loop does not grow too fast (the meniscus must be as thick as possible, so as to minimize the pressure excess ΔP), its shape has enough time to reach mechanical equilibrium (Fig. C.VIII.35). This shape is slightly flattened if the size of the loop is comparable to or larger than the gravitational capillary length created using the line tension E of the dislocation:

$$L_g = \sqrt{\frac{E}{\rho g a_o}} \qquad\qquad (C.VIII.55)$$

As an order of magnitude, $L_g \approx 400\ \mu m$ with $E \approx 5 \times 10^{-7}$ dyn, $g = 1\,000$ cm.s^{-2}, $\rho = 1$ g/cm^3 (liquid crystal density) and $a_o = 3 \times 10^{-7}$ cm. One must therefore use large enough loops (a fraction of a mm) for the flattening to be easily measurable.

To calculate the shape of the loop, let us write that, in each point of the loop, the line tension equilibrates the force due to the local pressure difference (eq. C.VIII.49). This yields, for a dislocation with Burgers vector b:

$$E\kappa = b\,(\Delta P_0 + \rho g z) \tag{C.VIII.56}$$

with ΔP_0 a constant giving the pressure difference imposed by the meniscus at the point of height $z = 0$ (taken, for instance, at the bottom edge of the frame) and κ the local curvature of the dislocation (its explicit expression is given by eq. C.VIII.10). Note that, if the film is thin, ΔP_0 includes the energy difference $f[Na_0)] - f[(N-1)a_0]$ originating in the interactions between the two free surfaces. Eq. C.VIII.56 is, in fact, a differential equation for the shape of the loop. It can be solved numerically, and the theoretical shape thus obtained can be used to fit the experimentally determined loop shape (Fig. C.VIII.35). In this case, the fit parameters are the line tension E and the constant ΔP_0. This method was used to measure the line tension of a dislocation in 8CB, as a function of film thickness, Burgers vector and temperature.

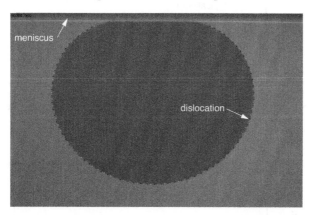

Fig. C.VIII.35 Dislocation loop with a Burgers vector $b = 2a_0$ in a vertical 8CB film 12 layers thick at 33°C. The loop, tending to ascend, is blocked by the edge of the upper meniscus, visible at the top of the photograph. The dotted line corresponds to the shape calculated by solving eq. C.VIII.56 numerically (from ref. [33b]).

The main results are that, at a fixed temperature, the line tension increases slightly as the film thickness decreases [8] (Fig. C.VIII.36). This increase is predicted by the theory of elasticity (see chapter C.III) and is largely a result of the deformation of the free surface of the film [9]. On the other hand, it can be checked that, in smectics, the energy of a dislocation varies linearly with its Burgers vector (in thick films, at least) (Fig. C.VIII.37) [8]. This result is in strong contrast with that of usual solids, where the energy of a dislocation is proportional to the Burgers vector squared. Experiment also shows that, in an infinite medium (or, equivalently, in very thick films of several hundred layers), the dislocation energy decreases as $(T_{NA} - T)^{-\varpi}$ with $\varpi \approx 0.45$ in 8CB [33b] (Fig. C.VIII.38a). This result is in keeping with the second-order character of the transition in 8CB. On the other hand, the line tension of the dislocations is finite at T_{NA} in thin films (Fig. C.VIII.38b). This is not directly related to the surface tension effect discussed above (as this correction vanishes in thick films at T_{NA} according to eq. C.III.40), but to an effect related to the surface increase in smectic order that leads to a finite value of B (and hence of E) in the film. This result confirms that the line tension should vanish at a temperature above T_{NA} in thin films (to be precise, at the thinning temperature T(N) already defined).

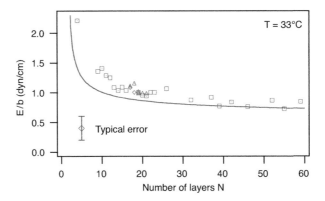

Fig. C.VIII.36 Line tension of an elementary edge dislocation as a function of the number of layers at 33°C in 8CB. The solid line is the theoretical prediction (calculated using the law C.III.38 and independent measurements for the coefficients K, B and γ) (from ref. [33b]).

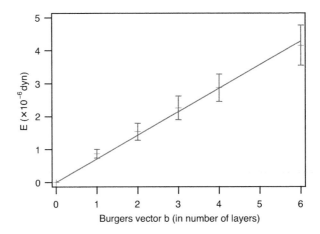

Fig. C.VIII.37 Line tension of an edge dislocation as a function of the Burgers vector, measured in thick films of the 8CB-10CB mixture. The dependence is linear, as predicted by the theory (from ref. [8]).

Unfortunately, it is impossible to determine the energy of the dislocations above T_{NA}, since the meniscus melts and the loops deform so fast that the quasi-static approximation upon which the shape analysis relies is never legitimate.

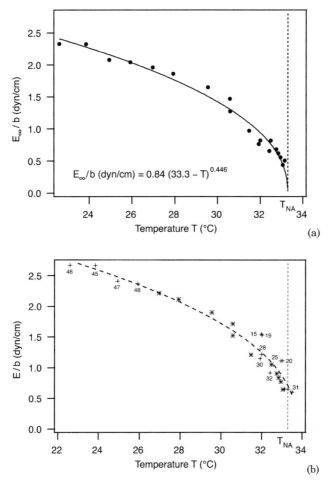

Fig. C.VIII.38 a) Line tension in an infinite medium (obtained by measuring the line tension in thick films, and then correcting the obtained value for the surface tension effects) as a function of temperature; b) line tension measured in films of variable thickness; the number of layers is indicated for each point. For the stars, the number of layers is close to 100. The dotted line is only a guide for the eye. This time, the line tension does not vanish at T_{NA}. The material is 8CB (from ref. [33b]).

In the following subsections, we shall analyze the dynamics of dislocations. We shall proceed in steps, first considering the case of an isolated loop, then that of two separate or concentric loops, and finally the more subtle situation of two crossing loops. We shall then discuss the problem of meniscus "impermeability." As we shall see, one can calculate the equilibration time of two menisci connected by a film, the nucleation time of a loop when the film is locally heated using the wire, as well as the drop in tension induced by suddenly stretching a film spanning a deformable frame.

VIII.4.d) Dynamics of an isolated loop

In this and the three following subsections, we neglect the interactions between free surfaces (f(H) = 0). In 8CB films, for instance, this assumption is legitimate as soon as their thickness exceeds ten layers. The general case will be discussed in subsection VIII.4.i.

This assumption being made, one can adopt two fairly different (but mathematically equivalent) points of view to study the dynamics of dislocations in films. A **global** point of view, consisting of using the dissipation theorem, then minimizing the dissipation, and a more local point of view, consisting of writing the balance of **local** forces and pressure fields acting on each dislocation.

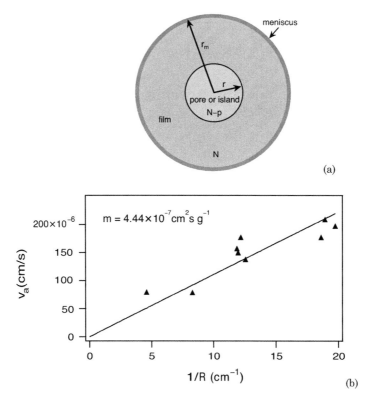

Fig. C.VIII.39 a) Loop of radius r containing N – p layers in an N-layer film; b) asymptotic velocity v_a of an elementary edge loop in a thick 8CB film as a function of 1/R. The temperature is 28°C (from ref. [18]).

The assumptions are nevertheless the same in both cases and they deserve some discussion.

First of all, we are reasoning at a **mesoscopic scale**, larger than the thickness of the film and smaller than its lateral dimensions. This allows us to ignore all that is happening in the immediate vicinity of the dislocation cores, where the stress and velocity fields are very complicated, or even singular. We also assume that, at this scale, the hydrodynamic velocity of the molecules v_N, for each domain of thickness N, is the same in all layers, including the surface ones (implicitly assuming there is no surface crystallization). Another assumption is that the pressure P_N is constant within each domain of thickness N, neglecting the pressure variations due to flow in the plane of the film (these variations can be calculated, but they are completely negligible with respect to the pressure jumps between domains of different thickness). Finally, we neglect all inertial effects (which is fully justified, considering that the effective mass of a dislocation is of the order of the mass of a molecular row [34]).

These remarks being made, let us consider the elementary problem of a circular loop of radius r, Burgers vector $b = pa_o$ (p is a relative integer), in a film of thickness $H = Na_o$ stretched on a circular frame of radius r_m, considered as much larger than the width and the curvature radius R of the meniscus (Fig. C.VIII.39a.) By convention, the loop corresponds to a pore if $b > 0$ and to an island if $b < 0$. The velocity of the loop (in the laboratory reference frame) is $v = dr/dt$. Finally, let V_m be the total volume of the meniscus. The aim of this subsection is to find the dynamical equation for a loop, to solve it, and then to compare the theoretical predictions with the observations (subsection VIII.4.a.)

i) Global approach

We already stated that the "motor force" of the growth is the global gain in surface energy for the system film + meniscus. The "brake," so to speak, is related to the increase in the energy of the loop and to the energy dissipated in the film and in the meniscus during its deformation. The global point of view consists in writing that the energy variation is equal to the dissipated energy (dissipation theorem, neglecting the inertial effects and assuming the temperature is constant). This theorem yields (per unit of time):

$$\frac{\gamma}{R}\frac{dV_m}{dt} - \frac{d}{dt}(2\pi rE) = \Phi_{disl} + \Phi_m \qquad (C.VIII.57)$$

The first term on the left-hand side is the gain in surface energy of the system. Taking into account the conservation of the total volume, it is equivalent to:

$$\frac{\gamma}{R}\frac{dV_m}{dt} = 2\pi\frac{\gamma}{R}rbv \qquad (C.VIII.58)$$

The second term on the left-hand side represents the variation in the line energy of the dislocation:

$$-\frac{d}{dt}(2\pi rE) = -2\pi Ev \qquad (C.VIII.59)$$

This term is negative (respectively positive) when the loop grows (respectively collapses), as its length increases (respectively decreases).

The first term on the right-hand side accounts for the dissipation occurring in the film by flow around the dislocation. Indeed, the dislocation moves at a velocity v with respect to the smectic at rest (inside the loop). In the first approximation, this motion can be compared with that of a ribbon of width $|b|$ moving at the velocity v in a film of width $H = Na_0$. This problem was solved exactly and leads to the following result for the dissipation (see eq. C.II.140):

$$\Phi_{disl} = 2\pi r \, \frac{|b|v_2^2}{m(1 - |b|/H)} \tag{C.VIII.60a}$$

where m is the mobility of the dislocation in an infinite medium ($m \approx \sqrt{\lambda_p/\mu}$). In a thick film, the b/H correction is negligible, hence:

$$\Phi_{disl} = 2\pi r \, \frac{|b|v_2^2}{m} \qquad \text{(in thick films)} \tag{C.VIII.60b}$$

We shall use this formula in the following.

The last term on the right-hand side corresponds to dissipation in the meniscus. Indeed, matter conservation requires that a matter flux enters the meniscus, at an average velocity:

$$v_m = \frac{b}{H}\frac{r}{r_m} v \tag{C.VIII.61}$$

This velocity is taken as positive when the matter enters the meniscus and as negative in the opposite case. This matter flux is accompanied by flow around the dislocations of the meniscus, considered as fixed in the first approximation. This results in an additional dissipation term, which can be written as:

$$\Phi_m = 2\pi r_m C \frac{H}{m} v_m^2 = 2\pi C \frac{1}{m}\frac{b}{H}\frac{r}{r_m} rb \, v_2^2 \tag{C.VIII.62}$$

where C is a dimensionless parameter related to the number of meniscus dislocations that participate in energy dissipation. This parameter (which depends in a non-trivial way on film thickness and on the curvature radius of the meniscus, and which varies from a few tens to a few thousands) will be estimated in subsection VIII.4.f.

Using eqs. C.VIII.58–62 and keeping in mind that $\Delta P = P_{air} - P_m = \gamma/R$, where P_m is the pressure inside the meniscus (see subsection VIII.4.g), after simplifying by $2\pi r v$ one can rewrite the fundamental relation C.VIII.57 under the following form:

$$\Delta Pb - \frac{E}{r} = \frac{|b|}{m^*} v \tag{C.VIII.63a}$$

with

$$m^* = \frac{m}{1 + C\dfrac{|b|}{H}\dfrac{r}{r_m}} \qquad \qquad \text{(C.VIII.63b)}$$

This equation has a simple interpretation in terms of **effective (or configurational) forces acting on the dislocation**. Indeed, **everything occurs as if the dislocation line was a material line, acted upon by real forces**. However, this is only a convenient representation, since the forces involved are not real.

The first of these forces, ΔPb, can be identified with the elastic (Peach and Koehler) force acting on the dislocation. This is the motor force. The force $-E/r$, pointing toward the inside of the loop, is related to the line tension of the dislocation. It opposes the extension of the loop, while the force $-(|b|v)/m^*$, opposed to the velocity, is identified with the viscous force acting on the dislocation. Note that m^* appears as an effective mobility accounting for dissipation effects in the meniscus.

Eqs. C.VIII.63a and b describe the dynamics of an isolated loop. Two types of loops must be distinguished, pores or islands.

For a pore, $b > 0$, such that there exists a critical nucleation radius $r_c = E/(b\Delta P)$ above which the loop grows ($v > 0$) and below which it collapses ($v < 0$). One recovers thus expression C.VIII.53b for the critical radius of a pore when the interactions between the free surfaces are negligible. Note that the critical radius does not depend on the Burgers vector of the dislocation in thick films, as E is proportional to b (Fig. C.VIII.37).

For an island, $b < 0$, such that $v < 0$ for all r. An island will collapse, irrespective of its initial radius.

The dynamical problem of a loop in very thick films (more than 100 layers) can also be analytically solved. In this case, dissipation in the meniscus is not important, as $C(|b|/H)(r/r_m) \ll 1$ such that $m^* \approx m$.

For a pore ($b > 0$), eq. C.VIII.63 becomes, introducing the critical radius r_c:

$$\frac{dr}{dt} = m\Delta P \left(1 - \frac{r_c}{r}\right) \qquad \qquad \text{(C.VIII.64)}$$

This equation has the first integral:

$$r - r_i + r_c \ln\left|\frac{r - r_c}{r_i - r_c}\right| = m\Delta P\, t \qquad \qquad \text{(C.VIII.65)}$$

with r_i the initial radius of the loop. The curve r(t) is plotted in figure C.VIII.30 for two initial radii, above and below r_c, respectively. Fitting the experimental points with the theoretical law C.VIII.65 yields the mobility and the line tension of the dislocation.

Note that, for $r_i > r_c$, the loop grows and tends asymptotically toward a constant velocity $v_a = m\Delta P = m\gamma/R$, inversely proportional to the radius of curvature of the meniscus. This simple, but non-trivial, dependence was checked experimentally in thick 8CB films (Fig. C.VIII.39b). The value of the mobility found in thick films is also in good agreement with that measured in bulk samples by microplasticity (see chapter C.III). This result confirms that the dissipation around a dislocation is very localized and that the mobility correction due to finite thickness effects (going as $1/(1 + b/H)$ in the ribbon model) is negligible. This is also confirmation that the Laplace law holds for a smectic meniscus, which is not an obvious result.

If, on the other hand, $r_i < r_c$, the loop collapses after a finite time, given by:

$$t_c \text{ (pore)} = \frac{r_c}{v_a} \left[\ln\left(\frac{1}{1-x}\right) - x \right] \qquad \text{with} \qquad x = \frac{r_i}{r_c} < 1 \qquad \text{(C.VIII.66)}$$

This time diverges logarithmically as $r_i \to r_c$.

The case of an island ($b < 0$) is treated in the same way. Its collapse time is given by (with r_c and v_a the critical radius and the asymptotic velocity of a pore with opposite Burgers vector):

$$t_c \text{ (island)} = \frac{r_c}{v_a}[x - \ln(1+x)] \qquad \text{with} \qquad x = \frac{r_i}{r_c} \qquad \text{(C.VIII.67)}$$

This collapse time is finite, irrespective of the initial radius of the island.

It is noteworthy that the collapse time of a pore or an island is independent of the Burgers vector of the dislocation (still in the thick film limit), since r_c and v_a are themselves independent of b.

ii) Local approach

To understand better the dynamics of a loop, let us now reason at a more local level, trying to guess the motion laws of a dislocation.

The first observation is that flow around a dislocation can only occur if there is a pressure difference across the dislocation. It is then reasonable to write that the dislocation acquires a velocity v proportional to this pressure difference. We are thus led to the first phenomenological law:

$$v = \mu_{N-p} (P_N - P_{N-p}) \qquad \text{(C.VIII.68)}$$

with μ_{N-p} a coefficient characterizing the mobility of the dislocation.

Similarly, one must have between the film and the meniscus a pressure difference proportional to the velocity v_m at which the molecules penetrate into the meniscus:

$$v_m = \mu_m (P_N - P_m) \qquad \text{(C.VIII.69)}$$

This pressure difference is directly related to dissipation in the meniscus, since one must have:

$$\Phi_m = 2\pi r_m (P_N - P_m) H v_m = 2\pi r_m H \frac{v_m^2}{\mu_m} \qquad \text{(C.VIII.70)}$$

Comparing with eq. C.VIII.62, we find that:

$$\mu_m = \frac{m}{C} \qquad \text{(C.VIII.71)}$$

Finally, the equilibrium of **(very real) pressure forces** acting on the matter surrounding a circular segment of dislocation of length $rd\theta$ (Fig. C.VIII.40) leads in addition to the following relation (in the $dr \to 0$ limit):

$$[P_N H - P_{N-p}(H - b) - P_{air} b] r d\theta + E d\theta = 0 \qquad \text{(C.VIII.72a)}$$

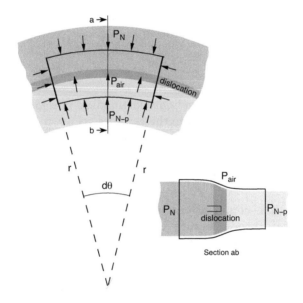

Fig. C.VIII.40 Top and side views of the box for which we write the force balance. The two vertical walls parallel to the dislocation line are taken far enough from the core so that the pressure and velocity fields are homogeneous over the thickness of the film. Note that the viscous stress vanishes at this distance from the core and does not contribute to the force balance. The arrows depict the pressure field.

The first term in brackets is the resultant of radial pressure forces exerted by the air and the two smectic domains on the box containing the dislocation segment. The second term, $Ed\theta$, is the radial component of the lateral pressure forces exerted on the box by the smectic. This equation can also be written as:

$$(P_N - P_{air}) b + (P_N - P_{N-p})(H - b) + \frac{E}{r} = 0 \qquad \text{(C.VIII.72b)}$$

Eliminating P_N and P_{N-p} between the three eqs. C.VIII.68, 69 and 72b, one obtains the following equation:

$$\Delta P b - \frac{E}{r} = \frac{v}{\mu_{N-p}}(H - b) + b\frac{V_m}{\mu_m} \qquad \text{(C.VIII.73a)}$$

where $\Delta P = P_{air} - P_m = \gamma/R$. Using eqs. C.VIII.61 and 71, one then obtains:

$$\Delta P b - \frac{E}{r} = \frac{v}{\mu_{N-p}}(H - b) + \frac{C}{m}\frac{b}{H}\frac{r}{r_m}bv \qquad \text{(C.VIII.73b)}$$

This equation is exactly identical to that (eq. C.VIII.63) found by applying the dissipation theorem, if we take:

$$\mu_{N-p} = m\frac{H - b}{|b|} = m\frac{N - p}{|p|} \qquad \text{(C.VIII.74)}$$

We can therefore obtain the pressure difference across the dislocation as a function of the mobility m of a dislocation:

$$P_N - P_{N-p} = \frac{v}{m}\frac{|p|}{N - p} \qquad \text{(C.VIII.75a)}$$

This pressure difference is inversely proportional to m and to the film thickness (inside the loop), and proportional to the Burgers vector of the dislocation. Note that this equation can also be written as:

$$(P_N - P_{N-p})(N - p) a_o v = \frac{|p|a_o v^2}{m} \qquad \text{(C.VIII.75b)}$$

Under this form, it becomes apparent that the energy dissipated around the dislocation is equal to the pressure difference across the dislocation multiplied by the matter flux crossing it (calculated in the reference frame of the dislocation).

In conclusion, the dynamical loop problem can be solved in one of two ways [35a,b]:

1. Either in a "**global**" way, applying the dissipation theorem to the entire system ("film + meniscus"). This procedure has the advantage of directly providing us with the evolution equation for a single loop. It is however incomplete if several loops are involved, the number of equations lacking being equal to that of additional loops. In this case, the global energy dissipation must be minimized to yield a complete set of equations. Some concrete examples will be given in the next subsections.

2. Or in a "**local**" way, by first writing the balance of pressure forces acting on the dislocation (far from the core):

$$(P_N - P_{air}) b + (P_N - P_{N-p})(H - b) + \frac{E}{r} = 0 \qquad \text{(C.VIII.76)}$$

and then writing the pressure jumps at the entrance of the meniscus and across the dislocation using the two equations:

$$P_N - P_m = \frac{C}{m} v_m \qquad\qquad (C.VIII.77)$$

$$P_N - P_{N-p} = \frac{v}{m} \frac{|p|}{N-p} \qquad\qquad (C.VIII.78)$$

It is then enough to eliminate P_N and P_{N-p} between these three equations to find the dynamical equation for the loop. The local method presents the advantage of being readily applicable in all cases, especially that of several competing loops. In the next subsections, we shall prove the equivalence of the two methods by successively analyzing the cases of two separate loops, of two concentric loops, and finally of two intersecting loops.

VIII.4.e) Dynamics of two separate loops

It is very easy to nucleate two separate dislocation loops in a film. We shall see that the evolution of these two loops is not completely independent, as they are coupled by the meniscus.

To write the equations of motion for the two loops, it is easier to use the local approach, so we employ it first.

Let r_1 and r_2 be the radii of the two loops, $b_1 = pa_o$ and $b_2 = qa_o$ their respective Burgers vectors (with p and q relative integers). The meniscus is taken as circular, with a radius r_m and the film has a thickness $H = Na_o$.

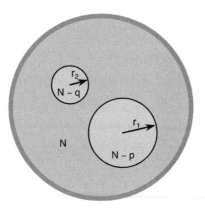

Fig. C.VIII.41 Two isolated loops in an N-layer film.

Applying the law C.VIII.76 (taking into account eqs. C.VIII.77–78) immediately yields:

$$\frac{v_1}{m}|b_1| + \frac{E_1}{r_1} = \left(\Delta P - \frac{C}{m}v_m\right)b_1 \qquad \text{(C.VIII.79a)}$$

$$\frac{v_2}{m}|b_2| + \frac{E_2}{r_2} = \left(\Delta P - \frac{C}{m}v_m\right)b_2 \qquad \text{(C.VIII.79b)}$$

while matter conservation imposes:

$$v_m = \frac{1}{N}\left(\frac{r_1}{r_m}p\,v_1 + \frac{r_2}{r_m}q\,v_2\right) \qquad \text{(C.VIII.80)}$$

These three equations form a complete equation set, completely describing the coupled dynamics of two loops. Before discussing them, let us show how they can be found using the global approach. Starting by applying the dissipation theorem to the entire system, we obtain (up to a prefactor of 2π):

$$\Delta PH v_m r_m - E_1 v_1 - E_2 v_2 = r_1 \frac{|b_1| v_1^2}{m} + r_2 \frac{|b_2| v_2^2}{m} + Cr_m \frac{H}{m} v_m^2 \qquad \text{(C.VIII.81)}$$

where the velocity v_m is given by eq. C.VIII.80.

To find the third equation, it is enough to write that, for each couple of values for the radii r_1 and r_2 of the two loops, the system chooses the velocities v_1 and v_2 that minimize the dissipation. This approach is legitimate (see, for instance, de Groot S.R. and Mazur P., *Non-Equilibrium Thermodynamics*, Dover, New York, 1984, p. 45) as the system is very close to thermodynamic equilibrium. It should also be noted that the motion of dislocations can be considered as stationary, since the Reynolds numbers associated with flow around the dislocation core are extremely small, below 10^{-6}). The global dissipation function $\Phi(v_1, v_2)$ is given by the right-hand side of equation C.VIII.81. Denoting by $E(v_1, v_2)$ the left-hand side of this equation, we need to minimize Φ with the condition $E - \Phi = 0$. Using in addition the fact that E is a linear form of the v_i and Φ a quadratic form of the v_i, we obtain the following equations:

$$\frac{\partial \Phi}{\partial v_i} = 2\frac{\partial E}{\partial v_i} \qquad i = 1,2 \qquad \text{(C.VIII.82a)}$$

In the case under consideration, this yields explicitly:

$$\Delta P b_i r_i - E_i = r_i \left(|b_i|\frac{v_i}{m} + \frac{C}{m}v_m b_i\right) \qquad i = 1, 2 \qquad \text{(C.VIII.82b)}$$

We recognize equations C.VIII.79 obtained using the local approach, thus proving the equivalence of the two methods.

In both calculations, we implicitly assumed that both loops remained circular. This is only an approximation (incidentally, very well verified in experiments) since each loop is immersed in the hydrodynamic velocity field created by the other loop. A small deformation ensues (infinitesimal, since the dislocations have a very strong line tension which favors their circular shape). Another consequence is the existence of a hydrodynamic interaction between the two loops, which generally tend to drift apart (this is the case, in particular, for two growing pores, or for a growing pore and a collapsing island).

This interaction aside, the growth of each loop is however not independent of the other, since they are coupled by the meniscus. This latter behaves as an imperfect reservoir, in that it resists being filled, leading to a pressure drop at the entrance of the meniscus. This effect is nevertheless weak in very thick films, where $v_m \ll v_1$ and v_2. In this case, the coupling term in $(C/m)v_m b$ in eqs. C.VIII.76a and b becomes negligible and the two loops evolve in a quasi-independent manner.

On the other hand, the coupling with the meniscus becomes important in thin or medium thickness films (less than 100 layers). To show this, assume that loop number 2 is a pore ($q > 0$) and let us calculate its critical nucleation radius in the presence of the other loop (which can be a pore or an island). To this purpose, it is enough to set $v_2 = 0$ in eqs. C.VIII.79b and 80, yielding:

$$r_{2c} = \frac{r_c}{1 - C\dfrac{p}{N}\dfrac{r_1}{r_m}\dfrac{v_1}{m\Delta P}} \qquad \text{(C.VIII.83a)}$$

with r_c the critical radius of an isolated pore. Using v_1 from eq. C.VIII.79a, we find in the end:

$$r_{2c} = r_c \frac{1 + C\dfrac{|p|}{N}\dfrac{r_1}{r_m}}{1 + C\dfrac{p}{N}\dfrac{r_c}{r_m}} \qquad \text{(C.VIII.83b)}$$

Two cases must then be distinguished:

1. Either loop "1" is a pore ($p > 0$). In this case, $r_{2c} > r_c$ for $r_1 > r_c$ (i.e., when loop "1" grows), while $r_{2c} < r_c$ when $r_1 < r_c$ (i.e., when loop "1" collapses).

2. Or loop "1" is an island ($p < 0$); in this case, loop "1" collapses and we always have $r_{2c} > r_c$.

These behaviors are easily explained in terms of pressure fields, loop "2" being immersed in the pressure field P_N of loop "1", different from P_m in the dynamical regime.

Thus, if loop "1" is an island, it collapses irrespective of its radius, leading to an increase in the pressure P_N, and hence in the dynamical critical radius r_{2c}.

If loop "1" is a pore, two cases must be distinguished:

1. Either its radius is smaller than r_c; in this case it collapses, leading to a decrease in the pressure P_N, and hence in the critical radius r_{2c};

2. Or its radius is larger than r_c; in this case, the loop grows and the pressure P_N, as well as the critical radius r_{2c}, increase.

We emphasize that these phenomena are directly related to the lack of permeability of the meniscus, characterized by the factor C, which leads to a change in film pressure. This effect is small in very thick films, above a few hundred layers, but can become important in films of less than 100 layers. For instance, one can estimate using eq. C.VIII.83b that in a film of 20 layers $r_{2c}/r_c \approx 18$, for $p = 1$, $r_m/r_c = 50$, $r_1/r_m = 0.5$ and $C \approx 2300$ (see subsection VIII.4.g). This considerable increase in the critical nucleation radius was observed by Picano [35a], who noticed that it is much more difficult to nucleate a loop in a film where another loop is already growing, than in a "fresh" film.

We finally point out that a loop can start by growing and end by collapsing as another – larger – loop grows in its vicinity. This singular behavior was found numerically by solving the coupled eqs. C.VIII.79a and b [35b].

Another interesting case (frequently encountered in experiments) is that of two concentric loops.

VIII.4.f) Dynamics of two concentric loops

The situation is slightly more complicated if the two loops are concentric. The notations are the same as in the previous subsection, but this time loop "2" is inside loop "1." It is clear that, on symmetry grounds and due to the flow they engender in the plane of the film, these loops will tend to center on each other. We shall assume this in the following.

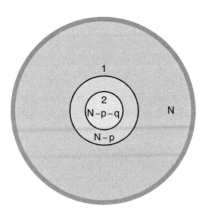

Fig. C.VIII.42 Two concentric loops.

The calculation in this configuration is analogous to the previous case. We start by noting that the law of matter conservation remains unchanged, the velocities being measured in the laboratory frame of reference:

$$V_m = \frac{1}{N}\left(\frac{r_1}{r_m}pv_1 + \frac{r_2}{r_m}qv_2\right)$$

(C.VIII.84)

This time, however, we need to account for the fact that loop "1" is immersed in the velocity field of loop "2." Consequently, the pressure jump across loop "1" will be proportional not to v_1, which is the velocity of dislocation "1" in the laboratory frame of reference, but rather to the velocity $v_1 - [qv_2/(N-p)](r_2/r_1)$ relative to the smectic flowing in the domain of thickness $N-p$. Keeping this in mind, we find the equations of motion by applying to each dislocation the laws C.VIII.76–78:

$$\left(\Delta P - \frac{C}{m}V_m\right)b_1 - \frac{E_1}{r_1} = \frac{|b_1|}{m}\left(v_1 - \frac{qv_2}{N-p}\frac{r_2}{r_1}\right)$$

(C.VIII.85a)

$$\left[\Delta P - \frac{C}{m}V_m + \left(v_1 - \frac{qv_2}{N-p}\frac{r_2}{r_1}\right)\frac{|p|}{(N-p)m}\right]b_2 - \frac{E_2}{r_2} = \frac{|b_2|}{m}v_2$$

(C.VIII.85b)

Again, it is easily checked that the global procedure gives exactly the same laws. One can also see that, if the film is very thick, the coupling terms are very small, so that the loops evolve in a quasi-independent manner. The interactions are however stronger than in the case of two isolated loops. This can be shown immediately if we proceed as before, assuming the inner loop "2" is a pore ($q > 0$) and determining its critical radius when the other loop "1" is much larger than the critical radius r_c. This time we find:

$$r_{2c} = r_c \frac{1 + C\dfrac{|p|}{N}\dfrac{r_1}{r_m}}{1 + C\dfrac{p}{N}\dfrac{r_c}{r_m} + \dfrac{p}{N-p}\left(1 - \dfrac{|p|}{p}\dfrac{r_c}{r_1}\right)}$$

(C.VIII.86)

Both at the numerator and the denominator of this formula we find the same corrections, proportional to the constant C, as in the case of two isolated loops (eq. C.VIII.83b). These terms can be interpreted in exactly the same way as in the previous case. In contrast, the denominator contains an additional term, proportional to $p/(N-p)$, which is independent of the meniscus, stemming from the fact that loop "2" feels the inner pressure field of loop "1," P_{N-p}. If loop "1" is an island ($p < 0$), this correction is always negative, leading to an increase in r_{2c}: this effect was expected, since the pressure P_{N-p} inside the island is always stronger than outside. If loop "1" is a pore, the sign of the correction depends on the value of its radius r_1 with respect to r_c. Thus, for $r_1 > r_c$, the correction is positive and r_{2c} decreases. This effect is also easy to understand, since loop "1" grows, leading to a decrease in the inside pressure. The reasoning is reversed when $r_1 < r_c$, leading to an increase in r_{2c}. Still, one must make sure that, in this last case, $r_{2c} < r_1$. It is easily seen, from eq. C.VIII.86, that this only occurs for very small r_1 (below r_c/N, typically), which is strongly limiting. In practice, it is almost impossible to nucleate a pore inside another – collapsing – one.

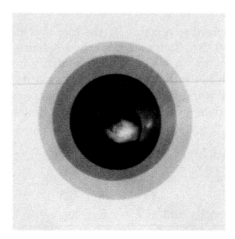

Fig. C.VIII.43 Nucleation using the heating wire (visible below the film) of several concentric pores (photograph by F. Picano, 2000).

It is also important to realize that when $r_1 \gg r_c$, the dominant term in eq. C.VIII.86 is that appearing at the numerator. This term is related to dissipation in the meniscus and always leads – as in the case of two separate loops – to an increase in the critical radius. This effect was observed experimentally by Picano, who noted a systematic increase of the critical radius as several pores are each nucleated within the previous one using the heating wire. This phenomenon invariably occurs if the thinning temperature $T(N-1)$ is notably exceeded, as shown in figure C.VIII.43. In this case, the central loops will collapse, although their radii are larger than the critical radius r_c, while the largest one will keep on growing. This phenomenon is, however, very difficult to describe quantitatively due to the thermal transients inherent to the nucleation method.

VIII.4.g) Dynamics of two crossing loops

In this subsection, we shall consider the more complicated case of two crossing loops (Fig. C.VIII.44).
     To simplify the problem as much as possible, let us assume the loops are circular pores of identical radii r and Burgers vectors $b = pa_o$ ($p > 0$) and let 2L be their "entanglement" distance (Fig. C.VIII.45a).
     Let us immediately point out that two microscopic dust particles must be present at the crossing points of the two loops. These particles, albeit invisible in the photograph, are, however, indispensable because otherwise the dislocations would tend to separate and form two concentric loops

(Fig. C.VIII.44b). Indeed, the total line length is smaller for two concentric loops than for two crossing loops, at equal surfaces for the domains of thickness $N - 1$ and $N - 2$. One should also note that the rule of nodes (namely, that the sum of line tensions around each node vanishes) is well verified experimentally. It can also be checked in figures C.VIII.1a and C.III.17, where the textures are more complex. These photographs also show that the steps are circular – to a very good approximation – as we always assumed in our calculations. Note that, in the lyotropic film in figure C.VIII.1b, the steps are not circular. These deformations are caused by flow in the film, induced by the flow of humid air that regulates the vapor pressure around the film.

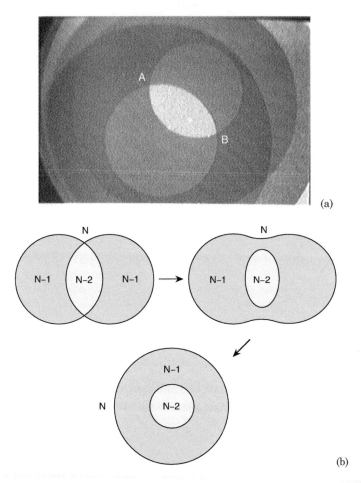

(a)

(b)

Fig. C.VIII.44 a) photograph of two crossing loops. One can check experimentally that they are circular and that the node rule is satisfied; b) in the absence of dust particles, the nodes should disappear, leaving the place to two concentric loops.

Let us now return to our problem. One question worth asking is the following: will the two loops tend to separate or to come closer together?

To answer it, let us look for the equations obeyed by the radius r and the distance L.

Matter conservation requires:

$$V_m = \frac{2}{N}\frac{r}{r_m}p\,v \tag{C.VIII.87}$$

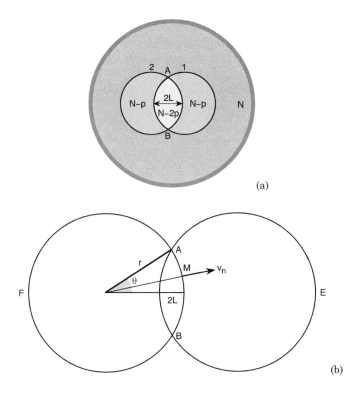

(a)

(b)

Fig. C.VIII.45 a) Crossing of two symmetric loops; b) definition of the angular aperture of the central domain.

The two other equations can be obtained using the global method. Set $v = dr/dt$ and $\dot{L} = dL/dt$. The energy gain is independent of L and is given by (per unit time):

$$E(v,\dot{L}) = 2\pi\,\Delta P\,H\,v_m\,r_m - 4\pi E\,v \tag{C.VIII.88}$$

The dissipation is more difficult to calculate, since three dissipation sources must be distinguished: the first one related to flow in the meniscus Φ_m, the second one, Φ_i, produced at the level of the dislocation segments delimiting the inner domain of thickness $N - 2p$ and the third one, Φ_e, occurring at the other segments, at the boundary between the domains of thickness $N - p$ and the domain of thickness N.

Dissipation in the meniscus writes as before:

$$\Phi_m = 2\pi C \, r_m \frac{H}{m} v_m^2 \tag{C.VIII.89}$$

To calculate Φ_i, one must first determine the normal velocity (with respect to the smectic at rest contained in the central domain) in each point of the curved dislocation segments AB bordering this area. Let θ be the angular aperture of this domain (Fig. C.VIII.45). If the angle $\theta \ll 1$, this velocity is given by (up to a second-order correction in θ):

$$v_n = \dot{L} \tag{C.VIII.90}$$

whence:

$$\Phi_i \approx 4 \frac{b}{m} r \theta \dot{L}^2 \tag{C.VIII.91}$$

To calculate Φ_e, one must account for the fact that the matter expelled from the central area must be found in the two lateral zones, generating a radial flow (with a velocity v_e with respect to the center of each loop, assumed uniform) in which grow the external dislocation segments AEB and AFB. In these conditions, one has:

$$\Phi_e = 2 \frac{b}{m} \int_{AEB} (v - v_e)^2 \, ds = 4 \frac{b}{m} (\pi - \theta) r \, (v - v_e)^2 \tag{C.VIII.92}$$

The drag velocity can be obtained from purely geometrical considerations, leading to the following result:

$$v - v_e = \frac{[\pi - 2\theta + \sin\theta \cos\theta + (L/r) \sin\theta] v - (\sin\theta) \dot{L}}{\pi - \theta} \tag{C.VIII.93a}$$

reducing, in the $\theta \ll 1$ approximation, to:

$$v - v_e \approx v - \dot{L} \frac{\theta}{\pi} \tag{C.VIII.93b}$$

which yields:

$$\Phi_e \approx 4\pi \frac{b}{m} r \left(v^2 - \frac{\theta}{\pi} v^2 - 2\theta \dot{L} v \right) \tag{C.VIII.94}$$

Finally, putting together all the terms gives:

$$\Phi = 8\pi \frac{C}{m} \frac{p}{N} \frac{r}{r_m} a_o r v^2 + 4\pi \frac{b}{m} r v^2 + 4\theta \frac{b}{m} r \left(\dot{L}^2 - v^2 - 2\pi v \dot{L} \right) \tag{C.VIII.95}$$

In particular, the case of two isolated loops is retrieved by setting $\theta = 0$. It is then enough to minimize Φ with respect to variables v and $\dot{L}$, recalling that

$E = \Phi$. As E does not explicitly depend on $\dot{L}$, one finds immediately that the dissipation is minimal for:

$$\frac{\partial \Phi}{\partial \dot{L}} = 0 \qquad\qquad (C.VIII.96)$$

resulting in the desired relation between v and $\dot{L}$:

$$\dot{L} = \pi v \qquad\qquad (C.VIII.97)$$

The velocity of the loop is given by writing that $E = \Phi$. In the limit of very thick films, and for $r \gg r_c$, we find once again $v \approx m\Delta P$.

In conclusion, the two loops tend to regroup since their centers are coming together (in the laboratory frame of reference) at a velocity $(\pi - 1)v$.

These principles must also apply to step foams, which are transiently formed immediately after creating the film (Fig. C.VIII.1a). Nevertheless, the analytical calculations quickly become unwieldy for more complicated configurations.

In the two following subsections, we discuss the problem of dissipation in the meniscus.

VIII.4.h) Dissipation and pressure drop in the meniscus

As loops or islands empty into the meniscus, a flow takes place at its entrance, with an average velocity v_m. Far inside the meniscus (strictly speaking, on the vertical wall of the frame that supports it), this velocity vanishes. The flow dissipates energy and leads to a pressure drop in proportion with the velocity v_m. Let us now try to quantify this result.

Let ϕ_m be the dissipation per unit time and per unit meniscus length and $P_N - P_m$ the pressure drop. These two quantities are connected by the following relation:

$$\phi_m = (P_N - P_m) H v_m \qquad\qquad (C.VIII.98)$$

with H the thickness of the film. It is then sufficient to calculate one of them to know the other.

Since the exact calculation cannot be performed, let us consider two extreme cases, which we shall combine in order to obtain an approximate solution for the general case.

We start by assuming that the dislocations within the meniscus are far enough from each other, giving the velocity and pressure fields enough time to become homogeneous over the entire thickness of the meniscus after crossing each dislocation. The meniscus being always very large with respect to the volume of the film, its deformation during filling can be neglected, so that the dislocations can be considered as immobile and immersed in a flow with a velocity that is $v = v_m$ at the entrance of the meniscus and decreases as it penetrates into the meniscus (Fig. C.VIII.46a). The dissipation can then be obtained simply by summing over the dissipated energies (per unit length) around each dislocation:

$$\phi_m = \phi_1 + \phi_2 + ...$$ (C.VIII.99)

For the first dislocation, one has:

$$\phi_1 = \frac{a_o v_m^2}{m}$$ (C.VIII.100)

The pth dislocation, immersed in the velocity field

$$v_{mp} = v_m \frac{N}{N + p - 1}$$ (C.VIII.101)

dissipates an energy:

$$\phi_p = \frac{a_o v_m^2}{m} \left(\frac{N}{N + p - 1} \right)^2$$ (C.VIII.102)

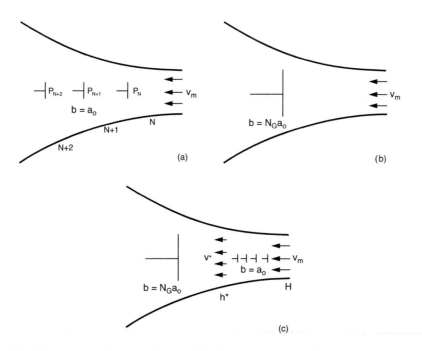

Fig. C.VIII.46 Extreme and intermediate models for calculating the dissipation in the meniscus. a) When the dislocations are very far from each other, the pressure and the velocity profile have enough time to become homogeneous between successive dislocations; b) when the dislocations are very close together, they act as a "giant" dislocation blocking the flow in the center of the meniscus; c) intermediate model. Close to the film (H < h < h*), the dislocations are far enough from each other and model "a" is valid. Far from the film (h > h*), model "b" applies.

Summing over all these contributions, we obtain the desired formula:

$$\phi_m = \frac{a_o v_m^2}{m} \sum_{p=1}^{\infty} \left(\frac{N}{N+p-1} \right)^2 \qquad \text{(C.VIII.103a)}$$

This series is convergent and its sum is N, leading to the following – very simple – result:

$$\phi_m = \frac{H}{m} v_m^2 \qquad \text{(C.VIII.103b)}$$

Comparing this formula with the phenomenological equation C.VIII.62, we find that C = 1.

To find the limit in which this model applies, let us first point out that the first 10N terms must be taken into account in the preceding series to ensure convergence. As a consequence, at the position of the 10 Nth dislocation (where the thickness h of the meniscus is about 10 H), the distance between two successive dislocations must be larger than $(10H)^2/l_p = 100N^2d^2/l_p$ (with $l_p = (\eta\lambda_p)^{1/2}$ the permeation length). This requires that the radius of curvature R of the meniscus be larger than $10^5N^5d^3/l_p^2$. Experimentally, this condition is never satisfied. This simple model under its present form must therefore be abandoned, although we can see that it is more adequate for thin films (small N) and at high temperature (l_p large).

This calculation also shows that the dislocations are very close to one another in the meniscus, for permeation purposes (this effect is even stronger when the contact angle is not zero). They are so close that we could even assume that they are almost touching, thus forming an obstacle similar to a giant dislocation with a Burgers vector $N_G a_o$ (Fig. C.VIII.46b). Modeling this dislocation as a ribbon perpendicular to the layers, of a width equal to the Burgers vector, one can estimate the friction force exerted upon it by a flow with velocity v_m using the formula C.II.154 (replacing $H = (N_G + N)a_o$ to take into account the confinement effect of the giant dislocation in the meniscus). Multiplying this force by the velocity v_m, we obtain a new expression for the dissipation:

$$\phi_m = \frac{1}{m} \frac{(N_G + N)N_G}{N} a_o v_m^2 \qquad \text{(C.VIII.104)}$$

Comparing this formula with the phenomenological equation C.VIII.62, we find that $C = N_G(N+N_G)/N^2$, where N_G is a constant characterizing dissipation in the meniscus. Experimentally, it seems reasonable to take for N_G the number of additional layers (with respect to the number of layers in the film) present in the meniscus in the zone where it is well oriented (Fig. C.VIII.16) (for a calculation of N_G, see Picano et al. [18]). Beyond this area, dissipation in the meniscus is negligible, since it is filled with focal conic defects, which certainly bring about a strong increase in permeability.

Following this line of reasoning, one expects the value of N_G to be of the order of some thousands, yielding values of $C \approx N_G^2/N^2$, generally much larger than 1, in contrast with the prediction of the first model.

The truth is in fact somewhere between these two models (Fig. C.VIII.46c). Indeed, close to the entrance of the meniscus the dislocations are far enough from their neighbors so the first model applies. Conversely, the second model becomes valid far from the entrance, where the dislocations are very close to one another. Let L* be the distance from the entrance of the meniscus where the crossover occurs from one model to the other. In this point, the thickness of the meniscus is h*, and the distance Λ between two dislocations is typically $(\Lambda l_p)^{1/2} \approx h*/2$. This occurs close to the dislocation of rank p*, which satisfies the equation:

$$p^*(p^* + N)^4 \approx \frac{4 \, l_p^2 \, R}{a_o^3} \tag{C.VIII.105}$$

These p* dislocations dissipate an energy:

$$\phi_{(1)} = \frac{a_o v_m^2}{m} \sum_{p=1}^{p^*} \left(\frac{N}{N+p-1} \right)^2 \approx \frac{p^*}{p^*+N} \frac{v_m^2}{m} N a_o \tag{C.VIII.106}$$

In the second part of the meniscus, formula C.VIII.104 still applies if we replace N by N+p* and take as velocity, instead of v_m, the average velocity v* calculated at the position of dislocation p*. This yields:

$$\phi_{(2)} = \frac{1}{m} \frac{(N_G + N + p^*)N_G}{N + p^*} a_o v^{*2} \tag{C.VIII.107a}$$

with

$$v^* = v_m \frac{N}{N + p^* - 1} \tag{C.VIII.107b}$$

Overall, we have:

$$\phi_m = \phi_{(1)} + \phi_{(2)} = \left[\frac{p^*}{p^* + N} + \frac{(N_G + N + p^*) N_G N}{(N + p^*)(N + p^* - 1)^2} \right] N a_o \frac{v_m^2}{m} \tag{C.VIII.108}$$

yielding for the constant C [36a]:

$$C = \frac{p^*}{p^* + N} + N_G \frac{N}{(N + p^* - 1)^2} \frac{N_G + N + p^*}{N + p^*} \tag{C.VIII.109}$$

where p* is obtained by solving eq. C.VIII.105.

The constant C is plotted as a function of N in figure C.VIII.47 with R = 1 mm, N_G = 2700, l_p = 190 Å and a_o = 30 Å. With these values (obtained by fitting the theoretical prediction to the experimental data presented in the next

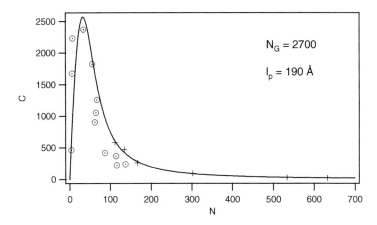

Fig. C.VIII.47 The constant C as a function of the number of layers N in the film. The solid line is the best fit of the experimental data obtained by analyzing either the slowing-down dynamics of a growing pore (open dots), or the equilibration process of two menisci of different size connected by a film (crosses) (see subsection VIII.4..i).

subsection [36a,b]), p* decreases more or less linearly from 33, for N = 3, to 0, for N ≈ 85. Thus, C can vary considerably, typically between 5 and 2500 in the thickness range probed by the experiments (where N takes values between 4 and 700 layers).

VIII.4.i) Measurement of the dissipation constant

Two complementary methods were used to measure the permeability constant of the meniscus. The first one is based on analyzing the growth of a pore and its slowing down over long time intervals, as its radius becomes sizable compared to that of the meniscus. The second one consists of monitoring pressure equilibration between two menisci with different curvatures connected by a film of known thickness.

i) Growth of a pore in thin films: analysis of its slowing down
  over long periods of time

During his doctorate work, Picano noticed that, in films of less than 150 layers, the growth velocity of pores tended to decrease with time [35a]. This effect, illustrated by the two graphs in figure C.VIII.48, becomes significant when the radius of the pore is much larger than its critical radius and approaches the radius of the meniscus. It is strongest in films of intermediate thickness, about thirty layers, as shown by the N = 34 curve shown in the first graph. One can also see in the second graph that the slowing down effect persists in very thin films, of less than 10 layers, in spite of the dramatic global increase in the growth velocity of the pores with respect to their asymptotic velocity v_a in thick films (Fig. C.VIII.39).

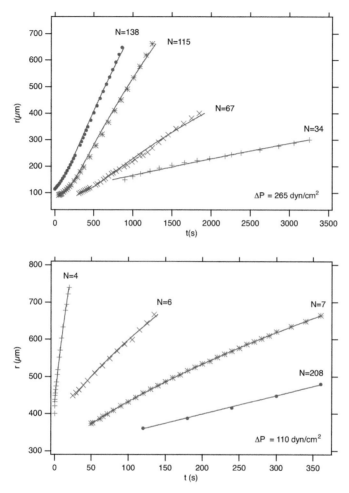

Fig. C.VIII.48 r(t) curves measured at long periods of time in films of various thicknesses. The solid lines are best fits using the theoretical law C.VIII.110. The values of C obtained from these fits are shown in the graph of figure C.VIII.47 (open dots) (from ref. [36b])

To analyze these curves, in particular those presented in the second graph, one must take into account not only dissipation in the meniscus, but also interactions between the free surfaces which, as we saw in subsection VIII.3.d, dominate the capillary depression imposed by the meniscus in films of less than 10 layers.

This work was performed in Oswald et al. [36b]. Here, we summarize the results.

The first step is to generalize the dynamical equation C.VIII.63 to account for the interactions between surfaces. This calculation immediately gives, for a pore with a Burgers vector of $b = a_o$ and preserving the previous notations:

$$\frac{1}{m}\frac{dr}{dt}\left(1 + \frac{Cr}{r_m N}\right) = \frac{\gamma}{R} + \frac{f[Na_o] - f[(N-1)a_o]}{a_o} - \frac{E}{ra_o} \qquad (C.VIII.110)$$

This equation can be solved analytically and leads to the following evolution law for the radius of the pore as a function of time:

$$\frac{t}{t_c} = \left(1 + \frac{Cr_c}{r_m N}\right)\left[\frac{r - r_i}{r_c} + \ln\left(\frac{r - r_c}{r_i - r_c}\right)\right] + C\frac{r^2 - r_i^2}{2r_m r_c} \qquad (C.VIII.111)$$

where r_i is the initial pore radius, r_c its – already calculated – critical radius (see eq. C.VIII.53a) and t_c a typical time:

$$t_c = \frac{a_o r_c}{m[a_o\Delta P + f(Na_o) - f((N-1)a_o)]} \qquad (C.VIII.112)$$

In these equations, all parameters are known, except for the constant C. The fits to the experimental curves (solid lines in the two graphs of figure C.VIII.48) yield the value of C in "thin" films (in practice, below 150 layers; the slowing down of elementary loops becomes undetectable over long time intervals in thicker films).

The results of these measurements are presented in figure C.VIII.47 (open dots). One can see the existence of a maximum value for C close to $N=20-30$ layers, in good agreement with the theory developed in the previous subsection.

ii) Equilibration of two menisci

Another way of testing the theoretical predictions for dissipation in the meniscus consists of measuring the equilibration time for two menisci of different sizes. This problem occurs, for instance, when the film is pierced by a fiber coated with the liquid crystal. How long does it take for the meniscus that forms around the fiber to reach equilibrium with the main meniscus, connecting the film to the frame?

The geometry of the setup is shown in figure C.VIII.49. The two menisci are circular, with radii r_1 and r_2. Experimentally, the outer meniscus has the same height h_o as the frame holding the film, while the meniscus surrounding the needle wets it completely, as shown by the photograph in the figure. In practice, after their initial phase of stabilization – which can take a few hours – the two menisci have different curvatures. Then comes a second phase, much slower than the first one, during which the two menisci equilibrate their inner pressures P_1 and P_2 by matter exchange through the film. In the experiment described in Caillier and Oswald [36a], the matter flows from the outer meniscus to the inner one, as shown in the diagram of figure C.VIII.49.

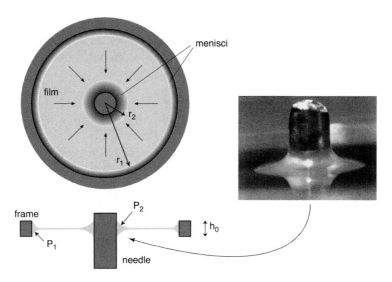

Fig. C.VIII.49 Equilibration of two concentric circular menisci. The arrows depict the radial flow in the film. The photograph shows the meniscus formed around the needle (from ref. [36a]).

Let us now see how to determine the constant C by monitoring in time the shape of the two menisci.

From the law C.VIII.77, we have:

$$P_N - P_1 = -\frac{C(N)}{m} v_1 \qquad \text{(C.VIII.113)}$$

with P_N the pressure within the film and P_1 the pressure in the outer meniscus "1." Note that the velocity v_1 is taken here as positive when the outer meniscus empties into the inner meniscus (imposing $P_1 > P_N$). For the inner meniscus, this law writes:

$$P_N - P_2 = \frac{C(N)}{m} v_2 \qquad \text{(C.VIII.114)}$$

with v_2 the velocity at its entrance. Total mass conservation (provided there is no leak along the needle or on the frame, which is well verified experimentally), requires $v_2 r_2 = v_1 r_1$, so that eq. C.VIII.114 can be rewritten as

$$P_N - P_2 = \frac{C(N)}{m} \frac{r_1}{r_2} v_1 \qquad \text{(C.VIII.115)}$$

Finally, after subtracting the two previous equations:

$$v_1 = \frac{m}{C(N)} \frac{r_2}{r_1 + r_2} \Delta P \qquad \text{(C.VIII.116)}$$

where $\Delta P = P_1 - P_2$ is the pressure difference between the two menisci.

In practice, the radii r_1 and r_2, as well as ΔP, can be measured by analyzing the shape of each meniscus. The velocity v_1 on the other hand is so small (below $1\,\mu m/hour$, usually) that it cannot be measured directly, for instance by seeding the film with small particles and following their trajectories. This quantity, on the other hand, can be obtained by measuring the time variation in the volume of one of the menisci, for instance, that formed around the needle. Indeed, let V_1 and V_2 be the volumes of the outer and inner menisci, respectively. From total mass conservation, $dV_2/dt = -dV_1/dt = 2\pi r_1 Na_o v_1$. Replacing in eq. C.VIII.117 gives:

$$C(N) = 2\pi m Na_o \frac{r_1 r_2}{r_1 + r_2} \left(\frac{dV_2}{dt} \right)^{-1} \Delta P \qquad \text{(C.VIII.117)}$$

This formula was used to measure the constant $C(N)$ in films thicker than about 100 layers (for thinner films, the equilibration times are so long – over one month – that the method can no longer be applied). The values obtained in this way are shown in the graph of figure C.VIII.47 (crosses). The results agree very well with the previous ones, as well as with the proposed model.

Note that, in the thick film range $(110 < N < 2000)$ the constant $C(N)$ is described with excellent precision by the following simplified law [36a]:

$$C(N) = \left(\frac{N_G}{N} \right)^2 \qquad \text{with} \qquad N_G = 2700 \qquad \text{(C.VIII.118)}$$

VIII.4.j) Nucleation dynamics of a loop

In this subsection, we examine the mechanisms of loop nucleation when the film is locally heated using the wire, or when it is stretched or compressed using a deformable frame.

i) Nucleation using a heating wire

We have already seen that, using the heating wire technique, one can nucleate an elementary dislocation loop (pore) in an N-layer film by locally heating it for a very short time (a few ms), to a temperature between $T(N)$ and $T(N-1)$.

Microscopy observations using a fast camera (500 images/second) show that the loop grows very fast during the pulse (Fig. C.VIII.50a), its radius growing initially as $\sqrt{t}$ (Fig. C.VIII.50b).

For a possible qualitative explanation of this phenomenon, let us assume that, during the initial nucleation phase, the film above the wire locally reaches a temperature T_N between $T(N)$ and $T(N-1)$ (Fig. C.VIII.51a). In these conditions, permeation (i.e., the molecules passing from the central layers to its neighbors) becomes very easy, so the mobility of dislocations strongly increases in this region. On the other hand, matter expelled from the center towards the edges (figure C.VIII.51b) must penetrate in the meniscus; this latter "resists," as it is not overheated. This flow in the meniscus controls loop dynamics. Indeed, direct application of the dissipation theorem yields, neglecting the dissipation in the film above the tip:

$$[\Delta P\, H + f(N, T_N) - f(N-1, T_N)]\, v_m r_m = r_m C\, \frac{H}{m}\, v_m^2 \qquad \text{(C.VIII.119)}$$

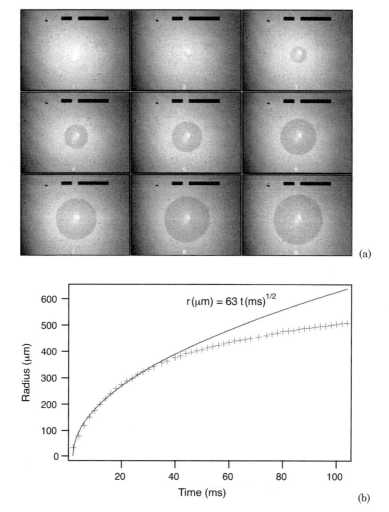

(a)

(b)

Fig. C.VIII.50 a) Loop nucleation in an 8CB film of 9 layers. The time is given in ms. The heating wire situated under the film can be seen in transparency (fuzzy white spot in the center) (photographs by F. Picano [35a]); b) loop radius as a function of time. The solid line is a fit of the first experimental points with a √t law. At long periods of time, the loop velocity strongly decreases, since the film cools down and the mobility of dislocations within the film is no longer "infinite."

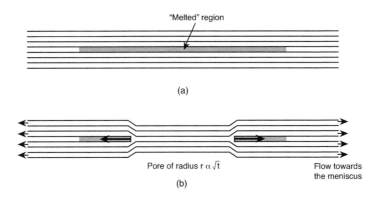

(a)

(b)

Fig. C.VIII.51 First nucleation mechanism. After melting of the central layer (a), a pore forms and its radius grows as √t (b).

In this expression, the mobility m and the constant C are taken at the temperature of the meniscus, T_m. As to ΔP, in this case it represents the pressure difference between the surrounding air and the inside of the film at the nucleation point. This pressure difference does not equal $P_{air} - P_m$, since the temperature is not uniform in the film, which is strongly heated locally. Its expression can be found by writing that film tension is the same everywhere (condition of mechanical film equilibrium), giving:

$$\Delta PH = (P_{air} - P_m)H + 2\delta\gamma + \delta f \qquad \text{(C.VIII.120)}$$

where

$$\delta\gamma = \gamma(T_m) - \gamma(T_N) \qquad \text{(C.VIII.121)}$$

$$\delta f = f(N, T_m) - f(N, T_N) \qquad \text{(C.VIII.122)}$$

Unfortunately, these two quantities are not well-known, but they will clearly count for the nucleation process.

  The other equation of the problem accounts for matter conservation and writes:

$$Hv_m r_m = va_0 r \qquad \text{(C.VIII.123)}$$

Eliminating v_m between eqs. C.VIII.119 and C.VIII.123 yields:

$$r\frac{dr}{dt} = \frac{m}{C}\frac{r_m}{H}N\Delta F \qquad \text{(C.VIII.124)}$$

with ΔF the motor force for nucleation:

$$\Delta F = \Delta PH + f(N, T_N) - f(N-1, T_N)$$

$$= (P_{air} - P_m)H + \gamma(T_m) - \gamma(T_N) + f(N, T_m) - f(N-1, T_N) \qquad \text{(C.VIII.125)}$$

Integrating eq. C.VIII.124 leads to a diffusive growth law in √t:

$$r(t) = \sqrt{2 \, N \frac{r_m}{H} \frac{m}{C} \Delta F \, t}$$ (C.VIII.126)

This type of law is indeed observed experimentally at the beginning of the growth (Fig. C.VIII.50b). Nevertheless, our lack of information as to the actual value of the ratio $\Delta F/C$ precludes any quantitative testing of this model in the case of strongly overheated thin films.

In conclusion, we point out that pore nucleation is still more complex in 8CB films of less than 8 layers. In this case, a drop of isotropic liquid is formed in the film immediately before pore nucleation. Clearly, this occurs as the thinning temperature exceeds the nematic-isotropic liquid transition temperature [35a] (Fig. C.VIII.52). The drop of isotropic liquid is transient, disappearing after a few ms as the film cools down. As to the pore, it keeps on growing, since its radius is always much larger than the critical nucleation radius (below 1 μm for films of less than 10 layers). This phenomenon was also observed in fluorinated materials exhibiting a first-order smectic A-isotropic transition [27e].

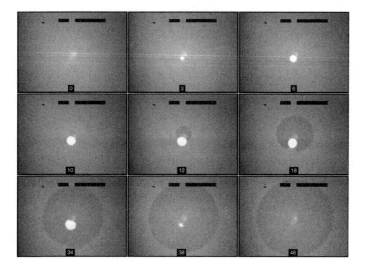

Fig. C.VIII.52 Sequence of photographs showing the nucleation of a drop of isotropic liquid (bright disk) followed by the nucleation of a pore near the drop (dark disk) in an 8CB film of 8 layers. As the temperature decreases, the drop quickly disappears, while the pore continues to grow, engulfing the drop. The oven is stabilized at 28°C. The heating wire is still visible in transparence (fuzzy spot in the center) (photographs by F. Picano [35a, 36b]).

ii) Nucleation in a film under tension or compression

In the Introduction, we showed that pores and islands can be nucleated by suddenly deforming a film stretched on a mobile frame (Fig. C.VIII.53a). This occurs by heterogeneous nucleation on dust particles of micrometric size, meaning that the critical nucleation radius, in the presence of the flow, is comparable with the size of the particles, namely in or below the micrometer range. But we already know that the critical nucleation radius is inversely proportional to the pressure difference $P_{air} - P_N$ between the air and the inside of the film (eq. C.VIII.53). We conclude that suddenly stretching (compressing) the film and its meniscus decreases (increases) the pressure within the film. Incidentally, this pressure variation was detected using the balance technique presented in subsection VIII.2.b).

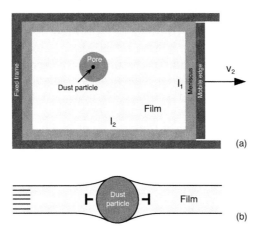

Fig. C.VIII.53 a) Nucleation of a pore (or of an island) on a dust particle in a film under tension (or compression); b) if the smectic wets the dust particle, this latter is surrounded by a dislocation forming a precursor island. The island grows if the radius exceeds the critical nucleation radius under compression.

To calculate this radius, we must account for the fact that, by deforming the frame, the surface area of the film changes. Matter flows from the meniscus towards the film or from the film toward the meniscus, leading to pressure variation in the film. This pressure variation $(\Delta P)_e$ was calculated previously (eq. C.VIII.77) as:

$$(\Delta P)_e = P_m - P_N = -\frac{C}{m} v_m \qquad\qquad (C.VIII.127)$$

where v_m is the velocity at the entrance of the meniscus (counted as positive when the meniscus fills up). From matter conservation, and assuming a constant velocity v_m over the contour of the frame, we obtain:

$$-2(l_1 + l_2) H v_m = H l_1 v_2 \qquad\qquad (C.VIII.128)$$

where $v_2 = dl_2/dt$ is the deformation velocity of the frame. These two relations yield the pressure variation:

$$(\Delta P)_e = \frac{C}{m} \frac{l_1}{2(l_1 + l_2)} v_2 \qquad\qquad (C.VIII.129)$$

and the value of the critical nucleation radius for a pore (b > 0), or for an island (b < 0):

$$r_c = \frac{E}{b[\Delta P + (\Delta P)_e]} = \frac{r_c(\text{eq.})}{1 + \left|\dfrac{(\Delta P)e}{\Delta P}\right|} \qquad \text{(C.VIII.130)}$$

In this formula, $\Delta P = \gamma/R$ is the pressure difference imposed by the meniscus and $r_c(\text{eq.})$ the critical pore radius in the film at rest. This formula shows that the critical radius for pore nucleation decreases when the film is stretched ($v_2 > 0$) and increases when the film is compressed ($v_2 < 0$), diverging for $\Delta P = -(\Delta P)_e$. In the case of islands, the critical radius remains infinite under stretching and becomes finite at fast enough compression, requiring $-(\Delta P)_e > \Delta P$. This flow effect can be quite important, as the $(\Delta P)_e$ term quickly becomes much larger than ΔP. Indeed, one has:

$$(\Delta P)_e = 1.5 \times 10^8 \, v_2 \qquad \text{(in CGS units)} \qquad \text{(C.VIII.131)}$$

for a square frame ($l_1 = l_2$) with $N = 100$, $C = 600$ (see Fig. C.VIII.47) and $m = 10^{-6}$ cm^{2}sg^{-1}. Taking $v_2 = 100$ μm/s, which is a typical value for the frame deformation velocity, yields $(\Delta P)_e = 1.5 \times 10^6$ dyn/cm^2. This pressure variation is considerable compared with that imposed by the meniscus, which is typically $\Delta P = 100$–3000 dyn/cm^2. Thus, film deformation leads to a decrease by a factor $1.5 \times 10^6/500 \approx 3000$ of the critical nucleation radius, which drops from 30 μm, typically (with $E/b = 1.5$ dyn/cm and $\Delta P = 500$ dyn/cm^2) in the film at rest to 0.01 μm in the film under flow. It is therefore not surprising that loops or islands spontaneously nucleate in the film. On this topic, note that island nucleation under compression is easier than pore nucleation under stretching (for the same absolute value of the deformation velocity). This can be very easily demonstrated by blowing cigarette smoke over a film before compression or stretching. In the first case (compression), islands nucleate in large numbers around the particles that settle on the film, while in the second case (stretching), only a few pores nucleate. The reason for this asymmetry is that each particle that falls onto the film forms a minute island, rather than a pore (Fig. C.VIII.53b), and this island can then grow if its size exceeds the critical nucleation radius.

We emphasize that this calculation implicitly assumes a constant value for C, at all times equal to that measured in the quasi-static flow regime. For this hypothesis to hold, the meniscus would have to preserve its equilibrium shape, which is obviously not the case in the experiments. Indeed, the flow (much more violent here than in all experiments described above) quickly destabilizes the meniscus, in particular in its well-oriented part, which plays the role of a "plug" in the quasi-static regime. As a result, the constant C very quickly decreases with time.

The scenario is thus the following: at the moment frame deformation begins, plug flow at the entrance of the meniscus (initially stable) leads to a sudden variation in film pressure. The critical nucleation radius of pores or islands (depending on the type of deformation) typically drops from 30 μm to below 1 μm, allowing defect nucleation on the dust particles (if any) contained in the film. Then nucleation quickly ceases as the meniscus becomes destabilized, increasing its permeability and allowing the pressure to reach its equilibrium value in the film.

We add that recent experiments of F. Caillier (still unpublished results) on the collapse dynamics of smectic bubbles inflated at the end of a hollow cylinder show that parameter C typically ranges between 0.3 and 1 when the meniscus is completely "destructured" (i.e., filled with focal conics). In this nonlinear regime, the value of C varies as $v_m^{-0.7}$ and is almost independent of the film thickness. This "shear-thinning" behavior contrasts with what happens in the quasi-static regime (at velocities $v_m \ll 1\mu m/s$) in which C is independent of v_m but varies considerably as a function of the film thickness (as can be seen in figure C.VIII.47).

The same arguments also explain why the smectic bubbles tend to thin and form pores that migrate to the top (Fig. C.VIII.28) when they are inflated too fast (see Fig. C.VIII.15a).

VIII.5 Film structure and phase transitions at a fixed number of layers

In subsection C.VIII.3.e, we already tackled the problem of phase transitions in the context of thinning transitions. We have shown that a so-called presmectic film of N layers can coexist with a nematic meniscus (Fig. C.VIII.54) as long as the temperature of the system "film + meniscus" is between T_{NA} and T(N). Below T_{NA}, the meniscus goes to the smectic A phase and the film remains stable, while above T(N) the film of thickness N destabilizes and loses one layer. The new film, containing N – 1 layers, is stable until the next thinning transition occurring at T(N – 1), and so on.

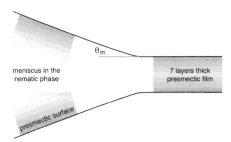

Fig. C.VIII.54 Coexistence of the meniscus in the nematic phase with a presmectic film of 7 layers. Film stability depends on the temperature and on the suction exerted by the meniscus limited by curved surfaces.

We also emphasized that the existence of presmectic films is related to the appearance of smectic order at the free surface of the nematic (or isotropic) phase. This order, vanishing exponentially over a distance $\xi(T)$, stabilizes the discrete thickness of presmectic films. These films are, however, under stress, bearing a pressure difference $\Delta P = \gamma/R$ due to the curvature $1/R$ of the meniscus surfaces. The thinning transitions of presmectic films can thus also be induced by an increase in temperature, as well as by an increase in ΔP.

From a thermodynamical point of view, the thinning transitions can be seen as first-order phase transitions, as the thickness of presmectic films changes discontinuously.

In this section, we shall be concerned with phase transitions in films at a **constant number of smectic layers**. In particular, we shall analyze, on concrete examples, the influence of confinement and surface effects on the transitions, and show that they can even induce new transitions.

VIII.5.a) The SmA → SmC* transition: optical measurements

As a smectic film undergoes a first-order phase transition **at a constant number of layers**, the optical path $\delta = <nH>$ (with n the optical index and H the film thickness) has a discontinuity $\Delta\delta$, which changes the light reflectivity of the film. If $\Delta\delta$ is large enough, the transition becomes clearly visible in reflecting microscopy, as were the thinning transitions (Fig. C.VIII.25). Four examples of first-order transitions (where the two phases can coexist at the transition point) are shown in the photographs of figure C.VIII.55.

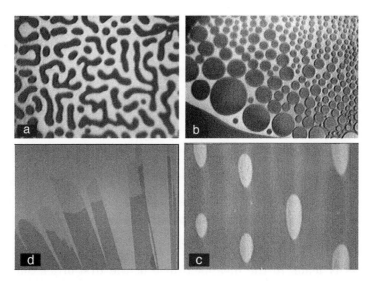

Fig. C.VIII.55 Examples of four first-order phase transitions in smectic films, directly visible in reflecting microscopy: a) $SmB_{cryst} \Rightarrow SmC$ transition in a 7O.7 film. The dark domains of the SmC phase are in equilibrium with the brighter matrix of the SmB_{cryst} phase; b) first-order SmA → SmC* transition in a thick ALLO-902C13M5T film. The circular domains of the SmC* phase are this time immersed in a matrix of the SmA phase; c) SmI → SmC* transition in a DOBAMBC film. The domains of SmC* phase coexist with a matrix of SmI phase, but, contrary to the previous photo, their shape is anisotropic and they are regularly placed; d) lamellar → cubic $(L_\alpha \rightarrow Ia3d)$ transition in a film of lyotropic mixture $C_{12}EO_6/H_2O$.

Clearly, as the films are thick enough, the transitions are the same as in the bulk: $SmB_{cryst} \rightarrow SmC$ in 7O.7 (Fig. C.VIII.55a), $SmA \rightarrow SmC^*$ in ALLO-902C13M5T (Fig. C.VIII.55b), $SmI \rightarrow SmC^*$ in DOBAMBC (Fig. C.VIII.55c) and $L_\alpha \rightarrow Ia3d$ in the binary mixture $C_{12}EO_6$/water (Fig. C.VIII.55d).

What happens when the film thickness N decreases?

To answer this question, Krauss et al. examined the case of ALLO films and determined very precisely their optical thickness as a function of temperature from the reflectivity spectra (Fig. C.VIII.56) [38a,b]. They noticed that, in a film of 126 layers (Fig. C.VIII.56a), the optical path δ undergoes a discontinuous variation at 82.5°C. This jump reflects the first-order character of the $SmA \rightarrow SmC^*$ transition, well established for this material. We should say that this discontinuity ($\Delta\delta \approx 50$ nm) can be measured directly in the film at the transition temperature, as circular domains of the SmC^* phase are in equilibrium with the SmA matrix (Fig. C.VIII.55b) (in this case, the measurement is performed by successively focusing the incident beam on a SmC^* domain, and then on the SmA matrix). Repeating the measurement on thinner and thinner films shows that, as N decreases, $\Delta\delta$ also decreases, and finally vanishes at $N_c = 90$ layers. Below this thickness no discontinuity occurs anymore, as illustrated by the curve in figure C.VIII.56b, determined for an 80-layer film. One can see that δ varies strongly around 85°C, but this variation is nevertheless continuous and the slope dδ/dT remains finite.

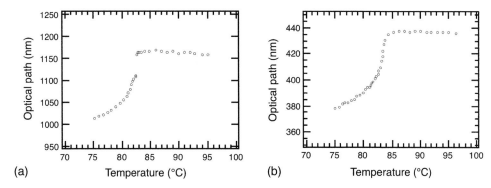

Fig. C.VIII.56 Variation of the optical thickness as a function of temperature in ALLO films: a) film thickness 126 layers, b) film thickness 80 layers (from ref. [38a,b]).

As a conclusion of this study, one is tempted to say that decreasing the film thickness N leads to the disappearance of the first-order $SmA \rightarrow SmC^*$ transition.

An identical phenomenon was demonstrated by Bahr and Fliegner [39] in free-standing films of the C7 compound (4-(3-methyl-2-chloropentanoyloxy)-4'-heptyloxybiphenyl), a material that also exhibits a $SmA \rightarrow SmC^*$ transition. The optical method used this time is ellipsometry; its principle is presented in figure C.VIII.57.

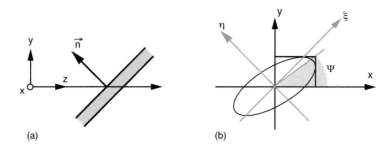

Fig. C.VIII.57 Ellipsometry applied to the study of phase transitions in smectic films.

In this experiment, the smectic film is crossed by the beam of a He-Ne laser, at a 45° angle with respect to the normal to the film **n**. The incident beam, directed along the z axis, is linearly polarized at a 45° angle with respect to the **x** axis, perpendicular to the (**n**,**z**) plane. Thus, in the (x,y,z) reference frame, the electric field of the incident beam appears as the superposition of two waves, polarized along x and y axes:

$$E_x = E_{xo}e^{i\varphi_x} \qquad \text{and} \qquad E_y = E_{yo}e^{i\varphi_y} \qquad\qquad \text{(C.VIII.132)}$$

with the same amplitude ($E_{xo} = E_{yo}$) and the same phase ($\varphi_x = \varphi_y$). If we characterize a generic wave by $\Psi = \arctan(E_{yo}/E_{xo})$ and $\Delta = \varphi_y - \varphi_x$, in our case the incident beam is described by $\Psi = 45°$ and $\Delta = 0°$. After crossing the film, however, the laser beam acquires an elliptic polarization described by a phase difference $\Delta \neq 0°$ and an angle $\Psi \neq 45°$. To measure these two angles, Bahr and Fliegner used the setup shown in figure C.VIII.58. It contains a polarizer and an analyzer, making with the x axis angles P and A, respectively. The sample is placed between the polarizer and the analyzer, preceded by a quarter-wave plate at a 45° angle with respect to the x axis. A computer adjusts the P and A angles to obtain extinction of the transmitted beam, whose intensity is measured using a photomultiplier, an optical chopper and a lock-in amplifier. Without the sample, extinction is obtained for P = 45° and A = 45° + 90°.

To simplify the analysis, let us assume that the sample does not absorb light, only inducing a phase shift Δ_s. We shall prove that, in the presence of the sample, extinction is achieved by turning the polarizer by an angle

$$dP = \Delta_s / 2 \qquad\qquad \text{(C.VIII.133)}$$

To prove this result, let us first notice that, by turning the polarizer by dP, the polarization of the incident beam becomes elliptical after crossing the quarter-wave plate. The principal axes of the polarization ellipse are then oriented along the slow and fast axes, ξ and η of the quarter-wave plate, themselves tilted by 45 ° with respect to the x and y axes (Fig. C.VIII.58b). Let E_o be the amplitude of the beam after the polarizer. After going through the quarter-wave plate, the beam will be described in the (ξ, η, z) reference frame by the polarization components:

$$E_{\xi o} = E_o \cos(\omega t) \cos(dP) = \frac{1}{2} E_o [\cos(\omega t - dP) + \cos(\omega t + dP)] \quad \text{(C.VIII.134a)}$$

$$E_{\eta o} = E_o \sin(\omega t) \sin(dP) = \frac{1}{2} E_o [\cos(\omega t - dP) - \cos(\omega t + dP)] \quad \text{(C.VIII.134b)}$$

yielding, after projection onto the x and y axes:

$$E_{xo} = \frac{\sqrt{2}}{2} E_o \cos(\omega t + dP) \quad \text{and} \quad E_{yo} = \frac{\sqrt{2}}{2} E_o \cos(\omega t - dP) \quad \text{(C.VIII.135)}$$

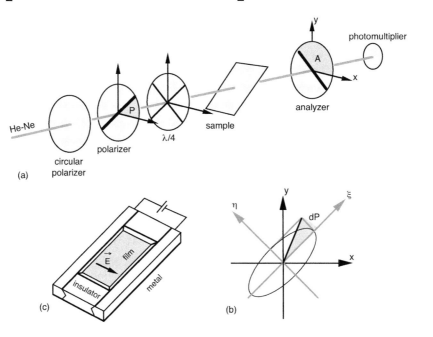

Fig. C.VIII.58 Ellipsometer used for the study of smectic films.

In conclusion, turning the polarizer induces a phase shift $\Delta_{pol} = -2dP$. Since the sample introduces the additional phase shift Δ_s, extinction is obtained after the analyzer for $\Delta_s + \Delta_{pol} = 0$, i.e., for $dP = \Delta_s/2$. This proves the relation C.VIII.133.

In the experiment of Bahr and Fliegner [39], the C7 films are stretched on a "semi-metallic" frame, permitting the application of a DC electric field parallel to the x direction (Fig. C.VIII.58c). The C7 molecules being homochiral, the SmC* phase exhibits a spontaneous polarization perpendicular to the molecular tilt plane (see chapter C.IV) and forms a helix along **n**. In films much thinner than the helix pitch, this results in a finite average polarization in the plane of the film and perpendicular to the molecular tilt plane. By applying an electric field, the molecules can all be aligned in the (y,z) plane perpendicular to the field.

Two configurations are then possible, according to the sign of the applied voltage, with different ellipsometric phase shifts Δ_+ and Δ_- (Fig. C.VIII.59a). The experimental diagram in figure C.VIII.59b confirms this

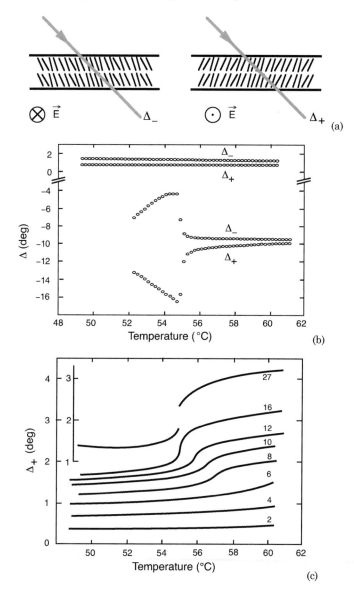

Fig. C.VIII.59 Ellipsometric phase shifts $\Delta_\pm$ measured as a function of temperature in C7 films of different thicknesses; a) schematic representation of the two configurations induced by the electric field, parallel and anti-parallel to the x axis; b) phase shifts Δ_+ and Δ_- measured in C7 films of 4 (upper curves) and 190 layers (lower curves); c) phase shift Δ_+ measured as a function of temperature in films of 27 ± 2, 16 ± 1, 12, 10, 8, 6, 4, and 2 layers (from ref. [39]).

reasoning and shows that the expected discontinuities of $\Delta_+(T)$ and $\Delta_-(T)$ at the SmA → SmC* transition are clearly visible in a film of 190 layers; in a 4-layer film, on the other hand, the $\Delta_+(T)$ and $\Delta_-(T)$ curves exhibit no anomaly between 48°C and 62°C, which would imply that the SmA → SmC* transition disappeared. To understand better this phenomenon, Bahr and Fliegner [39] performed systematic measurements of $\Delta_+(T)$ using films of different thicknesses (Fig. C.VIII.59c). These measurements clearly show that **the SmC* → SmA transition disappears below $N_c \approx 15$ layers.**

This result is **paradoxical** – to say the least – since, from a theoretical point of view, the SmA → SmC* transition cannot disappear, since it breaks the $D_\infty → C_2$ symmetry. Strictly speaking, the character of the transition could change, becoming second order above a tricritical point given by a certain thickness N_c. The only possible explanation of this paradox is that the **symmetry of smectic films is not truly broken** at this transition.

This lack of symmetry breaking is, in fact, obvious in the diagrams of figure C.VIII.59b, as the difference $\Delta_+ - \Delta_-$ never goes to zero, even above the SmC* → SmA transition temperature (55°C). This proves that **C7 films never have the D_∞ symmetry** of the SmA phase. The reason is that, in the surface layers of the film, the molecules are not perpendicular to the plane of the layers, as in the SmA phase, but rather tilted, like in the SmC phase. **It is of no consequence if the number of layers of the SmC type is small with respect to the total number of layers N, since a unique SmC layer suffices to break the D_∞ symmetry.**

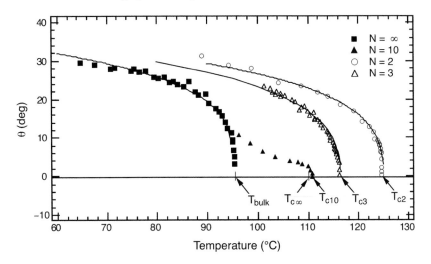

Fig. C.VIII.60 Variation of the average tilt angle as a function of temperature in DOBAMBC films (from ref. [40]). The solid lines are the predictions of the theoretical model.

Breaking of the D_∞ symmetry of the SmA phase, by molecular tilt in the surface layers of free-standing films was however observed by Heinekamp et al. [40] in another famous material, DOBAMBC (see chapter C.IV).

These authors determined the average tilt angle $<\theta>$ of the molecules with respect to the layer normal by ellipsometric measurements of the difference $\Delta_+(T) - \Delta_-(T)$. In contrast with C7 and ALLO, the SmC* → SmA transition in DOBAMBC is second order, so that in a bulk sample, or in a very thick film, the angle θ increases continuously from zero below $T_{bulk} = 95.15°C$ ($<\theta_\infty(T)>$ curve in figure C.VIII.60). The situation is, however, more complex in thin films, where the variation in the average tilt angle $<\theta_N>$ with temperature is strongly dependent on the number of layers N.

For instance, in a film of only 2 layers, the $<\theta_2(T)>$ curve is completely similar to that of $<\theta_\infty(T)>$, except that the temperature below which the angle θ starts to grow is higher ($T_{c2} = 125.11°C$).

In a three layers film, the behavior is similar, but the transition temperature T_{c3} is lower than T_{c2}. This decrease continues with increasing N, reaching $T_{c10} = 106°C$ for N = 10. This time, the shape of the $<\theta(T)>$ function below T_{c10} is not the same as for N = 2 or 3, since the average tilt angle $<\theta_{10}(T)>$ remains relatively small until T_{bulk}, then reaches the $<\theta_\infty(T)>$ curve below T_{bulk}.

These experiments led Heinekamp et al. to the conclusion that the SmA → SmC* transition occurs in the surface layers at a temperature T_{cN} larger than T_{bulk} and that depends on the film thickness N. A remarkable result is that, in the N → ∞ limit, the surface transition temperature $T_{c\infty} = 100°C$ is about 5° higher than the bulk transition temperature T_{bulk}. Thus, in the interval $T_{bulk} < T < T_{c\infty}$ and for very thick films, the local angle of molecular tilt θ decreases from a finite value at the surface toward zero when moving away from the surfaces.

Before presenting the theoretical analysis of these results, let us point out the similarity between these phenomena and the smectic order induced at the surface of nematic or isotropic phases, described previously in section VIII.3.

VIII.5.b) The SmA → SmC* transition: theoretical models

The results of the three experiments on the SmA → SmC* transition in films of DOBAMBC, C7 and ALLO are summarized in the diagrams of figure C.VIII.61.

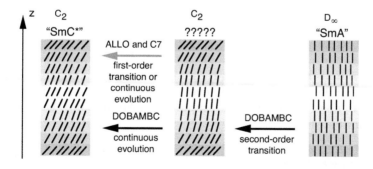

Fig. C.VIII.61 Structure of smectic films close to the SmA → SmC* transition.

To write a Landau model for the SmA → SmC* transition in films, we must admit that the tilt angle θ of the molecules depends on their position in the layer stack of each film. We shall consider successively the case of thick films and of materials exhibiting a first-order SmA → SmC* transition in the bulk (C7 and ALLO), and that of thin films and of materials such as DOBAMBC, where the transition is second order.

i) Thick films: the bulk SmA → SmC transition is first order*

In films of a few tens of layers it is more convenient to assume, as Kraus [38b] and Demikhov et al. [41] did, that the order parameter (here, the angle of local molecular tilt) is a continuous function θ(z) of the z coordinate. If the SmA → SmC* transition is first order, the free energy density per unit volume in the film can be expanded in the following series:

$$f_c(z) = \frac{1}{2}a(T)\theta^2(z) - \frac{1}{4}b\theta^4(z) + \frac{1}{6}c\theta^6(z) + \frac{1}{2}g\left(\frac{d\theta}{dz}\right)^2 \qquad \text{(C.VIII.136)}$$

where

$$a(T) = \alpha\frac{T - T^*}{T^*} \qquad \text{(C.VIII.137)}$$

and α, b, c and g are positive parameters. The last constant in this expression, g, is a "stiffness modulus" in the sense that it favors a uniform order parameter (see the analogy with eq. B.VI.14 in volume 1).

In an infinitely thick film, the role of the surfaces is negligible and the SmA → SmC* transition temperature is:

$$T_c^{bulk} = T^*\left(1 + \frac{3b^2}{16\alpha c}\right) \qquad \text{(C.VIII.138)}$$

At this temperature (different from the spinodal temperature T^* of the smectic A phase), the two phases coexist (first-order transition) and the tilt angle goes from zero (for $T > T_c^{bulk}$) to

$$\theta_c = \sqrt{\frac{3b}{4c}} \qquad \text{(C.VIII.139)}$$

for $T < T_c^{bulk}$.

In a film of finite thickness H, one must consider the effect of surfaces, where the molecules are tilted by θ_s, even above T_c^{bulk}. To simplify the analysis, assume that θ_s does not depend on temperature. Minimizing the free energy of the film per unit surface:

$$F_{/cm2} = \int_{-H/2}^{+H/2} f_c[\theta(z),\theta'(z)]\,dz \qquad \text{(C.VIII.140)}$$

leads to the following differential equation:

$$a(T)\theta - b\theta^3 + c\theta^5(z) - g\frac{d^2\theta}{dz^2} = 0 \qquad \text{(C.VIII.141)}$$

with the boundary conditions $\theta(z = H) = \theta(z = 0) = \theta_s$ at the surfaces (note that, by symmetry, $d\theta/dz = 0$ in the center of the film). The solutions $\theta(z)$ of this equation depend on the coefficients $a(T)$, b, c, and g, as well as on the thickness H, and must be calculated numerically. To simplify the numerical analysis, it is convenient to define the following three characteristic values:

$$f_c = \frac{3b^2}{4c}, \quad \xi_c = \sqrt{\frac{4gc}{3b^2}} \quad \text{and} \quad \theta_c = \sqrt{\frac{3b}{4c}} \qquad \text{(C.VIII.142)}$$

defining the units of energy, length and tilt angle. With respect to the dimensionless variable $\bar{z} = z/\xi_c$, eq. C.VIII.141 becomes:

$$\frac{1}{4}\eta(T)\,\theta - \frac{\theta^3}{\theta_c^2} + \frac{3\theta^5}{4\theta_c^4} - \frac{d^2\theta}{d\bar{z}^2} = 0 \qquad \text{(C.VIII.143)}$$

where we used

$$\eta(T) = \frac{T - T^*}{T_c^{bulk} - T^*} = \frac{\Delta T}{\Delta T_c^{bulk}} \qquad \text{(C.VIII.144)}$$

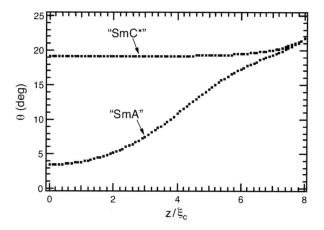

Fig. C.VIII.62 Two solutions $\theta(\bar{z})$ calculated numerically for a film of thickness $H = 16\xi_c$ at a temperature $\eta = 1.07$.

Let us now consider the example of a film of thickness $H = 16\xi_c$ and surface angle $\theta_s = 21.7° = 0.374$ rad. Let $\eta = 1.07$ (i.e., slightly above the bulk transition temperature T_c^{bulk}). Numerically solving the differential eq. C.VIII.143 by the Runge-Kutta method shows that two solutions exist at this temperature, shown in figure C.VIII.62. The first solution, labelled "SmA," is characterized by a lower value of the average tilt angle than the second solution, labelled, for this reason, "SmC*" (see also figure C.VIII.63a). Calculating the corresponding energies (integrated over the thickness z, Fig. C.VIII.63b) shows that the "SmC*" solution is energetically favorable with respect to the "SmA" solution, although the temperature already exceeds T_c^{bulk}. Nevertheless, at $\eta = 1.3$, a still higher temperature, the "SmA" solution, exhibiting the lower tilt angle, becomes energetically favorable. Indeed, the diagram in figure C.VIII.63b shows that the SmA → SmC* phase transition in a film of thickness $H = 16\xi_c$ takes place at the reduced temperature $\eta = 1.2$.

We also emphasize that the two solutions only coexist over a finite range of the temperature η. The two limits of this interval correspond to the spinodal temperatures (i.e., to the absolute stability limits of the competing "phases"). Systematic numerical analysis shows that the width $\Delta\eta$ of this interval decreases as the thickness H decreases and that, for $H < 8\xi_c$, eq. C.VIII.143 only admits a unique solution, irrespective of the temperature η; this signifies that the first-order phase transition SmA → SmC* disappeared.

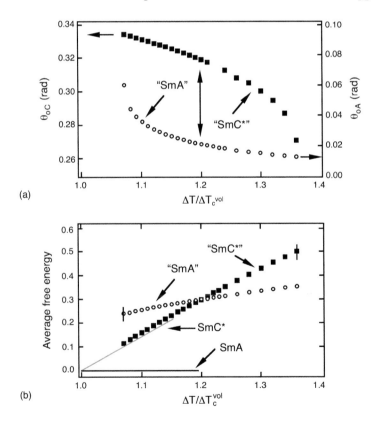

Fig. C.VIII.63 Comparison of the two solutions of equation C.VIII.143 for $H = 16\xi_c$; a) variation of the average tilt angle as a function of the temperature η; b) variation of the average free energy as a function of the temperature η.

ii) Thin films: the bulk SmA → SmC transition is second order*

The model we employed previously, built on a continuous order parameter, is inapplicable to the case of DOBAMBC. Indeed, in this material the "SmA → SmC*" transition is second order and the anomalies appear in films consisting only of a very small number of smectic layers. On the other hand, the molecular tilt angle in the surface layers is not constant; it depends on the temperature and must be calculated.

For these reasons, Heinekamp et al. [40] put forward a **discrete Landau model** whereby the order parameter takes values θ_i indexed by the position i (i = 1, ... ,N) of the layers in a stack of thickness N. According to this model, the free energy of a film can be written as the sum of three contributions:

$$F = F_S + F_B + F_C \tag{C.VIII.145}$$

where

$$F_S = \frac{1}{2} A'(T)\,(|\theta_1|^2 + |\theta_N|^2) + \frac{1}{4} B'\,(|\theta_1|^4 + |\theta_N|^4) + \frac{1}{6} C'\,(|\theta_1|^6 + |\theta_N|^6)$$

with $A'(T) = \alpha'(T - T_S)$,

$$F_B = \sum_{i=2}^{N-1} \left(\frac{1}{2} A(T)|\theta_i|^2 + \frac{1}{4} B|\theta_i|^4 + \frac{1}{6} C|\theta_i|^6 \right)$$

with $A(T) = \alpha\,(T - T_{bulk})$

$$F_C = \sum_{i=1}^{N-1} \frac{1}{2} g\,(|\theta_{i+1}| - |\theta_i|)^2 \tag{C.VIII.146}$$

In this model, F_S is the energy of the two surface layers and F_B the energy of the bulk layers. Hence, there is no *a priori* reason for the "surface coefficients" α', B' and C' to be equal to the bulk coefficients α, B and C. As to the third term, F_C, it accounts for the coupling between layers in the stack, the role of the g coefficient being similar to that of the stiffness modulus in the continuous model.

To discuss this model and its physical consequences, let us start by studying the two limiting cases N = 2 and N = ∞.

N = 2 case

In a film composed of only two layers, minimizing the free energy F_S with respect to $\theta = \theta_1 = \theta_2$ yields:

$$\theta_1 = \theta_2 = \sqrt{\frac{B'}{2C'}} \sqrt{\sqrt{1 + \frac{t}{t_o}} - 1} \tag{C.VIII.147}$$

with

$$t = T - T_S \quad \text{and} \quad t_o = -\frac{B'^2}{4\alpha'C'} \tag{C.VIII.148}$$

Fitting expression C.VIII.147 to the experimental results in figure C.VIII.60 results in: $T_S = 125.11°C$, $\alpha' = 0.0408$, $B' = 1.4126$ and $C' = 15.75$. The theoretical curve $\theta_2(T)$ corresponding to this set of parameters is plotted as a solid line in figure C.VIII.60.

N = ∞ case

A very thick film must behave as a bulk sample. In this limit, one can therefore neglect the surface term and assume that the order parameter θ does not depend on i. Minimizing the energy F_B/N with respect to θ naturally leads to an expression identical to C.VIII.147, where the coefficients α, B, C and T_{bulk} replace α', B', C', and T_S. The best fit to the experimental results (solid line in figure C.VIII.60) yields: $T_{bulk} = 95.15°C$, $\alpha = 0.0503$, $B = 0.2876$, $C = 17.55$.

N = 3 case

In both cases, $N = 2$ and $N = \infty$, the set of free energy coefficients (α,B,C) is known, up to a prefactor. Similarly, in the $N = 3$ case where $\theta_1 = \theta_3$ and the free energy writes:

$$F = F_S + F_B + F_C$$

$$F_S = A'(T)\,|\theta_1|^2 + \frac{1}{2}B'(T)\,|\theta_1|^4 + \frac{1}{3}C'\,|\theta_1|^6$$

$$F_B = \frac{1}{2}A(T)\,|\theta_2|^2 + \frac{1}{4}B(T)\,|\theta_2|^4 + \frac{1}{6}C\,|\theta_2|^6$$

$$F_C = g\,(|\theta_1| - |\theta_2|)^2 \tag{C.VIII.149}$$

the set of seven coefficients is only known up to a prefactor. If, on the other hand, we (arbitrarily) set $g = 1$, the values of the (α, α', B, B', C, C') coefficients are fixed. The values we gave previously correspond to this choice. Minimizing F with respect to θ_1 and θ_2 yields the system of equations:

$$[(A'(T) + 1 + B'(T)\theta_1^2 + C'\theta_1^4]\theta_1 - \theta_2 = 0 \tag{C.VIII.150a}$$

$$[(A(T) + 2 + B(T)\theta_2^2 + C\theta_2^4]\theta_2 - 2\theta_1 = 0 \tag{C.VIII.150b}$$

Its numerical solution, with the previously given values for the (α, α', B, B', C, C') coefficients, gives an average angle $<\theta_3(T)> = (2\theta_1 + \theta_2)/3$, in good agreement with the experimental data (solid line in figure C.VIII.60).

The analysis can be applied to a larger number of layers, but the calculations quickly become complicated and do not provide further insight into the problem. We shall therefore consider another example: the SmA → SmB$_{cryst}$ crystallization transition.

VIII.5.c) SmA → SmB$_{cryst}$ transition in 4O.8: mechanical measurements

All the experiments we described so far show that the **"surface" layers tend to be more ordered than those inside the film**. Incidentally, this order excess is not limited to the two surface layers; it propagates in the neighboring layers and is exponentially damped over a distance of the order of the correlation length. It is then natural to wonder whether there are materials and phase transitions for which **the excess of order can be limited to the surface layers**.

The first experiment suggesting this was indeed the case (for a more detailed review of this topic, see Bahr [42] and Huang and Stoebe [43]) was performed by Pindak et al. [44] on the 4O.8 liquid crystal at the SmA → SmB$_{cryst}$ transition. This experiment is a true tour de force, as it detects by purely mechanical means phase transitions, such as crystallization, in very thin films. The schematic of the experimental setup is shown in figure C.VIII.64a. Its most important element is a torsion oscillator, machined from a single piece of a Cu–Be alloy. It consists of a disk of radius $R_1 = 0.59$ cm, fixed at the extremity of a thin torsion rod, itself welded to a heavy holder. Owing to its high quality factor, of the order of 10^3, the resonance frequency of this oscillator (about 523 Hz) can be determined with a relative precision of 10^{-5}. The smectic film is stretched on a circular frame with radius $R_2 = 0.7$ cm, centered on the torsion rod and placed in the same plane as the oscillating disk. During stretching, the oscillator is lowered with respect to the film plane. Once the film is equilibrated, its thickness is determined optically, and then the oscillator is brought back in position. As the film makes contact with the disk, it adheres to it spontaneously, without breaking; an annular film is formed between the edge of the disk and its circular frame (Fig. C.VIII.64b).

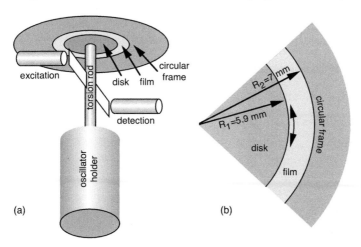

Fig. C.VIII.64 a) Torsion oscillator used to measure the viscoelasticity of smectic films. The excitation and detection of oscillations are based on electrostatics; b) close-up view of the smectic film.

The experiment consists of measuring the changes in the resonance frequency $f = \omega/2\pi$ and the quality factor Q of the oscillator in the presence of the film. These changes are directly related to the complex shear modulus of the film

$$\mu = \mu'(\omega) + i\omega\mu''(\omega) \tag{C.VIII.151}$$

where the real part μ' describes the elasticity of the film and the imaginary part μ'', the viscous response (dissipation).

To quantify this result, let us first calculate the torque exerted by the film on the disk of radius R_1. This torque is related to the two-dimensional stress tensor

$$\sigma_{ij} = 2\mu\,\varepsilon_{ij} \qquad (i,j = 1,2) \tag{C.VIII.152}$$

itself proportional to μ and to the two-dimensional strain ε_{ij}, associated with the displacement field **u** in the plane of the film:

$$\varepsilon_{ij} = \frac{1}{2}\left(\frac{\partial u_i}{\partial x_j} + \frac{\partial u_j}{\partial x_i}\right) \tag{C.VIII.153}$$

In polar coordinates (r,ψ), only the orthoradial component $\sigma_{\psi r}$ of the stress tensor contributes to the torque exerted on the disk, given by:

$$\Gamma = 2\pi R_1^2\,\sigma_{\psi r} = 2\pi R_1^2\,\mu\left(\frac{\partial u_\psi}{\partial r} - \frac{u_\psi}{r}\right) \tag{C.VIII.154}$$

To calculate $\sigma_{\psi r}$, let us neglect film inertia and write that $\mathrm{div}\underline{\sigma} = 0$. This yields the Navier equation obeyed by the orthoradial component $u_\psi(r)$ of the displacement field:

$$\mu\left(\frac{\partial^2 u_\psi}{\partial r^2} + \frac{1}{r}\frac{\partial u_\psi}{\partial r} - \frac{u_\psi}{r^2}\right) = 0 \tag{C.VIII.155}$$

supplemented by the boundary conditions:

$$u_\psi(R_1) = \theta(t)\,R_1 \qquad \text{and} \qquad u_\psi(R_2) = 0 \tag{C.VIII.156}$$

This equation admits as solution:

$$u_\psi(r) = ar + \frac{b}{r} \tag{C.VIII.157a}$$

with

$$a = -\theta(t)\,\frac{R_1^2}{R_2^2 - R_1^2} \qquad \text{and} \qquad b = \theta(t)\,\frac{R_1^2 R_2^2}{R_2^2 - R_1^2} \tag{C.VIII.157b}$$

This gives, using eq. C.VIII.154, the torque Γ exerted by the film on the oscillating disk:

$$\Gamma = -4\pi\mu \, \frac{R_1^2 R_2^2}{R_2^2 - R_1^2} \, \theta(t) = -K_F \theta(t) - K_F' \frac{d\theta}{dt} \qquad \text{(C.VIII.158a)}$$

with

$$K_F = 4\pi\mu' \, \frac{R_1^2 R_2^2}{R_2^2 - R_1^2} \qquad \text{and} \qquad K_F' = 4\pi\mu'' \, \frac{R_1^2 R_2^2}{R_2^2 - R_1^2} \qquad \text{(C.VIII.158b)}$$

Finally, the equation of motion for the oscillator in the presence of the film writes:

$$I_o \frac{d^2\theta}{dt^2} = -(K_o + K_F)\,\theta - K_F' \frac{d\theta}{dt} \qquad \text{(C.VIII.159)}$$

where I_o is the moment of inertia of the oscillator and K_o its torsion constant. Hence, the resonance frequency is (assuming $K_o \gg K_F$ and $K_F' \ll (K_o I_o)^{1/2}$, which can be checked a posteriori):

$$\omega \approx \sqrt{\frac{K_o + K_F}{I_o}} \approx \omega_o \left(1 + \frac{1}{2}\frac{K_F}{K_o}\right) \qquad \text{(C.VIII.160)}$$

with $\omega_o = (K_o/I_o)^{1/2}$. In the presence of the film, ω is shifted by

$$\frac{\Delta\omega}{\omega_o} = 2\pi \, \frac{\mu'}{K_o} \, \frac{R_1^2 R_2^2}{R_2^2 - R_1^2} \qquad \text{(C.VIII.161)}$$

with respect to the resonance frequency ω_o of the free oscillator. This relation gives access to the elastic shear modulus of the film:

$$\mu' = \frac{K_o}{2\pi} \, \frac{R_2^2 - R_1^2}{R_1^2 R_2^2} \, \frac{\Delta\omega}{\omega_o} \qquad \text{(C.VIII.162)}$$

Similarly, calculating the quality factor of the oscillator provides information on film viscosity, but this information is not so easily interpreted.

By changing the moment of inertia of the oscillator, Pindak et al. [44] measured the torsion constant of their oscillator, finding $K_o = 4.9 \times 10^6$ erg, which finally yields:

$$\mu'(\text{erg/ cm}^3) = 6.5 \times 10^5 \, \frac{\Delta\omega}{\omega_o} \qquad \text{(C.VIII.163)}$$

Let us now return to the experiments performed on 4O.8 (its chemical formula is given in figure A.III.9). In the bulk, this material exhibits a SmA $\rightarrow$ SmB$_{\text{cryst}}$ transition at about 49°C; it is therefore a prime candidate for stretching films in the smectic A phase. The experimental values of the resonance frequency and the quality factor of the oscillator as a function of temperature, for a 24-layer film, are shown in figure C.VIII.65. Their analysis reveals the presence of two events at decreasing temperature:

1. The first event, very clear-cut, occurs close to 49°C. It corresponds to the expected SmA → SmB$_{cryst}$ transition, and consists in a "strong" increase in the resonance frequency of the oscillator when the film crystallizes in the SmB$_{cryst}$ phase. Note that the relative shift $\Delta f/f_o$ is of the order of 1.6% a few degrees below the transition, amounting, from eq. C.VIII.163, to a two-dimensional shear modulus of the order of 10^4 erg/cm^2.

2. The second event occurs toward 56°C: it is also accompanied by an increase in the resonance frequency, but the effect is roughly 10 times weaker than below 49°C.

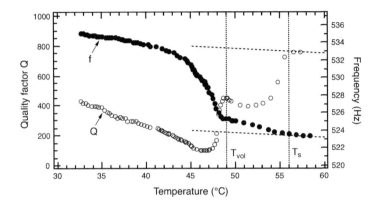

Fig. C.VIII.65 Quality factor and resonance frequency of the torsion oscillator in the presence of a 4O.8 film 24-layers thick. The dotted lines are extrapolations of frequency and quality factor measurements performed on the SmA phase (from ref. [44]).

Similar experiments performed on films of different thicknesses show that this small increase in the resonance frequency subsists in all films between 49°C and 56°C and that its relative amplitude $\Delta f/f_o$ does not depend on film thickness. On the other hand, the strong increase in the amplitude of the resonance frequency observed below 49°C is proportional to film thickness.

To explain these experimental results, Pindak et al. [44] suggested that the SmA → SmB$_{cryst}$ transition occurs in two steps: first in the two surface layers, between 49 and 56°C, then in the rest of the film below 49°C, which is the usual crystallization temperature.

Note that these experiments provide an estimate for the bulk shear modulus μ_{3D} of the SmB$_{cryst}$ phase in 4O.8, since below 49°C one must have $\mu_{3D} = \mu_{2D}/H$, where H is the film thickness. At 40°C, for instance, one obtains for a 24-layer film $\mu_{2D} \approx 10^4$ erg/cm^2 which, knowing that each layer is 28.6 Å thick, results in: $\mu_{3D} \approx 1.5 \times 10^9$ erg/cm^3. This is a typical value for the SmB$_{cryst}$ plastic crystal.

The sequence of phase transitions in 4O.8 films was recently refined by Chao et al. [45] who employed three techniques simultaneously, namely, optical reflectivity (see section C.VIII.1), electron diffraction, and alternative calorimetry (described in the next subsection). The scenario proposed by Chao et al. is presented in figure C.VIII.66. We can see that, besides the transitions detected mechanically by Pindak at 56°C and 49°C, three additional transitions exist. For instance, electron diffraction shows that, at 62°C, the outer layers of the film (those with indices i = 1 and N) go from the SmA phase to the hexatic SmB$_{hex}$ phase. A similar transition occurs at 51°C in the neighboring layers, of indices i = 2 and N – 1. It should be noted that these two transitions do not change the period of the oscillator since, as we have already shown in section C.VI.4, the hexatic phase does not exhibit "solid"-type elasticity ($\mu' = 0$). The SmA → SmB$_{hex}$ transition, on the other hand, is "visible" by electron diffraction, as well as by calorimetry [45], as we shall see in the following subsection.

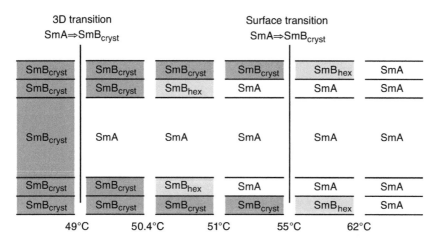

Fig. C.VIII.66 Sequence of the phase transitions in 4O.8 films according to Chao et al. [45]. Two of these transitions, at 55°C and 49°C, were discovered using mechanical measurements by Pindak et al. [44].

We would like to conclude by pointing out that the SmB$_{hex}$ phase that appears in the surface layers of the film is not present in the phase sequence of the three-dimensional material.

VIII.5.d) AC calorimetry and the "SmA-SmB$_{hex}$" transition

Historically, the first experimental demonstration of the transition "SmA → SmB$_{hex}$" was not in 4O.8 films, but in films composed of other materials of the mnOBC series. The technique used was calorimetry, and we shall start by presenting its general principle.

Differential scanning calorimetry is a method widely used in chemistry to detect phase transitions in bulk samples of thermotropic mesogenic substances. This method consists of measuring, as a function of time t, the heat flow between the sample under study and a reference sample, both of them placed in a chamber and subjected to a temperature ramp T(t) (upward or downward) at a controlled rate. From the record of the heat flow which has to be supplied to or withdrawn from the sample to maintain a zero temperature difference between the sample and the reference, one can determine the latent heat associated with first-order phase transitions as well as information on the specific heat of the sample.

AC calorimetry operates on a different principle, and is used for measuring the specific heat. In this case, the sample, held inside a temperature-regulated chamber, is in contact with a heat source (for instance, a metallic film traversed by an AC current), periodically modulated at a frequency ω. The AC component of the heat $Q(\omega)$ produced by the source induces a periodical modulation in sample temperature $\Delta T(\omega)$, which is detected very precisely using thermocouples and a lock-in amplifier. Thus, one can measure the specific heat C_p of the sample, which is inversely proportional to the temperature modulation amplitude.

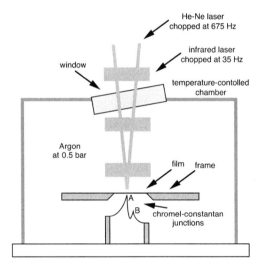

Fig. C.VIII.67 Diagram of the setup employed for measuring the specific heat of smectic films (from ref. [47]).

Applying these calorimetric methods to the study of phase transitions in smectic films faces a major challenge, especially in the most interesting case, that of thin films: the mass of the samples. Indeed, a 4-layer film with a surface of 1 cm² has a mass of about 10^{-6} g, which is at least three orders of magnitude lower than the typical mass of the samples used in commercial calorimeters. In these conditions, it is easy to understand the interest caused by the publication of the first paper by Geer et al. [46] reporting on measurements of the specific heat anomaly in a 65OBC film of only four layers. The setup used [47] is represented schematically in figure C.VIII.67. The crucial part is a differential thermocouple composed of two junctions A and B between chromel and constantan wires with a diameter of 13 μm. Junction A is placed 30 μm below the film, while the other one, B, is 5 mm lower. The whole system is held in a temperature-controlled chamber filled with argon at 0.5 bar. The experiment is performed by shining on the film an infrared laser beam ($\lambda = 3.4$ μm, $P = 7$ mW), chopped at 35 Hz. Part of this radiation (proportional to film thickness) is absorbed by the smectic film, heating it. As a result, the temperature of the film is modulated at 35 Hz, with an amplitude inversely proportional to the specific heat of the film. Argon being a good heat conductor, junction A – very close to the film – is sensitive to the temperature modulation, while junction B, much farther away, is not. Finally, the two junctions deliver a 35-Hz AC voltage (correlated with the chopped laser beam), which is then demodulated by a lock-in amplifier. As for the film thickness N, it is measured using a He-Ne laser.

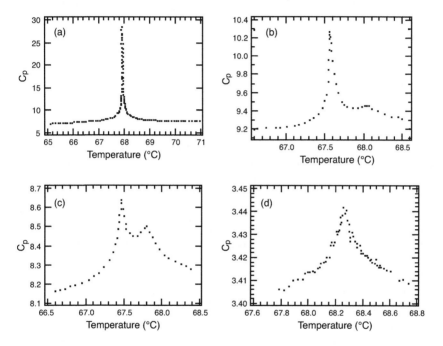

Fig. C.VIII.68 Specific heat C_p as a function of temperature in 65OBC films. In a 200-layer film (a) only one peak is seen at the SmA → SmB$_{hex}$ transition; in 10-layer (b) and 6-layer (c) films, two peaks can clearly be distinguished, while in a 4-layer film (d), only one peak is left [46].

The results obtained using this setup for 65OBC films of different thicknesses are shown in figure C.VIII.68. The first diagram was obtained for a relatively thick film of 200 layers. One can distinctly see a **specific heat anomaly**, attributed by the authors to the transition between the three-dimensional SmA and SmB$_{hex}$ phases. The first essential feature of this anomaly is the lack of thermal hysteresis, the shape and position of the peak being identical when the measurements are performed during heating and cooling. This indicates a continuous transition. The second feature of the measured anomaly is its shape: it is a quasi-symmetric peak, with its shape well represented by a formula of the type:

$$C_p = A_{\pm}|t|^{-\alpha} + B \qquad\qquad\qquad (C.VIII.164)$$

with $t = (T - T_c)/T_c$ the reduced temperature. The best fit to the experimental points gives the critical temperature $T_c = 67.930 \pm 0.003°C$, a critical exponent $\alpha = 0.59 \pm 0.03$, and a ratio $A_+/A_- = 0.99 \pm 0.08$. These values are very close to results obtained previously [48] by a different method, using three-dimensional 65OBC samples. One can therefore conclude that a free film of 200 layers behaves at the SmA-SmB$_{hex}$ transition like a bulk sample. Note (see chapter C.VI) that the shape of the specific heat anomaly does not agree with the Landau model for second-order transitions (mean-field model), which predicts a specific heat discontinuity at T_c (see, for instance, the SmA-SmC transition in 4O.7 [49]). The absence of such a discontinuity in the specific heat (which is, by definition, the second derivative of the free energy with respect to temperature) means that the transition is of higher than second order, in the sense of Ehrenfest. This result is in agreement with theory, which predicts that this defect-induced phase transition is of infinite order.

In the two following diagrams (Fig. C.VIII.68.b and c), obtained with thinner films (ten and six layers) one can see that a second peak appears at a temperature above T_c, its relative amplitude increasing as the film thickness decreases. According to the authors, this second peak corresponds to a liquid → hexatic phase transition in the surface layers, which precedes (on cooling) the liquid → hexatic transition in the inner layers of the film. This surface transition should also occur in the 200-layer film, but the associated specific heat anomaly is drowned in the signal of the bulk transition in diagram C.VIII.68a.

Finally, in the 4-layer 65OBC film, all layers undergo the liquid → hexatic transition at the same time so that only one peak appears in the corresponding diagram (Fig. C.VIII.68d).

In conclusion, in this first article Geer et al. [46] showed using calorimetric methods that the SmA → SmB$_{hex}$ transition in 65OBC films of more than four layers occurs in two steps: first in the surface layers and then in the inner film.

Performing similar measurements on films of 75OBC, another compound in the homologous mnOBC series, Huang and Stoebe [51] discovered a more complicated behavior, illustrated by the diagram in figure C.VIII.69, recorded for a 10-layer film. Let us start by analyzing the record obtained on cooling. Instead of two, this diagram exhibits four different transitions. The first three are similar to the anomalies observed in the 65OBC films and, according to the authors, correspond to a SmA → SmB$_{hex}$ transition occurring in three steps. Thus, the first small peak at 71°C would be due to the SmA → SmB$_{hex}$ transition in the two outermost layers of the film ($i = 1$ and N). The second small peak at 65.2°C would signal the transition of the adjacent layers ($i = 2$ and $N - 1$), while the third large peak would correspond to the bulk SmA → SmB$_{hex}$ transition. The fourth transition is different in nature, being characterized by a **specific heat discontinuity** at 64.5°C. The authors think that it corresponds to the SmB$_{hex}$ → SmE$_{cryst}$ transition in the outermost layers of the film.

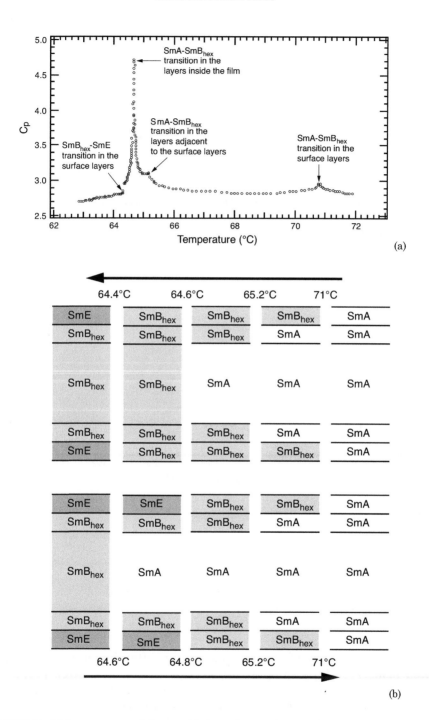

Fig. C.VIII.69 a) Measurements of the specific heat C_p as a function of temperature in a 75OBC film of ten layers. This diagram was obtained on cooling (from ref. [50]); b) sequence of phase transitions in 75OBC films, according to [50]. Note that the sequence is not the same on cooling and on heating.

It is noteworthy that, unlike the three other transitions, this surface crystallization of the film is a first-order transition, exhibiting *hysteresis*, as we can clearly see in the records of figure C.VIII.70. Indeed, the $C_p(T)$ curves measured on cooling and heating the films are not superposed. More precisely, the temperature of surface crystallization ($SmB_{hex} \rightarrow SmE_{cryst}$ transition) is shifted by about 0.3°C with respect to the surface melting temperature ($SmE_{cryst} \rightarrow SmB_{hex}$ transition). On the other hand, the $SmB_{hex} \rightarrow SmA$ transition inside the film occurs at the same temperature in both directions. The specific heat anomaly characterizing this transition is however much less intense in the presence of crystallized surface layers (on heating) than in their absence (cooling). This result shows that localized crystallization or melting of the two surface layers of a film can change the state of the inner layers of the film. Consequently, the diagrams in figures C.VIII.66 or C.VIII.69 should only be seen as qualitative results.

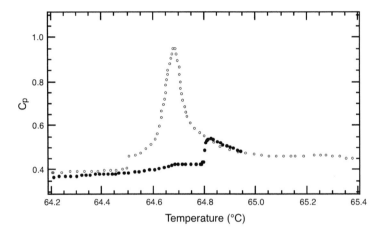

Fig. C.VIII.70 Measurements of the specific heat C_p as a function of temperature in a 75OBC film of 13 layers. The records, taken on cooling (open dots) and on heating (solid dots) the films, show strong hysteresis (from ref. [50]).

VIII.5.e) Polymorphism of 7O.7 films as revealed by X-ray diffraction

For the structural study of smectic phases, it is often imperative to use well-oriented monocrystalline samples. Bulk samples are most often prepared in capillary tubes, and then oriented under strong magnetic field. It is however obvious that the orientation of smectic layers in these samples is far from perfect, due to the geometrical imperfections of capillaries and the (often very large) viscosity of the phases under study. Using free films considerably increases the quality of the samples, as we shall show.

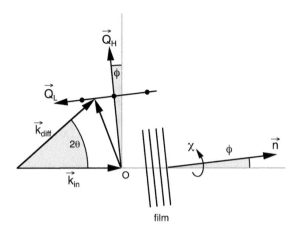

Fig. C.VIII.71 Typical X-ray diffraction geometry.

To begin with, we recall the typical geometry of a diffraction experiment (Fig. C.VIII.71), with the smectic layers perpendicular to the diffraction plane ($\mathbf{k}_{in}$, $\mathbf{k}_{diff}$), their normal $\mathbf{n}$ making a variable angle ϕ with the incident beam $\mathbf{k}_{in}$. The sample can also be turned around the normal $\mathbf{n}$ so that, by changing the rotation angle χ, an array of nodes in the reciprocal lattice can be brought into the diffraction plane. Scans along the $\mathbf{Q}_L$ axis are performed by simultaneously changing angles ϕ and 2θ according to the formula:

$$\frac{4\pi}{\lambda}\sin\theta\cos(\theta - \phi) = \frac{4\pi}{\lambda}\sin\theta_{(100)} \qquad\qquad (C.VIII.165)$$

with $\theta_{(100)}$ the Bragg angle of the (100) reflection.

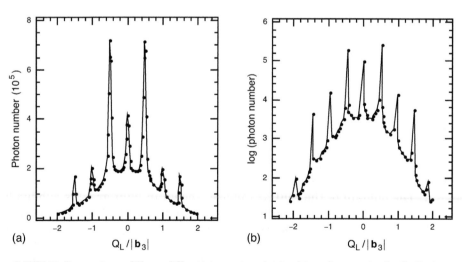

Fig. C.VIII.72 Comparison of X-ray diffraction spectra obtained in a Q_L scan in the SmB phase. a) Bulk 4O.8 sample (from ref. [52]); b) free 7O.7 film (from ref. [53]). Note that the y axis is in logarithmic scale in graph (b). $|\mathbf{b}_3| = 2\pi/a_0$, where a_0 is the thickness of a smectic layer.

Figure C.VIII.72 shows two Q_L-scans, the first one obtained using a bulk sample (Fig. C.VIII.72a), and the second using a free film (Fig. C.VIII.72b). In both diagrams, a series of sharp Bragg peaks can be distinguished, sitting on the background thermal noise. The height of the Bragg peaks with respect to the intensity of the thermal signal provides an estimate for the quality of the samples. For the bulk sample, this ratio is of about 3.5 for the highest peaks. The same peaks, obtained using a free smectic film, give a signal/noise ratio about 10 times larger (note that the y axis in figure C.VIII.72b is in logarithmic scale).

Owing to the exceptional quality of the spectra obtained with 7O.7 smectic films, Sirota et al. [54] could establish the phase diagram in figure C.VIII.73.

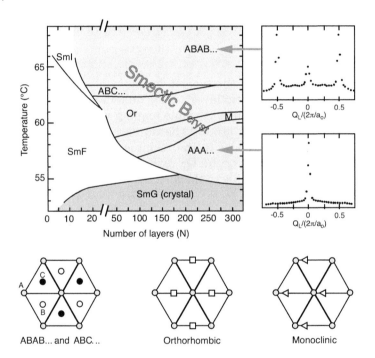

Fig. C.VIII.73 Polymorphism of 7O.7, determined by X-ray diffraction on smectic films (from ref. [54]).

Without going into the details of this remarkable diagram, let us point out two of its more salient features:

1. The domain of the SmB_{cryst} phase (shown in light gray) is subdivided into several domains, corresponding to different ways of stacking layers with a triangular structure. For instance, the ABAB... stacking corresponds to the hexagonal compact packing. In the Q_L spectrum, this type of alternating stacking gives Bragg peaks at $Q_L = \pm 2\pi/(2a_0)$, corresponding in real space to a period $2a_0$, twice the thickness a_0 of a smectic layer. At lower temperature, one observes successively the ABCABC..., orthorhombic (Or) and monoclinic (M)

stackings, and finally the simple AAA... stacking, of period a_o. In this latter case, the Bragg peaks at $Q_L = \pm 2\pi/(2a_o)$ are absent from the $\mathbf{Q}_L$ spectra.

2. In films of thickness N < 190, the hexatic phases SmI and SmF are intercalated between the SmB_{cryst} and SmG phases, **although they are absent in the 3D phase sequence**.

VIII.5.f) Can we hear the structure of the smectic films? Film vibrations and polymorphism of 7O.7 films

Is X-ray diffraction absolutely necessary to detect the multitude of phase transitions in 7O.7 films? According to Collett et al. [53], no latent heat or specific heat anomaly would be associated with stacking transitions, which also appear to be invisible in reflection or polarizing microscopy. In this context, it might seem far-fetched to pretend that one can simply "hear" them. This is nevertheless what we intend to show.

Miyano [55] was the first to propose the idea of exciting the vibration of smectic films to measure their tension and to detect their phase transitions. This idea was then used by Kraus et al. [56].

In the following, we shall distinguish the case of planar films from the more complex one of curved films.

i) Vibrations of planar films

For his experiment, Miyano used **planar** 4O.8 films, vibrating under vacuum (Fig. C.VIII.74). The vibrations were excited by a pointed electrode placed very close to the film and fed a voltage containing a direct component U_{DC} and an alternating one U_{AC} at a frequency ω:

$$U_{exc} = U_{DC} + U_{AC} e^{i\omega t} \tag{C.VIII.166}$$

The role of the direct voltage is to accumulate charges (ions) in the film close to the electrode. The total charge Q_{DC} is proportional to the direct component of the applied voltage and to the capacitance C of the capacitor formed by the film and the electrode: $Q_{DC} = CU_{DC}$. This capacitor charges very slowly (seconds to minutes), due to the weak conductivity of the liquid crystal. The charge Q_{DC} feels the electric field E_{AC}, proportional to the alternating component of the voltage, resulting in a localized force

$$f = Cst\ U_{DC}\ U_{AC} e^{i\omega t} \tag{C.VIII.167}$$

that acts on the film and excites its vibrations. These are detected using a laser beam, reflected by the film onto a position-sensitive photodiode. The experiment depends upon detecting the frequencies of the eigenmodes and in following their variation as a function of temperature or film thickness.

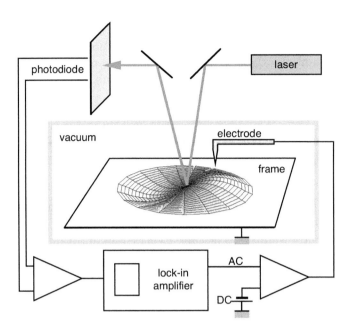

Fig. C.VIII.74 Setup employed by Miyano [55] to study the vibration of smectic films under vacuum.

Under vacuum, and assuming that dissipation is negligible, the equation of motion of the transverse displacement ζ of the film writes:

$$\rho_f \frac{\partial^2 \zeta}{\partial t^2} = \tau \Delta \zeta + \frac{EH^3}{12(1-\sigma^2)} \Delta^2 \zeta \quad \text{with} \quad \Delta^2 = \frac{\partial^4}{\partial x^4} + 2\frac{\partial^4}{\partial x^2 \partial y^2} + \frac{\partial^4}{\partial y^4} \quad \text{(C.VIII.168)}$$

where ρ_f is the mass per unit film surface and τ its tension. If the film is crystallized, an additional elastic force is added to the inertial force and to the restoring force due to film tension. The second term on the right-hand side of eq. C.VIII.168, corresponds to this additional force. It is calculated here within the approximation of isotropic elasticity and under the assumption that *the film is flat* (E is the Young modulus of the crystal and σ its Poisson coefficient). Before estimating the relative importance of the elastic term with respect to the tension term, we point out that the film is connected to the frame by a meniscus of negligible size, where the displacement ζ must vanish. The wave function ζ must therefore obey a boundary condition of the Dirichlet type:

$$\zeta(\text{contour}) = 0 \quad\quad\quad \text{(C.VIII.169)}$$

For a rectangular frame of size $L_x \times L_y$, the wave function, the solution of eq. C.VIII.168, which fulfills this boundary condition, has the following simple form:

$$\zeta(x,y) = \zeta_0 e^{i\omega t} \sin(k_x x) \sin(k_y y) \quad\quad\quad \text{(C.VIII.170)}$$

where

$$k_x = m \frac{\pi}{L_x} \qquad k_y = n \frac{\pi}{L_y} \qquad m, n = 1, 2, 3, \ldots \qquad \text{(C.VIII.171)}$$

In this case, the ratio between the restoring force due to bending elasticity and that due to film tension is given by

$$\frac{f^{bend}}{f^{cap}} = \frac{\dfrac{EH^3}{12(1-\sigma^2)} \Delta^2 \zeta}{\tau \Delta \zeta} \approx \frac{EH^3}{\tau L^2}(m^2 + n^2) \qquad \text{(C.VIII.172)}$$

With $E = 10^9$ erg/cm^3, $\tau = 50$ erg/cm^2, $H = 4 \times 10^{-6}$ cm ($N = 20$) and $L = 1$ cm, we find a ratio of the order of $10^{-9} \times (m^2 + n^2)$. It immediately ensues that, for *flat films* composed of several tens (or several hundreds) of layers, one can completely neglect solid-type elasticity, at least in the SmB$_{cryst}$ phase, so that the frequency of the eigenmodes only depends on film tension:

$$\omega_{mn} = \pi \sqrt{\frac{\tau}{\rho_f}} \sqrt{\frac{m^2}{L_x^2} + \frac{n^2}{L_y^2}} \qquad \text{(C.VIII.173)}$$

If the frame is circular, of radius R, the wave functions calculated in cylindrical coordinates have a more complex form:

$$\zeta_{mn}(r, \psi) = \zeta_0 e^{i\omega t} J_m(k_{mn} r) \sin(m\psi) \qquad \text{(C.VIII.174)}$$

with

$$k_{mn} R = \mu_{mn} \qquad \text{(C.VIII.175)}$$

where $m = 0, 1, 2 \ldots$ and $n = 1, 2, 3 \ldots$ are the indices of roots μ_{mn} of the Bessel functions of order m ($J_m(\mu_{mn}) = 0$).

If the ratio

$$\frac{EH^3}{\tau R^2} \mu_{mn}^2 << 1 \qquad \text{(C.VIII.176)}$$

the elasticity can be neglected and the frequencies of the eigenmodes are simply given by the formula:

$$\omega_{mn} = \frac{1}{R} \sqrt{\frac{\tau}{\rho_f}} \mu_{mn} \qquad \text{(C.VIII.177)}$$

Thus, it is not surprising that measuring the resonance frequency of the eigenmodes, as Miyano did (see Fig. C.VIII.75) for 4O.8 films, did not reveal any anomaly at the SmB$_{cryst}$ → SmA transition (49.5°C). However, these results show that the tension of 4O.8 films does not change until 55°C (i.e., as long as the film surface remains crystallized) (see Fig. C.VIII.65), then increases with temperature in the existence range of the SmA phase.

This observation shows that the excess of entropy per unit film surface

$$S_s = -\frac{\partial \tau}{\partial T} \qquad \text{(C.VIII.178)}$$

is negative and, as a consequence, that 40.8 films are better ordered than an equal thickness slab of the bulk SmA phase, even when their surface layers are liquid.

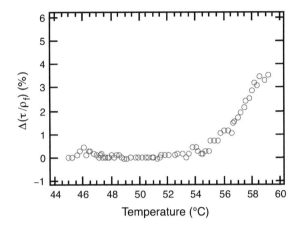

Fig. C.VIII.75 Temperature variation of the tension of 40.8 films, determined by measuring the frequency of the eigenmodes (from ref. [55]).

In conclusion, we emphasize that our demonstration of the negligible role of elasticity on the vibration spectrum of 40.8 films in the SmB phase relies heavily upon the *strong* assumption that the films are flat at equilibrium. Indeed, the expression of the restoring elastic force

$$f^{bend} = \frac{EH^3}{12(1-\sigma^2)} \Delta^2 \zeta \qquad \text{(C.VIII.179)}$$

that we employed is only valid if the films deviate from a perfectly plane shape by less than H. Obviously, for smectic films with surface areas of a cm², and a fraction of a micrometer thick, such flatness is almost impossible to achieve in practice. Thus, we must analyze the case of curved films.

ii) Vibrations of curved films

What happens when the vibrating film is not flat when at rest? Let us take the example of a film stretched on an octagonal non-planar frame, obtained by shifting vertically the two pairs of opposite sides of a flat L×L frame by an amount $L_z = \alpha L$. At equilibrium, the shape of the film stretched on this frame is described by a saddle-like minimal surface known as the "T" surface.

We introduce the orthogonal curvilinear coordinates (ξ_1, ξ_2, ξ_3), the first two parameterizing the surface. We assume that, as it vibrates in the fundamental mode, the points of the surface incur a displacement $\zeta(\xi_1, \xi_2)$ along the ξ_3 direction, perpendicular to the surface. In particular, when the film moves upward in the center, as shown in figure C.VIII.76a, it must also undergo compression along ξ_1 and elongation along ξ_2. According to the theory of elastic shells, this displacement engenders at a point of curvilinear coordinates (ξ_1, ξ_2) a deformation ε_{ij} given by:

$$\varepsilon_{ij} = \zeta \begin{bmatrix} 1/R_1 & 0 \\ 0 & 1/R_2 \end{bmatrix} = \frac{\zeta}{R_o} \begin{bmatrix} 1 & 0 \\ 0 & -1 \end{bmatrix} \qquad \text{(C.VIII.180)}$$

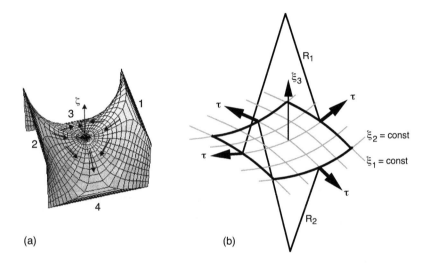

(a)   (b)

Fig. C.VIII.76 Vibration of a curved smectic film. a) Minimal "T" surface; b) definition of the orthogonal curvilinear coordinates.

where $1/R_1 = 1/R_o$ and $1/R_2 = -1/R_o$ are the principal curvatures of the minimal surface at this point. This deformation tensor is of zero trace, showing that the surface area is conserved. However, this deformation generates a shear stress in the film, of amplitude:

$$\sigma_{ij} = 2\mu H \frac{\zeta}{R_o} \begin{bmatrix} 1 & 0 \\ 0 & -1 \end{bmatrix} \qquad i,j = 1,2 \qquad \text{(C.VIII.181)}$$

where μ is the elastic shear modulus of the three-dimensional crystal in the plane of the layers and H the film thickness. In turn, this stress gives a restoring force perpendicular to the film:

$$f_3^{sh2D} = \frac{\sigma_{11}}{R_1} + \frac{\sigma_{22}}{R_2} \approx \frac{4\mu H}{R_0^2} \zeta \qquad \text{(C.VIII.182)}$$

which must be added to the elastic bend force already introduced in the equation of motion of the flat film:

$$f_3^{\,bend} \approx \frac{EH^3}{L^4}\,\zeta \qquad\qquad\qquad\qquad\qquad\qquad\text{(C.VIII.183)}$$

The ratio of these two forces is approximately:

$$\frac{f^{\,sh2D}}{f^{\,bend}} \approx \frac{L^4}{R_o^2 H^2} \qquad\qquad\qquad\qquad\qquad\text{(C.VIII.184)}$$

With $L = 1$ cm, $R_o = 20$ cm and $H = 4\times10^{-6}$ cm, this ratio is of the order of 10^8. This huge factor foretells a significant elastic contribution to the frequency of the eigenmodes, since the ratio between the elastic force and the capillary force now becomes:

$$\frac{f^{\,sh2D}}{f^{\,cap}} \approx \frac{\mu H L^2}{\tau R_o^2} \approx \frac{1}{5} \qquad\qquad\qquad\qquad\text{(C.VIII.185)}$$

with $\mu = 10^9$ erg/cm^3, $\tau = 50$ erg/cm^2 and the previously given values for L, R_o and H.

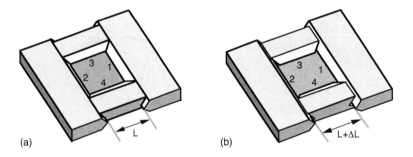

Fig. C.VIII.77 Curved films, shaped as the minimal "T" surface, are stretched on frame "b" obtained by increasing the span of rails 1 and 2 of the flat frame "a" by an amount ΔL. In this way, sides 3 and 4 are below sides 1 and 2. See Fig. C.VIII.76a.

After these theoretical considerations, it is now time to return to the experiments. To obtain curved smectic films of 7O.7, Kraus et al. [56] used a variant of the "classical" rectangular frame, composed of two mobile parts fitted on rails (Fig. C.VIII.77a). The modification consists of slightly increasing the span of the two rails 1 and 2, such that the two mobile pieces 3 and 4 end up being lower than the rails (Fig. C.VIII.77b). The shape of the SmA films stretched on this deformed frame is close to that of the minimal "T" surface (Fig. C.VIII.76a). This frame is placed in the temperature-controlled and airtight chamber of the experimental setup described in figure C.VIII.78. Vibrations are excited electrostatically and detected optically. The lock-in amplifier I allows a precise measurement of the amplitude and phase of eigenmodes. Since the tension of films in the SmA phase depends very little on thickness and temperature, their thickness is obtained by measuring, under vacuum, the frequency of the eigenmode (1,2).

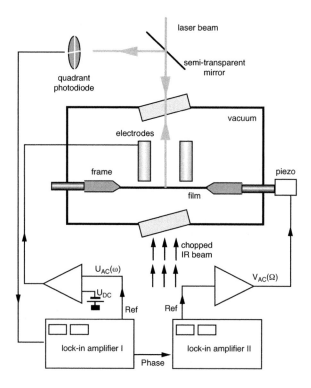

Fig. C.VIII.78 Setup used to study the vibrations of curved smectic films.

As shown by the diagrams in figure C.VIII.79, this frequency drops from 5700 Hz for a film of thickness H = 59 nm to 1 250 Hz for H = 1 240 nm. Hence, the method allows a precise determination of the thickness.

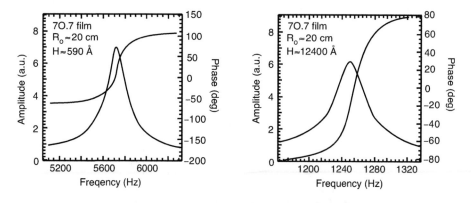

Fig. C.VIII.79 The frequency of the eigenmode (1,2) is measured under primary vacuum. Proportional to $H^{-1/2}$, it gives access to the thickness of 7O.7 films. The slope of the curve representing the phase as a function of frequency is proportional to the quality factor of the resonance and inversely proportional to the dissipation (from ref. [56]).

Afterward, in order to observe the phase transitions in the film, the vacuum inside the chamber is broken to facilitate the thermal exchange with the oven. The resonance frequency of the (1,2) mode is continuously monitored during a temperature sweep. Note that the presence of air considerably decreases the resonance frequency of the (1,2) mode, which is typically 800 Hz in the SmA phase, irrespective of film thickness. In this case, the dominant contribution is that of the mass of air set in motion by the vibrating film. During the temperature sweep, the resonance frequency is measured by slaving the excitation frequency to the phase, in order to remain on the resonance peak. The diagrams in figure C.VIII.80 show that in curved films – and in agreement with theory –, a jump in the resonance frequency of the (1,2) eigenmode can be seen at the SmC → SmB$_{cryst}$ transition (63°C). In this case, the restoring force f$_3^{sh2D}$ adds to the capillary force f$_3^{cap}$ and increases the frequency of the eigenmode. This frequency jump must depend on the thickness H of the film, since the force f$_3^{sh2D}$ is proportional to H. This effect is indeed observed in experiments (Fig. C.VIII.81).

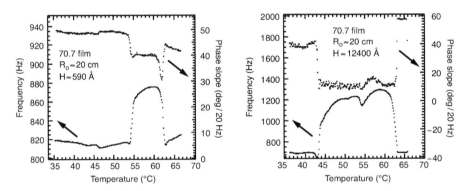

Fig. C.VIII.80 Variation of the parameters of the (1,2) eigenmode of curved 7O.7 films as a function of temperature. Cf. the similar diagram obtained with flat 4O.8 films (Fig. C.VIII.75) (from ref. [56]).

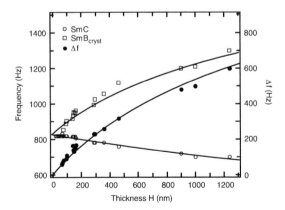

Fig. C.VIII.81 Variation of the resonance frequency of eigenmode (1,2) at the SmC → SmB$_{cryst}$ transition (from ref. [56]).

The two diagrams in figure C.VIII.80 also show that, at the SmC → SmB$_{cryst}$ transition the slope dφ/df, measuring the quality of mode (1,2), drops markedly. Even more important − in the context of smectic film physics − is that the width of the interval, which is influenced by the elasticity and the plasticity of the SmB phase, depends on film thickness. This interval, of about 20°C in thick films (H = 1 240 nm), drops to 7.5°C in a film of thickness H = 59 nm and disappears in films thinner than 30 nm; hence, the SmB$_{cryst}$ phase disappears for H < 30 nm. This result is in agreement with the phase diagram established by X-ray diffraction and given in figure C.VIII.73. This agreement between the results goes even further: A more careful analysis of the diagram of the thick film in figure C.VIII.80 reveals, inside the existence range of the SmB phase, some anomalies that seem to correspond to the stacking transitions. In the same way, the anomaly observed at 45°C in the diagram of the thin film (H = 59 nm) can be attributed to the SmF → SmG transition. The diagram in figure C.VIII.82 presents a map of all detected "acoustic" anomalies, as a function of thickness and temperature.

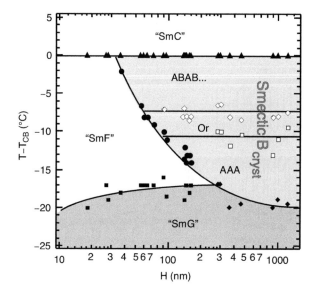

Fig. C.VIII.82 Diagram of acoustic anomalies in 7O.7 films. Phase identification relies on the comparison with the phase diagram obtained by X-ray diffraction (Fig. C.VIII.73) (from ref. [56]).

Clearly, the topology of this diagram is very close to that of the diagram obtained by X-ray diffraction, allowing identification of the domains corresponding to the different phases of films. The phase names "SmC," "SmB," and "SmG" are proposed by extrapolation from the thick film limit. These names are, however, given in inverted commas, as we know for sure that the structure inside a film is not always uniform. Beyond answering questions directly related to the physics of films, acoustics experiments on thick films also provide information on the rheological properties of three-dimensional smectic phases. For instance, in the diagram of the thick film in figure C.VIII.80, one can see that the frequency of eigenmode (1,2) reverts in the SmG phase (at T < 43°C) to the same value as in the SmC phase (at T > 63°C). This is surprising since the SmG phase, like the SmB phase, is crystalline in 3D; as such, its elasticity in the plane of the layers should contribute to the increase of the eigenfrequency. The only possible explanation of this paradox is that the shear modulus in the plane of the layers is much weaker in the SmG phase than in the SmB_{cryst} phase. How is this possible? One essential difference between these two phases is that, in the first one, the molecules are perpendicular to the plane of the smectic layers, while in the second one, they are tilted by about 30° with respect to the layer normal. In the SmB phase, any deformation ε_{ij} in the plane of the layers will therefore result in a change of the intermolecular distance or of the angle between nearest-neighbor molecules, requiring an additional energy, and thus resulting in the mechanical stress σ. The SmG phase, on the other hand, can respond to the deformation ε_{ij} by changing the angle of molecular tilt or by rotating the molecules around the layer normal, so that the lattice, as seen along a direction perpendicular to the molecules, remains unchanged. This concept is illustrated in figure C.VIII.83 for the simple case of uniaxial compression.

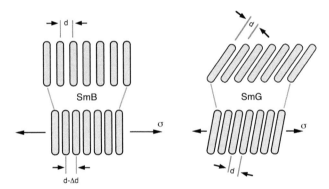

Fig. C.VIII.83 Uniaxial compression in the plane of the smectic layers induces a much larger stress in the SmB_{cryst} phase than in the SmG phase. In the latter case, the molecular tilt changes, so that the intermolecular distance can remain constant.

iii) Thermal expansion of smectic phases

The experimental system in figure C.VIII.78 also includes an infrared beam (generated by a halogen lamp) chopped at a very low frequency Ω, of the order of Hz. Its effect is identical to that of the infrared laser used in the AC calorimetry experiment (Fig. C.VIII.67), since it modulates the temperature of the film according to $T = T_o + \Delta T \sin(\Omega t)$. This periodical film heating induces a frequency modulation of the (1,2) eigenmode of the film, which can be written as:

$$\omega_{1,2} = \omega_o + \Delta\omega \sin(\Omega t + \Phi) \qquad\qquad (C.VIII.186)$$

As the eigenfrequency $\omega_{1,2}$ is shifted by $\Delta\omega(t)$ with respect to ω_o, the phase $\varphi(t) = \varphi_o + \Delta\varphi(t)$, detected by the lock-in amplifier I, is changed by

$$\Delta\varphi(t) = \frac{d\varphi}{d\omega}\Delta\omega(t) \qquad\qquad\qquad\text{(C.VIII.187)}$$

This phase, converted to an analog signal, is fed to the second lock-in amplifier, II, operating at a frequency Ω. In this way, the amplitude $\Delta\omega$ and the phase Φ are measured by lock-in amplifier II. The results of the measurements are shown in figure C.VIII.84. Two main conclusions can be drawn:

1. The first one concerns the phase transitions (including the stacking transitions) and the anomalies they produce in the $\Delta\omega(T)$ curves. These anomalies, clearly visible, were used to draw the diagram in figure C.VIII.82.

2. The second information concerns the thermal dilation coefficients of crystalline phases. It is obtained from the $\Phi(T)$ curves, which show that the phase of the $\Delta\omega(t)$ modulation changes by 180° between the SmB_{cryst} phase and the SmF phase (H = 59 nm) or the SmG phase (H = 1240 nm). Differently stated, heating the film decreases the frequency of the (1,2) mode in the SmB phase while, on the contrary, it increases the frequency of the (1,2) mode in the SmF and SmG phases.

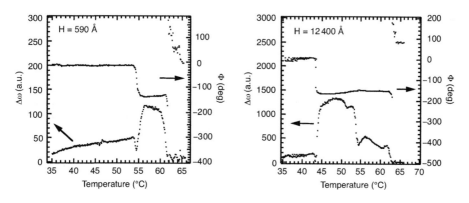

Fig. C.VIII.84 Thermomechanical response of 7O.7 films.

To analyze this phenomenon in more detail, let us assume that, in the SmB phase, the film tends to dilate upon heating in the (ξ_1,ξ_2) plane. As the meniscus is also in the SmB phase, this dilation can only relax by flow in the meniscus. There ensues a compressional mechanical stress σ_{ij}^{thm} (negative, by definition), adding to the film tension and that can be put under the form:

$$\sigma_{ij}^{tot} = \sigma_{ij}^{cap} + \sigma_{ij}^{thm} = (\tau - C\Delta T)\begin{bmatrix} 1 & 0 \\ 0 & 1 \end{bmatrix} \qquad\qquad\text{(C.VIII.188)}$$

where C is a thermo-mechanical coefficient (positive in the SmB_{cryst} phase). Overall, the effective tension of the film decreases, and so does the frequency of the (1,2) eigenmode.

In the SmF or SmG phases the effect is reversed (i.e., the film tends to contract along both directions (ξ_1, ξ_2) on heating). The explanation is simple, having to do with the molecular tilt with respect to the layer normal in these two phases. Indeed, it is well known that the tilt angle of the molecules decreases with increasing temperature. If, on the other hand, their equilibrium distance does not change (or very little) and, if the film does not exchange matter with the meniscus, this results in dilation, since the film has the natural tendency of contracting on heating. This dilation engenders a stress of the same sign as the intrinsic film tension. Finally, the film tension increases and the frequency of the eigenmode increases with it.

In conclusion, the sign of the C coefficient is positive in the SmB_{cryst} phase and negative in the SmF and SmG phases.

VIII.6 Smectic films as model systems

We already know that one can stretch smectic films of uniform thickness on frames of arbitrary shape, place them under vacuum and set in vibration by electrostatic excitation. In the previous section, their vibrations were employed for "hearing their crystalline structure" and for detecting their phase transitions.

Now, we shall create "smectic drums" of well-known smectic phases (A, C or C*), and inquire, paraphrasing Mark Kac [57], whether it is possible to "hear their shape." In particular, we shall see that these films, owing to their exceptional robustness, allow us not only to illustrate this proposition, but also to explore certain – still unsolved – related mathematical problems. The starting point of this study is the solution of the eigenvalue equation

$$\frac{\partial^2 \zeta}{\partial x^2} + \frac{\partial^2 \zeta}{\partial y^2} = -k^2 \zeta \qquad\qquad (C.VIII.189)$$

in the presence of Dirichlet boundary condition on a closed contour:

$$\zeta(\text{contour}) = 0 \qquad\qquad (C.VIII.190)$$

In the context of smectic films, this eigenvalue equation stems directly from the equation of motion of films:

$$\tau \left(\frac{\partial^2 \zeta}{\partial x^2} + \frac{\partial^2 \zeta}{\partial y^2} \right) = \rho H \frac{\partial^2 \zeta}{\partial t^2} \qquad\qquad (C.VIII.191)$$

Indeed, testing it for solutions of the form:

$$\zeta = e^{i\omega t} \, \zeta(x,y) \qquad\qquad (C.VIII.192)$$

gives eq. C.VIII.189, with

$$k^2 = \frac{\rho H}{\tau} \omega^2 \qquad\qquad\qquad (C.VIII.193)$$

To start with, we shall prove the existence of isospectral drums with different shapes.

VIII.6.a) Isospectral planar drums

The first problem we approach is that of the connection between the shape of the frame and the eigenvalue spectrum of k (or of the frequency ω). Indeed, it is clear – by considering, for instance, the very simple case of a rectangular frame – that its eigenfrequencies depend on the shape, namely on the ratio $r = L_x/L_y$ at constant frame area $L_x \times L_y$. This dependence is expressed by equation (see eq. C.VIII.173):

$$f_{mn} = \sqrt{\frac{\tau}{\rho H}} \frac{1}{2 \sqrt{L_x L_y}} \sqrt{m^2 \frac{L_y}{L_x} + n^2 \frac{L_x}{L_y}} \qquad\qquad (C.VIII.194)$$

where m and n are integers characterizing each mode. In figure C.VIII.85, we plotted the reduced frequency for the modes of order (1,1) (1,2) and (2,1) as a function of the aspect ratio r. One can immediately recognize that, in the case of rectangular frames, a univocal relation between the spectrum and the aspect ratio of the frame exists: to put it otherwise, the shape of the frame can be deduced from the spectrum of its eigenfrequencies.

Does this result hold for all possible contour shapes? In particular, can one find frames with different shapes but the same eigenfrequency spectrum?

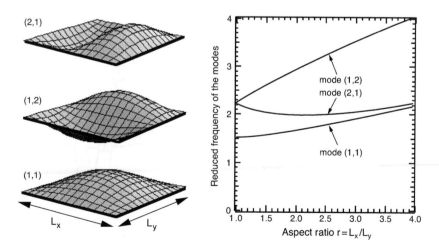

Fig. C.VIII.85 Eigenmode frequencies for a rectangular drum as a function of its shape.

This problem, formulated in 1966 by Mark Kac as the famous question "Can one hear the shape of a drum?" [57], was only solved in the Euclidian two-dimensional space in 1992 by Gordon, Webb and Wolpert [58]. The answer to this question turns out to be **negative**, as these three mathematicians were able to design two contours with different shapes but with an identical eigenvalue spectrum (Fig. C.VIII.86).

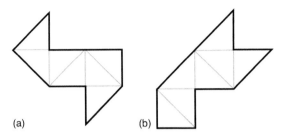

(a)　　　　(b)

Fig. C.VIII.86 The first pair of isospectral flat drums, invented by Gordon, Webb and Wolpert.

The discovery of flat isospectral drums took place in several steps, the penultimate one being the construction by Buser [59, 60] of a pair of isospectral surfaces obtained by assembling seven identical cross-shaped domains (Fig. C.VIII.87). It must be noted that, from a mathematical point of view, the eigenvalue equation C.VIII.189 applies to the Buser surfaces as well as to the flat domains of Gordon, Webb and Wolpert, for a very clear reason: in both cases, the metric is Euclidian, so that the Laplace operator remains well defined.

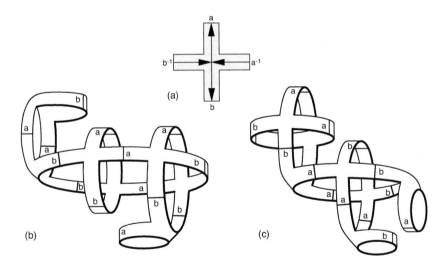

Fig. C.VIII.87 Isospectral surfaces (b) and (c) of Buser, obtained by assembling seven identical cross-shaped domains (a). Before assembling, the elementary domains (a) are flat. During the assembling of the surfaces, these elementary domains become curved, but this deformation is isometric, preserving the Euclidian metric.

However, the Buser surfaces cannot be seen as true drums, as they are not minimal surfaces, whereupon the total curvature vanishes in each point. Consequently, only the planar domains of Gordon, Webb and Wolpert rise to the challenge of Kac in its literal interpretation.

The isospectrality of Buser surfaces and of the domains of Gordon, Webb, and Wolpert is a result of assembling schemes obeying certain rules derived from group theory. Once these rules are known, other pairs of isospectral surfaces can be constructed. Stating the rules and understanding their origin is beyond the intent of this work (we refer the reader to references [7, 58, 59 ,60] for additional information on this aspect of isospectrality). Nevertheless, one can prove – more easily, and a posteriori – the isospectrality of the domains of Gordon, Webb and Wolpert using the so-called "eigenfunction transplantation" method. Below, we detail this demonstration, since it can be seen as an application of the superposition principle for the solutions of the wave equation.

To begin with, let us assume that the solutions (i.e., the eigenvalues and eigenfunctions), are already known for one of the contours of the GWW pair (for instance, shape "a" in Fig. C.VIII.88a). Let k be one of these eigenvalues and $\zeta(x,y)$ the corresponding wave function. This latter can be seen as an assembly of seven "partial" wave functions $\zeta_A(x,y)$, $\zeta_B(x,y)$, ... , $\zeta_G(x,y)$, each defined on one of the triangles composing the contour "a." These functions must obviously fulfill the Dirichlet conditions at the boundaries of contour "a."

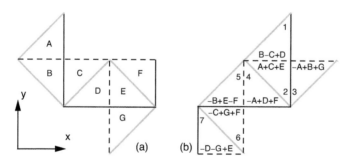

Fig. C.VIII.88 Eigenfunction transplantation from contour (a) onto contour (b) (see text).

Thus, for instance, the function $\zeta_A(x,y)$ must vanish on two sides of triangle A: the black and the gray ones. For the same reason, the function $\zeta_C(x,y)$ must vanish on the dotted side of triangle C. On the other hand, the partial wave functions belonging to adjacent triangles must join continuously along the common side. For functions $\zeta_A(x,y)$ and $\zeta_B(x,y)$, this implies that their values must be equal along the common dotted side:

$$\zeta_A(x,y) = \zeta_B(x,y) \qquad \text{on the dotted side} \qquad \text{(C.VIII.195)}$$

Moreover, for the junction to be smooth, their derivatives along the direction perpendicular to this common boundary must be equal

$$\frac{\partial \zeta_A}{\partial y} = \frac{\partial \zeta_B}{\partial y} \qquad \text{on the dotted side} \qquad (\text{C.VIII.196})$$

By the same rationale, for triangles B and C one must have:

$$\zeta_B(x,y) = \zeta_C(x,y) \quad \text{and} \quad \frac{\partial \zeta_B}{\partial x} = \frac{\partial \zeta_C}{\partial x} \quad \text{on the solid side} \qquad (\text{C.VIII.197})$$

Let us now construct seven new wave functions $\Psi_1(x,y)$, $\Psi_2(x,y)$, ..., $\Psi_7(x,y)$ defined on the elementary triangle by a linear superposition of the seven partial wave functions:

$$\Psi_1(x,y) = + \zeta_B(x,y) - \zeta_C(x,y) + \zeta_D(x,y)$$

$$\Psi_2(x,y) = + \zeta_A(x,y) + \zeta_C(x,y) + \zeta_E(x,y)$$

$$\Psi_3(x,y) = - \zeta_A(x,y) + \zeta_B(x,y) + \zeta_G(x,y)$$

$$\Psi_4(x,y) = - \zeta_A(x,y) + \zeta_D(x,y) + \zeta_F(x,y) \qquad (\text{C.VIII.198})$$

$$\Psi_5(x,y) = - \zeta_B(x,y) + \zeta_E(x,y) - \zeta_F(x,y)$$

$$\Psi_6(x,y) = - \zeta_C(x,y) + \zeta_G(x,y) + \zeta_F(x,y)$$

$$\Psi_7(x,y) = - \zeta_D(x,y) - \zeta_G(x,y) + \zeta_E(x,y)$$

These new functions fulfill the wave equation owing to the superposition principle, and they correspond to the same eigenvalue k as the initial function $\zeta(x,y)$.

Finally, let us "transplant" these seven new functions onto the triangles numbered from 1 to 7 in contour "b." As by a miracle, the new wave functions fulfill the Dirichlet conditions on the boundary of contour "b." At the same time, they join continuously at the common sides of neighboring triangles.

To check the first affirmation, let us consider for instance the function $\Psi_1(x,y) = \zeta_B(x,y) - \zeta_C(x,y) + \zeta_D(x,y)$ transplanted onto triangle 1. To respect the Dirichlet conditions, this function must vanish at the black and gray sides of this triangle. Let us start by the gray one. The wave function $\zeta_B(x,y)$ vanishes at this side, and we are left with the difference $\zeta_D(x,y) - \zeta_C(x,y)$. Inspection of contour "a" shows that it must vanish on the gray side, since functions $\zeta_D(x,y)$ and $\zeta_C(x,y)$ are equal at this boundary. In the exact same way, one can check that the function $\Psi_1(x,y) = \zeta_B(x,y) - \zeta_C(x,y) + \zeta_D(x,y)$ vanishes along the black side of triangle 1.

To check the second point, namely that the joining conditions between transplanted functions in two neighboring triangles are fulfilled, let us consider, for instance, the pair $\Psi_4(x,y) = - \zeta_A(x,y) + \zeta_D(x,y) + \zeta_F(x,y)$ and $\Psi_5(x,y) = - \zeta_B(x,y) + \zeta_E(x,y) - \zeta_F(x,y)$. Here, the junction must be smooth at the dotted side. We know that functions $\zeta_A(x,y)$ and $\zeta_B(x,y)$ are appropriately joined, as they are neighbors in the contour "a." The same holds for the pair $\zeta_D(x,y)$ and $\zeta_E(x,y)$. We are left with function $\zeta_F(x,y)$, taken with a + sign in triangle "4" and with a – sign in triangle "5." In contour "a," this function vanishes along the dotted side, ensuring its continuity at the boundary between triangles "4" and "5," regardless of the sign difference. The sign change is nevertheless necessary when dealing with the slope change across this boundary, in order to avoid a cusp. The joining conditions at the boundaries between the other triangles can be verified in the same way.

This verification concludes the isospectrality proof for the GWW drums.

On the other hand, we still ignore their eigenvalue spectrum and the shape of their wave functions. They can obviously be calculated numerically (using a computer), but we find it more edifying to determine them experimentally using smectic drums.

The setup used by Even et al. [7] (Fig. C.VIII.89) is similar to that of Miyano (see Fig. C.VIII.74): the smectic film (S2 in the SmA phase) is stretched on a frame precisely cut from a flat metal sheet. It is held in a chamber under primary vacuum, in order to suppress the effects related to air inertia.

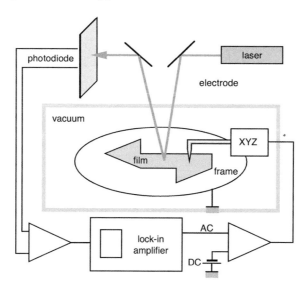

Fig. C.VIII.89 Setup used to study the eigenmodes of smectic drums.

The vibrations are excited electrostatically, using a sharp electrode, precisely positioned at coordinates $(x,y,z)_{exc}$ using a microtranslation system. Using this positioning system of the excitation electrode in the $(x,y)_{exc}$ plane of the film, the shape $\zeta(x,y)$ of the wave functions can be detected. To this end, it suffices to sweep the surface of the film very precisely with the excitation electrode, simultaneously measuring the vibration amplitude at a fixed point $(x,y)_{det}$ of the film using a laser beam. This method is based on the fact that the amplitude $\zeta(x,y)_{det}$ of vibrations detected at point $(x,y)_{det}$ is proportional to the value $\zeta(x,y)_{exc}$ of the wave function at the position $(x,y)_{exc}$ of the excitation electrode. To prove this result, we shall use a reasoning based on the energy.

Let

$$\zeta(x,y,t) = \zeta(x,y) \cos(\omega t + \phi) \tag{C.VIII.199}$$

be the displacement of the film, induced by the excitation force (considered as point-like on the scale of the film):

$$f(x,y,t) = C\, U_{DC} U_{AC}\, \delta(x - x_{exc}, y - y_{exc}) \cos(\omega t) \tag{C.VIII.200}$$

where C is a constant. As the excitation frequency ω coincides with one of the eigenfrequencies of the film, the phase shift ϕ of the vibration with respect to the excitation force (see subsection C.VIII.5.e) is $-\pi/2$, yielding:

$$\zeta(x,y,t) = \zeta(x,y) \sin(\omega t) \tag{C.VIII.201}$$

Let us now calculate the average power injected into the film by the excitation:

$$P = \frac{1}{T} \int_{film} \int_0^T \left(f(x,y,z) \frac{\partial \zeta(x,y,t)}{\partial t} \right) dx\,dy\,dt$$

$$= C\, U_{DC} U_{AC} \zeta(x,y)_{exc} \omega \frac{1}{T} \int_0^T \cos^2(\omega t)\, dt$$

$$= \frac{C\omega}{2} U_{DC} U_{AC} \zeta(x,y)_{exc} \tag{C.VIII.202}$$

We can see that it is proportional to $\zeta(x,y)_{exc}$. Now, the vibration amplitude $\zeta(x,y)_{det}$ detected with the laser at point $(x,y)_{det}$ is also proportional to the average power P injected into the film. As a consequence, amplitude $\zeta(x,y)_{det}$ detected with the laser is proportional to the vibration amplitude at the excitation point $\zeta(x,y)_{exc}$, QED.

Note that, in practice, the detected signal has a maximum (resp. a minimum) when the laser beam reflects on a vibration node (resp. a vibration antinode) of the film, since the photodiode detects the deflection of the reflected beam.

The experiments we are about to describe were performed by Even on S2 (BDH) films in the SmA phase. Under vacuum, the quality factor of the resonance peaks is of the order of 200, so that the eigenmodes of the two GWW drums can be easily detected. Their spectra are plotted in figure C.VIII.90 as a function of the reduced frequency f/f_{fund}, with f_{fund} the frequency of the fundamental mode, used here as reference. This normalization is required for a meaningful comparison, as the actual frequency of the modes depends on the thickness of the films, strongly variable between experiments. Once the normalization is performed, one can easily recognize the isospectrality of the two GWW contours. One might be surprised by the dispersion in height of the resonance peaks, but this is only an artifact due to the measurement being always performed at the same point $(x,y)_{det}$ during the frequency sweep; it can happen that the detection point is close to a node for one mode, giving a strong peak, and close to an antinode for a different mode, yielding a much weaker peak.

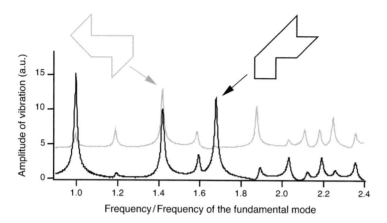

Fig. C.VIII.90 Eigenmode spectra of drums "a" and "b" of Gordon, Webb and Wolpert. The isospectrality of these two drums is obvious (from ref. [7a]).

Once the frequencies of the eigenmodes are known, one can then lock onto one of them and measure the corresponding wave function $\zeta(x,y)$ by sweeping the surface of the film with the excitation electrode. Figure C.VIII.91 shows the shape of the fundamental mode and of the ninth mode, determined by this method.

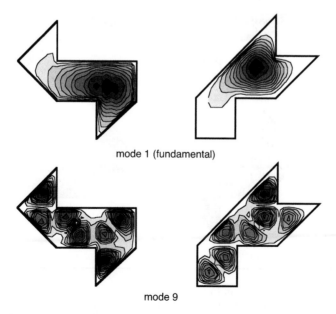

Fig. C.VIII.91 Shapes of the eigenmodes of the GWW drums, measured by sweeping the excitation point (from ref. [7a]).

The example of the fundamental modes is instructive because, in spite of their very different shapes, they vibrate at the same frequency, being related by the transplantation operation.

In the same vein, the ninth mode seems more trivial, as it results from the assembly of seven – identical – partial wave functions, vanishing at the sides of each triangle and changing sign from one triangle to the next. Clearly, the unique wave function used for building this mode on both drums corresponds to the fundamental mode of the elementary triangle.

However, the structure of the ninth mode can help us appreciate the subtlety of the assembly schemes of the drums. To this end, let us modify slightly frame "a" by "reversing" the last triangle G, as in figure C.VIII.92. Is this modified frame isospectral with frames "a" and "b" of GWW? To answer this question, Even recorded the shape of the ninth mode of the modified frame and showed that it is no longer an assembly of seven identical functions. Moreover, its frequency is lower than that of the ninth mode of the GWW frame. This occurs because the ninth mode of the modified frame cannot be obtained by assembling seven identical functions without introducing a discontinuity of the derivatives at the junction of triangles E and G (Fig. C.VIII.92).

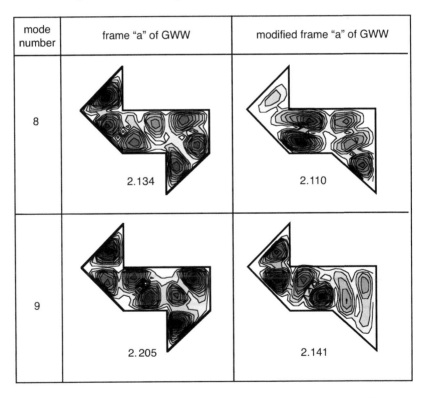

mode number	frame "a" of GWW	modified frame "a" of GWW
8	2.134	2.110
9	2.205	2.141

Fig. C.VIII.92 Effect of a modification in the assembling scheme for frame "a." The shape of the modes and the reduced eigenfrequencies (indicated next to each contour) are modified. For the modified frame, the ninth mode is no longer obtained by assembly of the fundamental mode of the elementary triangle (from ref. [7a]).

This counterexample provides some insight as to why, in isospectral frames, all pairs of adjacent triangles must exhibit mirror symmetry with respect to their common side. Although this symmetry is a necessary condition for isospectrality, it is however not sufficient. To illustrate this, consider frames "c" and "d" in figure C.VIII.93, which respect this symmetry. Obviously, one can construct for each of them an eigenmode with an associated frequency identical to that of the ninth mode of the GWW frames "a" and "b." On the other hand, these two frames have eigenmodes with frequencies different from those of the GWW frames. For instance, it is readily apparent that, due to its almost square-like shape, the frequency of the fundamental mode of frame "c" will be lower than that of the GWW frames, while for the frame "d," narrow and long, the frequency of the fundamental mode will be higher.

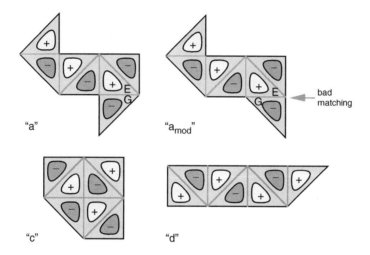

Fig. C.VIII.93 Eigenmodes of modified frames. The wave function constructed on frame "a_{mod}" is not an eigenmode, since triangles E and G are not properly matched. The wave functions constructed on frames "c" and "d" are eigenmodes, with a frequency identical to that of the ninth mode of frame "a". However, the frequencies of the other modes are different.

As Gottlieb noticed recently [61], the absence of isospectrality for the frames "a_{mod}," "c," and "d" in figure C.VIII.93 can be shown a priori using the asymptotic formula valid for polygonal frames

$$\sum_{n = 1}^{\infty} e^{-\lambda_n t} \propto \frac{A}{4\pi t} - \frac{\Lambda}{8\sqrt{\pi t}} + \sum_{i} \frac{\pi^2 - \theta_i^2}{\pi \theta_i} \quad \text{for} \quad t \to \infty \qquad \text{(C.VIII.203)}$$

where $\lambda_n = k_n^2$ are the eigenvalues of eq. C.VIII.189, A is the surface area of the frame, Λ its perimeter and θ_i the internal angles of the polygon. Indeed, frames "a" and "a_{mod}" have the same area and perimeter, but different angles. For the equal-area frames "c" and "d," the lack of isospectrality is even more obvious since not only the angles, but also the perimeters, are different.

VIII.6.b) Fractal planar drums

The asymptotic formula C.VIII.203 translates a global statistical property of the spectra and is related to the Weyl formula for the integrated density of eigenmodes of the scalar wave equation in two dimensions:

$$N(k) = \sum_{k_n < k} 1 + \sum_{k_n = k} \frac{1}{2} = \frac{A}{4\pi} k^2 - \frac{\Lambda}{4\pi} k \qquad \text{(C.VIII.204)}$$

Here, N(k) is the number of modes with an eigenvalue k_n below k, while A and Λ designate once again the area and the perimeter of the contour.

The signification of the Weyl formula is easy to comprehend in the case of a square drum with surface area $A = L^2$ and perimeter $\Lambda = 4L$. The eigenvalues of the drum are given by (see eq. C.VIII.194).

$$k_{mn}^2 = \left(\frac{\pi}{L}\right)^2 (m^2 + n^2) \qquad m, n = 1, 2, 3, \dots \qquad \text{(C.VIII.205)}$$

They can be represented graphically by a set of points in the wave vector space (k_x, k_y), as shown in figure C.VIII.94. In this reciprocal space, the eigenmode density is $1/(\pi/L)^2$, such that the number of modes within a circular sector of radius k is given by:

$$\frac{\pi k^2}{4} \frac{1}{(\pi/L)^2} = \frac{A}{4\pi} k^2 \qquad \text{(C.VIII.206)}$$

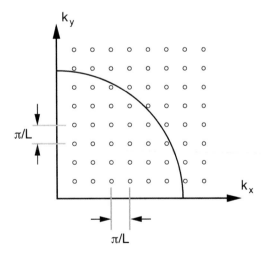

Fig. C.VIII.94 Calculating the integrated eigenmode density N(k) of a square drum. Due to the Dirichlet conditions, modes with k_x or $k_y = 0$ must be removed.

However, due to the Dirichlet boundary conditions, modes for which k_x or k_y are zero must be removed. In agreement with the Weyl formula, this yields:

$$N(k) = \frac{A}{4\pi} k^2 - \frac{k}{\pi/L} = \frac{A}{4\pi} k^2 - \frac{\Lambda}{4\pi} k \qquad \text{(C.VIII.207)}$$

If the contour of the drum is not an ordinary curve with dimension $D = 1$, but a fractal curve with a perimeter Λ that depends on the precision of the measurement, the Weyl-Berry-Lapidus conjecture [62] predicts that:

$$N(k) = \frac{A}{4\pi} k^2 - C(D) M(D) k^D \qquad \text{(C.VIII.208)}$$

with D the fractal Hausdorff-Besicovitch dimension, M(D) the measure of the contour and C(D) a constant. Since $1 < D < 2$, the number of modes to be removed is larger than for a contour with dimension $D = 1$. A heuristic argument for understanding this property is based on mode localization in fractal contour drums. More precisely, let us consider the contour in figure C.VIII.95. This figure shows how a square contour can be transformed into a quadratic Koch curve by successive iterations.

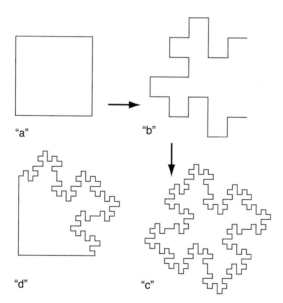

"a" "b"

"d" "c"

Fig. C.VIII.95 Construction by successive iterations of the quadratic Koch curve. The fractal dimension of this curve is D = log8/log4 = 1.5. For contour "d," the iterations were only applied to two sides of the initial square (from ref. [7a]).

By definition [63], the fractal dimension of the Koch curve is

$$D = \frac{\ln(N)}{\ln(1/r)} = \frac{\ln(8)}{\ln(4)} = \frac{3}{2} = 1.5 \qquad \text{(C.VIII.209)}$$

where $r = 1/4$ is the reduction factor applied to the generator at each iteration and N the number of segments that compose this generator. Figure C.VIII.95 shows clearly that the second iteration already leads to a very irregular contour. Using more graphic language, one might say that, if this contour represented a seacoast, this would have a multitude of bays or creeks of smaller and smaller size, separated by straits.

In our case, this contour represents a drum, and we are looking for its eigenmodes. Since the wave equation has no analytical solutions in this geometry, the eigenmodes can be found either numerically or experimentally, using smectic films. Once again, we choose the second approach.

Using the setup in figure C.VIII.89, Even measured the frequencies and wave functions of the eigenmodes for films of the liquid crystal S2, stretched on a frame with the contour shown in figure C.VIII.95d. Note that, in order to avoid mode degeneracy due to the symmetry of the Koch curve (Fig. C.VIII.95c), only two sides of the initial square underwent successive iterations (prefractal frame).

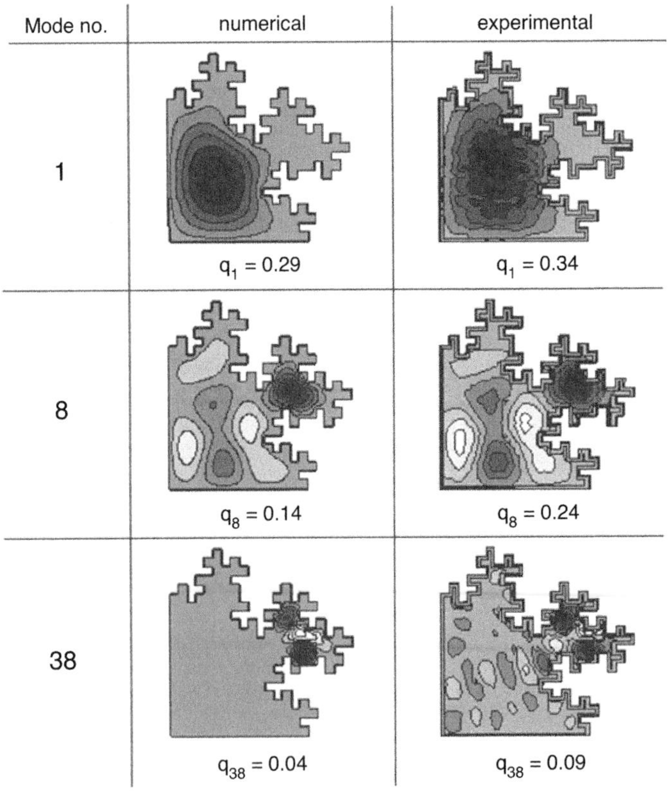

Mode no.	numerical	experimental
1	$q_1 = 0.29$	$q_1 = 0.34$
8	$q_8 = 0.14$	$q_8 = 0.24$
38	$q_{38} = 0.04$	$q_{38} = 0.09$

Fig. C.VIII.96 Eigenmodes of a smectic prefractal drum (from ref. [7b]).

Let us start by considering the wave function of the measured fundamental mode (Fig. C.VIII.96). Obviously, this mode does not penetrate into the crannies of the contour, remaining localized in the main "bay." To quantify this localization phenomenon, we define the *localization ratio* q_n defined by

$$q_n = \frac{1}{A \int |\zeta_n(x,y)|^4 dxdy} \qquad\qquad (C.VIII.210)$$

and calculated using the normalized wave function $\zeta_n(x,y)$:

$$\int |\zeta_n(x,y)|^2 dxdy = 1 \qquad\qquad (C.VIII.211)$$

Measurements of q_n performed on smectic films agree with the results of the numerical calculation, summarized in figure C.VIII.97.

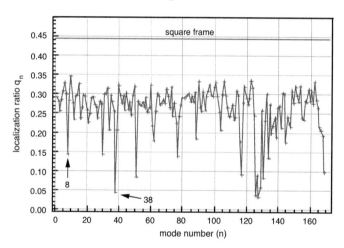

Fig. C.VIII.97 Localization ratio, calculated for the first 170 modes of the prefractal contour drum.

For comparison, this graph also shows the value of the analytically calculated localization ratio for the square frame (with dimension D = 1), where the wave functions are known (see eqs. C.VIII.170–171). This localization ratio is independent of the mode number and its value is exactly $q_{sq} = 4/9 = 0.44$. The situation is very different for the prefractal frame, where the localization ratio varies irregularly between 0.03 and 0.34 for the first 170 modes.

VIII.6.c) Quantum billiards and diabolic points

Continuing the exploration of the relation between the shape of a drum and its eigenmode spectrum, one can study the effects due to continuous modifications of the drum shape. For instance, one might ask what is the consequence of a change in the shape of the elementary triangles composing the isospectral drums of Gordon, Webb and Wolpert (GWW). In the initial drums (Fig. C.VIII.86), the basic element is an isosceles right angled triangle. What happens if it is deformed? The simplest deformation is to maintain an isosceles triangle and to change the value of the right angle. The experimental results in figure C.VIII.98 show that the GWW frames deformed in this manner remain isospectral, although the eigenfrequencies are changed.

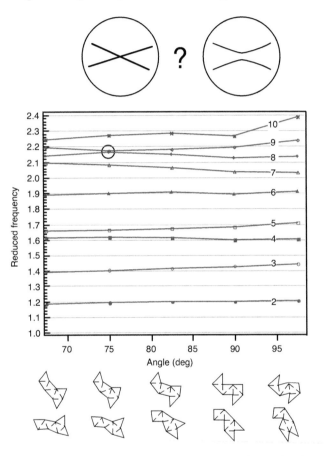

Fig. C.VIII.98 Spectra of modified GWW frames (from ref. [7]).

This result should not come as a surprise, once the isospectrality proof by the method of wave functions transplantation is understood (subsection C.VIII.6a). One can, however, wonder as to the nature of a small detail marked by a circle in the diagram of figure C.VIII.98. Do the curves of modes 8 and 9 cross, or do they avoid each other?

This question turns out to be related to the crossing problem for the energy levels of a quantum system described by a Hamiltonian depending on one or more parameters:

$$\hat{H}(a,b,\dots)\Psi = E\Psi \qquad\qquad \text{(C.VIII.212a)}$$

In the simplest case, this can be, as proposed by Berry and Wilkinson [64], a particle of mass m free to move inside a triangle-shaped potential well with infinitely high walls. The Schrödinger equation of this "quantum billiards" then writes:

$$-\frac{\hbar^2}{2m}\Delta\Psi = E\Psi(x,y) \qquad\qquad \Psi(\text{contour}) = 0 \qquad\qquad \text{(C.VIII.212b)}$$

Since the **shape** of the triangular well is determined by two of its angles, α and β, the energy levels of the particle also depend on these two parameters:

$$E_n = E_n(\alpha,\beta) \qquad\qquad \text{(C.VIII.213)}$$

Clearly, in the (α,β,E) space, the equations of adjacent levels n and n + 1, $E = E_n(\alpha,\beta)$ and $E = E_{n+1}(\alpha,\beta)$, define two surfaces. Let us assume that their intersection is not void.

If these were generic surfaces $E = S_1(\alpha,\beta)$ and $E = S_2(\alpha,\beta)$, they would intersect along a curve defined by the equation

$$S_1(\alpha,\beta) = S_2(\alpha,\beta) \qquad\qquad \text{(C.VIII.214)}$$

as in the example of figure C.VIII.99a, where one of the surfaces is a plane, the other one is an ellipsoid and their intersection is a circle.

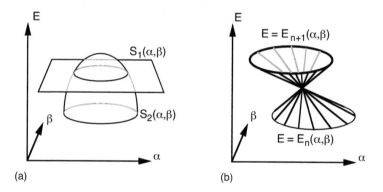

(a)   (b)

Fig. C.VIII.99 Examples of intersections between two surfaces: a) the common case of two generic surfaces; b) the particular case of two adjacent energy levels in a triangular quantum billiards: the intersection, when it is not void, is reduced to a "diabolic point."

In the case of energy levels of a quantum billiards depending on two parameters, Von Neumann, Wigner and Teller (see ref. [64]) showed that, in contrast with the generic case, the intersection is limited to isolated points. Close to these points, the surfaces $E = E_n(\alpha,\beta)$ and $E = E_{n+1}(\alpha,\beta)$ take the shape of cones, touching at the tip (Fig. C.VIII.99b). Due to their resemblance to the "diabolo" toy, Berry and Wilkinson [64] termed these points "diabolic points." In the special case of triangular quantum billiards, Berry and Wilkinson made the list of diabolic points: the first one brings together levels 5 and 6, when the angles of the triangular well are $\alpha = 130.57°$, $\beta = 30.73°$, and $\gamma = 18.70°$.

From the mathematical point of view, diabolic points reveal a feature of the eigenvalue equation:

$$\frac{\partial^2 \Psi}{\partial x^2} + \frac{\partial^2 \Psi}{\partial y^2} = -k^2 \Psi \qquad\qquad (C.VIII.215)$$

to which the Schrödinger equation reduces by the substitution:

$$k^2 = \frac{E}{\hbar^2 / 2m} \qquad\qquad (C.VIII.216)$$

and also the wave equation of smectic films C.VIII.191 with

$$k^2 = \frac{\rho H}{\tau}\, \omega^2 \qquad\qquad (C.VIII.217)$$

This analogy between quantum billiards and the vibrations of smectic films suggests that the approach of the frequencies of the eighth and ninth mode of the modified isospectral GWW drums (Fig. C.VIII.98) signals the presence of a diabolic point. To pinpoint this diabolic point, one should however vary not one, but two angles of the elementary triangle. Such an experiment is not easy to perform, requiring a large number of frames. The same difficulty appears when attempting to model the triangular billiards of Berry and Wilkinson with smectic films.

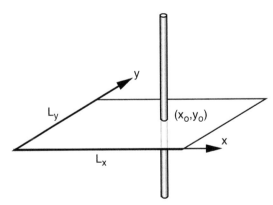

Fig. C.VIII.100 A smectic film pierced by a fiber is similar to a quantum billiards perturbed in one point.

To demonstrate the existence of diabolic points using smectic films and verify their mathematical properties, Brazovskaia [65a,b] studied a different theoretical system: the rectangular quantum billiard (or well) in the presence of a point-like perturbation. The Schrödinger equation now writes:

$$[- \Delta + s\, \delta(x - x_o)\, \delta(y - y_o)]\, \Psi(x,y) = \frac{E}{\hbar^2 / 2m}\, \Psi(x,y) \qquad \text{(C.VIII.218)}$$

The "perturbation potential," described by its "strength" s, acts pointwise at coordinates (x_o, y_o) inside the billiard. At s = const, the x_o and y_o coordinates of the perturbation are used to localize the diabolic points. Choosing a negative value for s, Cheon and Shigehara [66] numerically found several diabolic points in a billiard of aspect ratio $r = L_x/L_y = 1.155 \approx \pi/e$.

In the case of smectic drums, it is much easier to create a point-like perturbation of a positive and infinite strength s; to this end, one only needs to pierce the film with a very thin glass fiber, as shown in figure C.VIII.100. In this case, the film wets the fiber, forming a circular meniscus (see Fig. C.VIII.49). The fiber acts as a perturbation of infinite strength s, since the film is motionless at this point (the meniscus does not slide on the fiber). Thus, the wave function vanishes at point (x_o, y_o):

$$\xi(x_o, y_o) = 0 \qquad \text{(C.VIII.219)}$$

In her experiments, Brazovskaia measured the eigenfrequencies of a SCE4 film (SmC* at room temperature), stretched on a rectangular frame with dimensions $L_x = 1.15$ cm and $L_y = 1$ cm. The glass fiber, with a diameter D = 0.3 mm, is placed inside the frame with a precision of 0.01 mm. The diagram in figure C.VIII.101 shows the variation of the eigenfrequencies of the film as a function of the x_o coordinate of the fiber, the other coordinate remaining unchanged: $y_o = L_y/2$.

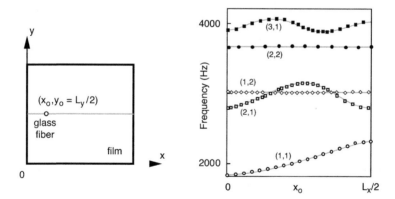

Fig. C.VIII.101 Eigenmode frequency for a rectangular film pierced by a fiber (from ref [65]).

First of all, one can notice that modes (1,2) and (2,2) of the rectangular frame – since they exhibit a nodal line coincident with the trajectory of the fiber – are not perturbed by it. The situation is completely different for the other modes, the perturbation vanishing only when the trajectory of the fiber crosses a nodal line: for instance, in the case of mode (2,1), this occurs when the fiber is at the edge of the film or in the center of the frame. Everywhere else, the wave function of this mode is modified by the fiber and its frequency increases with respect to that of the unperturbed (2,1) mode. If the frequencies of the unperturbed modes (1,2) and (2,1) are close enough, increasing the frequency of the (2,1) mode leads to two crossing points, visible in the diagram of figure C.VIII.101.

Are they diabolic points? There are three very convincing arguments for it.

The first, and most obvious, one is that the fiber positions for which the crossing of modes (1,2) and (2,1) occurs do not correspond to any obvious symmetry that could result in eigenmode degeneracy.

The second argument is provided by the shape of the $f_{1,2}(x_o,y_o)$ and $f_{2,1}(x_o,y_o)$ surfaces, measured close to the crossing points. Their plot in figure C.VIII.102 suggests that they are conic surfaces. It is therefore worthwhile to plot surfaces $f_{1,2}(x_o,y_o)$ and $f_{2,1}(x_o,y_o)$ (Fig. C.VIII.103) to see better the position of diabolic points. One can see in particular, by separating the two sheets, that there are four pairs of conic points corresponding, when they coincide, to the four diabolic points.

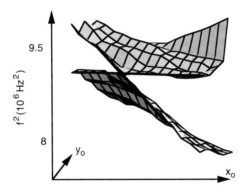

Fig. C.VIII.102 Conic shape of the $f_{1,2}(x_o,y_o)$ and $f_{2,1}(x_o,y_o)$ surfaces close to a diabolic point (point C in figure C.VIII.103) (from ref. [65]).

The third argument in favor of the "diabolic" character of points A, B, C and D involves the transformation of the wave function $\zeta(x,y;x_o,y_o)$, as the fiber describes a closed trajectory around one such point. Consider, for instance, the path shown in thick line around point B on the (1,2) sheet in figure C.VIII.103. Knowing the complete set of eigenmodes of the rectangular frame:

$$\xi_{mn}(x,y) = \sin\left(m\,\frac{\pi}{L_x}x\right)\sin\left(n\,\frac{\pi}{L_y}y\right) \qquad m,n = 1,2,3,... \qquad \text{(C.VIII.220)}$$

and their frequencies:

$$f_{mn} = \frac{1}{2}\sqrt{\frac{\tau}{\rho H}}\sqrt{\left(\frac{1}{L_x}\right)^2 m^2 + \left(\frac{1}{L_y}\right)^2 n^2} \qquad \text{(C.VIII.221)}$$

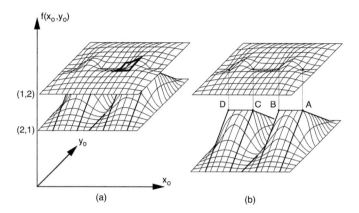

(a)   (b)

Fig. C.VIII.103 Global shape of surfaces $f_{1,2}(x_o,y_o)$ and $f_{2,1}(x_o,y_o)$. When the sheets are separated (b), one can clearly distinguish four pairs of conic points A, B, C and D, which coincide in figure (a). On sheet (1,2) in figure (a), we also represent a closed path around the conical point B. The wave function changes along this path as shown in figure C.VIII.104.

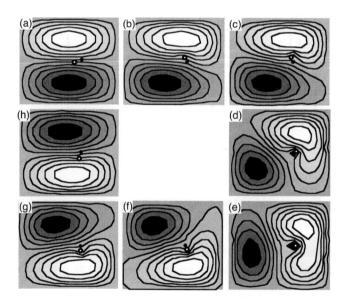

Fig. C.VIII.104 Transformation of the (1,2) mode (a) as a function of the fiber moving around the diabolic point B in figure C.VIII.103. In agreement with ref. [64], the wave function of the (1,2) mode changes sign as the fiber returns to its starting point (h) (from ref. [65]).

Brazovskaia calculated the wave function $\zeta(x,y;x_o,y_o)$ of the mode perturbed by the fiber:

$$\xi(x,y;x_o,y_o) = \sum_{m,n} \frac{\xi_{mn}(x_o,y_o)}{f^2(x_o,y_o) - f_{mn}^2} \xi_{mn}(x,y) \qquad \text{(C.VIII.222)}$$

where $f(x_0,y_0)$ is the measured frequency of the perturbed mode. Using this formula, one can calculate and trace (Fig. C.VIII.104) the wave function for eight positions of the fiber along this path. Clearly, in agreement with the definition of diabolic points, the wave function changes sign as the fiber returns to the initial point.

In the next subsection, we return to the problem of vibrations of curved films.

VIII.6.d) Stability and vibrations of a catenoid

Among all minimal surfaces, the catenoid is undoubtedly the simplest, being the only one endowed with revolution symmetry [67]. It can be obtained by rotation around the z axis of a curve, the catenary, described by equation:

$$r(z) = r_0 \cosh(z/r_0) \qquad\qquad (C.VIII.223)$$

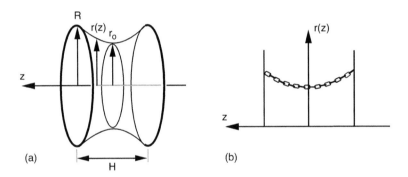

(a)   (b)

Fig. C.VIII.105 a) The catenoid is the minimal revolution surface obtained by rotation of the catenary around the z axis. b) The catenary.

This particular surface shape is the one adopted by a smectic film stretched between two circular frames facing each other, since it minimizes the surface energy:

$$F_{surf} = \tau \int_{-H/2}^{+H/2} 2\pi r \sqrt{1 + \left(\frac{dr}{dz}\right)^2}\, dz \qquad\qquad (C.VIII.224)$$

This integral is identical (up to a 2π prefactor) to that giving the potential energy of a chain (Fig. C.VIII.105b)

$$F_{grav} = \rho g \int_{-H/2}^{+H/2} r \sqrt{1 + \left(\frac{dr}{dz}\right)^2}\, dz \qquad\qquad (C.VIII.225)$$

with a mass per unit length ρ.

If the film is stretched between two circles of radius R separated by a distance H (Fig. C.VIII.105a), the catenoid must obey the boundary conditions:

$$\frac{H}{2} = r_0 \, \text{arcosh}\left(\frac{R}{r_0}\right) \tag{C.VIII.226}$$

In principle, this equation fixes the value of r_0. Using the radius R as length unit and introducing the reduced variables $\tilde{r}_0 = r_0/R$ and $\tilde{H} = H/R$, the previous equation writes, in its dimensionless form:

$$\tilde{H}(\tilde{r}_0) = 2 \, \tilde{r}_0 \, \text{arcosh}\left(\frac{1}{\tilde{r}_0}\right) \tag{C.VIII.227}$$

The function $\tilde{H}(\tilde{r}_0)$ is plotted in figure C.VIII.106. Returning to dimensional variables, one can see that, for $H < H_{max} = 1.3254 \, R$, there are two possible values of r_0, $r_{os}(H)$ and $r_{ou}(H)$, one larger and the other one smaller than $r^* = 0.5524 \, R$. The two branches of the curve meet at point $r^*(H_{max})$. For $H > H_{max}$, the equation C.VIII.227 no longer has solutions.

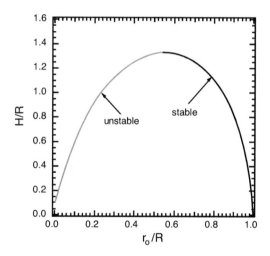

Fig. C.VIII.106 For $H < 1.3254 \, R$, **two** catenoids exist, one stable and the other one unstable.

Can the two catenoids be obtained in practice? Experiments performed using smectic liquid crystals show that one can stretch smectic films shaped as catenoids [67] using two cylindrical frames (Fig. C.VIII.107a). At the beginning, the frames are brought together and coated with the liquid crystal; a catenoid is obtained by progressively increasing the distance between them, with a minimal radius (at half-height) that follows the stable branch $r_{os}(H)$. When the distance H reaches H_{max}, the catenoid undergoes an instability and suddenly breaks, leaving behind two flat films, each one stretched on one of the two cylindrical frames. Additionally, a small bubble appears midway between the two frames the moment the film breaks [68].

To obtain the unstable solution of eq. C.VIII.227, one must use a cylindrical frame and a glass plate (Fig. C.VIII.107b). As before, the cylindrical frame and the glass plate are brought in contact, and then slowly separated. Using this method, one can stretch a "semi-catenoid," between $z = 0$ and $z = H/2$, since the film must be perpendicular to the glass plate in $z = 0$. When the distance $H/2$ is increased slowly, one obtains the stable catenoid, as with the first method, except that the evolution of the minimal radius $r_{os}(t)$ following a change in the distance $H/2$ is very long, since the meniscus connecting the film to the glass plate moves very slowly.

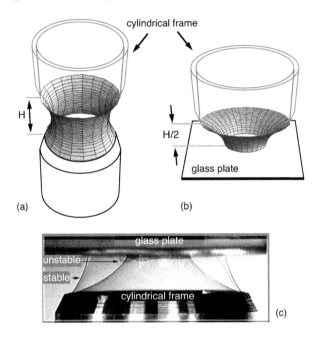

Fig. C.VIII.107 Two ways of obtaining catenoids: a) using two cylindrical frames; b) using a cylindrical frame and a glass plate. In both cases, the distance H is set using a micro-translation stage; c) the two possible shapes of the catenoid, observed experimentally for $H < H_{max}$ (the image is a superposition of two photos).

This very slow evolution enables one to go from $H_1 < H_{max}$ to $H_2 = H_1 + \Delta H > H_{max}$, without a significant change in r_o. If the system is subsequently left to evolve, the radius r_o varies progressively from $r_o(H_1)$ to zero, and the film ends by separating from the glass plate. It suffices then to return to the height H_1 at the precise moment when the radius r_o goes through the value $r_{ou}(H_1) < r^*$, to obtain the second solution $r_{ou}(H_1)$ of eq. C.VIII.227. This second shape of the catenoid is unstable because, for a fixed distance H_1, the radius r_o can either increase up to $r_{os}(H_1)$, or decrease to zero, resulting in film separation. These two processes being very slow, the radius r_o of the catenoid can be stabilized at its value $r_{ou}(H_1)$ by finely tuning the distance $H/2$.

The photograph of the unstable catenoid in figure C.VIII.107c could be taken using this "manual feedback" system.

When a film stretched between two cylindrical frames (Fig. C.VIII.107a) is deformed (due to a change in H), it maintains its revolution symmetry, whether it is stable (for $H < H_{max}$) or unstable (for $H > H_{max}$). This behavior suggests that, among the vibration eigenmodes of a catenoid, the fundamental mode with cylindrical symmetry plays a particular role. When the catenoid is stable, the frequency corresponding to this mode must be real, since any deformation with respect to the equilibrium shape must generate restoring forces. Nevertheless, these forces must decrease as the height H approaches its critical value H_{max}, vanish for $H = H_{max}$ and finally change sign above H_{max}. Consequently, the frequency of the fundamental mode must tend to zero as $H \rightarrow H_{max}$ and become imaginary beyond this value.

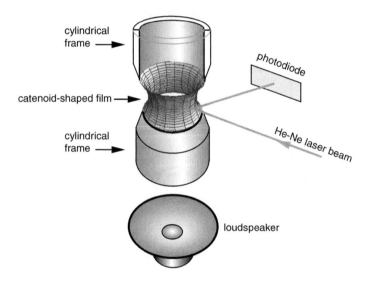

Fig. C.VIII.108 Setup for studying the fundamental vibration mode of the catenoid.

To detect the behavior of the fundamental mode on approaching the instability, Ben Amar et al. [67] used the experimental setup sketched in figure C.VIII.108. As for the experimental studies of vibrations in planar films, the vibrations are detected by a laser beam, which is reflected on the film before detection by a photodiode. For the excitation, on the other hand, a loudspeaker placed below the catenoid is more effective and better adapted to the geometry of the fundamental mode than electrostatic excitation. Catenoid vibrations induced by the loudspeaker are so strong that they can be visualized directly, by stroboscopy (Fig. C.VIII.109). In this example, the frequency of the fundamental mode is about 200 Hz for $H/R = 4.8$ mm/5 mm (Fig. C.VIII.110). The measurements show (Fig. C.VIII.111), in agreement with our heuristical analysis, that this frequency decreases as the height H increases, tending to zero for $H \approx 6.6$ mm.

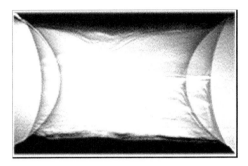

Fig. C.VIII.109 Stroboscopic observation of the fundamental vibration mode of the catenoid.

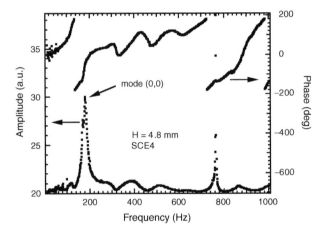

Fig. C.VIII.110 Amplitude and phase of the first two eigenmodes of the catenoid.

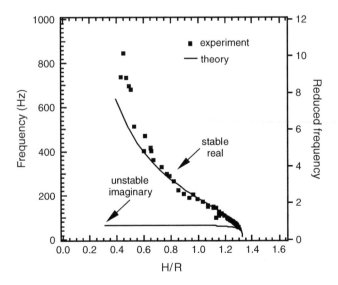

Fig. C.VIII.111 Frequency of the fundamental mode as a function of the reduced distance H/R.

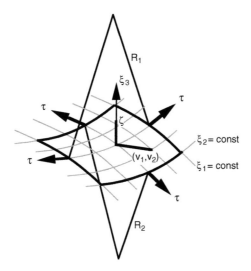

Fig. C.VIII.112 Definition of the (v_1, v_2, w) coordinates of a point on a minimal surface.

From a theoretical point of view, an exact analysis of the vibrations of minimal surfaces is more complex than in the case of planar membranes. Ben Amar, da Silva et al. [69] searched for the equations of motion using the Lagrange method. Let us start by introducing the local coordinates (ξ_1, ξ_2, ξ_3), where ξ_1 and ξ_2 are two curvilinear coordinates parameterizing the surface and ξ_3 the coordinate along the direction normal to the surface (Fig. C.VIII.112). In this reference system, the displacement of a point in the surface can be decomposed into a displacement (v_1, v_2) in the (ξ_1, ξ_2) plane tangent to the surface and a displacement ζ along the ξ_3 direction orthogonal to the surface (Fig. C.VIII.112). We assume that the film is perfectly liquid in two dimensions, such that the displacement (v_1, v_2) does not change the energy of the film. The displacement ζ, on the other hand, changes the shape of the surface and leads to an increase in capillary energy:

$$E^{cap} = \frac{1}{2}\tau \int (D^\alpha \zeta D_\alpha \zeta + 2G\zeta^2)\, dA \tag{C.VIII.228}$$

where D^α and D_α are respectively the covariant and contravariant derivatives with respect to the ξ_1 and ξ_2 coordinates on the surface, and G is the Gaussian curvature at point (ξ_1, ξ_2). As for the kinetic energy, its expression is

$$E^{kin} = \frac{1}{2}\rho H \int \left[\frac{\partial v^\alpha}{\partial t}\frac{\partial v_\alpha}{\partial t} + \left(\frac{\partial \zeta}{\partial t}\right)^2\right] dA \tag{C.VIII.229}$$

where v^α and v_α are the covariant and contravariant vector fields. Knowing the Lagrange function $L = E^{kin} - E^{cap}$, one can obtain the equation of motion for the displacement ζ:

$$\tau(D^\alpha D_\alpha - 2G)\zeta = \rho H \frac{\partial^2 \zeta}{\partial t^2} \tag{C.VIII.230}$$

The left-hand side of this equation is the restoring force along ξ_3 generated by the appearance of a total net curvature of the film under the action of the tension τ. To identify the first term of this capillary force let us note that, for a planar film, the covariant and contravariant derivatives simply reduce to the partial derivatives, so that this term corresponds to the usual Laplacian (see eq. C.VIII.191). The second term of the capillary force, on the other hand, has no equivalent in planar films, where the Gaussian curvature $G = 0$. Its interpretation is nevertheless very simple since we know, from the differential geometry of surfaces [70], that a surface placed at a distance ζ from a minimal surface acquires a total net curvature $2G\zeta$.

Applied to the case of the catenoid, eq. C.VIII.230 becomes:

$$\tau \frac{1}{r_o^2 \cosh^2\theta}\left(\frac{\partial^2 \zeta}{\partial\theta^2} + \frac{\partial^2 \zeta}{\partial\phi^2}\right) + 2\tau \frac{1}{r_o^2 \cosh^4\theta}\zeta = \rho H \frac{\partial^2 \zeta}{\partial t^2} \tag{C.VIII.231}$$

where $\theta = z/r_o$ and ϕ is the azimuthal angle. The θ and ϕ coordinates parameterize the surfaces drawn in figure C.VIII.107.

Writing the eigenmodes under the form

$$\zeta = e^{i\omega t}\zeta(\theta)\cos(n\phi) \tag{C.VIII.232}$$

yields

$$\frac{\partial^2\zeta(\theta)}{\partial\theta^2} + \left(\frac{2}{\cosh^2\theta} - n^2 + \tilde\omega^2\cosh^2\theta\right)\zeta(\theta) = 0 \tag{C.VIII.233}$$

where

$$\tilde\omega = \frac{\omega}{\omega_u} \tag{C.VIII.234a}$$

is the reduced frequency, measured in units of

$$\omega_u = \frac{1}{r_o}\sqrt{\frac{\tau}{\rho H}} \tag{C.VIII.234b}$$

The solutions $\zeta(\theta)$ of eq. C.VIII.233 must obey the Dirichlet conditions at the edges of the catenoid:

$$\zeta\left(\pm\frac{\Theta}{2}\right) = 0 \quad \text{with} \quad \Theta = \frac{H}{r_o} \tag{C.VIII.235}$$

These solutions can be found by numerical integration of eq. C.VIII.233, with the reduced frequency ω/ω_u as the adjustable parameter. For a given n several solutions, indexed by $m = 0,1,2\ldots$ are possible. The frequencies of eigenmodes (n,m) found by this method are plotted as a function of the reduced height of the catenoid in figure C.VIII.113. The most important result of this calculation is that only the fundamental mode (0,0) becomes unstable (the frequency ω then becomes imaginary), when the reduced height of the catenoid exceeds the critical value

$$(H/r_o)^* = 2.4 \qquad\qquad\qquad (C.VIII.236)$$

Note that this "dynamic" critical value is in agreement with the ratio $H^*/r_o^* = 1.3254/0.5524$ calculated from the "static" critical values $H^* = H_{max}/R$ and $r_o^* = r_o/R$ found previously (Fig. C.VIII.106).

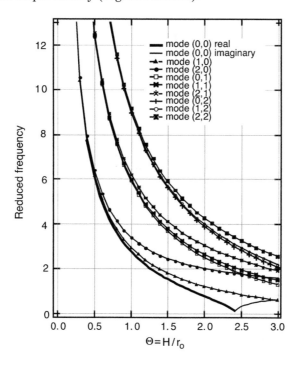

Fig. C.VIII.113 Frequencies of the eigenmodes of the catenoid.

To conclude this analysis of the eigenmodes of the catenoid, it seems important to point out a more general – and mathematically more elegant – result, concerning the families of surfaces related by isometric transformations. For instance, a catenoid can be changed into a helicoid by applying the isometric Bonnet transformation. By definition, an isometric transformation conserves the metric of the surface. On the other hand, Gauss's theorema egregium states that, if the metric is conserved, then so is the intrinsic curvature G. Now, the equation of motion C.VIII.231 only involves the Gaussian curvature and the metric. Consequently, the solutions of this equation also apply to the entire family of surfaces related to the catenoid. In other terms, all surfaces of this family are isospectral in the sense of Kac.

VIII.6.e) Nonlinear phenomena in vibrating films

In all the experiments described so far, the vibration amplitude was small enough so that the system remains in the linear regime, where the linear wave equations apply. For instance, in experiments involving electrostatically excited isospectral drums, the deflection of the laser beam is of the order of 10^{-3} rad, corresponding to a vibration amplitude of a few µm for a film size of a few cm. The vibration amplitude of the films is however very easily increased by a few orders of magnitude using acoustic excitation by a loudspeaker (Fig. C.VIII.114).

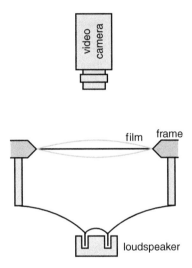

Fig. C.VIII.114 Exciting high-amplitude vibrations using a loudspeaker.

Leaving the linear regime has spectacular effects, such as the meniscus instability shown in the two photographs in figure C.VIII.115.

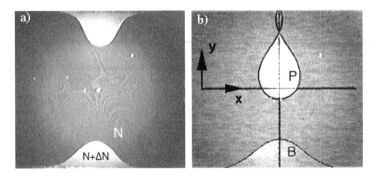

Fig. C.VIII.115 The meniscus instability (Beg-Rohu effect) is one of the nonlinear effects triggered by high-amplitude vibrations.

In the experiments of Brazovskaia et al. [71], this instability was studied in a SCE4 film (SmC* at room temperature) of thickness $N = 10$, stretched on a frame of size 5.3 mm × 5.3 mm. In air, the frequency of the fundamental mode of such a film is about 380 Hz. As the amplitude A of this mode exceeds 0.1 mm, one can observe a thick step ($\Delta N \approx 20$) that leaves the meniscus, forming a domain of thickness $N + \Delta N$ and a progressively increasing surface area S. On the photographs in figure C.VIII.115, this instability occurs simultaneously on two opposite edges of the frame. At a fixed amplitude A, the area S increases slowly. This slow evolution is undoubtedly related to the difficulty encountered by the molecules when leaving the meniscus (see subsection C.VIII.4.g). As the area S of the domain increases, the shape of the step changes and, in the end, it resembles a peninsula, connected to the meniscus by a narrow isthmus.

In the quasi-static approximation, neglecting permeation and viscous stress in the film, the shape of the step $y = y(x)$ is determined by the equilibrium of the forces acting upon it:

$$\frac{E}{R(x,y)} = \Delta\tau(x,y) + f_{int} \qquad \text{(C.VIII.237)}$$

with

$$\frac{1}{R(x,y)} = \frac{\dfrac{d^2y}{dx^2}}{\left[1 + \left(\dfrac{dy}{dx}\right)^2\right]^{3/2}} \qquad \text{(C.VIII.238)}$$

the local curvature of the step at the point (x,y) and

$$\Delta\tau(x,y) = \tau(x,y;N) - \tau(x,y;N + \Delta N) \qquad \text{(C.VIII.239)}$$

the local tension difference across the step between the N and $N + \Delta N$ domains. As for the last term f_{int}, it represents a short-range repulsive force (not very well understood) that becomes relevant when two dislocation segments of opposite signs come very close together, as in the case of the isthmus of step P in figure C.VIII.115b.

Before any calculations, let us analyze the shape of the step B in the photograph of figure C.VIII.115b: its curvature is clearly positive at the base (close to the meniscus) and negative on top of the bump. It is therefore obvious that the tension difference $\Delta\tau(x,y)$ varies along the step. On the other hand, we know that, if the interaction energy f(H) between the film surfaces is negligible (this assumption is not required here, but it does help simplify the problem), the tension difference is simply (see eq. C.VIII.22b):

$$\Delta\tau(x,y) = [P_{air} - P(x,y)]b \qquad \text{(C.VIII.240)}$$

where $b = \Delta N a_0$ is the Burgers vector of the step and P_{air} the atmospheric pressure. Clearly, the pressure P(x,y) inside a vibrating film must vary as a function of the position (x,y). When we studied the shape of steps in vertical films, this pressure variation was due to gravity. In a horizontal vibrating film, and in the absence of any flow (an experimental fact), this pressure variation can only be due to film vibrations. The experiments show that the relevant parameter for the meniscus instability and for the shape of steps is the product $(A\omega)^2$. This result suggests that the pressure $P(x,y)$ varies as:

$$P(x,y) = \rho \frac{\dot\zeta^2}{2} + P_m \qquad\qquad (C.VIII.241)$$

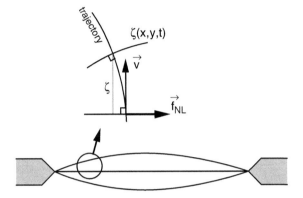

Fig. C.VIII.116 Mechanism responsible for the non-linear force in a smectic film undergoing high-amplitude vibrations.

To prove this expression, let us consider a volume element inside the film (Fig. C.VIII.116) and assume that its trajectory is at all times orthogonal to the film surface, described by the equation $\zeta = (x,y,t)$. The velocity along this trajectory has the components:

$$\mathbf{v} = \frac{d\mathbf{r}}{dt} \approx \frac{\partial\zeta}{\partial t}\left(-\frac{\partial\zeta}{\partial x}, -\frac{\partial\zeta}{\partial y}, 1\right) \qquad\qquad (C.VIII.242)$$

yielding an acceleration

$$\mathbf{a} = \frac{d\mathbf{v}}{dt} \approx \frac{\partial^2\zeta}{\partial t^2}\left(-\frac{\partial\zeta}{\partial x}, -\frac{\partial\zeta}{\partial y}, 1\right) + \frac{\partial\zeta}{\partial t}\left(-\frac{\partial^2\zeta}{\partial x\partial t}, -\frac{\partial^2\zeta}{\partial y\partial t}, 0\right) \qquad (C.VIII.243)$$

The first term of this expression represents the acceleration component perpendicular to the film. This is the term appearing in the linear wave equation for the film to first order in ζ, and that is balanced by the tension term. The second term only appears to second order in ζ, and corresponds to the centripetal acceleration. It is equilibrated by a pressure gradient P in the plane of the film:

$$\rho \frac{\partial \zeta}{\partial t}\left(-\frac{\partial^2 \zeta}{\partial x \partial t}, -\frac{\partial^2 \zeta}{\partial y \partial t}, 0\right) = -\rho \vec{\nabla}\left(\frac{\dot{\zeta}^2}{2}\right) = -\vec{\nabla}P \qquad \text{(C.VIII.244)}$$

This expression allows us to calculate the time-averaged value of the film pressure for the fundamental mode. Writing that for this mode, in the first approximation:

$$\zeta(x,y,t) \approx A \sin(\omega t) \sin\left(\frac{\pi}{L_x}x\right) \sin\left(\frac{\pi}{L_y}y\right) \qquad \text{(C.VIII.245)}$$

we obtain, using eq. C.VIII.244 and the fact that $\langle P(x,y) = P_m \rangle$ at the edge of the meniscus:

$$\langle P(x,y) \rangle = \frac{1}{4}\rho(A\omega)^2 \sin^2\left(\frac{\pi}{L_x}x\right)\sin^2\left(\frac{\pi}{L_y}y\right) + P_m \qquad \text{(C.VIII.246)}$$

Finally, combining eqs. C.VIII.237, 240 and 245 yields the differential equation describing the shape of the steps:

$$E\frac{\dfrac{d^2 y}{dx^2}}{\left[1+\left(\dfrac{dy}{dx}\right)^2\right]^{3/2}} = (P_o - P_m)H - \frac{\rho(A\omega)^2 H}{4}\sin^2\left(\frac{\pi}{L_x}x\right)\sin^2\left(\frac{\pi}{L_y}y\right) \qquad \text{(C.VIII.247)}$$

The numerical solution of this equation is given by the two black curves in the photograph of figure C.VIII.115b. Except for the isthmus, where the interaction between steps should be taken into account, these numerical solutions follow very closely the contour of the real steps.

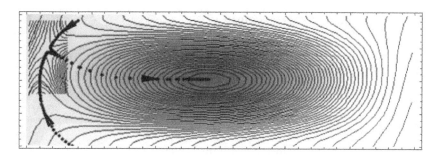

Fig. C.VIII.117 Trajectory of a small island in a vibrating film slightly tilted with respect to the horizontal.

The pressure field induced by the vibrations of a smectic film was also demonstrated by Dumoulin et al. [72] using a strongly elongated rectangular frame, slightly tilted along the azimuthal direction φ by an angle θ with respect to the horizontal. The pressure field is then given by:

$$<P(x,y)> = \frac{1}{4}\rho(A\omega)^2 \sin^2\left(\frac{\pi}{L_x}x\right)\sin^2\left(\frac{\pi}{L_y}y\right)$$

$$+ \rho g\theta\left[(L_x - x)\sin\varphi + (L_y - y)\cos\varphi\right] + P_m \qquad \text{(C.VIII.248)}$$

The "map" in figure C.VIII.117 (showing the isobaric lines) gives its profile in the plane (x,y) of the film.

Consider now that a small island, with surface area S and a height excess of ΔH appears in this film. It will be subjected to the buoyancy

$$\mathbf{f}_A = S\,\Delta H\,\mathbf{grad}\,P(x,y) \qquad \text{(C.VIII.249)}$$

Under the action of this force, the island (of negligible inertia) will move along a trajectory that follows the field lines of the vector field **grad** P, as shown in figure C.VIII.117.

VIII.6.f) Backflow and topological flows in SmC films

When discussing the hydrodynamic behavior of nematics (volume 1, chapter B.III), we emphasized the importance of the **mutual** coupling between the flow $\mathbf{v}(\mathbf{r},t)$ and the dynamics of the director field $\mathbf{n}(\mathbf{r},t)$. In the angular momentum equation

$$\mathbf{n}\times(\mathbf{h} + \mathbf{h}^M - \gamma_1\mathbf{N} - \gamma_2\underline{A}\mathbf{n}) = 0 \qquad \text{(C.VIII.250)}$$

the influence of flow on the orientation of the nematic is due to the hydrodynamic torque (the last two terms) exerted by the flow on the director. We recall that here,

$$\mathbf{N} = \frac{D\mathbf{n}}{Dt} - \Omega\times\mathbf{n} \qquad \text{where} \qquad \Omega = \frac{1}{2}\mathbf{curl}\,\mathbf{v} \qquad \text{(C.VIII.251)}$$

is the difference between the angular velocities of the director and the fluid, while

$$A_{ij} = \frac{1}{2}\left(\frac{\partial v_i}{\partial x_j} + \frac{\partial v_j}{\partial x_i}\right) \qquad \text{(C.VIII.252)}$$

is the symmetric part of the velocity gradient tensor.

The reciprocal effect, describing the influence of director orientation on the flow, appears in the momentum equation

$$\rho\frac{D\mathbf{v}}{Dt} = \text{div}\,\underline{\sigma} + \mathbf{F} \qquad \text{(C.VIII.253)}$$

via the stress tensor

$$\underline{\sigma} = -P\underline{I} + \underline{\sigma}^E + \underline{\sigma}^v \qquad \text{(C.VIII.254)}$$

where the last two terms, depending on the director field, are the viscous stress tensor:

$$\sigma_{ij}^v = \alpha_1 (n_k A_{kl} n_l) n_i n_j + \alpha_2 N_i n_j + \alpha_3 n_i N_j + \alpha_4 A_{ij} + \alpha_5 n_k A_{ki} n_j + \alpha_6 n_i A_{jk} n_k \qquad \text{(C.VIII.255)}$$

and the Ericksen elastic stress tensor:

$$\sigma_{ij}^E = -\frac{\partial f}{\partial n_{k,j}} \frac{\partial n_k}{\partial x_i} \qquad \text{(C.VIII.256)}$$

respectively, where f is the Frank-Oseen deformation energy.

The mutual couplings between the director and velocity fields are summarized in figure C.VIII.118.

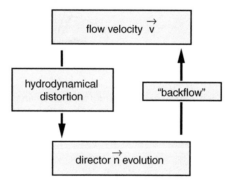

Fig. C.VIII.118 Diagram summarizing the coupling between the director and velocity fields.

To illustrate the influence of flow on the orientation of the nematic, we already cited the experiment of Wahl and Fischer (subsection B.III.1c, volume 1). In this experiment, a shear flow induces a change in the – initially homeotropic – director orientation. This flow → distortion coupling is essential for understanding the mechanisms responsible for the thermal, hydrodynamic and electrohydrodynamic instabilities of nematics (see chapter B.III in the first volume).

The reciprocal effect, known as "backflow," corresponds to a flow induced by the motion of the director. This effect exists, but directly demonstrating it in a nematic is much less easy to do, and a direct illustration was never achieved (see subsection B.III.1d, volume 1). We shall show here that smectic films in the SmC phase behave like two-dimensional nematics and, as such, allow a spectacular illustration of the Ericksen elastic stress and of the backflow effect. On this occasion, we shall see that the topology of the director field plays a fundamental role in nematic hydrodynamics.

i) c-director field in SmC films

We remember that, in the smectic C phase, the molecules are tilted with respect to the layer normal (see also chapter C.IV this volume, and subsection

B.IV.4, volume 1). The polar tilt angle θ is a thermodynamic parameter, depending on temperature, but is almost unchanged by the flow. Thus, one only needs to consider the projection of the molecules onto the plane of the layers to construct a two-dimensional vector order parameter $c(x,y,t)$. By definition $|c| = 1$, so that c is completely defined by the angle φ it makes with the x axis:

$$c = (\cos\varphi, \sin\varphi) \tag{C.VIII.257}$$

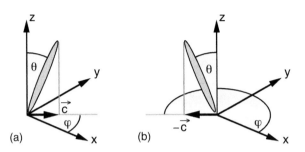

(a)　　　　(b)

Fig. C.VIII.119 A film in the SmC phase behaves as a two-dimensional nematic. The projection of the molecules onto the plane of the layers (x,y) gives the order parameter c. Contrary to the ordinary nematic phase, c (a) is not equivalent to $-c$ (b). This vector is therefore equivalent to a spin.

This order parameter, of amplitude $|c|$, considered fixed, and **phase** φ, bestows on the SmC phase the same elastic and hydrodynamic properties as for two-dimensional nematics, with the distinction that the c and $-c$ orientations are not equivalent (Fig. C.VIII.119). One particular consequence is that the "minimal" rank of disclinations of the director field c is $m = \pm 1$ (see chapter B.IV).

Pindak et al. [73] showed that the director field c of a film in the SmC phase is clearly visible in polarized microscopy between crossed polarizers (in transmission or reflection) if the thickness N of the film is large enough. For very thin films, the contrast can be enhanced by slightly uncrossing the polarizers. These observations showed that the appearance of the director field $c(x,y)$ depends on the film preparation method. If the film is stretched in the SmA phase, and then cooled rapidly to the SmC phase, the director field can exhibit several disclinations with individual charges $m_i = \pm 1$ and total topological charge

$$m = \sum_i m_i \tag{C.VIII.258}$$

usually equal to +1. This total charge originates in the preferred anchoring of the c director on the meniscus delimiting the film. Indeed, if the c director maintains the same local orientation over the contour of the meniscus, the result is a defect of angle 2π when following the Burgers circuit enclosing the film, which therefore bears a global topological charge of +1 (Fig. C.VIII.120a).

One can however create a director field that is everywhere continuous, without any defect, by slowly stretching the film directly in the SmC phase; in this case, a $+2\pi$ localized wall appears, which crosses the edge of the film (Fig. C.VIII.120c). In this point, the preferred anchoring of the **c** director is broken.

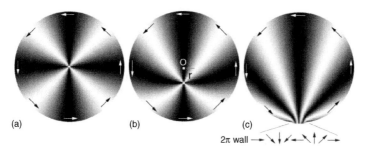

(a)   (b)   (c)

2π wall

Fig. C.VIII.120 Some possible configurations when the **c** director prefers to orient parallel to the meniscus bordering the film. In a) the anchoring is respected over the entire contour of the meniscus: a +1 disclination appears inside the film; in b) the director field is deformed as the disclination moves a distance r from the center of the film; in c) the **c** director field is everywhere continuous (no topological charge): the anchoring is respected at the meniscus, with the exception of the wall of angle 2π crossing the edge of the film.

ii) Brownian motion of a disclination

Muzny and Clark [74] studied the behavior of the +1 disclination in circular 10.O7* films. When the film is very thin (N = 3), the disclination, although it remains localized close to the center of the film, undergoes an intense erratic motion (Fig. C.VIII.121a).

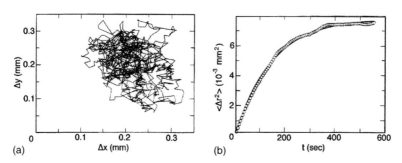

(a)   (b)

Fig. C.VIII.121 Brownian motion of a +1 disclination in a 10.O7* film of three layers at 74.6°C. a) the trajectory of the disclination; b) mean square distance from the center of the film as a function of time (from ref. [74]).

Such behavior is typical of a Brownian particle evolving in a potential well

$$U(r) = \frac{Cr^2}{2} \tag{C.VIII.259}$$

with r the distance with respect to the center of the well. In the present case of a +1 disclination, this potential corresponds to the deformation energy (integrated over the entire surface of the film) associated with a shift r of the disclination with respect to the center of the film (Fig. C.VIII.120b). In the approximation of isotropic elasticity ($k_1 = k_3 = k$), the Frank-Oseen energy for a two-dimensional nematic writes:

$$F = \frac{k}{2} \int (\mathbf{grad}\varphi)^2 \, dxdy \qquad\qquad\qquad \text{(C.VIII.260)}$$

where k is a two-dimensional elastic constant (measured in erg).

At mechanical equilibrium, the function $\varphi(x,y)$ must fulfill the Laplace equation

$$\Delta\varphi = 0 \qquad\qquad\qquad \text{(C.VIII.261)}$$

signifying that the elastic torque per unit surface vanishes.

When the disclination is at the center O of the film (Fig. C.VIII.120a), the director field exhibits revolution symmetry and the expression of the phase $\varphi(x,y)$ is (in the first quadrant $x > 0$ and $y > 0$):

$$\varphi = \arctan\left(\frac{y}{x}\right) + \frac{\pi}{2} \qquad\qquad\qquad \text{(C.VIII.262a)}$$

As the disclination moves from the center of the film to the point of coordinates $(0,-r)$ (Fig. C.VIII.120b), the director field is deformed. Pleiner [75] showed that it is enough to add a second disclination at the point with coordinates $(0,-R^2/r)$ to satisfy automatically the tangential anchoring conditions at the edge of the film. In these conditions, the angle φ of the director field becomes:

$$\varphi = \arctan\left(\frac{y - r}{x}\right) + \arctan\left(\frac{y - R^2/r}{x}\right) + \frac{\pi}{2} \qquad\qquad\qquad \text{(C.VIII.262b)}$$

Calculating the associated energy of elastic deformation, using eq. C.VIII.260, gives the sought-after potential $U(r)$:

$$U(r) = \pi k \ln\left(\frac{1}{1 - (r/R)^2}\right) \approx \pi k \left(\frac{r}{R}\right)^2 \qquad\qquad\qquad \text{(C.VIII.263a)}$$

finally yielding the sought-for constant $C = 2\pi k/R^2$.

To determine the elastic constant k, Muzny and Clark [74] measured the mean square value of the disclination shift $<r^2>$ as a function of time (Fig. C.VIII.121b). In the $t \rightarrow \infty$ limit, this shift tends toward the value $<r^2>_\infty = 0.8\times10^{-2}$ mm^2. Knowing that for a Brownian particle in a two-dimensional harmonic potential the equipartition theorem gives:

$$<U(r)> = \pi k \frac{<r^2>_\infty}{R^2} = k_B T \qquad\qquad\qquad \text{(C.VIII.263b)}$$

one can deduce $k \approx 6 \times 10^{-12}$ erg, using the film diameter $2R = 3.7$ mm. This value of the two-dimensional elastic constant is in good agreement with that calculated using the following estimate:

$$k \approx K H \sin^2\theta \qquad\qquad\qquad (C.VIII.264)$$

with $K \approx 10^{-5}$ dyn (which is typically one order of magnitude larger than in usual nematics), $H \approx 2 \times 10^{-6}$ cm and $\sin^2\theta \approx 0.25$. This formula shows that the elastic constant k increases in direct proportion to the film thickness H. Thus, the Brownian motion of the disclination becomes undetectable in thick films.

Finally, the initial slope of the "$<r^2>$ versus t" graph in figure C.VIII.121b yields the diffusion constant: $D \approx 1.8 \times 10^{-7}$ cm^2/s. This constant is related to the mobility μ of the defect by the Einstein relation:

$$D = k_B T \mu \qquad\qquad\qquad (C.VIII.265)$$

This mobility is itself related to the two-dimensional effective rotational viscosity $\gamma = \gamma_1 H \sin^2\theta$. It can be obtained by integrating the local dissipation $\gamma(d\varphi/dt)^2$ over the surface of the film, assuming that the director field is the same as in the static case (this is undoubtedly a very rough approximation). The calculation yields:

$$\mu^{-1} = 2\pi\gamma \ln(R/r_c) \qquad\qquad\qquad (C.VIII.266)$$

where r_c is the radius of the disclination core. With $r_c \approx 300$ Å and the previously obtained value for D, one finds $\gamma_1 \approx 0.01$ poise, a reasonable – although certainly too small – value. It is therefore very likely that other dissipation sources exist aside from that taken into account by this model, in particular in the disclination core.

iii) Phase winding and backflow

The director field **c** is very easily perturbed by flow or by the action of an external torque (Fig. C.VIII.122). This was shown by Cladis et al. [76, 77] and by Dascalu et al. [78] in a series of experiments that we shall present in the following paragraphs.

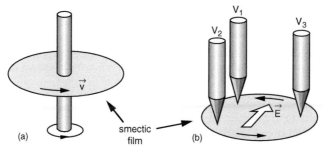

Fig. C.VIII.122 Experiments of Cladis et al. [76, 77]: a) the smectic film is pierced by a rotating fiber; b) a rotating electric field exerts a torque on the **c** director. This field can be produced by a three-phase voltage (V_1,V_2,V_3) applied to three electrodes forming an equilateral triangle.

In their first experiments, Cladis et al. [76] used a fiber to pierce TB9A films in their center (see Fig. C.VIII.122a). The TB9A (terephthalylidene-bis-(4-nonylaniline)) exhibits a SmC phase between 157.5°C and 192.7°C. In their experimental setup, the fiber can be set in rotation and the evolution of the **c** director field observed in polarizing microscopy. The most remarkable outcome of their experiments was showing that **the distortions induced by fiber rotation are strongly dependent on the topology of the director field and on the presence (or absence) of a +1 disclination.**

Let us start by analyzing the case of films **without topological defects**, such as that in figure C.VIII.123a. In this case, the phase φ(x,y) is "wound" by the rotation of the fiber, meaning that, with each complete turn, the angle φ along the circumference of the fiber increases by 2π [76]. With W the number of turns, the phase on the surface of the fiber after rotation stops is given by

$$\varphi(r_o,\psi) = \varphi_o(r_o,\psi) + W2\pi \tag{C.VIII.267}$$

where $\varphi_o(r_o,\psi)$ is the initial phase, in polar coordinates (r,ψ), such that it appears at the fiber surface as it is inserted in the center of the film. After the fiber rotation stops, the φ(r,ψ) field relaxes to its equilibrium configuration given, in isotropic elasticity, by the Laplace equation:

$$\frac{1}{r^2}\frac{\partial^2\varphi}{\partial\psi^2} + \frac{1}{r}\frac{\partial}{\partial r}\left(r\frac{\partial\varphi}{\partial r}\right) = 0 \tag{C.VIII.268}$$

and the boundary conditions at the outer edge and at the fiber surface. We already specified that boundary condition on the fiber (eq. C.VIII.267).

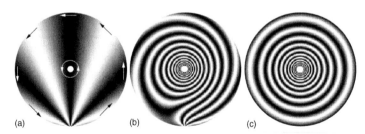

Fig. C.VIII.123 Phase winding by fiber rotation. a) Initial configuration of the **c** director field; b) equilibrium configuration after eight turns of the fiber; c) "wound" component of the phase field (crossed polarizers).

If we assume that at the outer edge (i.e., at the junction with the meniscus), the phase remains unchanged with respect to its initial configuration:

$$\varphi(R,\psi) = \varphi_o(R,\psi) \tag{C.VIII.269}$$

then the solution to the Laplace equation is obvious and reads:

$$\varphi(r,\psi) = \varphi_0(r,\psi) + \Phi_0 \frac{\ln(r/R)}{\ln(r_0/R)} \qquad \text{(C.VIII.270)}$$

where $\varphi_0(r,\psi)$ is the initial phase field. The corresponding **c** director field, as seen between crossed polarizers, is shown in figure C.VIII.123b. Note that here, the final equilibrium state of the director field does not depend on the angular velocity of the fiber since, although the flow accompanies phase winding, it does not produce it.

Let us now see how the distortion produced by fiber rotation can be relaxed. The most obvious way is to turn the fiber eight times the opposite way. Note that, if the fiber were free to turn, it would spontaneously perform these eight turns under the action of the elastic torque

$$\Gamma^{elast} = 2\pi r_0 k \left(\frac{\partial \varphi}{\partial r}\right)_{r=r_0} = 2\pi k \frac{\Phi_0}{\ln(r_0/R)} \qquad \text{(C.VIII.271)}$$

exerted upon it by the **c** director field. Similarly, the distortion could relax if the film holder were free to turn around the axis of the system. Calculating the torque applied upon it by the director field yields exactly the same result – up to a sign change – as for that on the fiber. This should not come as a surprise, since the smectic film transmits the torque between the fiber and the film holder. We emphasize the analogy between this experiment and that of Grupp, described in chapter B.1 of volume 1. However, in the experiment of Cladis et al. [76], neither the fiber nor the holder is free to turn. Thus, one must either turn the fiber the opposite way or withdraw it from the film to relax the distortion.

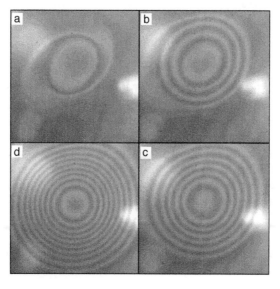

Fig. C.VIII.124 Winding of the phase φ of the **c** director field in a SmC film under the action of a rotating electric field. The three bright spots are the – very fuzzy – images of the electrodes that produce the rotating electric field.

The second way of winding the phase φ is by applying to the molecules a torque **Γ** produced by a rotating electric field (Fig. C.VIII.122b). The effect of this torque on a SmC film is shown in figure C.VIII.124. These images, taken at the beginning of the winding process, show the progressive formation of a "target" texture, composed of dark concentric rings corresponding to equiphase lines φ = Nπ (integer N). These rings are created in the center of the texture and then migrate outward, accumulating at the periphery. This observation reveals that the angular velocity dφ/dt is stronger in the center than at the periphery, due to the torque **Γ** exerted by the rotating field on the molecules being non-uniform in the plane of the film, with a maximum in the center of the triangle formed by the electrodes. When the torque **Γ** is suppressed, the distortion of the "target" texture relaxes by the reverse process, the rings converging towards the center, where they disappear.

Another remarkable feature is that the two processes, the winding and the relaxation of the director field **c**(r,t) (or the phase φ(r,t)), are accompanied by a circular flow v_ψ(r,t) tangent to the rings. This flow is easily detectable by following the motion of small dust particles contained in the films. In both cases, this flow is brought about by the evolving director field. **This is therefore precisely the backflow we wanted to demonstrate.** The experiment in figure C.VIII.124 shows that *backflow*:

1. Occurs as long as the **c** director field evolves.

2. Ceases when the distortion satisfies the static torque balance equation:

$$- k \left[\frac{1}{r^2} \frac{\partial^2 \varphi}{\partial \psi^2} + \frac{1}{r} \frac{\partial}{\partial r} \left(r \frac{\partial \varphi}{\partial r} \right) \right] = \Gamma \qquad \text{(C.VIII.272)}$$

where Γ is the constant applied torque, finite as long as the phase is kept wound up and zero when the winding has relaxed. Note that the angular velocity of this flow, defined by ω = v_ψ/r, has the same sign as the angular velocity of director rotation dφ/dt.

For a theoretical interpretation of backflow, we shall use the simplified equations of nematodynamics. The first simplification is related to the **c** and **v** fields being two dimensional, and to the cylindrical symmetry. A second simplification resides in using the isotropic elasticity approximation $k_1 = k_3 = k$. The last simplification concerns the expression of viscous stress, and consists in considering only two linearly independent viscosity coefficients instead of six (five, if we take into account the Parodi relation). In the following, we shall therefore set:

$$\alpha_1 = \alpha_5 = \alpha_6 = 0, \qquad -\alpha_2 = \alpha_3 = \alpha \quad \text{and} \quad \alpha_4 = \beta \qquad \text{(C.VIII.273)}$$

where the viscosity coefficients α and β describe dissipation due to rotational and irrotational flows, respectively. With this notation:

$$\gamma_1 = \alpha_3 - \alpha_2 = 2\alpha \neq 0, \quad \gamma_2 = \alpha_6 - \alpha_5 = 0, \text{ and } \quad \eta_b = \eta_c = \frac{1}{2}(\alpha + 2\beta) \quad \text{(C.VIII.274)}$$

This last relation implies that the film behaves as an isotropic liquid under simple shear. Note that the viscosities used here are not bulk values, but rather effective two-dimensional viscosities integrated over the thickness of the film, accounting for the molecular tilt angle θ with respect to the layer normal ($\gamma_1 \equiv \gamma_1 H \sin^2\theta$, etc.).

We also assume that the radial velocity is zero ($v_r = 0$) and that the angle φ between the **x** axis and the **c** director, in principle a generic function of r and ψ, can be written as:

$$\varphi = m\psi + \bar{\varphi}\,(r) \tag{C.VIII.275}$$

with integer m. The first part $m\psi$, represents a disclination of rank m, while the second one $\bar{\varphi}(r)$ describes the non-singular part of the director field. Note that, for this particular choice, $\partial^2\varphi / \partial\psi^2 = 0$, further simplifying the equations.

In this "minimalist" framework and using cylindrical coordinates, the equations of angular momentum and momentum have a simple form (for the moment, we do not assume m = 0).

Thus, the angular momentum equation, translating the equilibrium of internal torques, reads:

$$\Gamma - \gamma_1 \left(\frac{D\varphi}{Dt} - \frac{1}{2r}\frac{\partial(vr)}{\partial r} \right) + k\,\frac{1}{r}\frac{\partial}{\partial r}\left(r\,\frac{\partial\varphi}{\partial r} \right) = 0 \quad \text{where} \quad \frac{D\varphi}{Dt} = \frac{\partial\varphi}{\partial t} + \frac{v}{r}\frac{\partial\varphi}{\partial\psi} \tag{C.VIII.276}$$

We recognize in this expression the external torque Γ, the viscous torque proportional to γ_1, and the elastic torque proportional to k. Note that these are torques integrated over the film thickness, so they are expressed per unit surface.

To obtain the momentum equation, we must first calculate the viscous stress tensor. With our approximations, this tensor reduces to only two terms, one of them symmetric, as for an isotropic liquid, and the other one antisymmetric:

$$\sigma^v_{ij} = 2\beta A_{ij} + \alpha(n_i N_j - n_j N_i) \tag{C.VIII.277a}$$

In cylindrical coordinates $(\mathbf{e}_r, \mathbf{e}_\psi)$, this tensor is:

$$\sigma^v_{ij} = \beta r\,\frac{\partial}{\partial r}\left(\frac{v}{r}\right)\begin{bmatrix} 0 & 1 \\ 1 & 0 \end{bmatrix} + \alpha\left(\frac{D\varphi}{Dt} - \frac{1}{2r}\frac{\partial(vr)}{\partial r}\right)\begin{bmatrix} 0 & 1 \\ -1 & 0 \end{bmatrix} \tag{C.VIII.277b}$$

Its antisymmetric part is a result of director rotation with respect to the fluid and corresponds to the viscous torque already encountered in the angular momentum equation. The symmetric part is related to shear flow.

Let us now write the dynamic balance of the torques acting on a – virtually delimited – disk of radius r in the film (Fig. C.VIII.45).

First of all, the disk experiences a torque Γ exerted by the electric field on the molecules within the disk:

$$\Gamma^{ext} = \int_0^r 2\pi r \Gamma(r) \, dr \qquad\qquad (C.VIII.278)$$

This torque must be balanced by the actions exerted on the boundary of the disk by the outer region of the smectic film. First, the edge of the disk bears an elastic torque per unit length $C = k \partial\varphi/\partial r$. At equilibrium, this torque is the only contribution so in this case (and only in this case) we have:

$$\int_0^r 2\pi r \Gamma(r) \, dr + 2\pi r k \frac{\partial\varphi}{\partial r} = 0 \qquad\qquad (C.VIII.279)$$

Taking the derivative of this equation with respect to r gives back the equilibrium equation of internal torques C.VIII.276:

$$\Gamma(r) + k \frac{1}{r} \frac{\partial}{\partial r}\left(r \frac{\partial\varphi}{\partial r}\right) = 0 \qquad\qquad (C.VIII.280)$$

Note that eqs. C.VIII.276 and C.VIII.280 are only equivalent in the static case.

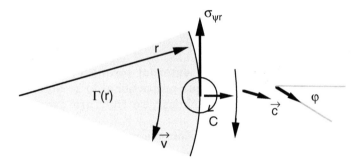

Fig. C.VIII.125 Balance of torques acting on a disk with radius r.

Out of equilibrium, i.e., in the dynamic regime, one must also take into account the viscous stress $\sigma_{\psi r}^V$ and the Ericksen stress $\sigma_{\psi r}^E$ exerted by the outer region of the film on the boundary of the disk. These stresses result in a global torque $2\pi r^2 (\sigma_{\psi r}^V + \sigma_{\psi r}^E)$, where we recognize in the $2\pi r^2$ factor the circle perimeter $2\pi r$ and the lever arm r. The Ericksen stress can be determined from its general expression, and its value is:

$$\sigma_{\psi r}^E = -k \frac{1}{r} \frac{\partial\varphi}{\partial r} \frac{\partial\varphi}{\partial\psi} \qquad\qquad (C.VIII.281)$$

Overall, using expressions C.VIII.277b and C.VIII.281 for the viscous stress and the Ericksen stress, we obtain:

$$\int_0^r 2\pi r \Gamma(r)dr + 2\pi rk\frac{\partial \varphi}{\partial r} + 2\pi r^2 \left[-k\frac{1}{r}\frac{\partial \varphi}{\partial \psi}\frac{\partial \varphi}{\partial r} - \frac{\gamma_1}{2}\left(\frac{D\varphi}{Dt} - \frac{1}{2r}\frac{\partial(rv)}{\partial r}\right) + \beta r\frac{\partial}{\partial r}\left(\frac{v}{r}\right) \right] = 0$$

$$(C.VIII.282)$$

This equation is nothing else than the momentum conservation equation. Adding the internal angular momentum equation:

$$\Gamma - \gamma_1\left(\frac{\partial \varphi}{\partial t} + \frac{v}{r}\frac{\partial \varphi}{\partial \psi} - \frac{1}{2r}\frac{\partial(vr)}{\partial r}\right) + k\frac{1}{r}\frac{\partial}{\partial r}\left(r\frac{\partial \varphi}{\partial r}\right) = 0 \qquad (C.VIII.283)$$

provides a system of coupled equations for the velocity v and the phase φ. Note that this system is valid in the general case, whether the film contains a topological defect in its center or not (eq. C.VIII.275).

These equations once established, let us return to the simple case of a film without topological defect, that we are considering in this subsection. In this case, the Ericksen stress (eq. C.VIII.281) vanishes since the φ field only depends on r and $D\varphi/Dt = \partial\varphi/\partial t$ from eq. C.VIII.276. Furthermore, we assume that the total torque Γ^{ext} is applied in a small region of radius r_0 at the center of the film. Thus, outside this region ($r > r_0$), eq. C.VIII.282 gives:

$$\frac{\Gamma^{ext}}{\pi r^2} + 2k\frac{1}{r}\frac{\partial \varphi}{\partial r} + 2\left[-\frac{\gamma_1}{2}\left(\frac{\partial \varphi}{\partial t} - \frac{1}{2r}\frac{\partial(rv)}{\partial r}\right) + \beta r\frac{\partial}{\partial r}\left(\frac{v}{r}\right) \right] = 0 \qquad (C.VIII.284)$$

while eq. C.VIII.283 becomes:

$$-\gamma_1\left(\frac{\partial \varphi}{\partial t} - \frac{1}{2r}\frac{\partial(rv)}{\partial r}\right) + k\frac{1}{r}\frac{\partial}{\partial r}\left(r\frac{\partial \varphi}{\partial r}\right) = 0 \qquad (C.VIII.285)$$

By subtracting these two equations the viscous term is cancelled, and we find:

$$\frac{\Gamma^{ext}}{\pi r^2} + 2k\frac{1}{r}\frac{\partial \varphi}{\partial r} - k\frac{1}{r}\frac{\partial}{\partial r}\left(r\frac{\partial \varphi}{\partial r}\right) + 2\beta r\frac{\partial}{\partial r}\left(\frac{v}{r}\right) = 0 \qquad (C.VIII.286a)$$

or, equivalently:

$$\frac{\Gamma^{ext}}{\pi r^2} - kr\frac{\partial}{\partial r}\left(\frac{1}{r}\frac{\partial \varphi}{\partial r}\right) + 2\beta r\frac{\partial}{\partial r}\left(\frac{v}{r}\right) = 0 \qquad (C.VIII.286b)$$

If the external torque Γ^{ext} has just been applied, the deformation does not yet exist. Hence, at first, the only action of the Γ^{ext} torque is to create the cylindrical Poiseuille flow, obtained by solving the preceding equation (neglecting the elastic restoring term):

$$V = \frac{\Gamma^{ext}}{4\pi\beta}\left(\frac{1}{r} - \frac{r}{R^2}\right)$$

(C.VIII.287)

The velocity of this flow, vanishing on the outer edge at $r = R$, goes in the same direction as the applied torque. This flow slows down progressively as the phase is wound up and finally stops when the elastic torque is strong enough to compensate the applied torque: equilibrium has been reached. The deformation, obtained by solving eq. C.VIII.286b with $v = 0$, assumes the already encountered logarithmic form (see eq. C.VIII.280):

$$\varphi = \varphi_0(r) - \frac{\Gamma^{ext}}{2\pi k}\ln\left(\frac{r}{R}\right) \qquad (r > r_0)$$

(C.VIII.288)

with $\varphi_0(r)$ the initial phase field.

If, after winding up the phase, the external torque Γ^{ext} is suppressed, the elastic term in eq. C.VIII.286, given by $-\Gamma^{ext}/\pi r^2$, induces a backflow of expression identical to that in eq. C.VIII.287, up to a sign change. To find out the appearance of the deformation and of the velocity field at the end of the relaxation, we assume that

$$V = \frac{k}{2\beta}\frac{\partial\varphi}{\partial r} \qquad \text{and that} \qquad \varphi = e^{-t/\tau}\,\tilde{\varphi}(r) + \varphi_0(r)$$

(C.VIII.289)

This first relation shows that it is the $\partial\varphi/\partial r$ deformation – when it is not balanced by an external torque – that is responsible for the backflow (note that, for the initial phase field $\varphi_0(r)$, $\partial\varphi_0/\partial r = 0$). With this choice, eq. C.VIII.286b is automatically fulfilled, while the torque equation C.VIII.285 yields:

$$-\gamma_1\left[-\frac{1}{\tau}\tilde{\varphi}(r) - \frac{1}{2r}\frac{\partial}{\partial r}\left(\frac{k}{2\beta}r\frac{\partial\tilde{\varphi}(r)}{\partial r}\right)\right] + k\frac{1}{r}\frac{\partial}{\partial r}\left(r\frac{\partial\tilde{\varphi}(r)}{\partial r}\right) = 0$$

(C.VIII.290)

Using R as the length unit ($\tilde{r} = r/R$), this equation can also be written as:

$$\frac{\gamma_1 R^2}{k}\frac{1}{\tau}\tilde{\varphi}(r) + \left(\frac{\gamma_1}{4\beta} + 1\right)\frac{1}{\tilde{r}}\frac{\partial}{\partial\tilde{r}}\left(\tilde{r}\frac{\partial\tilde{\varphi}(r)}{\partial\tilde{r}}\right) = 0$$

(C.VIII.291)

Taking

$$\tau = \frac{(\gamma_1 R^2)/k}{1 + \dfrac{\gamma_1}{4\beta}}$$

(C.VIII.292)

one can see that the deformation $\tilde{\varphi}(\tilde{r})$, solution of this equation, is given by the Bessel function $\mathrm{Re}[Y_0(i\tilde{r})]$. The important point here is that the backflow reduces the rotational viscosity γ_1 by a factor $1 + (\gamma_1/4\beta)$ and thus speeds up the deformation relaxation.

iv) The paradox of the magic spiral

Let us now consider the case of the director field for a +1 disclination placed in the center of the film and containing in its core a solid object ensuring homeotropic anchoring. In the experiment of Cladis et al. [76], this object is a fiber that pierces the film, and its surface induces homeotropic anchoring. In the experiment of Chevallard et al. [79], the anchoring is ensured by a solid inclusion placed in the core of the disclination. Figures C.VIII.126a and b illustrate the fact that the core of a "fresh" disclination, recently created during film stretching, is devoid of solid impurities, so that the anchoring effect on the phase φ does not yet exist. As a result, the phase of the director field can be changed at will by application of a rotating electric field. One can thus obtain the "wound up" director field shown in figure C.VIII.126b. As long as the disclination remains "clean," the phase change at the disclination core is reversible; as the rotating electric field is removed, the director field relaxes towards an unwound phase configuration (Fig. C.VIII.126a). If, on the other hand, the wound-up state in figure C.VIII.126b is maintained for about 24 hours, dust particles and other impurities accumulate in the disclination core (Fig. C.VIII.126c) and block the phase such that, after removing the rotating field, the director field relaxes toward the logarithmic configuration in figure C.VIII.126d.

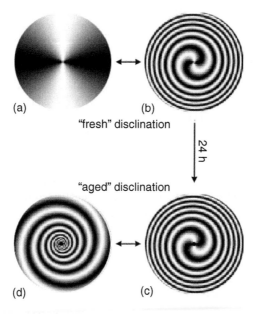

Fig. C.VIII.126 Preparation of a disclination containing a solid inclusion at the core. Due to this inclusion, a stable orientation defect can be created. The film is observed between slightly uncrossed polarizers.

Due to the director being anchored on the edge of the film and on the surface of the inclusion with radius r_0, one can thus create a **stable** phase difference of $\Delta\varphi = \varphi(r_o) - \varphi(R) = W\pi$ between the film edge and the disclination center. At equilibrium, the configuration of the director field has a logarithmic profile, as we have just seen:

$$\varphi(r,\psi) = \psi + W\pi \left(\frac{\ln(r/R)}{\ln(r_o/R)} - 1 \right)$$
(C.VIII.293)

ensuring that the elastic torque per unit surface vanishes. The appearance of such a director field, observed in polarizing microscopy, is represented in figure C.VIII.126d. When the orientation defect is $\Delta\varphi = \pi/4$ (W = 1/4), the director field lines are spirals and the director field exhibits revolution symmetry, as shown in diagram (a) of figure C.VIII.127.

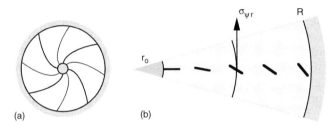

(a)   (b)

Fig. C.VIII.127 Magic spiral configuration. Under the effect of the elastic torque exerted by the director field, one would expect the inclusion at the center of the film to turn. However, it is clear – due to revolution symmetry – that it does not turn. The Ericksen elastic stress (non-zero in the presence of a +1 defect) allows us to solve this paradox.

R.B. Meyer noticed (see ref. [32]) that, in this **magic spiral** configuration, the elastic torque given by eq. C.VIII.271 and accounted for by the second term in eq. C.VIII.282, still acts on the fiber. This finite torque clearly tends to turn the fiber. Yet, the fiber should not turn, as the director field exhibits revolution symmetry (rotating the fiber does not change the director field). Therefore, another elastic torque must compensate this torque exactly, otherwise the paradox cannot be solved. This elastic torque is given by the third term in eq. C.VIII.282: it stems directly from the Ericksen elastic stress tensor given by eq. C.VIII.281. If no defect is present, this stress vanishes. However, it does not vanish in the presence of a rank m defect, since this time $\partial\varphi/\partial\psi = m$:

$$\sigma^E_{\psi r} = -k \frac{\partial\varphi}{r\,\partial\psi}\frac{\partial\varphi}{\partial r} = -km \frac{1}{r}\frac{\partial\varphi}{\partial r}$$
(C.VIII.294)

This term must therefore be considered in eq. C.VIII.282. One must also consider the director advection term appearing in the convective derivative so that, in the presence of a rank m defect (see eq. C.VIII.276):

$$\frac{D\varphi}{Dt} = \frac{\partial\varphi}{\partial t} + m \frac{v}{r}$$
(C.VIII.295)

Finally, in the presence of a rank m disclination, eq. C.VIII.282 can be explicitly written as:

$$\int_0^r 2\pi r \Gamma(r) dr + 2\pi r k \frac{\partial \varphi}{\partial r}(1 - m) + 2\pi r^2 \left[-\frac{\gamma_1}{2}\left(\frac{\partial \varphi}{\partial t} + m\frac{v}{r} - \frac{1}{2r}\frac{\partial(rv)}{\partial r}\right) + \beta r \frac{\partial}{\partial r}\left(\frac{v}{r}\right) \right] = 0$$

$$(C.VIII.296)$$

For m = +1, the second term of this equation can be seen to vanish. In other terms, the total elastic torque exerted by the director field on the perimeter of the fiber, or of a virtually delimited disk of any radius r, is strictly zero for m = +1 (the only case exhibiting revolution symmetry). Thus, the "*perpetuum mobile*" paradox raised by Meyer is solved.

υ) Topological flows

We have just seen that, in the case of a +1 defect, the elastic torque in eq. C.VIII.296 vanishes. As a consequence, the torque Γ^{ext} applied to the fiber, or directly to the film by the rotating field, can only be balanced by the viscous stress. In these conditions, the system can only tend to a **stationary state** characterized by static deformation (as in the equilibrium state), and also by a persistent flow v(r). These permanent flows are termed **topological**, their existence being directly related to the topology of the director field. They are easily detected by following the motion of dust particles suspended in the film, as shown in figure C.VIII.128.

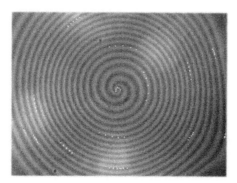

Fig. C.VIII.128 Experimental demonstration of topological flow induced by a rotating field in a SmC film. (from ref. [79]). The white dots indicate the position of dust particles, plotted every 20 seconds.

Let us now investigate the equations more closely. In the stationary state, $\partial\varphi / \partial t = 0$ and eq. C.VIII.296 becomes:

$$\int_0^r 2\pi r \Gamma(r) dr + 2\pi r^2 \left[\beta\left(1 + \frac{\gamma_1}{4\beta}\right) r \frac{\partial}{\partial r}\left(\frac{v}{r}\right) \right] = 0 \qquad (C.VIII.297)$$

As before, let us assume that the torque is applied directly to the fiber or to a small area of radius r_0 in the center of the film. In this case, for $r > r_0$ one finds again a cylindrical Poiseuille flow identical to that given by eq. C.VIII.287, provided the viscosity β is replaced by the effective viscosity

$$\beta_{eff} = \beta \left(1 + \frac{\gamma_1}{4\beta} \right)$$

(C.VIII.298)

larger than β. The reason for this viscosity increase is that the molecules turn as they move, due to the revolution symmetry of the director field. As to the director field, it can be calculated from the equation of internal angular momentum C.VIII.283, including the advection term. In the stationary regime, this equation writes:

$$\frac{\gamma_1}{4\beta} r \frac{\partial}{\partial r} \left(\frac{v}{r} \right) + k \frac{1}{r} \frac{\partial}{\partial r} \left(r \frac{\partial \varphi}{\partial r} \right) = 0$$

(C.VIII.299)

Replacing the velocity by its expression C.VIII.287 (with β replaced by β_{eff}), we obtain:

$$\frac{\Gamma^{ext}}{\pi r^2} \frac{\gamma_1}{4\beta_{eff}} = k \frac{1}{r} \frac{\partial}{\partial r} \left(r \frac{\partial \varphi}{\partial r} \right)$$

(C.VIII.300)

yielding by integration, and taking into account the boundary conditions in figure C.VIII.127:

$$\varphi = \psi + a \frac{\ln(r/R)}{\ln(r_0/R)} + b \ln^2(r/R) - W\pi$$

(C.VIII.301a)

with

$$a = W\pi - b\ln^2(r_0/R) \qquad \text{and} \qquad b = \frac{\Gamma^{ext}}{\pi k} \frac{\gamma_1}{4\beta_{eff}}$$

(C.VIII.301b)

Another interesting case is that when the torque applied to the film is homogeneous ($\Gamma(r) = \Gamma = $ const). The velocity field, calculated from eq. C.VIII.297, is then:

$$v = -\frac{\Gamma}{2\beta_{eff}} r \ln(r/R)$$

(C.VIII.302)

while the general torque equation C.VIII.276 becomes:

$$\tilde{\Gamma} + k \frac{1}{r} \frac{\partial}{\partial r} \left(r \frac{\partial \varphi}{\partial r} \right) = 0$$

(C.VIII.303)

where

$$\tilde{\Gamma} = \Gamma \left(1 - \frac{\gamma_1}{4\beta_{eff}} \right) \qquad \text{(C.VIII.304)}$$

can be seen as an effective torque acting on the director. Integrating this equation, with the same boundary conditions as in figure C.VIII.127, yields:

$$\varphi(r,\psi) = m\psi + W\pi \left[(1 - \alpha) \frac{\ln(r/R)}{\ln(r_o/R)} + \alpha \frac{R^2 - r^2}{R^2 - r_o^2} - 1 \right] \quad m = 1 \qquad \text{(C.VIII.305a)}$$

with

$$\alpha W\pi = \frac{\tilde{\Gamma}}{4k} (R^2 - r_o^2) \qquad \text{(C.VIII.305b)}$$

In the absence of an external torque, the magic spiral has a purely logarithmic shape, since $\alpha = 0$. If, on the other hand, an external torque is applied, one can say that a fraction α of the deformation is converted into a quadratic configuration. In figure C.VIII.128, this transformation is partial; $\alpha < 1$ (see also the photos in figure C.VIII.130).

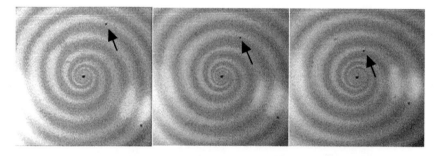

Fig. C.VIII.129 Deformation relaxation for the magic spiral. The dust particle shown by the arrow is dragged by the deformation toward the disclination center.

As the external torque is removed, the deformation relaxes to its logarithmic configuration. This process is illustrated by the series of three photos in figure C.VIII.129. In these images, one can see a dust particle, animated by almost radial motion, dragged toward the center of the disclination.

VIII.6.g) Orbital motion of Cladis disclination lines

The transformation of the logarithmic deformation configuration under the action of the rotating electric field is illustrated in more detail in the series of photos in figure C.VIII.130 [77, 80].

In the first photo, the deformation is due to the phase difference between the center of the disclination and the film edge. In the absence of an external torque, the deformation assumes a purely logarithmic configuration, evidenced in photo 1 by a strong concentration of interference fringes in the vicinity of the disclination core. The effect of the external torque is to transform this logarithmic configuration into a quadratic one, in agreement with formula C.VIII.305a. Configurations 2-5 are stationary states, discussed previously, where the phase does not vary and the topological flows persist. In photo 5, the applied torque Γ is such that the coefficient α is of the order of 1 (defining the critical torque Γ$_{crit}$).

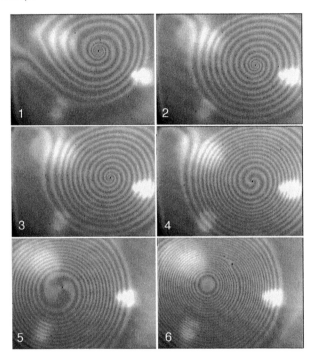

Fig. C.VIII.130 Transformation of the deformation around a disclination under the effect of the rotating electric field. The phase difference Δφ = Wπ between the center of the disclination and the edge of the film remains constant. 1) The logarithmic deformation in the absence of field is very "concentrated" around the core. 2–5) The logarithmic deformation is progressively transformed into a quadratic deformation. 6) Above a critical value of the applied torque, the disclination leaves the center of the film and orbits around the "target" central texture (from ref. [80]).

Cladis et al. [77] noticed that, if the torque Γ is stronger than Γ$_{crit}$, the disclination leaves the center of the film and starts moving along an orbit whose radius increases with the difference ΔΓ = Γ − Γ$_{crit}$. For a constant ΔΓ > 0, the system tends to a new stationary state where the radius of the orbit remains fixed, but the phase of the central "target" texture within the orbit is no longer constant. More precisely, one can observe experimentally that extinction rings are constantly created at the center of the texture and migrate outward, exactly as in the experiment performed on the film with no topological defect (see Fig. C.VIII.124). In the present case, however, the rings do not accumulate in the texture since, with each turn of the disclination, two extinction rings annihilate at the outer edge of the texture. Analytically, this is expressed by the equation:

$$\frac{\partial \varphi (r = 0, t)}{\partial t} - \omega m \, 2\pi = 0 \qquad\qquad (C.VIII.306)$$

where the first term represents the rate of phase generation at the center of the texture and the second term is the rate of phase annihilation by the disclination moving on its orbit at an angular velocity ω (Fig. C.VIII.131).

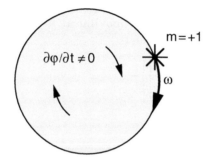

Fig. C.VIII.131 Phase creation and annihilation in the "target" texture.

VIII.6.h) Molecular diffusion in films

It is an experimental fact that a micrometer-sized particle immersed in an isotropic liquid diffuses more slowly in the vicinity of a solid wall than in the bulk [81]. This effect is due to a decrease in the hydrodynamic mobility μ of the particle as it approaches the wall and to the fact that the diffusion coefficient D of the particle is directly proportional to its mobility via the Einstein formula [82]:

$$D = \mu \, k_B T \qquad\qquad (C.VIII.307)$$

We recall that (in the stationary regime) the mobility is defined by the relation $v = \mu F$, where v is the velocity acquired by the particle under the action of an external force F. In an infinite medium, and for a sphere of radius r immersed in a Newtonian fluid of dynamic viscosity η, $\mu_\infty = 1/(6\pi\eta r)$ (Stokes formula [83]). If the fluid is replaced by a smectic A, de Gennes [84] showed that $\mu_\infty = 1/(8\pi\eta r)$ for a sphere moving parallel to the layers (see subsection II.5.b in chapter C.II). The situation is more complicated as the particle moves at a distance z from the free surface of the smectic; in this case, the hydrodynamic boundary conditions at the free surface (vanishing normal velocity and shear stress) are satisfied by placing an image sphere on the other side of the interface. The problem is then reduced to that of two spheres, at a distance 2z, moving parallel to the layers (Fig. C.VIII.132) at the same velocity **v**. This problem is well known by the specialists of suspensions in ordinary liquids. The main result is that, in the first approximation, each particle creates around itself a hydrodynamic flow, which surrounds the other particle and drags it along. Thus, two particles move faster than a single one, due to this dragging effect quantified by the Faxen formula [85] (given here in its simplified form):

$$v \approx v_{Stokes} + v(2z)$$ (C.VIII.308)

where v_{Stokes} is the velocity that the particle would have in an infinite medium, and $v(2z)$ the hydrodynamic velocity produced at its position by the image particle, a distance 2z away and moving at a velocity v_{Stokes}. In the smectic case, this velocity is approximately:

$$v(2z) \approx v_{Stokes} \left(\frac{r}{2z} + O\left[\left(\frac{r}{2z} \right)^2 \right] \right)$$ (C.VIII.309)

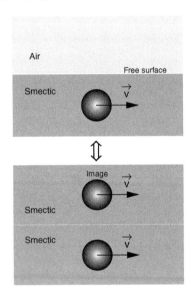

Fig. C.VIII.132 Hydrodynamic equivalent of a sphere moving close to a free surface.

Thus, the mobility of a particle at a distance z from the free surface increases, since, for an equal applied force, its velocity increases. In particular, for the smectic case this mobility $\mu(z)$ is:

$$\mu(z) = \mu_\infty \left(1 + \frac{r}{2z} \right) = \frac{1}{8\pi\eta r} \left(1 + \frac{r}{2z} \right)$$ (C.VIII.310)

In a free film of thickness Na_o, the same reasoning leads (only taking into account the first two images with respect to the two free surfaces) to the following formula:

$$\mu(z) = \mu_\infty \left(1 + \frac{r}{2 \left(\dfrac{Na_o}{2} + z \right)} + \frac{r}{2 \left(\dfrac{Na_o}{2} - z \right)} \right) \qquad \text{(C.VIII.311)}$$

this time z being the distance (z > 0) from the particle to the center of the film.

It turns out that the Stokes-Einstein formula remains applicable in the bulk [82] even when the particle is a simple molecule, for instance a fluorescent probe dissolved in the liquid crystal. On the other hand, Bocquet and Barrat showed, by theoretically studying the problem of self-diffusion close to a solid wall using mode-coupling models [86a,b], that the hydrodynamic arguments invoked above remained valid down to very small scales, of the order of a few molecular layers. In these conditions, we can admit that formula C.VIII.311 remains valid for a fluorescent molecule dissolved in a smectic film. It is then very easy to determine its average diffusion coefficient in the plane of a film with thickness $H = Na_o$:

$$D(N) = 2k_B T \int_0^{Na_o/2} \rho(z)\, \mu(z)\, dz \qquad \text{(C.VIII.312)}$$

where $\rho(z)$ is the normalized profile of probe concentration in the film $(2\int_0^{Na_o/2} \rho(z)\, dz = 1)$.

This formula can also be written as:

$$D(N) = D_\infty \left(1 + \frac{N_h}{N} \right) \qquad \text{(C.VIII.313)}$$

where D_∞ is the bulk diffusion coefficient and N_h the following quantity:

$$N_h = 2N \int_0^{Na_o/2} \rho(z) \left(\frac{\mu(z)}{\mu_\infty} - 1 \right) dz \qquad \text{(C.VIII.314)}$$

Two limiting cases can arise:
1. Either the concentration is uniform throughout the film (i.e., $\rho(z) = 1/Na_o$). In these conditions:

$$N_h = \frac{2}{a_o} \int_0^{Na_o/2} \frac{\sigma_h}{2\left(\dfrac{Na_o}{2} + z \right)} + \frac{\sigma_h}{2\left(\dfrac{Na_o}{2} - z \right)} \, dz = \frac{\sigma_h}{a_o} \ln \left(\frac{Na_o}{\sigma_c} \right) \qquad \text{(C.VIII.315)}$$

In this formula, σ_h is the "hydrodynamic radius" of the molecule (of the order of its size) and $\sigma_c \approx a_o$ is a cutoff distance, necessary so that the mobility does not diverge at the interface.

2. Or the molecules concentrate in the center of the film, being repelled by the surfaces. In this extreme case:

$$N_h = \frac{2\sigma_h}{a_o} \tag{C.VIII.316}$$

In any case, the diffusion coefficient increases as the film thins. The thickness dependence depends on the density profile, but it essentially varies as $1/N$. This slow dependence is a direct consequence of the $1/r$ decrease of hydrodynamic interactions between particles.

To test these theoretical predictions, Bechhoefer et al. [87], and then Picano [35a], measured the diffusion coefficients of two fluorescent probes of different size in SmA films of 8CB as a function of thickness.

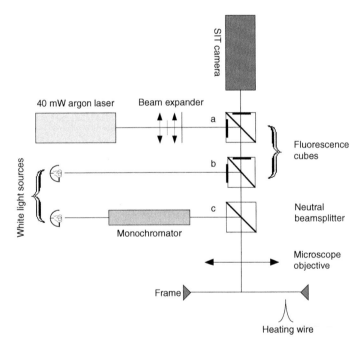

Fig. C.VIII.133 Setup used to measure the diffusion coefficient of fluorescent molecules in smectic A films. a) Thickness measurement; b) photobleaching; c) observation of the diffusion spot by fluorescence microscopy (from ref. [88]).

The setup employed is shown in figure C.VIII.133. It consists of three optical paths: the first one (a) is for focusing the beam of a 40 mW Argon laser on the film for a few seconds. In this way, one can bleach most of the fluorescent probes within the lit area. The second line (b) uses a very sensitive SIT camera, a fluorescence cube and a low-intensity white light source for visualizing the fluorescence profile of the spot as a function of time. It can be shown that the profile very quickly becomes Gaussian (irrespective of its initial profile) (Fig. C.VIII.134) and that its half-maximum surface varies in time as cst+4Dt (Fig. C.VIII.135). Line (c), equipped with a monochromator, allows a precise measurement of film thickness. Finally, a heating wire placed under the film is used to thin it by one layer after each measurement.

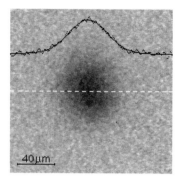

Fig. C.VIII.134 Image of the film after photobleaching. The black dots represent the concentration profile measured along the dotted line. The solid curve is the best fit with a Gaussian profile (from ref. [87]).

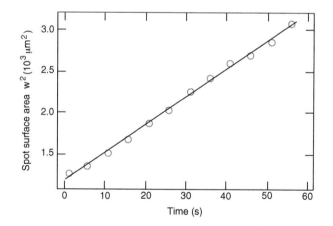

Fig. C.VIII.135 Half-maximum surface area of the diffusion spot as a function of time. A linear interpolation gives $D = 6.1 \times 10^{-8}$ cm^2/s (DIA, N = 7) (from ref. [87]).

The two probes employed are shown in figure C.VIII.136. The first one, DIA (4-(4-dihexadecylaminostyryl)-N-methylpyridinium iodide), is typically twice as large as the other, NBDDOA (4-(N,N-dioctyl)amino-7-nitrobenz-2-oxa-1,3-diazole). The experimental results for these two probes are shown in figure C.VIII.137. As expected, the smaller probe diffuses faster but, more significantly, the diffusion coefficients of both probes increase slowly, following a 1/N law as the film thickness decreases. Moreover, the measured coefficient N_h

is 2.3 times larger for NBDDOA than for DIA. This result is in agreement with the theory, which predicts that N_h is proportional to the molecular size.

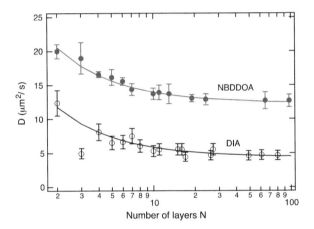

DIA

NBDDOA

Fig. C.VIII.136 The two fluorescent probes used for measuring the diffusion coefficients.

Fig. C.VIII.137 Diffusion coefficients as a function of thickness in 8CB films. The solid lines correspond to the best fits with the law C.VIII.313; a) DIA, $D_\infty = 4.4 \pm 0.1 \times 10^{-8}$ cm^2/s, $N_h = 3.4 \pm 0.3$; b) NBDDOA, $D_\infty = 12.2 \pm 0.1 \times 10^{-8}$ cm^2/s, $N_h = 1.46 \pm 0.1$ (from ref. [87] and the experimental data of F. Picano [35a]).

In conclusion, let us point out the anomalous behavior of the DIA probe in 3-layer films of the liquid crystal 8CB: the diffusion coefficient is almost twice as small as expected by interpolating the experimental values obtained for $N = 2$ and $N = 4$. This is not an experimental error, as this value was measured several times, in different films (ergo the small error bar on the experimental point). An explanation could be that the DIA molecules form dimers in 3-layer films, but this interpretation remains hypothetical.

BIBLIOGRAPHY

[1] Georges Friedel, "Mesomorphic states of matter," *Annales de Physique*, **18** (1922) 273.

[2] a) Young C.Y., Pindak R., Clark N.A., Meyer R.B., *Phys. Rev. Lett.*, **40** (1978) 773.
b) Chiang H.-T., Chen-White V.S., Pindak R., Seul M., *J. Phys. II (France)*, **5** (1995) 835.

[3] Boys C.V., *Soap Bubbles: Their Colors and Forces Which Mold Them*, Dover Publications, Inc., New York, 1959.

[4] Smith G.S., Sirota E.B., Safinya C.R., Plano R.J., Clark N.A., *J. Chem. Phys.*, **92** (1990) 4519.

[5] Bergeron V., *J. Phys. Condens. Matter*, **11** (1999) R215.

[6] Pershan P.S., *Structure of Liquid Crystal Phases*, World Scientific, Singapore, 1988, p. 45.

[7] a) Even C., "Vibrations of smectic films on a fractal frame," Solid-state physics M.Sc. internship report, University of Paris XI, Orsay 1996.
b) Even C., Russ. S., Repain V., Pieranski P., Sapoval B., *Phys. Rev. Lett.*, **83** (1999) 726.
c) Even C., Pieranski P., *Europhys. Lett.*, **47** (1999) 531.

[8] Geminard J.-C., Laroche C., Oswald P., *Phys. Rev. E*, **58** (1998) 5923.

[9] Lejcek L., Oswald P., *J. Phys. II (France)*, **1** (1991) 931.

[10] Pieranski P., Beliard L., Tournellec J.-Ph. et al., *Physica A*, **194** (1993) 364.

[11] Eberhardt M., Meyer R.B., *Rev. Sci. Instr.*, **67** (1996) 2846.

[12] a) Schneider F., *Rev. Sci. Instr.*, **73** (2002) 114.
b) Jacquet R., Schneider F., *Phys. Rev. E*, **67** (2003) 021707.

[13] Oswald P., *J. Physique (France)*, **48** (1987) 897.

[14] a) Stannarius R., Cramer C., *Liq. Cryst.*, **23** (1997) 371.
b) Stannarius R., Cramer C., *Europhys. Lett.*, **42** (1998) 43.
c) Stannarius R., Cramer C., Schüring H., *Mol. Cryst. Liq. Cryst.* A, **329** (1999) 1035 (Part 2).
d) Stannarius R., Cramer C., Schüring H., *Mol. Cryst. Liq. Cryst.*, **350** (2000) 297.
e) Schüring H., Thieme C., Stannarius R., *Liq. Cryst.*, **28** (2001) 241.

[15] Mach P., Grantz S., Debe D.A., Stoebe T. , Huang C.C., *J. Phys. II (France)*, **5** (1995) 217.

[16] Mach P., Huang C.C., Nguyen H.T., *Phys. Rev. Lett.*, **80** (1998) 732.

[17] Miyano K., *Phys. Rev. A*, **26** (1982) 1820.

[18] Picano F., Holyst R., Oswald P., *Phys. Rev. E*, **62** (2000) 3747.

[19] a) Picano F., Oswald P., Kats E., *Phys. Rev. E*, **63** (2001) 021705.
b) Poniewierski A., Holyst R., Oswald P., *Langmuir*, **18** (2002) 1511.

[20] de Gennes P.-G., *Langmuir*, **6** (1990) 1448.

[21] Gorodetskiï E.E., Pikina E.S., Podnek V.E., *JETP*, **88** (1999) 35.

[22] Richetti P., Kekicheff P., Barois P., *J. Phys. II (France)*, **5** (1995) 1129.

[23] Safran S. A., *Statistical Thermodynamics of Surfaces, Interfaces, and Membranes*, Addison-Wesley Publishing Company, New York, 1994.

[24] Rowlinson J. S., Widom B., *Molecular Theory of Capillarity*, Clarendon Press, Oxford, 1982.

[25] Mol E.A.L., Wong G.C.L., Petit J.M., Rieutord F., de Jeu W.H., *Physica B*, **248** (1998) 191.

[26] a) Dolganov V.K., Demikhov E.I., Fouret R., Gors C., *Phys. Lett. A*, **220** (1996) 242.
b) Demikhov E.I., Dolganov V.K., Meletov K.P., *Phys. Rev. E*, **52** (1995) R1285.

[27] a) Stoebe T., Mach P., Huang C.C., *Phys. Rev. Lett.*, **73** (1994) 1384.
b) Johnson P.M.,Mach P., Wedell E.D., Lintgen F., Neubert M., Huang C.C., *Phys. Rev. E*, **55** (1997) 4386.
c) Pankratz S., Johnson P.M., Nguyen H.T., Huang C.C., *Phys. Rev. E*, **58** (1998) R2721.
d) Pankratz S., Johnson P.M., Holyst R., Huang C.C., *Phys. Rev. E*, **60** (1999) R2456.
e) Pankratz S., Johnson P.M., Paulson A., Huang C.C., *Phys. Rev. E*, **61** (2000) 6689.
f) Veum M., Kutschera E., Voshell N., Wang S.T., Wang S.L., Nguyen H.T., Huang C.C., *Phys. Rev. E*, **71** (2005) 020701.

[28] a) Bubbles of isotropic liquid were also observed in "thick" films of SmC or SmC* phases as they are heated above their melting temperature. These

bubbles, more similar to flat lenses than to spheres, are stable over the temperature range where the two phases coexist. This observation emphasizes the importance of impurities in these systems. For a detailed study of their shape, see Schüring H., Stannarius R, *Langmuir*, **18** (2002) 9735.

b) It should also be said that, in these films, the bubbles interact elastically via the Schlieren and self-organize. These problems were studied in the following references: Cluzeau P., Dolganov V., Poulin P., Joly G., Nguyen H.T., *Mol. Cryst. Liq. Cryst.*, **364** (2001) 381; Cluzeau P., Joly G., Nguyen H.T., Dolganov V., *JETP Lett.*, **75** (2002) 482; *ibid.*, *JETP Lett.*, **76** (2002) 351.

[29] a) Moreau L., Richetti P., Barois P., *Phys. Rev. Lett.*, **73** (1994) 3556.
b) Richetti P., Moreau L., Barois P., Kekicheff P., *Phys. Rev. E*, **54** (1996) 1749.

[30] Davidov D., Safinya C.R., Kaplan M., Dana S.S., Schaetzing R.J., Birgeneau R.J., Lister J.D., *Phys. Rev. Lett.*, **31** (1979) 1657.

[31] Sprunt S., Solomon L., Lister J.D., *Phys. Rev. Lett.*, **31** (1979) 1657.

[32] de Gennes P.G., Prost J., *The Physics of Liquid Crystals*, Clarendon Press, Oxford, 1995.

[33] a) Geminard J.-C., Holyst R., Oswald P., *Phys. Rev. Lett.*, **78** (1997) 1924.
b) Zywocinski A., Picano F., Oswald P., Geminard J.-C., *Phys. Rev. E*, **62** (2000) 8133.

[34] Friedel J., *Dislocations*, Pergamon Press, Oxford, 1967.

[35] a) Picano F., "Edge dislocations and thinning transitions in free-standing films of smectic liquid crystal," Ph.D. Thesis, Ecole Normale Supérieure de Lyon, Lyon, 2001.
b) Oswald P., Pieranski P., Picano F., Holyst R., *Phys. Rev. Lett.*, **88** (2002) 15503.

[36] a) Caillier F., Oswald P., *Phys. Rev. E*, **70** (2004) 031704.
b) Oswald P., Picano F., Caillier F., *Phys. Rev. E*, **68** (2003) 061701.

[37] Fournier J.-B., "Bulk faceting of smectic liquid crystals," D.Sc. Thesis, University of Paris XI, Orsay, 1991.

[38] a) Kraus I., Pieranski P., Demikhov E.I., Stegemeyer H., Goodby J.W., *Phys. Rev. E*, **48** (1993) 1916.
b) Kraus I., "Studies of free-standing films of thermotropic liquid crystals in the smectic phase," Ph.D. Thesis, University of Paris XI, Orsay, 1995.

[39] Bahr Ch., Fliegner D., *Phys. Rev. A*, **46** (1992) 7657.

[40] Heinekamp S., Pelcovits R., Fontes E., Chen E. Yi, Pindak R., Meyer R.B., *Phys. Rev. Lett.*, **52** (1984) 1017.

[41] Demikhov E.I., Hoffmann U., Stegemeyer H., *J. Phys. II (France)*, **4** (1994) 1865.

[42] Bahr Ch., *Intern. J. Mod. Phys.*, **8** (1994) 3051.

[43] Huang C.C., Stoebe T., *Adv. in Phys.*, **42** (1993) 343.

[44] Pindak R., Bishop D.J., Sprenger W.O., *Phys. Rev. Lett.*, **44** (1980) 1461.

[45] Chih-Yu Chao, Chia-Fu Chou, Ho J. T., Hui S.W., Anjun Jin, Huang C.C., *Phys. Rev. Lett.*, **77** (1996) 2750.

[46] Geer R., Huang C.C., Pindak R., Goodby J.W. , *Phys. Rev. Lett.*, **63** (1989) 540.

[47] Geer R., Stoebe T., Pitchford T., Huang C.C., *Rev. Sci. Instr.*, **62** (1991) 415.

[48] Huang C.C., Viner J.M., Pindak R., Goodby J.W., *Phys. Rev. Lett.*, **46** (1981) 1289.

[49] Birgeneau R.J., Garland C.W., Kortan A.R., Litster J.D., Meichle M., Ocko B.M., Rosenblatt C., Yu L.J., Goodby J.W., *Phys. Rev. A*, **27** (1983) 1251.

[50] Geer R., Stoebe T., Huang C.C., Goodby J.W., *Phys. Rev. A*, **46** (1992) 6162.

[51] Huang C.C., Stoebe T., *Adv. in Phys.*, **42** (1993) 343.

[52] Aeppli G., Litster J.D., Birgeneau R.J., Pershan P.S., *Mol. Cryst. Liq. Cryst.*, **67** (1981) 205.

[53] Collett J., Sorensen L.B., Pershan P.S., Litster J.D., Birgeneau R.J., Als-Nielsen J., *Phys. Rev. Lett.*, **49** (1982) 553.

[54] Sirota E.B., Pershan P.S., Sorensen L.B., Collett J., *Phys. Rev. A*, **36** (1987) 2890.

[55] Miyano K., *Phys. Rev. A*, **26** (1982) 1820.

[56] Kraus I., Bahr Ch., Chikina I.V., Pieranski P., *Phys. Rev. E*, **58** (1998) 610.

[57] Kac M., *Am. Math. Monthly*, **73** (1966) 1.

[58] Gordon C., Webb D., Wolpert S., *Bull. Am Math. Soc.*, **27** (1992) 134.

[59] Buser P., *Lect. Notes Math.*, **1339** (1988) 63.

[60] Buser P., *Geometry and Spectra of Compact Riemann Surfaces*, Birkhäser, Basel, 1992.

[61] Gottlieb H.P.W., *Europhys. Lett.*, **50** (2000) 280.

[62] see, for instance, Lapidus M.L., *Trans. Am. Math. Soc.*, **323** (1991) 465.

[63] Mandelbrot B., *Fractal Objects* (in French), Flammarion, Paris, 1975.

[64] Berry M.V., Wilkinson M., *Proc. Roy. Soc. Lond. A*, **392** (1984) 14.

[65] a) Brazovskaia M., "Vibrations of freely-suspended smectic films: nonlinear effects, diabolic points and self-tuning oscillators," Ph.D. Thesis, University of Paris XI, Orsay, 1998.
b) Brazovskaia M., Pieranski P., *Phys. Rev. E*, **58** (1998) 4076.

[66] Cheon T., Shigehara T., *Phys. Rev. Lett.*, **76** (1996) 1770.

[67] Ben Amar M., da Silva P.P., Limodin N., Langlois A., Brazovskaia M., Even C., Chikina I.V., Pieranski P., *Eur. Phys. J.*, **3** (1998) 197.

[68] Geminard J-Ch., private communication, 2001.

[69] Ben Amar M., da Silva P.P., Brazovskaia M., Even C., Pieranski P., *Phil. Mag. B*, **78** (1998) 115.

[70] Struik D.J., *Lectures on Classical Differential Geometry*, Dover Publications Inc., New York, 1950.

[71] Brazovskaia M., Dumoulin H., Pieranski P., *Phys. Rev. Lett.*, **76** (1996) 1655.

[72] Dumoulin H., Brazovskaia M., Pieranski P., *Europhys. Lett.*, **35** (1996) 505.

[73] Pindak R., Young C.Y., Meyer R.B., Clark N.A., *Phys. Rev. Lett.*, **45** (1980) 1193.

[74] Muzny C.D., Clark N.A., *Phys. Rev. Lett.*, **68** (1992) 804.

[75] Pleiner H., *Phys. Rev. A*, **37** (1988) 3986.

[76] Cladis P.E., Couder Y., Brand H.R., *Phys. Rev. Lett.*, **55** (1985) 2945.

[77] Cladis P.E., Finn P.L., Brand H.R., *Phys. Rev. Lett.*, **75** (1995) 1518.

[78] Dascalu C., Hauck G., Koswig H.D., Labes U., *Liq. Cryst.*, **21** (1996) 733.

[79] Chevallard C., Gilli J.-M., Frisch T., Chikina I.V., Pieranski P., *Mol. Cryst. Liq. Cryst.*, **328** (1999) 595.

[80] Chevallard C., Gilli J.-M., Frisch T., Chikina I.V., Pieranski P., *Mol. Cryst. Liq. Cryst.*, **328** (1999) 589.

[81] Faucheux L.P., Libchaber A.J., *Phys. Rev. E*, **49** (1994) 5158.

[82] Hansen J.P., McDonald I.R., *Theory of Simple Liquids*, Academic Press, London, 1986.

[83] Guyon E., Hulin J.-P., Petit L., *Hydrodynamique Physique*, Inter Editions/CNRS Editions, Les Ulis, 1991.

[84] de Gennes P.G., *The Physics of Fluids*, **17** (1974) 1645.

[85] Happel J., Brenner H., *Low Reynolds Number Hydrodynamics*, Martinus Nijhoff Publishers, Dordrecht, 1983.

[86] a) Bocquet L., Barrat J.-L., *Europhys. Lett.*, **31** (1995) 455.
b) Bocquet L., Barrat J.-L., *J. Phys.: Condens. Matter*, **8** (1996) 9297.

[87] Bechhoefer J., Geminard J.-C., Bocquet L., Oswald P., *Phys. Rev. Lett.*, **79** (1997) 4922.

Chapter C.IX

Columnar phases

Soap makers have known for a very long time that the consistency of a soap and water mixture depends very strongly on its composition. From fluid and optically isotropic when it is rich in water (micellar phase), the mixture suddenly becomes very viscous when the soap concentration exceeds 40 to 50% by weight. This new phase, birefringent and very pasty, was first named "middle phase," as it always appears in the center of the phase diagrams. X-ray studies showed that the amphiphilic molecules form tubular aggregates (or columns) that organize on a hexagonal lattice, hence the name hexagonal phase (denoted by H_1 in the literature) (Fig. C.IX.1).

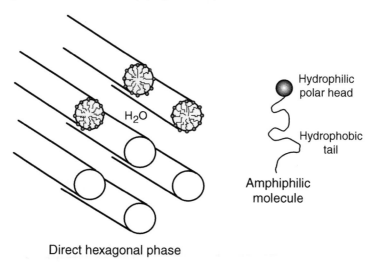

Direct hexagonal phase

Fig. C.IX.1 Schematic representation of an amphiphilic molecule and of the direct hexagonal phase H_1 (the water is outside the tubes).

At higher soap concentration (above 60%), another phase is encountered, which used to be called "neat phase"; also birefringent, this phase flows much more easily than the previous one, which might seem surprising since it is more concentrated in soap. It is the lamellar phase L_α. In certain systems, a very viscous isotropic phase, of cubic symmetry, appears between these two phases [1]. A typical phase diagram, for a mixture of water and the non-ionic surfactant $C_{12}EO_6$ (hexaethylene glycol mono-n-dodecyl ether), is shown in figure C.IX.2.

Although the hexagonal phase has been long known in lyotropic mixtures [1, 2] and in dry soaps [3], its discovery in thermotropic systems (of non-ionic molecules) is more recent, dating from the end of the 1970s [4]. Generally, the molecules that exhibit this type of organization have a disk-like shape (sometimes incorrectly called "discotic") with a stiff central core surrounded by flexible chains. In the hexagonal phase, these molecules stack like dishes into columns arranged on the sites of a two-dimensional hexagonal lattice (Fig. C.IX.3). The columns are usually fluid, in the sense that the distance between two molecules within the same column fluctuates more or less strongly with no long-range order. On the other hand, there is no positional correlation between molecules belonging to two neighboring columns. Thus, the columns must slide easily one over the other, like the layers of a smectic A phase. In the literature, this phase is denoted by D_h, where "D" points to the presence of disk-like molecules, while the subscript "h" specifies a hexagonal lattice. One sometimes encounters the additional subscripts "o" (for ordered) and "d" (for disordered), indicating that the distance between molecules within a column fluctuates more or less strongly. We would like to point out that there is no fundamental reason to distinguish between the D_{ho} and D_{hd} phases (the former ones being presumably stiffer than the latter).

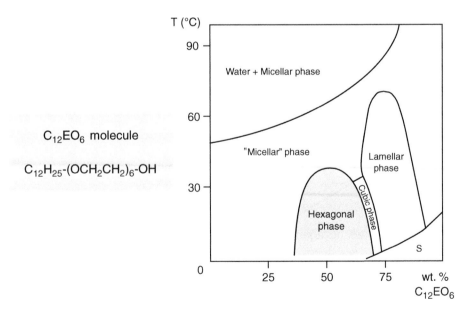

Fig. C.IX.2 The $C_{12}EO_6$ amphiphilic molecule and its phase diagram in aqueous solution (from ref. [2]).

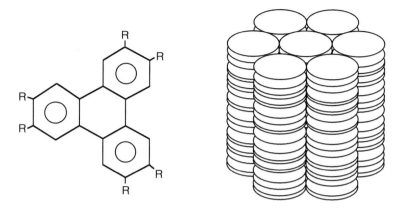

Fig. C.IX.3 Example of a "discotic" molecule (a derivative of triphenylene) exhibiting a hexagonal phase. R represents here an ether or ester function decorated with an aliphatic chain of varying length.

The plan of the chapter is as follows: in section IX.1 we describe several columnar phases and their optical properties. We shall see that the columns do not always form a hexagonal lattice, and that rectangular, oblique or triangular phases also appear. We shall also see that the molecules comprising these phases are not always disk-like as, for instance, the "phasmids," which are elongated molecules. In section IX.2 we shall establish, by analogy with smectics, the expression of the elastic free energy of a hexagonal phase. We shall then study its defects and textures. First, we shall describe the construction of "developable domains," similar to the focal domains of smectic phases (section IX.3), and then describe the dislocations (section IX.4). We shall conclude by a description of column buckling instabilities (section IX.5), one of which has no equivalent in smectics. The rheology of these phases will be only briefly discussed, as it is not yet very well known.

IX.1   Structure and optical properties

Since giving a list of all known molecules and of all possible columnar structures would be quite tedious, we shall mainly describe the phases obtained using certain triphenylene derivatives (Fig. C.IX.3), and mention a few other typical examples at the end of this section.

Figure C.IX.4 shows the two most commonly encountered types of lattices and their space groups as a function of the nature and the length of the R chains attached to the triphenylene core. The solid line disks represent a molecule perpendicular to the axis of the column. An ellipse shows a molecule tilted with respect to the axis of the cylinder. One can see in table 1 that all ethers give the hexagonal phase, while esters favor the rectangular phase (denoted by D_r in the literature).

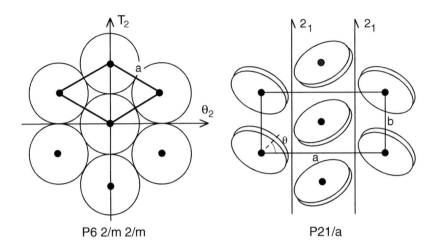

Fig. C.IX.4 The two most frequently encountered two-dimensional lattices and the corresponding space groups [5]. The arrows represent binary axes, while the half-arrows represent helical binary axes 2_1.

R	Space group	Lattice parameters
$C_5H_{11}O$	P6 2/m 2/m	a = 19 Å (80°C)
$C_8H_{17}O$	P6 2/m 2/m	a = 23.3 Å (80°C)
$C_7H_{15}COO$	P 2_1/a	a = 37.8 Å, b = 22.2 Å (100°C)
$C_{11}H_{23}COO$	P 2_1/a	a = 44.9 Å, b = 22.2 Å (100°C)
	P6 2/m 2/m	a = 26.3 Å (105°C)

Table 1 Columnar structures formed by some derivatives of triphenylene (from ref. [5]).

The hexagonal phase is optically uniaxial, the optical axis being parallel to the columns. Its birefringence Δn is not always easy to determine, as it is sometimes impossible to obtain monocrystals. An indirect method, which does not require the phase to be aligned, takes advantage of the textures formed while slowly cooling down from the isotropic phase a free droplet of the D_h phase placed on a glass slide. The molecular columns usually orient parallel to the glass and to the interface with air. Let us look at such a droplet using transmission microscopy between crossed polarizers (Fig. C.IX.5). If the droplet is thick enough, one can see in monochromatic light (for instance, in the green at λ = 546 nm) several dark fringes that follow the molecular columns. Since extinction occurs when the difference in optical path between the ordinary and the extraordinary rays is an integral multiple of the wavelength, the sample thickness varies between two dark fringes by $\Delta e = \lambda / |\Delta n|$. One only needs to determine Δe in order to obtain $|\Delta n|$. This can be done very precisely by mapping the height profile of the droplet using an interference Michelson objective. This method was used to determine the birefringence of hexapentyloxytriphenylene (C5HET in the following): $|\Delta n| = 0.143$ [6a]. As to the sign of Δn, it can be deduced by analyzing the deviations induced by the defects (see Fig. C.IX.13 in section 3). In triphenylene ethers, Δn is negative [6b].

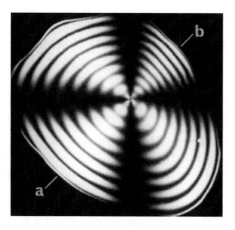

Fig. C.IX.5 Observation between crossed polarizers and using green light ($\lambda = 546$ nm) of a free C5HET droplet placed on a glass slide. The molecular columns are parallel to the glass surface and form circles centered on the center of the droplet. Measuring the profile of the droplet using Michelson interferometry shows that the difference in thickness between two dark fringes is 3790 nm.

The birefringence of lyotropic hexagonal phases is usually very weak. For instance, $|\Delta n| \approx 1.5 \times 10^{-3}$ at room temperature in the $C_{12}E O_6 +$ water mixture close to the azeotropic concentration (Fig. C.IX.2) [7]. We recall that at this concentration (50% of surfactant by weight in the $C_{12}EO_6 +$ water system), the solidus and the liquidus meet and the mixture behaves as a pure substance. Note that one can produce planar monodomains of this phase in directional growth.

The rectangular phases D_r are biaxial due to their symmetry. In phases with the space group $P2_1/a$ (Fig. C.IX.4), the refractive index ellipsoid has an axis (A1) along the helical axis, a second one (A2) at the intersection of the planes parallel to the two kinds of molecule and a third one (A3) perpendicular to the first two [8]. We shall see in section 3 that the angle Φ made by the (A2) axis with the **c** axis of the columns is easily determined using polarizing microscopy. It is related to the two structural angles θ and φ by the formula [8]:

$$\tan\Phi = \tan\varphi \cos\theta \qquad\qquad (C.IX.1)$$

where φ is the tilt angle of the molecules with respect to the (**a**,**b**) plane and θ the angle made by the **a** axis with the projection of the director (perpendicular to the molecule) onto the (**a**,**b**) plane (Fig. C.IX.4). One can also write the principal polarizabilities α_i of the phase as a function of the angles θ and φ and of the intrinsic polarizabilities of the molecule, along ($\alpha_{//}$) and perpendicular to ($\alpha_{\perp}$) its symmetry axis (itself perpendicular to the plane of the molecule) [8]:

$$\alpha_1 = \alpha_\perp - (\alpha_\perp - \alpha_{//}) \sin^2\varphi \sin^2\theta \qquad\qquad \text{(C.IX.2a)}$$

$$\alpha_2 = \alpha_\perp \qquad\qquad \text{(C.IX.2b)}$$

$$\alpha_3 = \alpha_\perp - (\alpha_\perp - \alpha_{//}) (1 - \sin^2\varphi \sin^2\theta) \qquad\qquad \text{(C.IX.2c)}$$

These quantities are directly related to the principal indices n_i by the Vuks formula:

$$\frac{n_i^2 - 1}{\bar{n}^2 + 2} = \frac{N\alpha_i}{3\varepsilon_0} \qquad\qquad \text{(C.IX.3)}$$

where $\bar{n}$ is the average index and N the number of molecules per unit volume. Measuring the n_i indices and the Φ angle one should then be able to obtain angles θ and φ. This could be useful, as the θ angle is difficult to measure using X-rays.

An example of the concerted use of X-ray and optical techniques is given in ref. [9]. This paper describes the two columnar phases of a homochiral triphenylene hexaester. The chains R have the chemical formula:

$$O - \underset{\underset{O}{\|}}{C} - CH_2 - \overset{*}{\underset{\underset{CH_3}{|}}{CH}} - C_6H_{13}$$

where C* stands for the asymmetric carbon.

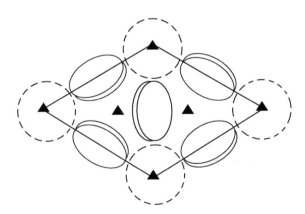

Fig. C.IX.6 Low-temperature phase observed in a chiral triphenylene hexaester. The lattice is triangular, with four columns per unit cell. Note the coexistence of a disordered column (dashed line) with three ordered columns. The solid triangles represent the ternary axes of the columnar lattice.

This material exhibits two columnar phases, one of them being rectangular, of apparent symmetry P2$_1$/a (high-temperature phase, Fig. C.IX.4), while the other (low-temperature phase, Fig. C.IX.6) has P321 symmetry. In this new phase, the lattice is triangular and the area of the lattice cell is double that of the high-temperature phase. Thus, there are four columns per unit cell (instead of two in the rectangular phase and one in the hexagonal phase). As for the rectangular phase, the molecules are tilted with respect to the axis of the columns, making an angle φ with the plane of the lattice. The original character of this phase is related to the presence of a disordered column (represented in dashed line in the Fig. C.IX.6), where the rotation axis of the molecules has no fixed orientation, and of three ordered columns, with well-defined rotation axes, related by a threefold symmetry. From the optical point of view, this phase is optically uniaxial, its optical axis being parallel to the columns. Its principal polarizabilities are given by [9]

$$\alpha_1 = \alpha_2 = \alpha_\perp - \frac{1}{2}(\alpha_\perp - \alpha_{//})\sin^2\varphi \qquad\qquad (C.IX.4a)$$

$$\alpha_3 = \alpha_{//} + (\alpha_\perp - \alpha_{//})\sin^2\varphi \qquad\qquad (C.IX.4b)$$

Optical measurements determined that $\theta = 41°$ in the rectangular low-temperature phase. The φ angle does not change significantly between the two phases, being close to $\varphi = 35°$.

We note that the birefringence of the hexagonal phase of the triphenylene ester with $R = C_{11}COO$ (see table 1) is abnormally low, indicating that, on the average, the molecules are tilted with respect to the axis of the columns (in this case, formulae C.IX.4 still apply). This result was confirmed by the X-ray studies of threads (see subsection IX.9.b and Safinya et al. [52]). On the other hand, the molecules are perpendicular to the columns in the hexagonal phases of triphenylene ethers.

It is also noteworthy that even molecules that are not disk-shaped can give columnar phases. This is the case for some molecules containing a cone-shaped core [10a–c] and that for "phasmids" (these molecules bear six chains, so they are named after the six-legged stick insects) exemplified in figure C.IX.7. The phasmids (Φ) can yield oblique or hexagonal phases, denoted by Φ_{ob} and Φ_h, respectively [11].

Another interesting family of compounds is that of substituted aromatic diamides [12a,b], some of which (such as that shown in figure C.IX.7) exhibit nematic and columnar phases. The mesomorphic properties of these compounds stem from their aptitude to form threads (thus resembling polymers), the amide groups belonging to two neighboring molecules being joined by a hydrogen bond. These columnar phases contain numerous defects of the column-end type (π defects) and their proliferation seems to play an important role in the transition towards the nematic phase (where the molecules still form columns, locally arranged on a hexagonal lattice) [13a,b]. At this point, we should mention the existence of columnar phases of metallic complexes derived from Schiff bases, or of molecules forming disk-like dimers with the metal atom in the center [14a,b].

Chain R	Cr		Φ_{ob}		Φ_h		I
C_8H_{17}	•	80	•		–	82	
$C_{12}H_{25}$	•	70	•	81.5	•	92	•

(• : the phase exists; – : the phase does not exist; Cr: crystal; I: isotropic liquid) [11]

Fig. C.IX.7 Phasmids and the phases obtained with two chains R of different length.

$H_{17}C_8COO$ — NHCOC$_{15}$H$_{31}$ / CH$_3$ / NHCOC$_{15}$H$_{31}$

$$Cr_1 \xrightarrow{102.5°C} Cr_2 \xrightarrow{135°C} \text{Hex.} \xrightarrow{147°C} \text{Nem.} \xrightarrow{166°C} \text{Iso.}$$

Fig. C.IX.8 Example of a substituted aromatic diamide exhibiting a nematic and a columnar phase (from ref. [12a,b]).

Finally, rectangular phases are encountered in certain ternary lyotropic systems such as the mixture of SDS (sodium dodecyl sulphate), water and decanol. The addition of a cosurfactant (an alcohol, in this case) leads to the transformation of cylinders into ribbons, which are positioned on the sites of a centered rectangular lattice [15] (Fig. C.IX.9).

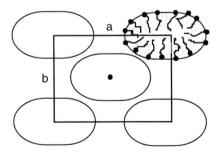

Fig. C.IX.9 Rectangular phase observed in certain ternary lyotropic mixtures. The lattice is centered rectangular. Inside the ribbon, the cosurfactant concentration (gray molecules in the drawing) is higher in the more planar regions than in the curved ones.

IX.2 Elasticity

In this section, we shall only deal with the hexagonal phase. In the static case, the bulk energy of elastic deformation ρf consists of two terms: the first one ρf_e describes the strain of the hexagonal lattice, while the second one ρf_c describes the deformation of the director field $\mathbf{n}$ (unit vector parallel to the axis of the columns). In the reference state, the columns are aligned along the z direction and the lattice is not distorted.

Let u and v be the x and y components of the displacement vector of the columns. To lowest order, and taking into account the hexagonal symmetry, we have, denoting by U_{ij} the deformation tensor components:

$$\rho f_e = \frac{1}{2} B (U_{xx} + U_{yy})^2 + \frac{1}{2} C (U_{xy}^2 - U_{xx} U_{yy}) \qquad \text{(C.IX.5)}$$

In this formula, U_{xx} and U_{yy} designate the unit elongation along the x and y axes and U_{xy} the shear deformation. The constant B is the compressibility modulus of the lattice and C its shear modulus. Note that the C modulus only comes into play when the lattice is stretched at constant density $(U_{xx} + U_{yy} = 0)$. As in the case of smectics, the anharmonic corrections to the deformations can be determined exactly [16]:

$$U_{xx} = \frac{\partial u}{\partial x} - \frac{1}{2} \left(\frac{\partial u}{\partial z} \right)^2 + \frac{1}{2} \left(\frac{\partial v}{\partial x} \right)^2 \qquad \text{(C.IX.6a)}$$

$$U_{yy} = \frac{\partial v}{\partial y} - \frac{1}{2} \left(\frac{\partial v}{\partial z} \right)^2 + \frac{1}{2} \left(\frac{\partial u}{\partial y} \right)^2 \qquad \text{(C.IX.6b)}$$

$$U_{xy} = \frac{1}{2} \left(\frac{\partial u}{\partial y} + \frac{\partial v}{\partial x} \right) - \frac{1}{2} \left(\frac{\partial u}{\partial x} \frac{\partial v}{\partial x} + \frac{\partial u}{\partial y} \frac{\partial v}{\partial y} + \frac{\partial u}{\partial z} \frac{\partial v}{\partial z} \right) \qquad \text{(C.IX.6c)}$$

These corrections ensure the rotational invariance of the energy. They are important for predicting the column buckling instabilities under mechanical stress (see section IX.6).

One must add to the elastic distortion energy of the hexagonal lattice the deformation energy of the director field. As in the case of nematics, it reads:

$$\rho f_c = \frac{1}{2} K_1 (\text{div } \mathbf{n})^2 + \frac{1}{2} K_2 (\mathbf{n} \cdot \mathbf{curl } \mathbf{n})^2 + \frac{1}{2} K_3 (\mathbf{n} \times \mathbf{curl } \mathbf{n})^2 \qquad \text{(C.IX.7)}$$

The K_1 and K_2 terms describe the splay and twist deformations of the columns, respectively. These two terms are irrelevant, since they are necessarily coupled to a deformation of the hexagonal lattice. The term K_3 describes the bending of the columns: it is relevant, since the columns can be bent without distorting the hexagonal lattice. In the following, we set $K_3 = K$. If the displacement of the columns remains small, the bending term can be expressed in terms of the displacement components u and v:

$$\rho f_c = \frac{1}{2} K \left[\left(\frac{\partial^2 u}{\partial z^2} \right)^2 + \left(\frac{\partial^2 v}{\partial z^2} \right)^2 \right] \tag{C.IX.8}$$

Finally, in the static case:

$$\rho f = \frac{1}{2} B \left(U_{xx} + U_{yy} \right)^2 + \frac{1}{2} C \left(U_{xy}^2 - U_{xx} U_{yy} \right) + \frac{1}{2} K \left[\left(\frac{\partial^2 u}{\partial z^2} \right)^2 + \left(\frac{\partial^2 v}{\partial z^2} \right)^2 \right] \tag{C.IX.9}$$

In the dynamic regime, one must add to the list of hydrodynamic variables the bulk dilation $\theta = -\delta\rho/\rho$. In this case, the elastic energy ρf_e is given by the more general expression:

$$\rho f_e = \frac{1}{2} A \, \theta^2 + \frac{1}{2} \bar{B} \left(U_{xx} + U_{yy} \right)^2 + \frac{1}{2} C \left(U_{xy}^2 - U_{xx} U_{yy} \right) - D \, \theta (U_{xx} + U_{yy}) \tag{C.IX.10}$$

The static case is recovered by writing that, at rest, the hydrostatic pressure P is constant (under the assumption that the sample boundaries are free):

$$P = - \frac{\partial \rho f_e}{\partial \theta} = 0 \qquad \text{(static case)} \tag{C.IX.11}$$

This relation yields

$$\theta = \frac{D}{A} (U_{xx} + U_{yy}) \tag{C.IX.12}$$

Replacing θ by its expression in eq. C.IX.10, eq. C.IX.9 is recovered, with

$$B = \bar{B} - \frac{1}{2} \frac{D^2}{A} \tag{C.IX.13}$$

In the following, we pose

$$\lambda = \sqrt{\frac{K}{B}} \qquad \text{and} \qquad A = \frac{1}{4} \frac{C}{B} \tag{C.IX.14}$$

As for smectics, λ has the dimensions of a length. λ is also related to the penetration distance of a perturbation along the direction perpendicular to the columns. If the columns are free to slide with respect to each another, λ must be of the order of the distance "a" between columns.

The dimensionless number A characterizes the elastic anisotropy of the hexagonal lattice. Note that $0 < A < 1$ for thermodynamic stability reasons (the energy is a positively defined quadratic form). For the same reason, one must have A > 0, $\bar{B} > 0$, C > 0, and B > 0 (which requires $D^2 < 2\bar{B}A$).

IX.3 Developable domains

IX.3.a) General construction

We know that the smectic phases exhibit configurations that are curved over large scales while the layer thickness is everywhere constant. In these configurations, known as confocal structures (one speaks of focal domains when the singularities are reduced to lines), the layers form a family of parallel surfaces (Dupin cyclides in the case of focal domains). Do columnar phases have similar structures? More specifically, can the columns be bent on a large scale without deforming the two-dimensional lattice, since this would be energetically very costly?

Polarizing microscopic observation of fan-shaped textures proves that this is indeed possible [17a,b] (Fig. C.IX.10). In this case, each fan is a domain where, in the first approximation, the columns assume a circular form. These domains are a particular example of more general structures known as **"developable domains."** The concept was introduced by Bouligand [17b], but it was Kléman who showed the unicity of this solution [18].

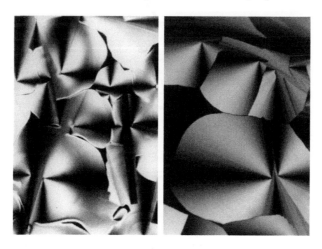

Fig. C.IX.10 Fan-shaped textures observed between crossed polarizers a) in a thin C5HET layer, a few μm thick, placed on an untreated glass slide in free air; b) in a lyotropic mixture of aerosol OT and water (the 75-μm thick sample is sandwiched between two glass slides, without being sealed on the sides). In the first case, the texture appears by slowly cooling to the liquid crystal from its isotropic phase. In the second case, the hexagonal phase grows slowly in the micellar phase by drying at the edges of the sample.

Let us delineate Kléman's proof. The problem is finding the curved configurations that preserve the (generic) column lattice. This requires that the distance and angles between columns be locally preserved. Let us analyze the implications of these two conditions:

i) Metric properties

Let S be the surface generated by a family of cylinders Γ, adjacent two by two. On this surface, two adjacent cylinders must remain parallel; this implies that the trajectories G orthogonal to them are geodesics of S and that

div **n** = **0**   (C.IX.15)

with **n** the director tangent to the cylinders (Fig. C.IX.11).

ii) Angular properties

We require that two generic surfaces S_1 and S_2 intersect along a line Γ at a constant angle. From Joachimsthal's theorem [19], Γ can either be a curvature line on both S_1 and S_2, or it is not a curvature line, but its geodesic torsion $1/\tau$ is constant and **non**-zero. We analyze these two cases separately.

First case: Γ is not a curvature line
The geodesic torsion $1/\tau \neq 0$ of Γ is the same on all surfaces S containing Γ. Since it is equal (within a factor 2) to the twist **n.curln** of the director field ($2/\tau$ = **n.curln**), the medium is twisted, so the lattice is necessarily deformed. Consequently, this solution does not satisfy the required conditions.

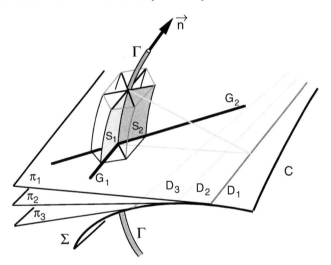

Fig. C.IX.11 Developable domain. The molecular cylinders Γ are perpendicular to the half-planes π that enclose the developable surface Σ with generatrices D tangent to the cuspidal edge C. The curvature centers in the π_1 plane (and in the π_2, π_3, etc. planes) of the columns are found on the line D_1 (respectively D_2, D_3, etc). This proves that the curvature radius of a column goes to zero as it crosses the developable surface Σ.

Second case: Γ is a curvature line.
Hence, it is also a curvature line on all surfaces S containing it. This is also true for all lines G that are orthogonal to it. Since these lines are also geodesics, they have no torsion and, consequently, they are plane.

Therefore, there exist planes π perpendicular to all the cylinders, yielding

$$\mathbf{n}.\mathbf{curl\ n} = 0 \qquad\qquad (\text{C.IX.16})$$

As a consequence, the curvature energy of the medium is reduced to the column bending term

$$\rho f_c = \frac{1}{2}K\,(\mathbf{n} \times \mathbf{curl\ n})^2 \qquad\qquad (\text{C.IX.17})$$

Kléman then showed that, in each plane π, the curves Γ have their curvature centers along the same straight line D. These lines generate a surface Σ, developable by definition, and are tangent to a curve C, called cuspidal edge (Fig. C.IX.11).

To summarize, one can construct a configuration that is curved on a large scale such that the lattice of the columns is not deformed by considering a generic developable surface Σ and the family of tangent half-planes π (Fig. C.IX.11). The molecular cylinders then describe the lines Γ perpendicular to these planes. It can also be shown that the reticular surfaces of the cylinders are developable. For this reason, Bouligand gave these configurations the name "developable domains" [17b].

It is noteworthy that the radius of curvature of the cylinders goes to zero on Σ (Fig. C.IX.11). This surface is therefore singular, as the bending energy diverges close to it. In particular, no cylinder can cross it.

As in smectics, it is therefore reasonable to consider configurations where the surface Σ is reduced to a line. In this limiting case, the line is necessarily straight, the cylinders forming circles centered on this line. The physical reality is very close to this solution. We shall nevertheless show that the actual solution is sometimes slightly different.

IX.3.b) Fan-shaped textures

These textures (Fig. C.IX.10) are frequently encountered in the hexagonal phases of thermotropic and lyotropic systems. They look very much like certain focal conic textures observed in thin planar samples of smectics A. One must therefore be cautious when concluding that a phase is columnar based solely on its texture.

As already stated, the columns (parallel to the glass surfaces) are approximately circular and centered on the convergence point of the black branches observed between crossed polarizers. The center of each fan thus contains the surface Σ (a straight line in the circular case) where the energy diverges: this is the domain core. From a topological point of view, each domain corresponds to a wedge line of rank 1 or 1/2, if the cross is complete or incomplete, respectively (Fig. C.IX.12a).

In many lyotropic materials, such as the $C_{12}EO_6$ + water mixture, the domain cores (grayed areas in the diagrams of Fig. C.IX.12a) are too small to be resolved in optical microscopy. In this case, the surface Σ is (on the scale of optical microscopy) degenerated into a straight line.

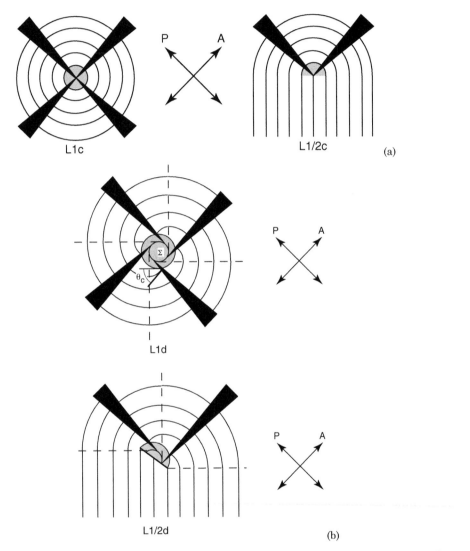

Fig. C.IX.12 a) Circular wedge lines of rank 1 and 1/2. The columns are centered on the convergence point of the black branches observed between crossed polarizers. Let us call these lines L1c and L1/2c; b) wedge lines of rank 1 and 1/2 in involutes of a circle. The generating circle (of radius r_0) is necessarily contained within the core (of radius r_c, shown in gray).

The situation is different in certain thermotropic materials such as triphenylene ethers. Careful observation shows that the cores of the wedge lines are large enough to be resolved in microscopy. Moreover, their radius, in the μm range, does not significantly change with temperature, which goes against a model where the core is filled by the isotropic liquid. Finally, and more importantly, the two dark branches of the 1/2 lines seem not to converge to the same point, which is not compatible with a circular structure of the columns. These observations, difficult to make since the observed features are at the resolution limit of optical microscopy, suggest that the developable surface S is a small cylinder of finite radius r_0. In this case, the columns have the shape of involutes of a circle (Fig. C.IX.12b).

This model was confirmed by the observation of macro-steps formed at the free surface of a thin liquid crystal layer; the steps follow very closely the molecular columns (this observation also demonstrates the coherence of the columns over very long distances, several tens of μm in this case). Their eccentricity πr_0 with respect to the core of the lines yields an estimate for r_0 (typically, $r_0 \approx 0.7$ μm in C5HET [20]). The model in involutes of a circle also explains the asymmetric deviations of the extraordinary rays induced by a line of rank 1/2 adhering to the lower glass plate of a thick C5HET sample [6b] (Fig. C.IX.13).

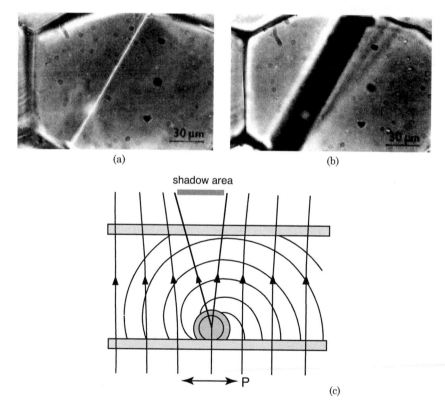

(a) (b)

shadow area

P

(c)

Fig. C.IX.13 Observation in extraordinary light of an L1/2d line adhering to the lower plate of a thick (75 μm) C5HET sample. a) The lower plate is in focus; note the "imposing" size of the line core. b) The upper plate is in focus. One can see a large shadow area, not centered on the defect core. By superposing the two photos, it can be seen that the rays are deflected twice as much on one side compared with the other. c) Model of involutes of a circle explaining this asymmetry [6b].

Finally, let us emphasize that the involutes of a circle were recently observed in a lyotropic mixture close to an L_β smectic phase [21].

Determining the energy of a wedge line is quite easy, since the energy is reduced here to the bending term C.IX.17. In the general case of a line of rank 1, the energy per unit length is given by [20]:

$$E = \frac{\pi K}{2} \ln\left(\frac{R^2 - r_0^2}{r_c^2 - r_0^2}\right) + E_{core} \qquad (C.IX.18)$$

with R the outer radius of the domain (from a few tens of μm to a few hundred μm, in practice). To determine the core radius r_c and the radius of the generating circle r_0, one needs a model for the core (note that $r_c > r_0$ necessarily). The hypothesis of a melted core, filled by the isotropic liquid, is not realistic in thermotropic columnar materials, as it leads (under the assumption that the penetration length λ is of the order of the distance between the columns, to be justified later) to radii of a few tens of Å, in disagreement with experiment (for a more detailed discussion see Sallen [22a] and Sallen et al [22b]). Another hypothesis is that the core is filled by the hexagonal phase, the columns being parallel to the defect axis. In this case, the core energy per unit length is simply given by:

$$E_{core} = 2\pi r_c E_s \qquad (C.IX.19)$$

where E_s is a surface energy depending on the angle θ_c between the cylinders and the tangent to the core (Fig. C.IX.14).

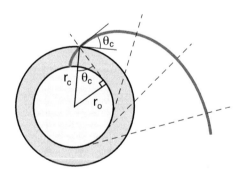

Fig. C.IX.14 Definition of the anchoring angle at the core θ_c.

Let us write [20]:

$$E_s = E_0 + E_2 \sin^2\theta_c \qquad (C.IX.20)$$

with $r_0/r_c = \sin\theta_c$. Clearly, the nature of the line depends on the sign of E_2, the L1c or L1d lines being favored for $E_2 > 0$ or $E_2 < 0$, respectively. More precisely, minimizing the total energy as a function of r_0 and r_c provides two distinct solutions, the first one describing an L1d line:

$$r_c = \frac{K}{4(E_o + E_2)} \qquad \text{and} \qquad r_o = r_c \sqrt{\frac{E_o + 2E_2}{E_2}} \qquad \text{(C.IX.21a)}$$

and the second one corresponding to an L1c line:

$$r_c = \frac{K}{2E_o} \qquad \text{and} \qquad r_o = 0 \qquad \text{(C.IX.21b)}$$

	Conditions on E_o and E_2	Number of solutions	Types of solutions	Energetic stability
Case 1	$E_2 > 0$ or $E_2 < 0$ and $-\dfrac{E_o}{E_2} > 2$	1	L1c line	Stable
Case 2	$E_2 < 0$ $1 < -\dfrac{E_o}{E_2} < 2$	2	L1c line L1d line	Unstable (saddle point of the energy) Stable
Case 3	$E_2 < 0$ $0 < -\dfrac{E_o}{E_2} < 1$	1	L1c line	Unstable (saddle point of the energy)

Table 2 Stability criteria for the lines of rank 1. Case 3 is unrealistic due to a calculation artifact.

The energetic stability of these two solutions depends on the constants E_o and E_2. Table 2 summarizes the calculations.

Clearly, the stability of rank 1 lines is strongly related to the anisotropy of the surface energy E_s and to the sign of E_2. Thus, only the L1c lines are stable when the core favors planar anchoring ($E_2 > 0$) or when the homeotropic tendency ($E_2 < 0$) is weak (case 1). If the homeotropic tendency increases, without becoming too strong (case 2), only the L1d lines in involutes of a circle are stable. Finally, if the homeotropic tendency becomes very strong (case 3), rank 1 lines disappear. This disappearance is only an artifact of the calculation, as in this extreme case $E_s(\theta_c = \pi/2) = E_o + E_2 < 0$, which is unphysical. To avoid the energy becoming negative, one must add an $E_4(\sin\theta_c)^4$ term to its power series expansion over $\sin\theta_c$ with $E_4 > 0$ and $4E_oE_4 > E_2^2$. The core energy C.IX.19 now has a minimum for $\theta_c^* = \arcsin[(-E_2/2E_4)^{1/2}]$, and the solution in involutes of a circle reappears, with an angle at the core θ_c close to θ_c^*.

Similar calculations can be performed for lines of rank 1/2. Assuming that their cores are filled by the hexagonal phase and their shapes are as shown in figure C.IX.12, and only considering expression C.IX.20 for the core energy, two solutions are found. The first one:

$$r_c = \dfrac{K}{8\left[E_o\left(\dfrac{1}{2}+\dfrac{1}{\pi}\right)+\dfrac{E_2}{2}\right]} \quad , \quad r_o = r_c \sqrt{\dfrac{E_o\left(\dfrac{1}{2}+\dfrac{1}{\pi}\right)+E_2\left(1-\dfrac{1}{\pi}\right)}{E_2\left(\dfrac{1}{2}-\dfrac{1}{\pi}\right)}} \qquad (C.IX.22a)$$

corresponds to an L1/2d line and the second one

$$r_c = \dfrac{K}{E_o\left(1+\dfrac{2}{\pi}\right)+\dfrac{2}{\pi}E_2} \quad , \quad r_o = 0 \qquad (C.IX.22b)$$

to an L1/2c line. The stability conditions of these lines are summarized in table 3.

As for rank 1 lines, case 3 is unrealistic, and the $E_4(\sin\theta_c)^4$ term must be taken into account to retrieve a stable solution of the L1/2d type.

Experimentally, the lines observed in C5HET are of the L1/2d type. These lines, sometimes very close to each other, as seen in figure C.IX.5, rarely associate, except in extremely thin samples [20]. These observations suggest that $E_2 < 0$ and that $0 < -E_o/E_2 < 0.83$. With the experimental values $r_o = 0.7\,\mu m$ and $r_c = 1\,\mu m$, we find, from eqs. C.IX.22a, $-E_o/E_2 \approx 0.72$ and $K/E_o \approx 1\,\mu m$.

To date, there is no direct evidence for the presence of the hexagonal phase within the disclination core.

Lyotropic liquid crystals are likely in case 1 described in table 3.

	Conditions on E_o and E_2	Number of solutions	Types of solutions	Energetic stability
Case 1	$E_2 > 0$ or $E_2 < 0$ and $-\dfrac{E_o}{E_2} > 0.83$	1	L1/2c line	Stable
Case 2	$E_2 < 0$ $0.61 < -\dfrac{E_o}{E_2} < 0.83$	2	L1/2c line L1/2d line	Unstable (saddle point of the energy) Stable
Case 3	$E_2 < 0$ $0 < -\dfrac{E_o}{E_2} < 0.61$	1	L1/2c line	Unstable (saddle point of the energy)

Table 3 Stability criteria for the lines of rank 1/2. Case 3 is unrealistic due to a calculation artifact.

We emphasize that these calculations are only approximate, as they do not take into account the bending energy of the columns, neglecting all deformations of the lattice. Indeed, developable domains, like the focal domains in smectics, are not exact solutions of the equilibrium equations (see section III.1.b), although they are obviously very close.

IX.3.c) Association rules for wedge lines

Along a Burgers circuit enclosing a wedge line of rank 1/2, the lattice of the columns turns through 180°. As such, the axis of the line must be parallel to a twofold symmetry axis of the hexagonal lattice, otherwise a disorientation wall appears (very costly in energy). Because the binary axes T_2 or θ_2 (Fig. C.IX.4) make angles of 0, 30, 60 or 90°, the same applies each time that two 1/2 lines associate.

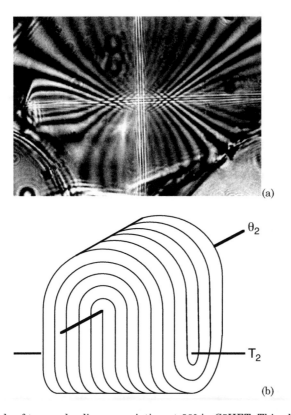

Fig. C.IX.15 Example of two wedge lines associating at 90° in C5HET. This observation confirms the hexagonal symmetry of the phase (from ref. [6a,b]). a) Micrograph taken between crossed polarizers; b) sketch of the column arrangement.

An example of association at 90° is given in figure C.IX.15. Also, the radius r_o of the developing circle is not arbitrary, but rather takes discrete values, since the lattice undergoes a translation by $\pi m r_o$ on a complete circuit around the line (with m = 1 or 1/2, depending on the type of the line). For instance, in the case of an L1/2d line of axis δ, one has:

$$r_0 = n\,\frac{a}{\pi} \qquad \text{for} \qquad \delta \mathbin{/\!/} T_2 \tag{C.IX.23a}$$

and

$$r_0 = n\,\frac{a\sqrt{3}}{\pi} \qquad \text{for} \qquad \delta \mathbin{/\!/} \theta_2 \tag{C.IX.23b}$$

with integer n. This discretization was never demonstrated experimentally and does not seem to play an important role.

IX.3.d) Fan-shaped textures in the rectangular phases of the P2₁/a type

Figure C.IX.16 shows a photograph (taken for crossed polarizers) of a thin layer of a rectangular phase spread on a glass plate in free air. One can recognize yet another fan-shaped texture where the molecular columns are parallel to the glass and to the air interface. The distinctive feature here is given by the presence of two types of fans. There are those (S) where the two extinction branches are parallel to the polarizers and those (T), much more common, for which the two extinction branches make an angle Φ with the polarizers. If the product is achiral, the two tilt angles ±Φ are equiprobable, while with a homochiral product, only one tilt is observed [23]. We analyze the two types of fans separately.

Fig. C.IX.16 Fan-shaped texture in the rectangular phase of C11HET. The polarizers are parallel to the edges of the photograph. Note the presence of two types of fans (S and T in the photo). The tilt angle Φ is close to 28° for the T fans.

In the tilted case (T fan), the $(\mathbf{a}, \mathbf{c})$ plane is parallel to the glass surface (Fig. C.IX.4). A neutral line of the sample makes an angle Φ with the $\mathbf{c}$ axis, while the other one is perpendicular to it. This angle is related to the structural angles θ and φ by the formula C.IX.1. The two extinction branches are therefore tilted by an angle Φ with respect to the polarizers. For this orientation, the birefringence of the medium is

$$(\Delta n)_1 = n_2 - n_3 \tag{C.IX.24}$$

where the n_2 and n_3 indices are given by formulae C.IX.2 and 3. It should be noted that, in this case, the axis of the wedge line is parallel to the 2_1 helical axis. It is therefore likely that the line has a screw component with Burgers vector $a/2$. Then, no wall can start at the line, which is well confirmed by microscopic observation.

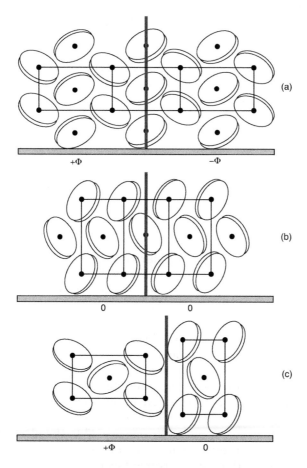

Fig. C.IX.17 The three main types of longitudinal walls observed in the rectangular phases. Walls (a) are observed in T fans; across such a wall, the tilt angle of the black crosses with respect to the polarizers goes from $+\Phi$ to $-\Phi$. Walls (b) are observed in S fans. They are less visible, since the extinction branches remain parallel to the polarizers on either side of the wall. Walls (c) are energetically more costly. They mark the passage from a T fan to an S fan.

In the straight case (S fan), the $(\mathbf{b}, \mathbf{c})$ plane is parallel to the glass plate. A neutral line is superposed with the $\mathbf{c}$ axis of the columns, while the other one is perpendicular to it. The birefringence of the sample is now

$$(\Delta n)_2 = n_1 - n \qquad \text{with} \qquad n = \frac{1}{\dfrac{\cos^2\Phi}{n_3^2} + \dfrac{\sin^2\Phi}{n_2^2}} \qquad\qquad \text{(C.IX.25)}$$

This value being different from $(\Delta n)_1$, the two types of fans should exhibit, for the same sample thickness, different chromatic polarization colors, which is indeed the case. Note that a longitudinal wall parallel to the cylinders is necessarily attached to the line, since there is no axis binary along the $\mathbf{b}$ direction. The most probable structure of this wall is shown in figure C.IX.17b.

Other longitudinal walls also exist. They follow the molecular columns and separate domains with different molecular tilt or different crystallographic orientation. The most frequently encountered types of walls are drawn in figure C.IX.17, along with their most probable structures.

One can also observe walls that are perpendicular to the cylinders. In this case, the molecular orientation changes within the column. These transverse walls are however much less frequent than the previous ones.

IX.4 Dislocations

IX.4.a) Direct observation

Dislocations are not visible using the polarizing microscope. However, they can be revealed by electron microscopy and freeze-fracture in lyotropic systems. This technique amounts to vitrifying the sample by plunging it into liquid propane. The quench must be fast enough so that the water does not have enough time to crystallize, requiring cooling rates of several thousand degrees per second. Glycerol is sometimes added to the mixture to avoid water crystallization. An alternative technique consists of using a non-aqueous polar mixture, such as formamide. The frozen sample is then broken under ultra-high vacuum, and immediately replicated to avoid contamination of the fracture surface. The replica is obtained by evaporating a few tens of Å of platinum under oblique incidence (shadowing), followed by a few hundreds of Å of carbon at normal incidence (cement "transparent" to the electrons). It is then peeled off the sample and observed using transmission electron microscopy.

Experience shows that the lyotropic hexagonal phases are particularly well cleaved along the dense planes of molecular cylinders. An example is shown in figure C.IX.18. It presents the hexagonal phase of DNA in concentrated aqueous solution [24]. Several steps separating different cylinder planes are visible. Some of them stop abruptly, signaling the emergence points of screw dislocations perpendicular to the cylinders. These dislocations are very frequent in the samples and can regroup as in figure C.IX.19 forming twist walls.

Fig. C.IX.18 Electron microscopy observation of a replica of the cleavage surface of the hexagonal phase in a concentrated DNA solution. Screw dislocations (indicated by arrows) emerge from the sample at the places where the steps end abruptly (photograph by F. Livolant [24]).

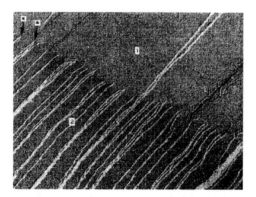

Fig. C.IX.19 Screw dislocations sometimes regroup, forming twist walls. (photograph by F. Livolant [24]).

Edge dislocations are much more difficult to see using this technique, due to its limited resolution. Some of them were observed in the hexagonal phase of the SDS-formamide system [25].

IX.4.b) Classification of the dislocations of a hexagonal phase

We shall distinguish two types of dislocations [17b]: those perpendicular to the molecular cylinders, called transverse dislocations, and those parallel to them, known as longitudinal.

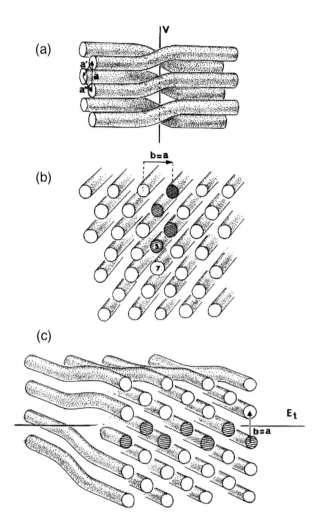

Fig. C.IX.20 Perfect dislocations with Burgers vector **b** = **a**. a) Transverse screw dislocation parallel to a T_2 axis; in the core region, the molecular cylinders are probably not broken up. b) Longitudinal edge dislocation. c) Transverse edge dislocation parallel to a θ_2 axis. The cylinders marked by cross-hatched ends were added during the Volterra process (from ref. [17b]).

In the general case, their Burgers vector **b** is a linear combination of the type:

b = l**a** + m**a**'   (C.IX.26)

where **a**, **a**', and **a**" are the three base vectors of the hexagonal lattice (with **a** + **a**' + **a**" = 0), and l and m two positive or negative integers.

Since **b** is normal to the director **n**, the screw dislocations are necessarily transverse. For instance, the screw dislocations of Burgers vectors **a**, **a**' or **a**" follow the T_2 principal axes of the hexagonal lattice (Fig. C.IX.20a).

The edge dislocations can be either transverse or longitudinal. Two edge dislocations, one of them longitudinal and the other transverse, with Burgers vectors **b** = **a**, are drawn in figures C.IX.20b and c. Generally speaking, any dislocation line can be replaced by a series of segments, parallel to **b**, **n**, or **b** × **n**. These three kinds of segments are depicted in figure C.IX.20 for **b** = **a**.

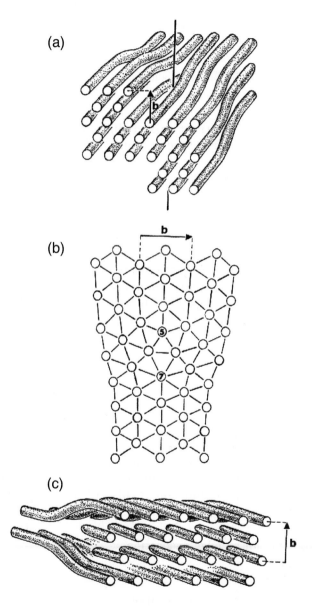

(a)

(b)

(c)

Fig. C.IX.21 Perfect dislocations with Burgers vector **b** = **a** − **a**'. a) Transverse screw dislocation parallel to a θ_2 axis; in the core region, the molecular cylinders are probably not broken up. b) Longitudinal edge dislocation. c) Transverse edge dislocation parallel to a T_2 axis (from ref. [17b]).

Another significant Burgers vector is $\mathbf{b} = \mathbf{a} - \mathbf{a}'$. The three main situations corresponding to this vector are represented in figure C.IX.21.

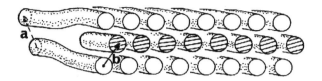

Fig. C.IX.22 Mixed transverse dislocation parallel to a θ_2 axis, with Burgers vector $\mathbf{b} = \mathbf{a}$ (from ref. [17b]).

Finally, one can construct mixed dislocations (of edge and screw character), such as the one represented in figure C.IX.22.

IX.4.c) Energy of the dislocations

The energy of the dislocations is calculated as in solids for the longitudinal dislocations and as in smectics for the transverse dislocations [16]. For this reason, we shall not go into the calculation details.

In the case of a longitudinal edge dislocation, one finds that the elastic stress falls off as $1/r$ in the plane of the hexagonal lattice and that the energy diverges logarithmically with the size R of the sample (in practice, R is the average distance between dislocations):

$$E_{el} = \frac{Cb^2}{4\pi}(1 - A)\ln\left(\frac{R}{r_c}\right) + E_{core} \qquad\qquad (C.IX.27)$$

where r_c is the core radius.

In the case of transverse dislocations, the calculations are performed as in smectics A. With the x axis along the dislocation and the z axis along the columns, one can show that the stress varies like $\exp(-z^2/4\lambda x)$ with $\lambda = (K/B)^{1/2}$ around an edge dislocation and like $\exp[-z^2(2A)^{1/2}/4\lambda x]$ around a screw dislocation. Their energies are given by:

$$E_{et} = \frac{Kb^2}{2\lambda r_c} + E_{core} \qquad \text{for an edge dislocation} \qquad\qquad (C.IX.28a)$$

$$E_{st} = \frac{\sqrt{A}Kb^2}{\sqrt{2}\lambda r_c} + E_{core} \qquad \text{for a screw dislocation} \qquad\qquad (C.IX.28b)$$

where r_c is again the core radius.

IX.4.d) Walls in hexagonal phases

Let us first consider longitudinal walls (i.e., parallel to the columns) separating two grains with different crystallographic orientations. These walls are frequent and well visible in homeotropic samples. Their energies can be measured by the "grain boundary method," the principle of which is presented in figure C.IX.23a. The homeotropic sample is placed in a linear temperature gradient G. A groove appears when a grain boundary meets the hexagonal-isotropic interface (Fig. C.IX.23b). Analyzing its equilibrium shape yields the surface tension and the energy of the grain boundary.

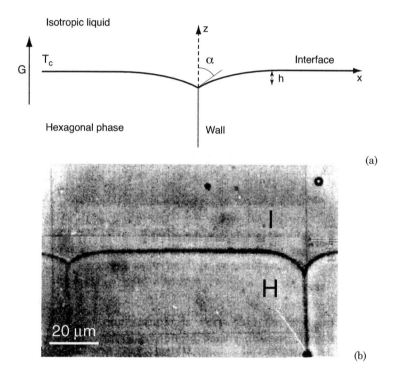

(a)

(b)

Fig. C.IX.23 a) Schematic of the groove formed as a grain boundary meets the hexagonal-isotropic interface. b) Grooves observed in C8HET for G = 16.6 K/cm (from ref. [26]).

We start by writing the Gibbs-Thomson relation between the temperature of the interface and its curvature 1/R:

$$T = T_c + \frac{d_o T_c}{R} \qquad\qquad (C.IX.29)$$

where $d_o = \gamma/L$ is the capillary length, γ the surface tension considered isotropic and L the latent heat per unit volume. This relation is also the differential equation determining the shape of the interface in the temperature gradient, as

$T_c - T = Gz$. In particular, its first integral yields an exact value for the depth h of the groove as a function of the angle α between the interface and the z axis at the bottom of the groove:

$$h = \sqrt{\frac{2d_o(1 - \sin\alpha)T_c}{G}} \qquad\qquad\qquad\qquad (C.IX.30)$$

The angle α is given by the equilibrium of the triple point at the bottom of the groove:

$$E_{wall} = 2\gamma \cos\alpha \qquad\qquad\qquad\qquad (C.IX.31)$$

Hence, measuring the depth h and the angle α is in principle enough to determine the surface tension and the energy of the grain boundary.

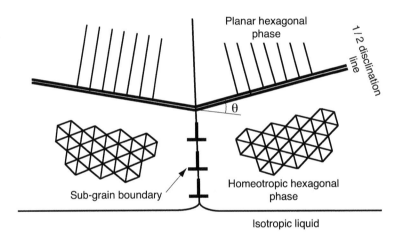

Fig. C.IX.24 Principle of the method for determining the disorientation angle θ of a subgrain boundary separating two homeotropic domains. Each 1/2 line separates a homeotropic region from a planar one. In this diagram, the two lines are taken as parallel to a θ_2 axis of the hexagonal lattice.

Measurements of this type were performed for C8HET [26] and for the lyotropic mixture $C_{12}EO_6$ + water [22a,b]. The surface tension at the hexagonal-isotropic interface is comparable in the two systems, although their nature is quite different:

$\gamma \approx 0.52$ erg/cm^2 in C8HET

$\gamma \approx 0.26$ erg/cm^2 in $C_{12}EO_6$ + water (at the azeotropic concentration)

Experience also shows that the energy of the grain boundaries varies with the disorientation angle θ of the hexagonal lattice as $|\theta \ln\theta|$, as long as its value remains low (below 15–20°) [26]. In practice, this angle can be measured if a wedge line of rank 1/2 is present in each of the two grains, since these lines always follow a twofold axis of the hexagonal lattice (Fig. C.IX.24). This $|\theta \ln\theta|$ variation is predicted by the theory for walls consisting of transverse screw dislocations a distance b/θ apart, which is only meaningful for small θ (these walls are often termed sub-grain boundaries). The θ factor simply stems from the fact the number of dislocations per unit length of wall is proportional to θ. As to the $|\ln\theta|$ dependence, this is due to the energy of each dislocation varying as $\ln(R/r_c)$, where R is typically the distance b/θ between two dislocations (further away from the wall, the lattice is essentially undistorted), and r_c a core radius that we take as equal to b. In these conditions, and neglecting the core energy of the dislocations, one obtains from eq. C.IX.27 [27]:

$$E_{sgb} \approx \frac{Cb(1-A)}{4\pi}|\theta \ln\theta| \qquad\qquad (C.IX.32)$$

Experimentally, $|\theta \ln\theta|/\cos\alpha \approx 0.38$ in C8HET, yielding $C \approx 10^8$ erg/cm^3 as $\gamma \approx 0.52$ erg/cm^3. This value of the elastic modulus is reasonable for this type of material.

Let us now consider transverse walls (i.e., all those that are not parallel to the columns). They belong to one of two kinds.

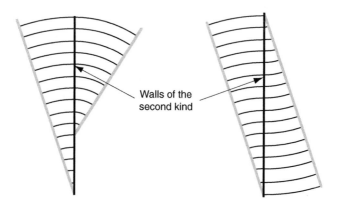

Walls of the second kind

Fig. C.IX.25 Wall of the second kind separating two fans. Only the curvature of the columns undergoes a jump across the wall.

There are those where the columns do not change their orientation, only the curvature undergoing a jump. These walls are frequently encountered in fan-shaped textures (Fig. C.IX.25). Bouligand named them "walls of the second kind" [28]. Obviously, their energetic cost is very low (but the energy calculation has not yet been performed).

Then there are the "walls of the first kind," more costly in energy, since the orientation of the columns is discontinuous. Their energy (here, we only consider symmetric walls) can be measured as a function of the disorientation angle of the columns $2\theta_o$ by the already presented "grain boundary method" (Fig. C.IX.23). The results of measurements in C8HET are shown in figure

C.IX.26 (very similar results were obtained for the lyotropic mixture $C_{12}EO_6$ + water). From the graph, we conclude that at least three types of walls must be distinguished:

1. Those corresponding to $10° < \theta_0 < 25°$, for which the energy varies as θ_0^2 (solid line parabola).

2. Those with $20° < \theta_0 < 80°$, where the energy is constant, close to 2γ;

3. Finally, the energy decreases linearly (dotted straight line) for those where $80° < \theta_0 < 90°$.

These experimental variations can be explained using very simple models inspired by those proposed for smectics A [6a, 20].

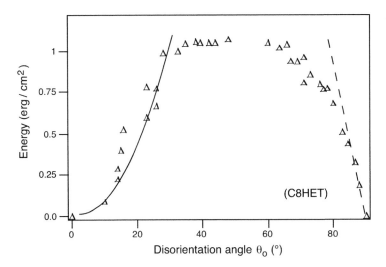

Fig. C.IX.26 Energy of a symmetric wall of the first kind as a function of the disorientation angle θ_0.

The most obvious is the one of "bend walls," shown in figure C.IX.27a. As the columns are continuously curved, the hexagonal lattice is necessarily dilated at the center of the wall. Let θ be the local tilt angle of the columns with respect to the x axis perpendicular to the wall.

The local energy density is easily determined:

$$\rho f = \frac{1}{2}[K_1(\sin\theta)^2 + K_3(\cos\theta)^2]\left(\frac{d\theta}{dx}\right)^2 + \frac{1}{2}B\left(1 - \frac{\cos\theta}{\cos\theta_0}\right)^2 \qquad (C.IX.33)$$

yielding, after minimization with respect to θ and setting $K = K_1 = K_3$:

$$\lambda^2\frac{d^2\theta}{dx^2} = \left(1 - \frac{\cos\theta}{\cos\theta_0}\right)\frac{\sin\theta}{\cos\theta_0} \qquad (C.IX.34)$$

where $\lambda = \sqrt{K/B}$. After integration, we obtain:

$$\frac{\sin\left(\frac{\theta + \theta_o}{2}\right)}{\sin\left(\frac{\theta - \theta_o}{2}\right)} = \exp\left[\tan\left(\theta_o\frac{x}{\lambda}\right)\right] \tag{C.IX.35a}$$

In the θ small limit, this equation simplifies:

$$\theta = \theta_o \tan\left(\frac{\theta_o}{2}\frac{x}{\lambda}\right) \tag{C.IX.35b}$$

yielding the energy of the wall:

$$E_{bend} = 2\sqrt{KB}\,(\tan\theta_o - \theta_o) \approx \frac{2}{3}\sqrt{KB}\,\theta_o^3 \tag{C.IX.36}$$

This formula shows that the energy increases very steeply with the θ_o angle, whence the idea of introducing pairs of edge dislocations at regular intervals, as shown in figure C.IX.27b.

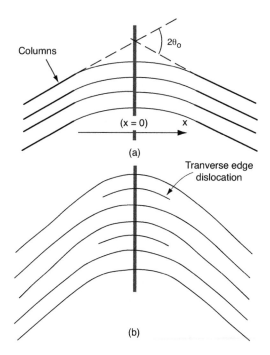

Fig. C.IX.27 a) Bend wall; b) mixed wall.

In this case, termed a "mixed wall," the energy is mainly localized in the vicinity of the dislocations. By summing over the individual energies of the dislocations, one has

$$E_{mixed} \approx \frac{\theta_o^2}{2b} 2E_{et} = \frac{E_{et}}{b} \theta_o^2 \qquad \text{(C.IX.37)}$$

where E_{et} is the energy of a transverse edge dislocation, given by equation C.IX.28a. The θ_o^2 dependence favors this type of wall for a large enough value of the angle θ_o (in practice, above 10°). Experimentally, this dependence is observed in the interval $10° < \theta_o < 25°$ (Fig. C.IX.26). Fitting this law to the experimental points yields a first estimate for the energy of transverse edge dislocations. For C8HET, this method yields $E_{et} \approx 10^{-5}$ dyn, while for the $C_{12}EO_6$ + water mixture, $E_{et} \approx 5\times10^{-6}$ dyn. Once more, the orders of magnitude are the same.

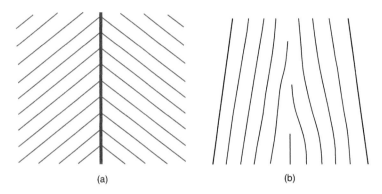

(a)   (b)

Fig. C.IX.28 a) Discontinuity wall; b) transverse sub-grain boundary.

As the angle increases, the distance between dislocations becomes so small that the system can no longer be envisioned in terms of separate dislocations. In this case, one speaks of discontinuity walls (Fig. C.IX.28a). Experiment shows that their energy varies very little, remaining close to 2γ (the core of the wall can be considered as "melted"):

$$E_{disc} \approx 2\gamma \qquad \text{(C.IX.38)}$$

The situation changes as θ_o tends toward 90°, since the wall must disappear for this value. In this limit, one can again design a model of a sub-grain boundary composed of transverse edge dislocations (Fig. C.IX.28b). This model is well adapted to the problem, especially since the dislocations interact very weakly, being in the same climb plane. Summing over their energies yields:

$$E_{sgb} \approx \frac{E_{et}}{b} (\pi - 2\theta_o) \qquad \text{(C.IX.39)}$$

Fitting this law to the experimental points (see, for instance the dashed line in Fig. C.IX.26) yields in both systems values of E_{et} that are compatible with the previous ones (up to a 50% variation).

To conclude this section, let us note that the discontinuity walls sometimes split into a sequence of stacked developable domains (Fig. C.IX.29).

These walls of a new kind are similar to the Grandjean walls in smectics (see section C.III.1c) and are only formed in the samples after annealing during several hours, or even days for the more viscous materials. They were observed in thermotropic as well as in lyotropic systems.

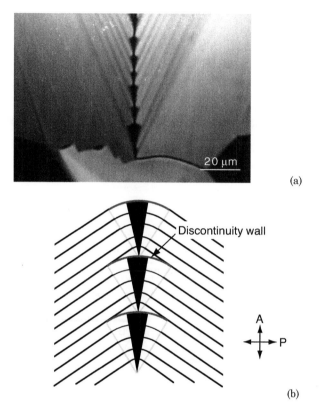

(a)

(b)

Fig. C.IX.29 a) Wall of "developable domains" in C8HET; b) schematic representation.

As an order of magnitude, figure C.IX.29b shows that the energy of such a wall is:

$$E_{dd} \approx E_{disc}\,\theta \qquad\qquad (C.IX.40)$$

where we neglected the bending energy of the fans, taking into account only the energy (considered constant) of the discontinuity walls between the fans. This relation shows that the formation of fans becomes favorable for $\theta_0 < 1$ rd $\approx 60°$, in good agreement with the experimental observations.

IX.4.e) Twist kink and motion of wedge lines

In thick samples one can sometimes notice twist kinks on the wedge lines (Fig. C.IX.30a). They are however rather infrequent, since their energy is high.

Indeed, screw dislocations are generated radially from each kink, forming a wall perpendicular to the disclination line. The number of dislocations is fixed by the requirement that the kink width AB be equal to the sum of all Burgers vectors (Fig. C.IX.30b):

$$AB = \sum b_i \qquad\qquad\qquad (C.IX.41)$$

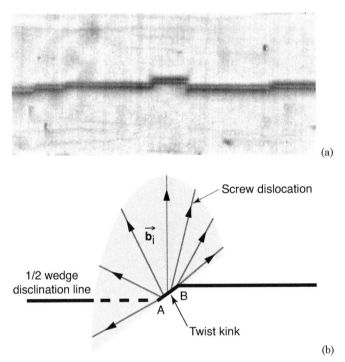

(a)

(b)

Fig. C.IX.30 a) Twist kink observed on a wedge line of rank 1/2 in a free drop of C5HET; b) schematic representation of the kink and of the screw dislocations attached to it.

The wedge lines of rank 1/2 can also move very slowly within the samples. Their movement was observed in C5HET during prolonged annealing of fan-shaped textures [6a]. They are intimately related to the presence of walls of the second kind and to the glide motion of transverse edge dislocations along these walls (the Peierls barrier is probably lower). In figure C.IX.31, we resolve the motion of a wedge line of rank 1/2. The different steps are as follows:

1. Formation of a streak AOBC (Fig. C.IX.31b); in this configuration, frequently encountered in the samples, the molecular cylinders are centered on points A, B, and C in triangles AOC, BOC, and ACB, respectively. Walls AO, BO, AC and BC are thus of the second kind, while the OC wall is of the first kind.

2. Glide motion of points A and B on the domain boundary; as the disorientation angle of the columns across the OC wall increases, the anchoring of the columns on the domain boundary (between A and B) progressively

changes (incidentally, this is probably the origin of the process that sets the line in motion). Note that during this step the line is no longer perfect, having an angle $\phi < \pi$ (so a rank $m = \phi/2\pi < 1/2$)

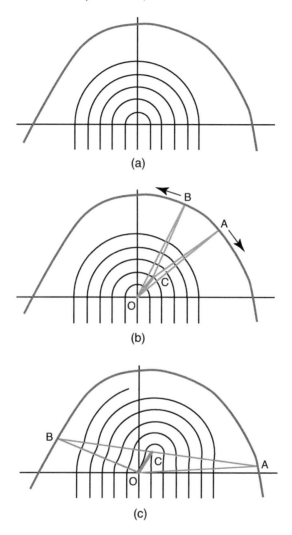

(a)

(b)

(c)

Fig. C.IX.31 Motion of a wedge line. a) Initial configuration; b) nucleation of a streak; c) as points A and B glide on the domain boundary, the 1/2 line moves from O to C, leaving behind a wall of the first kind: OC.

3. As points ACB are once again aligned (Fig. C.IX.31c), the disclination line is again perfect, of rank 1/2. On the other hand, its motion is accompanied by the formation of a wall of the first kind (the OC segment).

The result obtained in this particular case is in fact quite general, the motion of a wedge line being always accompanied (for topological reasons) by the "emission" of edge dislocations [29].

IX.5 Measuring the elastic constants

The compressibility modulus B of the hexagonal lattice can be determined by measuring the mechanical impedance of planar samples exhibiting a fan-shaped texture. These measurements yielded values close to $10^8 erg/cm^3$ for C5HET and for the $C_{12}EO_6$ + water mixture [22a,b].

By compressing (or dilating) a homeotropic sample and by measuring its instantaneous elastic response, one can obtain the compressibility modulus A of the medium. This measurement must be performed very quickly (over less than a ms) to avoid sample flow by permeation or any other plastic mechanism. The experiments of Cagnon et al. yield $A \approx 5 \times 10^8$ erg/cm^3 for C8HET [30], but these authors point out that this value is probably underestimated due to the presence of a few slightly disoriented grains in the samples. Hence, they propose that a value of 10^9 erg/cm^3 would be closer to reality.

Estimating K is an even more difficult task, as this coefficient is not directly accessible. On the other hand, one can estimate the penetration length $\lambda = (K/B)^{1/2}$ by indirect methods, for instance by using the aforementioned results for the energy of transverse edge dislocations [22a,b]. To this end one needs an explicit formula for their energy as a function of λ and other parameters, known or easily measurable. Formula C.IX.28a does not fit the bill, since the core radius r_c and its energy E_c are still unknown. To proceed, one therefore needs a model for the core. The simplest one assumes that the core is dissociated in the glide plane and that its energy is of the order of $\gamma\, r_c$, where γ is the energy of the hexagonal-isotropic interface, well known experimentally. Minimizing the total energy C.IX.28a of the dislocation as a function of r_c yields:

$$E_{ct} \approx \sqrt{2B\lambda\gamma}\, b \qquad\qquad (C.IX.42)$$

Using this formula and available results for B, γ and E_{et}, one obtains $\lambda \approx 10$ Å in the lyotropic mixture $C_{12}EO_6$ + water and $\lambda \approx 30$ Å in C8HET. These orders of magnitude indicate that, in the static regime and at the length scale of the dislocations (less than 1 µm), the penetration length λ is, as for the case of smectics, comparable with a molecular distance.

This result was confirmed by X-ray studies, from the spreading of the Bragg spot along the direction of the columns [31]. This diffuse scattering is due to the thermal undulations of the columns owing to their considerable flexibility. This method yields $\lambda \approx 20$ Å for the lyotropic mixture $C_{12}EO_6$ + water, in reasonable agreement with the preceding estimate. It was also used to determine λ in the columnar mesophases of other materials than the triphenylenes. For each and all of them, the measured values are microscopic [32].

The analysis of the diffuse scattering around the Bragg peaks in a plane perpendicular to the columns also yields some information on the elasticity of the lattice, more precisely on the anisotropy ratio A between the shear modulus and the compressibility modulus. Attributing this scattering to transverse elastic phonons, Clerc found $0.025 < A < 0.075$ for the lyotropic mixture $C_{12}EO_6$ + water [31]. This implies that it is easier to shear or stretch the hexagonal lattice at constant density than to change its density by compression or dilation.

IX.6 Mechanical instabilities

This is a delicate topic, the experimental results being very few and sometimes contradictory.

Two types of experiments can be distinguished. Those resorting to "gentle" methods to dilate or compress the samples, methods that we might term quasi-static (thermal methods), and the more violent ones, involving brutal deformation of the samples, using piezoelectric ceramics, for instance. Let us start with the first kind.

IX.6.a) Striations of thermomechanical origin

Many authors have noticed that fan-shaped textures are often striated [30–38]. These striations, developing spontaneously in the samples as they are heated or cooled, are then almost impossible to eliminate (unless perhaps after lengthy annealing over a few days). Their quantitative study was only recently performed [39] (Fig. C.IX.32).

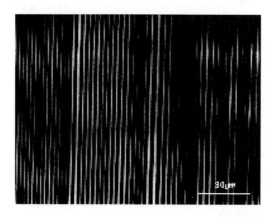

Fig. C.IX.32 Striations observed between crossed polarizers in a 100-μm thick sample of the lyotropic mixture $C_{12}EO_6$ + water. The contrast of the striations can be reversed by rotating the polarizers.

The main experimental results are the following:

1. The striations are due to a zigzag configuration of the molecular columns in the horizontal plane [30–33].

2. The angle of the zigzag (i.e., the maximum angle of the columns with respect to their average direction), is usually of the order of a few degrees.

3. The wavelength Λ of the striations varies as the square root of the sample thickness d:

$$\Lambda \propto \sqrt{d} \tag{C.IX.43}$$

Such a dependence is reminiscent of the layer undulation instability in a smectic A under dilation normal to the layers. This analogy suggests that the striations are due to the thermal dilation of the columnar lattice.

To prove this, let us determine the critical value γ_c of the dilation γ for which the columns start to undulate. We assume that the columns are initially straight and parallel to the two glass plates limiting the sample. We also assume they are strongly anchored at the surfaces and that the sample thickness remains constant. As the temperature changes, so does the distance between columns. The result is a deformation γ of the columnar lattice, of the type

$$u = \gamma x, \quad v = \gamma y \tag{C.IX.44}$$

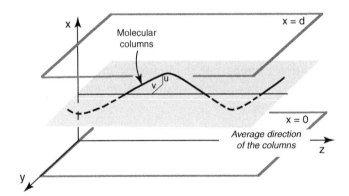

Fig. C.IX.33 Column undulations in a planar sample.

If the distance between columns increases γ is positive, corresponding to dilation. In the opposite case, the lattice is compressed. To study the stability of the deformed state, let us look for the value γ_c of the dilation above which a solution of the type

$$u = \gamma x + u_0 \sin(q_z z) \sin(q_x x) \tag{C.IX.45a}$$

$$v = \gamma y + v_0 \sin(q_z z) \sin(q_x x) \tag{C.IX.45b}$$

becomes energetically favorable. In this expression, $q_z = 2\pi/\Lambda$ is the wave vector of the undulation and $q_x = \pi/d$ enforces the strong anchoring condition at the plates. The coordinate system is defined in figure C.IX.33. Calculating the average energy per unit area of the sample

$$<\rho f> = \frac{1}{\Lambda} \int_0^\Lambda \int_0^d \rho f \, dz dx \tag{C.IX.46a}$$

yields, to second order in the amplitude and taking into account the anharmonic corrections (eqs. C.IX.6 and C.IX.9):

$$\frac{1}{Bd} <\rho f> = 2 \, (1 - A) \gamma^2 + \frac{1}{8} [\; \lambda^2 q_z^4 + q_x^2 - 2\gamma (1 - A) q_z^2] \; u_o^2$$

$$+ \frac{1}{8} [\; \lambda^2 q_z^4 + q_x^2 \, (2\gamma (1 - A) + A \, (1 - \gamma)^2) - 2\gamma (1 - A) \; q_z^2] \; v_o^2 + o \, (u_o^4, v_o^4)$$

$$\tag{C.IX.46b}$$

Note the absence of a cross-term in $u_o v_o$. This shows that the instabilities in the vertical plane and the horizontal plane are decoupled. To obtain the corresponding instability thresholds, one only needs to find the minimal values of γ for which the prefactors of the u_o^2 and v_o^2 terms go to zero. This yields:

$$\Lambda_{uc} = 2\sqrt{\pi \lambda d} \quad \text{and} \quad \gamma_{uc} = \frac{\lambda q_x}{1 - A} \qquad \text{in the vertical plane} \tag{C.IX.47a}$$

and

$$\Lambda_{vc} = \frac{2\sqrt{\pi \lambda d}}{\sqrt[4]{A}} \quad \text{and} \quad \gamma_{vc} = \frac{\lambda q_x}{1 - A} \sqrt{A} \qquad \text{in the horizontal plane} \tag{C.IX.47b}$$

Since $A < 1$, one always has $\gamma_{vc} < \gamma_{uc}$ and $\Lambda_{vc} > \Lambda_{uc}$. The columns first become unstable in the horizontal plane. In the following, we set $\gamma_{vc} = \gamma_c$ and $\Lambda_{vc} = \Lambda_c$ since these values define the true instability threshold in the system. In practice, this critical threshold γ_c is extremely low. For instance, one gets $\gamma_c \approx 10^{-5}$ for the lyotropic mixture $C_{12}EO_6$ + water using $\lambda = 15$ Å, $A = 0.05$ and $d = 100$ μm. This value explains why the striations appear so easily on cooling the sample. Since on the other hand we know that the value of the thermal dilation coefficient is close to 10^{-3} K^{-1}, both in lyotropic [33, 34, 36, 37] and thermotropic systems [38], it becomes apparent that the striations are always observed far from the threshold, in the strongly nonlinear regime. This prompts the very relevant question of whether formula C.IX.47 for the wavelength of the striations as a function of the sample thickness is still valid.

To answer this question, a nonlinear model was developed in Oswald et al. [39]. Inspired by a calculation of Singer [40] for lamellar phases, this model is technically difficult, so that we shall present a simplified version, yielding however the same results far from the threshold.

The main assumption of this model is that the columns form a zigzag rather than a sinusoid, a result that can be rigorously proven far from the threshold (Fig. C.IX.34a). In a "zig" (or a "zag"), the columns are straight, making an angle θ with the z axis and an angle α with the horizontal plane (y, z) (these angles are defined in the median plane, x = d/2). If the columns are not strongly disoriented with respect to their average direction (10° at most), a bend wall (Fig. C.IX.27a) forms between each zig and zag. This model is only meaningful if the width of this wall, of the order of λ/θ, is small with respect to the width L = $\Lambda/2$ of a stripe (whether zig or zag). This condition is generally fulfilled in practice since it is enough to have, for instance, $\theta >> 10^{-3}$ rd $\approx 0.06°$ if $\lambda \approx 20$ Å and L = 2 μm (in practice, θ is of the order of a few degrees).

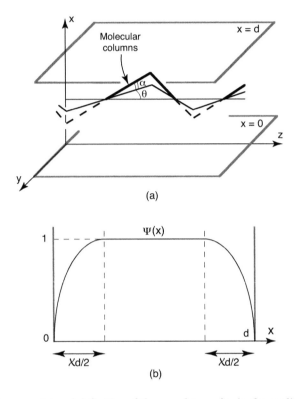

Fig. C.IX.34 a) Zigzag model and definition of the α and θ angles in the median plane; b) the $\Psi(x)$ function specifying the way that the columns turn in the thickness of the sample.

To determine the wavelength of the zigzag, one still needs to specify how the cylinders turn in the thickness of the sample. We choose to write the strain between two bend walls (i.e., within a zig or zag stripe) under the form:

$$u = \gamma x + \Psi(x)\alpha z \qquad\qquad\qquad \text{(C.IX.48a)}$$

$$v = \gamma y + \Psi(x)\theta z \qquad\qquad\qquad \text{(C.IX.48b)}$$

with:

$$\Psi(x) = \sin\left(\frac{\pi}{\chi}\frac{x}{d}\right) \qquad \text{if} \qquad 0 < x < d\frac{\chi}{2} \qquad\qquad \text{(C.IX.49a)}$$

$$\Psi(x) = 1 \qquad\qquad \text{if} \qquad d\frac{X}{2} < x < d\left(1 - \frac{X}{2}\right) \qquad\qquad \text{(C.IX.49b)}$$

$$\Psi(x) = \sin\left[\frac{\pi}{X}\left(1 - \frac{x}{d}\right)\right] \quad \text{if} \qquad d\left(1 - \frac{X}{2}\right) < x < d \qquad\qquad \text{(C.IX.49c)}$$

and $0 < X < 1$. This function, shown in figure C.IX.34b, fulfills the boundary conditions.

The next step is calculating the free energy per unit surface. This calculation, tedious but straightforward, leads to the following result [39]:

$$\frac{4}{Bd}<\rho f> = Cst - 2\gamma(2 - X)(1 - A)(\alpha^2 + \theta^2) + \frac{1}{12}(\alpha^2 + A\theta^2)\frac{q_x^2 L^2}{X}$$

$$+ \frac{1}{2}\left(1 - \frac{5}{8}X\right)(\alpha^2 + \theta^2)^2 + \frac{32}{9\pi}\frac{\lambda}{L}\left(X + \frac{3\pi}{4}(1 - X)\right)(\alpha^2 + \theta^2)^{3/2} \quad \text{(C.IX.50)}$$

Note that the last term of this expression corresponds to the energy of the bend walls separating the zigs from the zags (across such a wall, the columns are disoriented by an angle of $2(\theta^2 + \alpha^2)^{1/2}$). Minimization with respect to L can be performed analytically. It yields successively:

$$L = \left(\frac{16\lambda d^2}{\pi^2}\right)^{1/3} X^{1/3}\left[1 - X\left(1 - \frac{4}{3\pi}\right)\right]^{1/3}\frac{(\alpha^2 + \theta^2)^{1/2}}{(\alpha^2 + A\theta^2)^{1/3}} \qquad\qquad \text{(C.IX.51)}$$

and

$$\frac{4}{Bd}<\rho f> = Cst - 2\gamma(2 - X)(1 - A)(\alpha^2 + \theta^2) + \frac{1}{2}\left(1 - \frac{5}{8}X\right)(\alpha^2 + \theta^2)^2$$

$$+ (2\pi)^{2/3}\left(\frac{1}{X}\right)^{1/3}\left(\frac{\lambda}{d}\right)^{2/3}\left[1 - X\left(1 - \frac{4}{3\pi}\right)\right]^{2/3}(\alpha^2 + A\theta^2)^{1/3}(\alpha^2 + \theta^2) \quad \text{(C.IX.52)}$$

by substituting for L in the energy. Minimization with respect to the other variables X, θ and α is more difficult and can only be performed numerically, using for instance the *Mathematica* software. In normal conditions, the numerical calculation always yields $\alpha = 0$, meaning that *the columns prefer to stay in a plane parallel to the plates*. This somewhat surprising result was confirmed by X-ray measurements of the orientation of zigs and zags in a striated structure of a lyotropic hexagonal phase [41]. On the other hand, the values of θ, X and Λ depend on γ and on the thickness d of the samples. It can be shown numerically that, taking $\lambda = 10$ Å and $A = 0.05$, the following formulae apply for $\gamma > 10^{-3}$:

$$\Lambda \approx (1.020 \pm 0.007)\Lambda_c \qquad\qquad \text{(C.IX.53)}$$

$$\theta \approx (1.92 \pm 0.01)\sqrt{\gamma} \qquad\qquad\qquad (C.IX.54)$$

$$X \approx (1.071 \pm 0.006)\sqrt{\frac{\gamma_c}{\gamma}} \qquad\qquad (C.IX.55)$$

Of all these formulae the first one is the most surprising, showing that the wavelength of the zigzag is practically independent of the dilation, remaining equal to its threshold value Λ_c:

$$\Lambda \approx \Lambda_c = \frac{2\sqrt{\pi\lambda d}}{\sqrt[4]{A}} \qquad \text{(to within 5\%)} \qquad\qquad (C.IX.56)$$

The formula for θ also deserves some comment; from the experimental point of view, since it provides an estimate for γ once the angle θ – measurable from the textures – is known (in practice, θ takes values of a few degrees or fractions of a degree for values of γ between 10^{-4} and 10^{-2}), and from the theoretical point of view, since it can be found by writing that, within each stripe, the density in the plane normal to the columns remains almost unchanged (yielding $U_{xx} + U_{yy} = 2\gamma - \theta^2/2 \approx 0$, whence $\theta = 2\gamma^{1/2}$).

Finally, one may note that X is generally small (of the order of 0.1 in usual conditions) meaning that two well-defined twist boundary layers appear close to the plates.

Measurements of the wavelength as a function of the sample thickness performed for $C_{12}EO_6$ confirm the $d^{1/2}$ dependence (Fig.C.IX.35). They yield $\lambda/A^{1/2} \approx 190$ Å, resulting in $\lambda \approx 20$ Å for the smallest value of A compatible with the X-ray results ($A = 0.025$). This value of λ is in good agreement with the previously mentioned values.
This result leads to $\lambda/A^{1/2} \approx 100$ Å. Once again, we see that λ is close to a molecular distance.

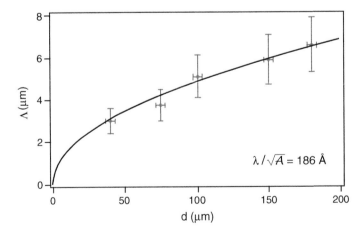

Fig. C.IX.35 Wavelength of the thermal striations as a function of the sample thickness in the mixture $C_{12}EO_6$ + water [39].

This experiment confirms once again that the penetration length λ is close to a molecular length, as in smectics A. Note that in this case the determination of λ (quasi-static, since the dilation is imposed by gently cooling the samples) is performed over a length scale of several μm.

The wavelength of the thermal striations was also measured for direct hexagonal phases where the cylinders are swollen with oil. The system is a mixture of SDS (sodium dodecyl sulphate), pentanol, cyclohexane and brine. These measurements showed that the penetration length λ increases like the diameter of the cylinders to the power 2/3, showing that the phase becomes softer upon swelling (it is mostly B that decreases). These swollen H_α phases can also be doped with magnetic particles, resulting in a slight decrease of λ, not yet well explained [42a–e].

The thermal striations are also visible in columnar mesophases of disk-like molecules, such as the triphenylene derivatives. In these materials, they appear on heating the samples, suggesting that the lattice parameter of the hexagonal lattice decreases with increasing temperature. The first measurements [43], performed on a 100-μm thick C8HET sample, show that $\Lambda \approx 2$ μm a few degrees below the transition temperature to the isotropic phase.

IX.6.b) Mechanical instabilities induced by sudden deformation

The main difference with respect to the instability of thermal origin described in the previous paragraph is that, this time, the deformation is imposed suddenly, rather than in a gradual way. In smectics A (we refer here to the layer buckling instability), this distinction was not very important as the measured wavelengths were (within experimental error), unaffected by the specific way of dilating the samples.

In columnar phases, this distinction becomes fundamental, as shown by the experiments of the Durand group in Orsay.

Let us first consider the layer buckling instability under uniaxial dilation. This instability, similar to that observed in smectics, develops when a planar sample is abruptly dilated. The difference with respect to the thermal case is that, here, the dilation is sudden and uniaxial. The calculation of the critical dilation for the appearance of undulations in the vertical plane and its wavelength can be performed as in the previous paragraph. It yields [16]:

$$\gamma_c = \left(\frac{\delta d}{d}\right)_c = 2\pi\lambda \quad \text{and} \quad \Lambda_c = 2\sqrt{\pi\lambda d} \tag{C.IX.57}$$

The experiment was performed by Gharbia et al. [43] using C8HET samples of variable thickness. As expected, the $d^{1/2}$ dependence is clearly verified; on the other hand, the value of λ found by this method is abnormally high: $\lambda \approx 1000$ Å. This result suggests that, at high frequency, the columns are correlated over similar distances.

To date, there is no entirely convincing explanation of this anomaly. One possibility would be that the system is less flexible at high frequency due to the presence of a high defect density (for instance, dislocation loops or column ends, which have been clearly shown to appear in the columnar phases of certain substituted aromatic diamides [13a,b]). In this case, Prost [44] showed that the bending modulus of the medium must be renormalized, taking the value K* (in practice, 1000 to 10,000 times larger than K when the defects are frozen within the matrix). Note that formulae C.IX.57 remain valid, provided that K is replaced by K*.

A second experiment, performed by the Orsay group, deals with the buckling instability of the columns under compression along their axis. This instability, predicted in 1982 [16], has no equivalent in smectics.

The principle of the experiment is presented in figure C.IX.36. The sample, in homeotropic anchoring (columns perpendicular to the plates), is brutally compressed. Let δd be its thickness variation and γ the compression ($\gamma = \delta d/d < 0$).

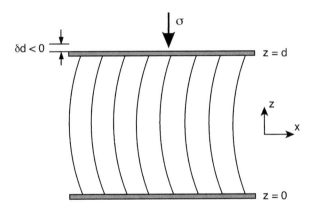

Fig. C.IX.36 Buckling instability of the molecular columns under compression in a hexagonal phase.

We assume that the buckling occurs in the (x, z) plane ($\partial/\partial y = 0$, $v = 0$), that the sample is thin (more precisely, that its thickness is much smaller than its lateral size, imposing $\partial u/\partial x = 0$, [45]) and that the molecules do not slip at the surfaces, nor do they flow by permeation during the experiment. By definition of the volume dilation θ, in the absence of flow one has:

$$\theta = \frac{\delta d}{d} \tag{C.IX.58}$$

From equation C.IX.10, the density of elastic energy is

$$\rho f = \frac{1}{2}A\theta^2 + \frac{1}{2}\bar{B}(U_{xx})^2 - D\theta U_{xx} + \frac{1}{2}K\left(\frac{\partial^2 u}{\partial z^2}\right)^2 \tag{C.IX.59}$$

where $U_{xx} = - (1/2) (\partial u/\partial z)^2$ (note that the volume dilation θ does not change to second order in the deformation [16]). To find the instability threshold, let us consider that a solution of the type

$$u = u_o \sin\left(\frac{\pi z}{d}\right)$$

(C.IX.60)

becomes energetically favorable. The total energy $< \rho f >$ per unit surface is, to second order in u_o:

$$<\rho f> = \frac{\pi^2}{4} \frac{u_o^2}{d} \left(D \frac{\delta d}{d} + \pi^2 \frac{K}{d^2} \right) + o(u_o^4)$$

(C.IX.61)

It ensues that the instability under compression [46] develops when

$$\frac{\delta d}{d} < \left(\frac{\delta d}{d}\right)_c = - \pi^2 \frac{K}{Dd^2}$$

(C.IX.62)

This instability threshold is extremely low if one takes for K its static value (this hypothesis is a priori not valid, considering the initial assumptions). Indeed, $(\delta d/d)_c \approx -10^{-9}$ with $D \approx A$ [47], $(K/A)^{1/2} \approx 10$ Å and $d = 100$ μm! How does this compare with the experimental results?

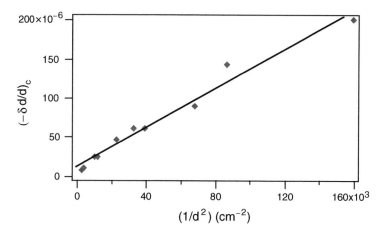

Fig. C.IX.37 Critical dilation as a function of the reciprocal of the squared thickness (from ref. [30]).

To test these theoretical predictions, Cagnon et al. [30] performed the experiment using C8HET. They demonstrated the existence of a buckling threshold varying as $1/d^2$ (Fig. C.IX.37), in agreement with the theoretical law C.IX.62 (up to a residual dilation, of the order of 10^{-5}). On the other hand, they find an abnormally high value for the penetration length, giving for K a renormalized value K* such that $K^{*1/2}/A \approx 1\,500$ Å (typically, $(\delta d/d)_c \approx 2.4 \times 10^{-5}$ for d = 100 μm).

This result, coherent with the measurements performed on planar samples [43], seems to confirm the idea that the bending constant does not have the same value at zero frequency ($K \approx 10^{-5}$ dyn) and at high frequency ($K^* \approx 0.2$ dyn). This difference is however considerable (4 to 5 orders of magnitude) and prompted us to study in more detail the dynamics of the buckling instabilities. Our conclusions are summarized in the next paragraph.

IX.7 Dynamics of the buckling instabilities

In the previous paragraph, we assumed the buckling instabilities to be instantaneous. In practice, they develop over finite times, in the ms range.

Let us start by considering the buckling instability of the columns under uniaxial dilation. Its development time can be calculated as for smectics [48], yielding:

$$\tau = \frac{2\pi d\eta}{B(\delta d - \delta d_c)} \qquad \text{with} \quad \delta d_c = 2\pi\lambda \qquad (\text{C.IX.63a})$$

where η is the viscosity under shear parallel to the columns. At twice the critical threshold, one typically has

$$\tau \approx \frac{d\eta}{\sqrt{KB}} \qquad (\text{C.IX.63b})$$

which, taking d = 100 μm, η = 10 poise (this value is hypothetical), $K \approx 10^{-5}$ dyn, and $B \approx 10^8$ erg/cm³, amounts to:

$$\tau \approx 3 \text{ ms} \qquad (\text{C.IX.63c})$$

Replacing K by a value $K^* \approx 0.1$ dyn yields an even faster development time, of the order of 30 μs. This calculation shows that, irrespective of the value of K, the buckling process is very fast. Our previous assumptions are thus justified as long as the applied mechanical stress does not relax significantly over the few ms during which the instability develops. This condition is satisfied in the experiments, as the planar samples (even the less well oriented) behave elastically above 20 Hz [22a,b].

Let us now analyze the case of column buckling under uniaxial compression (Fig. C.IX.36).

To determine its development time, let us write the dynamical equations of the problem, by analogy with the laws already known for smectic systems (for a detailed discussion, see Prost and Clark [49]).

The first two equations translate momentum conservation. They write simply:

$$\rho \frac{Dv_x}{Dt} = -\frac{\partial P}{\partial x} + \eta \, \Delta v_x + G_x \qquad \text{(C.IX.64a)}$$

$$\rho \frac{Dv_z}{Dt} = -\frac{\partial P}{\partial z} + \eta \Delta v_z \qquad \text{(C.IX.64b)}$$

within a simplified model described by a unique viscosity η. In the first equation, G_x is the elastic stress conjugated to the strain u. With the same hypotheses as in the previous paragraph ($v = 0$, $\partial/\partial x = 0$, $\theta = \delta d/d < 0$) it is given by:

$$G_x = -\frac{\delta \rho f}{\delta u} = \frac{\partial}{\partial z} \frac{\partial \rho f}{\partial u_{,z}} - \frac{\partial^2}{\partial z^2} \frac{\partial \rho f}{\partial u_{,zz}} = D \frac{\delta d}{d} \frac{\partial^2 u}{\partial z^2} - K \frac{\partial^4 u}{\partial z^4} \qquad \text{(C.IX.65)}$$

As for smectics, one can write a permeation law of the type:

$$\frac{Du}{Dt} - v_x = \lambda_p G_x \qquad \text{(C.IX.66)}$$

Finally, one needs the mass conservation law:

$$\text{div } \mathbf{v} + \frac{D\theta}{Dt} = 0 \qquad \text{(C.IX.67)}$$

To find the instability threshold and determine its growth rate, we assume that the permeation and inertial effects are negligible (to be justified a posteriori). Linearizing the equations yields:

$$v_z = 0, \qquad v_x = \frac{\partial u}{\partial t} \qquad \text{(C.IX.68a)}$$

$$P = \text{Constant} \qquad \text{(C.IX.68b)}$$

$$D \frac{\delta d}{d} \frac{\partial^2 u}{\partial z^2} - K \frac{\partial^4 u}{\partial z^4} + \eta \frac{\partial^2 v_x}{\partial z^2} = 0 \qquad \text{(C.IX.68c)}$$

We expect a solution of the type $u = u_o \sin(\pi z/d) \exp(t/\tau)$. Replacing in eq. C.IX.68c, one obtains:

$$\frac{\eta}{\tau} = -D \left(\frac{\delta d}{d} + \frac{K \pi^2}{D d^2} \right) \qquad \text{(C.IX.69)}$$

meaning that the instability develops if

$$\frac{\delta d}{d} < \left(\frac{\delta d}{d} \right)_c = -\frac{K \pi^2}{D d^2} \qquad \text{(C.IX.70)}$$

This is the already calculated threshold value (eq. C.IX.62). One can also estimate the developing time of the instability at twice the threshold value:

$$\tau \approx \frac{d^2\eta}{\pi^2 K} \qquad\qquad\qquad (C.IX.71)$$

With $K = K^* = 0.1$ dyn, $\eta = 10$ poise and $d = 100$ µm, one gets $\tau = 1$ms. This time, albeit 1 000 times longer than for the buckling instability (eq. C.IX.63b), remains very short compared to the time for plastic relaxation of the applied stress.

In conclusion, only the assumption of a renormalized K constant at high frequency can explain the abnormally high buckling threshold in compression parallel to the columns.

IX.8 Light scattering

Recently, Gharbia et al. [50] described the Rayleigh scattering from a homeotropic C8HET sample in the hexagonal phase. We shall see that this experiment raises new questions concerning the elasticity of the columnar phase. Indeed, one of the conclusions of this study, at variance with the experiments and interpretations discussed above, is that the column buckling is controlled by an elasticity of the three-dimensional solid type, rather than by bending elasticity. We shall show that this conclusion is not the only one consistent with the data and that the current theory of columnar phases can be reconciled with these light scattering results. Let us start by summarizing the experimental facts.

The geometry of the scattering setup is presented in figure C.IX.38. Let **k** be the wave vector of the incident beam (He-Ne laser, at a wavelength $\lambda = 6\,328$ Å), **k'** that of the scattered beam and $\mathbf{q} = \mathbf{k} - \mathbf{k'}$ the scattering vector (it will be denoted by **Q** in backscattering). The **k** and **k'** vectors can be taken either as the wave vectors of the ordinary or of the extraordinary beam.

The experiment shows that the scattering signal is depolarized, the allowed couples being $(\mathbf{k}_e, \mathbf{k}_o')$ and $(\mathbf{k}_o, \mathbf{k}_e')$. This observation is in agreement with the selection rules found for nematics (scattering by the orientation fluctuations of the director, see paragraph B.II.7 of volume 1).

The experiment also shows that scattering only occurs for discrete values of the angle α between the wave vector **k** and the z axis parallel to the columns. This observation shows that the column undulations scatter light and that the components of the scattering vectors **q** and **Q** along the z axis are quantified:

$$q_{//} = p\,\frac{\pi}{d} \qquad \text{and} \qquad Q_{//} = r\,\frac{\pi}{d} \qquad \text{with integral values for p and r} \qquad (C.IX.72)$$

Experimentally, modes p = 1, 2, 3, and 4 are observed in transmission, while modes r = 115, 116, 117, and 118 are visible in backscattering (with a 12.5 μm thick sample).

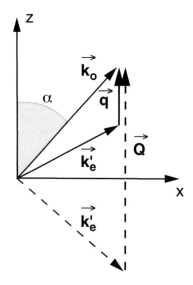

Fig. C.IX.38 Geometry of the light scattering setup. Vectors **q** and **Q** correspond to forward and back scattering, respectively. The **z** axis is parallel to the molecular columns.

Finally, and this is the main result of the study, the angular width at half-maximum $\delta\theta \approx q_\perp/q_{//}$ of each light scattering peak is **constant**, of the order of 4×10^{-2} rad, **from p = 1 to r = 118**. This observation is incompatible with an elasticity of the column bend type. Indeed, the width $\delta\theta$ defines the wave vector $q_\perp$ of the compression mode of the two-dimensional lattice ($B_\perp$ modulus) with the same energy as an undulation of the columns with a wave vector $q_{//}$ (K modulus). Denoting by u the strain of the columns, this yields $1/2\ B_\perp q_\perp^2 u^2 \approx 1/2\ Kq_{//}^4 u^2$ resulting in $\delta\theta \approx (K/B_\perp)^{1/2}q_{//}$, in striking disagreement with the experiment.

To explain this new anomaly, Gharbia et al. propose that an undulation with wave vector $q_{//}$ costs an energy $1/2\ B_{//}q_{//}^2u^2$. In these conditions, $\delta\theta \approx (B_{//}/B_\perp)^{1/2} = \text{Const}$, in agreement with the experiment.

Must one then infer that the hexagonal phase of C8HET exhibits an elasticity of the three-dimensional solid type, as proposed by these authors? Not necessarily.

To show this, let us first recall that a residual dilation θ_o persists within the sample and does not appear to relax easily. Its value, of the order of 10^{-5}, can be determined by the experiments of column buckling under compression, showing that they must first be compressed by a constant quantity, independent of the thickness, in order to induce the instability (see Fig. C.IX.37). In these conditions, the distortion free energy reads to second order in the distortion $\partial u/\partial x_i$, and taking into account the coupling between θ_o and the compression of the lattice:

$$\rho f = Cst + \frac{1}{2} B_\perp \left(\frac{\partial u}{\partial x}\right)^2 + \frac{1}{2} D\theta_o \left(\frac{\partial u}{\partial z}\right)^2 + \frac{1}{2} K \left(\frac{\partial^2 u}{\partial z^2}\right)^2 \qquad \text{(C.IX.73)}$$

where we set $B_\perp = \bar{B}$. This expression (that includes the anharmonic corrections and where we set $\partial/\partial y = 0$ for simplicity) contains the missing term of Gharbia et al. with $B_{//} = D\theta_o$. As such, it seems unnecessary to resort to a new type of elasticity to explain the results obtained in Rayleigh scattering, but this point deserves experimental confirmation.

IX.9 Threads of columnar phases

IX.9.a) Preliminary observation in polarizing microscopy

One cannot conclude this chapter without discussing threads, which are to thermotropic columnar phases as films are to the smectic phases.

Van Winkle and Clark [51] were the first to show that one could draw very thin and perfectly oriented threads of hexagonal or rectangular columnar phases. In this case, the molecular cylinders orient along the axis of the thread. The photographs in figure C.IX.39 show the appearance of a few threads drawn from the rectangular phase D_r of hexa-n-dodecanyloxytriphenylene (C11HAT) (the molecule is that shown in figure C.IX.3 with $R = OCOC_{11}H_{23}$). In the first photograph (a), the analyzer was removed. This thread, with a diameter of 1 µm, appears uniform under the microscope. Another, slightly thicker, thread (1.5 µm), is shown in photographs (b) and (c). In this case, the contrast varies continuously along the thread, when observed in polarized light without the analyzer (photo b). Between crossed polarizers (photo c), with the polarizer parallel to the axis of the thread, it is however possible to achieve extinction in any point along the thread by turning it around its axis. Based on this observation, the authors conclude that the thread locally behaves as a uniaxial medium, the optical axis making a finite angle with the axis of the thread. This observation is not entirely compatible with the $P2_1/a$ symmetry of the D_r phase, which confers upon it a biaxial character. Nevertheless, the biaxiality of the phase is weak, as shown by formulae C.IX.2a–c if angles θ and φ are not too large, and it behaves in the first approximation as a uniaxial phase, with an optical axis tilted with respect to the axis of the columns. On the other hand, the continuous rotation of the optical axis shows that the thread is twisted and that the columns take a helical shape (the columnar lattice turns by 2π from one end of the thread to the other). In photograph (d), the thread is even thicker (of diameter 3 µm). One can see this time two localized walls separating a dark region in the center from two brighter regions at the sides. Across each wall the optical axis, and hence the columnar lattice, abruptly turn by a $\pi/3$ angle.

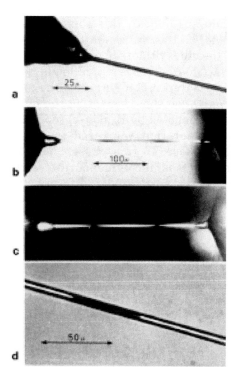

Fig. C.IX.39 A few threads of the rectangular phase of hexa-n-dodecanoyloxytriphenylene (C11HAT) observed in the polarizing microscope (from ref. [51]).

These first experiments showed the possibility of drawing threads and opened the way for more detailed structural studies of these phases using X-rays. The advantage is that the columns are perfectly aligned, which is rarely the case in bulk samples. However, even in threads a few microns in diameter, a certain mosaicity still persists, as shown by these preliminary observations.

IX.9.b) Study of the threads by X-ray diffraction

The first X-ray study of these threads was performed two years later, in 1984, by Safinya et al. [52] using the same material (C11HAT) and it revealed the rectangular and hexagonal structures of the low- and high-temperature columnar phases. This experiment also confirmed that, in the hexagonal phase of C11HAT, the molecules are tilted with respect to the axis of the columns, without a preferred orientation, in agreement with the birefringence measurements [9] (see also section IX.I.1). Other studies, on different materials from the series of triphenylene or truxene derivatives, followed this work and could specify the detailed structure of these phases. In particular, they confirm that, in the D_{hd} phase, the hydrocarbon chains are strongly disordered (as in a liquid), while the rigid cores exhibit orientational order, but are positionally disordered within the columns [53]. At this point, we should also mention the work of Davidson et al. [54] on the diffuse scattering around the Bragg peaks in the D_{hd} phase of HHTT (hexahexylthiotriphenylene). This diffuse scattering can be due to the phonons (this interpretation leads to a particularly small penetration length in this material, $\lambda \approx 3$ Å) or to the presence of static defects such as dislocations. Finally, we should mention the discovery in this material of a helical columnar phase denoted by H [55a–c] (no relation whatsoever to the D_{ho} phase, although this notation was used in the first reference [55a]) appearing below the D_{hd} phase. In this phase, the molecules form a helical structure with a pitch that is incommensurate with the intermolecular distance. The columns are also correlated and form a super-lattice with three columns per lattice cell, such that this phase resembles more a crystal than a liquid crystal.

IX.9.c) On the stability of threads

The experiments we have just described show the possibility of obtaining threads that are stable over long periods. These threads can have a wide range of diameters, ranging from the μm to several hundreds of μm, exactly as the thickness of smectic films can vary from a few layers to a few hundred layers.

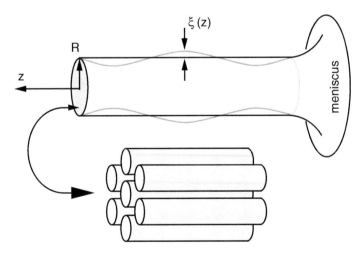

Fig. C.IX.40 Hexagonal phase thread and its meniscus.

In films, it is the elasticity of the layers that blocks the flow of material toward the meniscus (see chapter C.VIII).

In the case of threads, this effect exists, but one must also take into account their natural tendency to form droplets. Indeed, it has been known, ever since the work of Lord Rayleigh in 1879, that a cylindrical liquid jet, of axis z and radius R, is always unstable with respect to a sinusoidal deformation of its surface [R $\propto$ sin(qz)] since the deformation decreases its surface energy at a constant volume. This instability, purely geometrical in origin, has no equivalent in the case of films, which are planar.

In this paragraph, we shall show that the elasticity of the columnar lattice blocks the flow of material towards the meniscus and suppresses the jet instability, provided that the radius of the thread exceeds a critical radius that we shall determine [51, 56].

Let us consider a perfect cylindrical thread (i.e., without defects) of a hexagonal phase, with radius R and axis z (Fig. C.IX.40). This thread ends with two menisci that determine the pressure P_m inside the thread (we assume that one can employ the same kind of theory as for the menisci of smectic films). Let P be the pressure inside the thread and P_a the ambient pressure. Let u_r be the radial strain of the columns, σ_r the radial elastic stress and G_r the associated bulk elastic force. Because of the cylindrical geometry, $\partial/\partial\theta = 0$, so that:

$$G_r = \frac{\partial \sigma_r}{\partial r} \tag{C.IX.74}$$

with:

$$\sigma_r = B\left(\frac{\partial u_r}{\partial r} + \frac{u_r}{r}\right) = B\frac{1}{r}\frac{\partial}{\partial r}(ru_r) \tag{C.IX.75}$$

where B is the compressibility modulus of the columnar lattice. In the static case, the permeation equation in the bulk gives

$$G_r = 0 \tag{C.IX.76}$$

so that

$$\frac{\partial P}{\partial r} = \frac{\partial P}{\partial z} = 0 \tag{C.IX.77}$$

according to the equation of momentum conservation. From these two equations, one obtains

$$P = P_m \tag{C.IX.78}$$

in the thread. Thus, the meniscus imposes the pressure within the thread.
We still need the boundary conditions for the strain in the center of the film:

$$u_r(r = 0) = 0 \tag{C.IX.79}$$

and for the normal stress at the surface of the thread:

$$P - \sigma_r = P_a + \frac{\gamma}{R} - \frac{\gamma}{R^2}\left(\xi + R^2\frac{\partial^2\xi}{\partial z^2}\right) \qquad \text{in} \qquad r = R \qquad \text{(C.IX.80)}$$

where γ is the surface tension and $\xi = u_r(r = R)$ the radial strain at the surface. Note that, due to the cylindrical symmetry, the orthoradial stress $\sigma_{r\theta} = 0$.

As a first step, let us check if this problem has a solution $u_r(r)$ independent of z. We shall then study its stability with respect to a deformation in $\sin(qz)$.

Solving eq. C.IX.76 with the boundary condition C.IX.79 yields

$$u_r = cr \qquad \text{(C.IX.81)}$$

where c is a constant. This equation allows the determination of σ_r, using eqs. C.IX.78 and C.IX.80:

$$P_m - 2Bc = P_a + \frac{\gamma}{R} - \frac{\gamma}{R^2}cR \qquad \text{(C.IX.82a)}$$

resulting in

$$c = -\frac{\Delta P + \dfrac{\gamma}{R}}{2B - \dfrac{\gamma}{R}} \qquad \text{(C.IX.82b)}$$

with $\Delta P = P_a - P_m$ the pressure difference imposed by the meniscus. All equations are thus satisfied. A static solution to the problem of the thread exists in the hexagonal phase, unlike the case of an isotropic liquid (in which case, the elastic stress σ_r in eq. C.IX.80 must be replaced by the viscous stress).

Nevertheless, eq. C.IX.82b shows that the constant c diverges for $R = R_c = \gamma/2B$. We shall show that this is precisely the critical radius below which the thread becomes unstable with respect to a deformation of the type

$$\xi = cR + \xi_0 \sin(qz) \qquad \text{(C.IX.83)}$$

In this case, the strain field is of the form

$$u_r = cr + \xi_0\frac{r}{R}\sin(qz) \qquad \text{(C.IX.84)}$$

This field obeys the bulk equilibrium equation $G_r = 0$. By replacing in equation C.IX.80, one obtains

$$2B\xi_0 = \frac{\gamma}{R}(\xi_0 - R^2q^2\xi_0) \qquad \text{(C.IX.85a)}$$

or similarly

$$(1 - R^2 q^2)\xi_o = \frac{R}{R_c}\xi_o \qquad\qquad \text{(C.IX.85b)}$$

If $R > R_c$, this equation has no solution other than the trivial one $\xi_o = 0$. The thread is stable.

If on the other hand $R < R_c$, another solution appears, with an amplitude $\xi_o \neq 0$ and a wave vector

$$q = q_c = \frac{1}{R}\sqrt{1 - \frac{R}{R_c}} \qquad\qquad \text{(C.IX.86)}$$

The thread is then unstable and tends to form droplets. Note that, for this wave vector, the growth rate of the instability vanishes (which is normal, since we performed a marginal stability calculation). In fact, below R_c all wave vectors between 0 and q_c are unstable. Among these wave vectors there is one, q_{max}, where the growth rate is a maximum. This is the one chosen by the system. To determine q_{max}, one must solve the hydrodynamic equations. This calculation can be found in Kamenskii and Kats [56].

In fact, the critical radius calculated by this method is not very realistic. Indeed, typical values $\gamma = 70$ erg/cm^2 [57] and $B = 10^8$ erg/cm^3 [26] yield $R_c = 3.5\times10^{-7}$ cm $= 35$ Å. It should then be possible to draw threads as thin as a few molecular diameters. However, Van Winkle and Clark pointed out that it is impossible to produce threads with a diameter below 400 molecular distances [51]. This defines a new critical radius $R_c^* \approx 200a$ of the order of 0.5 μm, if we take $a \approx 25$ Å (see table 1), much larger than the calculated value for R_c.

To explain this disagreement between the calculated value R_c and the measured value R_c^*, several possibilities can be explored.

The first one is that the elastic modulus B is strongly overestimated. This could well be the case, as we took $B \approx 10^8$ erg/cm^3, a value found for the hexagonal phase of C8HET, a material where the molecules are perpendicular to the columns, unlike in the hexagonal or rectangular phases of C11HAT, where the molecules make with the axis of the columns an angle different from 90°. One could think that, by analogy with smectics, the compressibility modulus of the lattice is weaker in C11HAT than in C8HET. In this case, one would need to assume $B \approx 10^6$ erg/cm^3 in order to find the expected order of magnitude. This value was actually used in the two cited references [51, 56] to justify the calculation of R_c.

A different explanation, that we consider more plausible, would be to consider that the thread breaks, not when the elastic stress diverges (this is what happens at R_c in the previous calculation, where only the linear terms in ξ are taken into account), but rather as it reaches a finite threshold value σ^*. Let R_c^* be the corresponding radius of the thread. If $2B \gg \gamma/R_c^* \gg \Delta P$, formula C.IX.82b gives (in absolute value)

$$\sigma^* = \frac{\gamma}{R_c^*} \qquad\qquad \text{(C.IX.87)}$$

Taking $R_c^* = 0.5$ μm, one obtains $\sigma^* \approx 1.4 \times 10^6$ dyn/cm². It is probable that the columns break when the lattice is compressed by about $\gamma^* \approx 10\text{--}20\%$ (this is the value usually taken in metallurgy for the threshold of spontaneous nucleation, ex nihilo, of defects). From $\sigma^* = B\gamma^*$, one gets $B \approx 1.6 \times 10^6/0.15 \approx 10^7$ erg/cm³, which seems a more reasonable value for B in C11HAT. Note that, with this interpretation, $R_c^* \approx 10 R_c$.

All these calculations neglect the possible presence of defects in the thread, which could render it more fragile and decrease the plasticity limit σ^* even further. This might be the reason why it is so difficult to draw threads of certain materials such as C8HET (personally, we never managed to do it in the temperature range of the hexagonal phase).

IX.9.d) Vibration of threads

We have already seen, in chapter C.VIII, that one could very easily excite the vibrations of a smectic film, either mechanically using a loudspeaker, or electrically by placing an electrode in close proximity to it. This latter method was successfully employed by Gharbia et al. [57] to induce the vibration of C5HET threads. We recall that, for a thread, the resonance frequency of the fundamental mode is given by

$$\nu_1 = \sqrt{\frac{\gamma}{\rho d L^2}} \qquad\qquad (\text{C.IX.88})$$

where γ is the surface tension between the liquid crystal and air, ρ its density, d the diameter of the thread and L its length. Gharbia et al. checked that this formula applies very well to discotic threads by changing their length and diameter. They obtained a surface tension $\gamma \approx 70 \pm 10$ dyn/cm, typically twice as large as for the smectic A-air interface.

Note that, in the case of threads, air inertia is much less important than in the case of films. This result is not surprising, the air flowing much more easily around a thread than around a film. Therefore, it is not necessary to work in a vacuum in order to measure the true resonance frequency of a columnar thread.

Finally, we emphasize that formula C.IX.88 only applies if the hydrostatic pressure term $\Delta P \pi d^2/4$, contributing to the tension of the thread (with $\Delta P = P_{air} - P_m$), is negligible in comparison with the surface term $\pi d \gamma$:

$$\Delta P << \frac{4\gamma}{d} \qquad\qquad (\text{C.IX.89})$$

This is indeed the case in completely annealed threads where formula C.IX.88 is applicable. On the other hand, the pressure term can increase strongly during the drawing process or immediately afterward, if the thread did not have enough time to relax (this relaxation can take several minutes, the discotic phase being very viscous, much more than a smectic A phase). In this case, the thread will vibrate at higher frequency. This effect, also encountered in films (see subsection C.VIII.2b), was mentioned by Gharbia et al. [57], who observe increases in the resonance frequency of up to 30% (such an increase in a thread 1 μm in diameter would correspond to a pressure variation $\Delta P \approx 0.3 \times 8\gamma/d \approx 10^6$ dyn/cm^2, a very significant value, certainly very close to the limit of spontaneous thinning of the thread).

BIBLIOGRAPHY

[1] Ekwall P., "Liquid Crystalline Phases in Systems of Amphiphiles," in *Advances in Liquid Crystals*, Vol. 1, Ed. Glenn H. Brown, Academic Press, New York, 1975.

[2] Mitchell D.J., Tiddy G.J.T., Waring L., Bostock T., McDonald M.P., *J. Chem. Phys. Faraday Trans.*, **79** (1983) 975.

[3] Spegt P.A., Skoulios A.E., *Acta Cryst.*, **16** (1962) 301; ibid., **17** (1964) 198; *ibid.*, **21** (1966) 892.

[4] Chandrasekhar S., Sadashiva D.K., Suresh K.A., *Pramana*, **9** (1977) 471.

[5] Levelut A.M., *J. Chim. Phys.*, **88** (1983) 149.

[6] a) Oswald P., "Etude de quelques défauts observables dans les cristaux liquides hexagonaux," M.Sc. Thesis, University of Paris XI, Orsay, 1981.
b) Oswald P., *J. Physique Lett. (France)*, **42** (1981) L171.

[7] Sallen L., Sotta P., Oswald P., *J. Phys. Chem. B*, **101** (1997) 4875.

[8] Frank F.C., Chandrasekhar S., *J. Physique (France)*, **41** (1980) 1285.

[9] Levelut A.M., Oswald P., Ghanem A., Malthête J., *J. Physique (France)*, **45** (1984) 745.

[10] a) Malthête J., Collet A., *Nouv. J. Chim.*, **9** (1985) 151.
b) Levelut A.M., Malthête J., Collet A, *J. Physique (France)*, **47** (1986) 351.
c) Zimmermann H., Poupko R., Luz Z., Billard J., *Naturforsch. A: Phys. Chem., Kosmaphys.*, **40A** (1985) 149; ibid., **41A** (1986) 1137.

[11] Malthête J., Levelut A.M., Nguyen H.T., *J. Physique Lett. (France)*, **46** (1985) L875.

[12] a) Malthête J., Levelut A.M., Liébert L., *Adv. Mater.*, **4** (1992) 37.
b) Pucci D., Veber M., Malthête J., *Liq. Cryst.*, **21** (1996) 153.

[13] a) Albouy P.A., Guillon D., Heinrich B., Levelut A.M., Malthête J., *J. Phys. II (France)*, **5** (1995) 1617.
b) Levelut A.M., Deudé S., Megtert S., Petermann D., Malthête J., *Mol. Cryst. Liq. Cryst.*, **362** (2001) 5.

[14] a) El-ghayoury A., Douce L., Skoulios A., Ziessel R., *Angew. Chem. Int. Ed.*, **37** (1998) 2205.
b) Douce L., El-ghayoury A., Skoulios A., Ziessel R., *Chem. Commun.*, **20** (1999) 2033.

[15] Hendrikx Y., Charvolin J., *J. Physique (France)*, **42** (1981) 1427.

[16] Kléman M., Oswald P., *J. Physique (France)*, **43** (1982) 655.

[17] a) Saupe A., *J. Coll. Int. Sci.*, **58** (1977) 549).
b) Bouligand Y., *J. Physique (France)*, **41** (1980) 1297.

[18] Kléman M., *J. Physique (France)*, **41** (1980) 737.

[19] Hilbert D., Cohn-Vossen S., *Geometry and the Imagination*, Chelsea, New York, 1952.

[20] Oswald P., Kléman M., *J. Physique (France)*, **42** (1981) 1461.

[21] McGrath K.M., Kékicheff P., Kléman M., *J. Physique (France)*, **3** (1993) 903.

[22] a) Sallen L., "Elasticité et croissance d'une phase hexagonale lyotrope," Ph.D. Thesis, Ecole Normale Supérieure of Lyon, Lyon, 1996.
b) Sallen L., Oswald P., Géminard J.C., Malthête J., *J. Phys. II (France)*, **5** (1995) 937.

[23] Malthête J., Jacques J., Nguyen H.T., Destrade C., *Nature*, **298** (1982) 46.

[24] Livolant F., *J. Mol. Biol.*, **218** (1991) 165.

[25] Abiyaala M., "Etude en microscopie électronique par cryofracture des phases lyotropes du système SDS-formamide," Ph.D. Thesis, Rouen University, Rouen, 1994.

[26] Oswald P., *J. Physique (France)*, **49** (1988) 1083.

[27] Friedel J., *Dislocations*, Pergamon Press, Oxford, 1964.

[28] Bouligand Y., *J. Physique (France)*, **41** (1980) 1307.

[29] Kléman M., *Points, Lines and Walls: In Liquid Crystals, Magnetic Systems and Various Ordered Media*, John Wiley & Sons, Chichester, 1983.

[30] Cagnon M., Gharbia M., Durand G., *Phys. Rev. Lett.*, **53** (1984) 938.

[31] Clerc M., "Etude de transition de phase vers les phases cubiques des systèmes eau-surfactant," Ph.D. Thesis, University of Paris XI, Orsay, 1992.

[32] Davidson P., Clerc M., Ghosh S.S., Maliszewskyj N.C., Heiney P.A., Hynes J., Smith A.B. III, *J. Phys. II (France)*, **5** (1995) 249.

[33] Gilchrist C.A., Rogers J., Steel G., Vaal E.G., Winsor P.A., *J. Coll. Int. Sci.*, **25** (1967) 409.

[34] Rogers J., Winsor P.A., *J. Coll. Int. Sci.*, **30** (1969) 500.

[35] a) Livolant F., Levelut A.-M., Doucet J., Benoit J.P., *Nature*, **339** (1989) 724.
b) Livolant F., Bouligand F., *J. Physique (France)*, **47** (1986) 1813.
c) Livolant F., Leforestier A., *Mol. Cryst. Liq. Cryst.*, **215** (1992) 47.
d) Durand D., Doucet J., Livolant F., *J. Phys. II (France)*, **2** (1992) 1769.

[36] Clunie J.S., Goodman J.F., Symons P.C., *Trans. Faraday Soc.*, **65** (1969) 287.

[37] Constantin D., Oswald P., Impéror-Clerc M., Davidson P., Sotta P., *J. Phys. Chem. B*, **105** (2001) 668.

[38] Buisine J.M., Cayuela R., Destrade C., Nguyen H.T., *Mol. Cryst. Liq. Cryst.*, **144** (1987) 137.

[39] Oswald P., Géminard J.C., Lejcek L., Sallen L., *J. Phys. II (France)*, **6** (1996) 281.

[40] Singer S.J., *Phys. Rev. E*, **48** (1993) 2796.

[41] Impéror-Clerc M., Davidson P., *Eur. Phys. J. B*, **9** (1999) 93.

[42] a) Ramos L., "Phases hexagonales dopées," Ph.D. Thesis, University of Paris VI, Paris, 1997.
b) Ramos L., Fabre P., *Langmuir*, **3** (1997) 682.
c) Ramos L., Fabre P., *Prog. Colloid. Polym. Sci.*, **110** (1998) 240.
d) Ramos L., Molino F., *Europhys. Lett.*, **51** (2000) 320.
e) Ramos L., Fabre P., Nallet F., Lu C.-Y.D., *Eur. Phys. J. E*, **1** (2000) 285.

[43] Gharbia M., Cagnon M., Durand G., *J. Physique Lett. (France)*, **46** (1985) L683.

[44] Prost J., *Liq. Cryst.*, **8** (1990) 123.

[45] For a discussion of edge effects, see Palierne J.F., Durand G., *J. Physique Lett. (France)*, **45** (1984) L355.

[46] Note that we find a result different from that given by P.G. de Gennes and J. Prost in *The Physics of Liquid Crystals* (2nd ed.), Clarendon Press,

Oxford, 1993. The formula we find is identical to that of Kléman and Oswald [16], although the demonstration is different.

[47] The A and D coefficients are presumably of the same order of magnitude. A way of showing this is to rewrite the energy taking as variable the density within the column $\theta_{//}$ rather than θ. Since $\theta = \theta_{//} + U_{xx} + U_{yy}$, one can write the elastic energy under the equivalent form:

$$\rho f_e = \frac{1}{2}A\theta_{//}^2 + \frac{1}{2}(\bar{B} + A - 2D)(U_{xx} + U_{yy})^2 + \frac{1}{2}C(U_{xy}^2 - U_{xx}U_{yy})$$

$$+ (A - D)\,\theta_{//}\,(U_{xx} + U_{yy})$$

The deformations $\theta_{//}$ and $U_{xx} + U_{yy}$ being independent, one can reasonably assume that $A - D \ll A$ or, equivalently, that $A \approx D$. Similarly, one can assume that it is easier to compress the lattice than an individual column (as confirmed by the experiments). In this case, one has $\bar{B} + A - 2D \ll A$, resulting in $\bar{B} \approx A$.

[48] Ben Abraham S.I., Oswald P., *Mol. Cryst. Liq. Cryst.*, **94** (1983) 383.

[49] Prost J., Clark N.A., in *Liquid Crystals*, Proceedings of the International Conference, Bangalore, India, 1979, Ed. S. Chandrasekhar, Heiden, Philadelphia, 1980, p. 53.

[50] Gharbia M., Othman T., Gharbi A., Destrade C., Durand G., *Phys. Rev. Lett.*, **68** (1992) 2031.

[51] Van Winkle D.H., Clark N.A., *Phys. Rev. Lett.*, **48** (1982) 1407.

[52] Safinya C.R., Liang K.S., Varady W.A., Clark N.A., Andersson G., *Phys. Rev. Lett.*, **53** (1984) 1172.

[53] Fontes E., Heiney P.A., Ohba M., Haseltine J.N., Smith III A.B., *Phys. Rev. A*, **37** (1988) 1329.

[54] Davidson P., Clerc M., Ghosh S.S., Maliszewskyj N.C., Heiney P.A., Hynes J. Jr., Smith III A.B., *J. Phys. II (France)*, **5** (1995) 249.

[55] a) Fontes E., Heiney P.A., de Jeu W.H., *Phys. Rev. Lett.*, **61** (1988) 1202.
b) Fontes E., *J. Am. Chem. Soc.*, **113** (1991) 7666.
c) Idziak S., Fontes E., *Mol. Cryst. Liq. Cryst.*, **237** (1993) 271.

[56] Kamenskii V.G., Kats E., *Soviet Phys. JETP Lett.*, **37** (1983) 261.

[57] Gharbia M., Gharbi A., Cagnon M., Durand G., *J. Physique (France)*, **51** (1990) 1355.

Chapter C.X

Growth of a columnar hexagonal phase

In the previous chapters, we described the behavior of the nematic-isotropic liquid and smectic B-smectic A interfaces in directional growth. Now we shall consider the instabilities that develop during the **free growth** of the interface between a columnar hexagonal phase and its isotropic liquid. In this type of experiment, the growth temperature is fixed, while in directional growth it is the velocity of the interface that is imposed, the sample being placed in a temperature gradient. These two techniques are therefore complementary.

The material used is the columnar liquid crystal C8HET (hexaoctyloxy-triphenylene). We chose this substance because it is well known (see the previous chapter) and because, using it, one can observe the growth of a germ in the diffusive or in the kinetic regime. On the other hand, it allows a quantitative study of several morphological transitions and of the associated mechanisms. We shall see, in particular, that it is possible to reach the absolute restabilization threshold of the front at high velocity, a phenomenon extremely difficult to observe and study in metals or in the usual plastic crystals.

The structure of the chapter is as follows. First (section X.1), we give the phase diagram of C8HET, as well as the values of the main physical constants relevant for the growth process (surface tension and its anisotropy, diffusion coefficients, etc.). We then describe the growth of an isolated germ in the diffusive regime, at a supersaturation lower than 1 (section X.2). We shall see how a germ – circular in the beginning – becomes destabilized and then develops dendrites that can be two- or three-dimensional, depending on the imposed supersaturation. In this part we shall also deal with such topics as side branching and the transition towards the dense branching regime. Finally, we describe the growth of a germ in the kinetic regime, at a supersaturation larger than 1 (section X.3). We shall see that the front becomes stable again at high velocity, a phenomenon known as absolute restabilization. We shall show how one can measure the kinetic coefficient of molecular attachment at the interface, as well as its anisotropy using a kinetic Wulff construction.

Let us emphasize that many of the formulae used in this chapter have been proven in chapter B.IX of the first volume (overview of the theory of growth phenomena).

X.1 Phase diagram and physical constants of the material

In figure C.X.1 we present the molecular structure of C8HET, as well as the structure of the columnar hexagonal phase. This material exhibits a hexagonal phase between 68°C and 86°C and is isotropic above 86°C (see Fig. A.III.14 in the first volume).

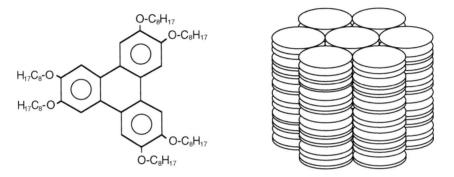

Fig. C.X.1 C8HET molecule and structure of the columnar hexagonal phase.

In this section, we detail the phase diagram close to the melting point and give the values of a few physical constants that will be useful in interpreting the growth experiments.

X.1.a) The phase diagram

This is the first thing to determine precisely when studying crystalline growth. It is not always an easy task, because organic materials decompose easily under the influence of light or heat. The liquid crystal C8HET is no exception, although it can be considered as very stable chemically (it degrades much more slowly than 4O.8 or MBBA, for instance). Purifying it by chromatography results in an impurity concentration [1] close to 10^{-3}. Since the material is only available in very small quantities (less than one gram of product, synthesized by J. Malthête, was used to perform the experiments described in this chapter), it is difficult to mix it with a known impurity, especially as it slowly degrades with time. We therefore chose to let the samples "age" naturally, and to measure at regular intervals of time the temperatures T_l and T_s of the liquidus and solidus. If the impurity concentrations C are very low, these two curves are straight lines, their slopes being given by the law of dilute mixtures (van't Hoff law). Strictly speaking, one cannot verify this law experimentally, the absolute impurity concentration not being directly accessible. Assuming that the liquidus is a straight line, one can however check experimentally that the same holds true for the solidus. This procedure leads to the phase diagram in figure C.X.2. In this diagram, the concentration scale is arbitrary, but we can estimate it using the van't Hoff law:

$$T_l = T_c - \frac{k_B T_c^2}{L v_s}(1 - K)\, C_\infty \qquad\qquad (C.X.1)$$

This formula yields the liquidus temperature T_l as a function of the melting temperature of the pure substance T_c, the partition coefficient of the impurities K, the transition latent heat per unit volume L, the average impurity concentration C_∞ and the volume v_s of a molecule in the solid. Using the following experimental values:

$$L \approx 4{,}37 \text{ J/cm}^3 \text{ [1]}, \quad T_c \approx 86.1°C \text{ [2]}, \quad K \approx 0.33 \text{ [2, 3]}, \quad v_s \approx 1{,}65 \times 10^{-21} \text{ cm}^3$$

one has $C_\infty \approx 3\%$ for $\Delta T = T_l - T_s = 5°C$, which is small enough to justify the dilute mixture assumption.

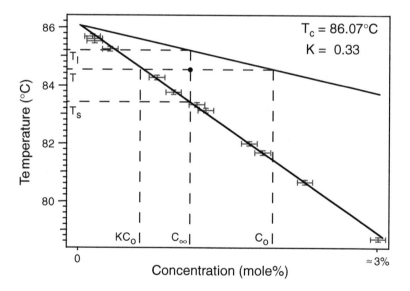

Fig. C.X.2 Experimental phase diagram (from ref. [2]).

In the following, we shall denote by m (< 0) the slope of the liquidus, by m/K the slope of the solidus and by ΔT the temperature difference $T_l - T_s$.

In the diffusive regime ($T_s < T < T_l$), we take as control parameter the chemical supersaturation Δ_c expressed by:

$$\Delta_c = \frac{C_0 - C_\infty}{C_0\,(1 - K)} \qquad\qquad (C.X.2a)$$

This quantity, lying between 0 and 1, gives the bulk fraction of solid phase at thermodynamical equilibrium.

In the kinetic regime ($T < T_s$), it is more convenient to use the thermal supersaturation, given by:

$$\Delta_t = \frac{T_I - T}{T_I - T_s} \tag{C.X.2b}$$

In this case, $\Delta_t > 1$.

In the following, we shall denote the supersaturation by Δ, with the understanding that it stands for Δ_c in the diffusive regime and for Δ_t in the kinetic regime.

X.1.b) Surface tension and anisotropy

These two parameters were measured by using the grain boundary method and by analyzing the equilibrium shape of a germ of the hexagonal phase in the isotropic phase. The experimental setups employed are the same as in chapter C.VII.

Let us first consider the equilibrium shape of a germ, when the molecular columns are perpendicular to the glass plates. This homeotropic orientation (the optical axis is perpendicular to the plates) appears spontaneously for a liquid crystal layer a few microns thick, the disk-like molecules preferring to align parallel to the glass surface. In figure C.XI.2, we show a germ after equilibrating for a few hours. It is almost circular in shape, implying that the anisotropy of the surface free energy in the plane of the hexagonal lattice is extremely small. Closer inspection shows that its shape is actually slightly hexagonal and that its radius can be expanded in polar coordinates as [4]

$$R(\theta) = R_0 + R_6 \cos(6\theta) + R_{12} \cos(12\theta) + \dots \tag{C.X.3}$$

Let ε_6^* be the shape anisotropy of the germ:

$$\varepsilon_6^* = \frac{R_{max} - R_{min}}{R_{max} + R_{min}} \tag{C.X.4}$$

and ε_6 the anisotropy of the surface free energy:

$$\gamma(\theta) = \gamma_0 \left[1 + \varepsilon_6 \cos(6\theta) \right] \tag{C.X.5}$$

One can show that, in the cylindrical geometry (when the wetting angle between the interface and the glass surfaces equals $\pi/2$), the Gibbs-Thomson law yields rigorously

$$\varepsilon_6^* = \varepsilon_6 \tag{C.X.6}$$

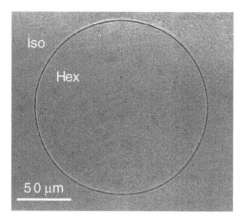

Fig. C.X.3 Germ of the columnar hexagonal phase in equilibrium with its isotropic liquid (from ref.[3]).

However, this result is no longer valid if the wetting angle is different from $\pi/2$ and a meniscus appears in the thickness of the sample. This is almost always the case, especially for C8HET, where the wetting angle is 30°, typically [2]. It is nevertheless possible to calculate numerically (and even analytically, in certain asymptotic limits) the three-dimensional shape of the germ as a function of the contact angle and of the anisotropy ε_2 of the surface tension in the thickness of the sample [4]. These calculations show that the error due to taking $\varepsilon_6 = \varepsilon_6^*$ is always below 10%. In conclusion, measuring the shape anisotropy of a germ provides a reasonable estimate for the anisotropy of the surface energy in the plane of the hexagonal lattice. For C8HET, the experiment gives [2, 3]:

$$\varepsilon_6 \approx 1.5 \pm 0.5 \times 10^{-3} \tag{C.X.7}$$

Fig. C.X.4 Two grain boundaries crossing the hexagonal-isotropic liquid interface in C8HET (from ref. [3].

The grain boundary method is well adapted for determining the surface tension. It consists of measuring the depth h of the grooves that appear each time a grain boundary crosses the hexagonal-isotropic interface (Fig. C.X.4). The depth h of the groove depends on the thermal gradient G and on the energy of the grain boundary, setting the angle α between the interface and the temperature gradient G at the crossing point. This depth has a simple analytical expression for a quasi-isotropic surface tension:

$$h = \sqrt{\frac{2\gamma_0 T_c}{GL}}(1 - \sin \alpha)$$
(C.X.8)

and we shall make this assumption here. The $h \propto 1/\sqrt{G}$ law is well verified experimentally, yielding the surface energy (with cylinders parallel to the interface) [2]:

$$\gamma_0 \approx 0.52\,erg/cm^2$$
(C.X.9)

The corresponding capillary length is:

$$d_0 = \frac{\gamma_0}{L} \approx 1.2\,\text{Å}$$
(C.X.10)

X.1.c) Diffusion coefficients

The diffusion coefficients of the impurity in the two phases (denoted by D_L in the isotropic phase and by D_S in the hexagonal phase) are also essential parameters in the growth process.

One can find a first relation between these coefficients by measuring the destabilization velocity V_c of the hexagonal-isotropic front in directional growth. Indeed, the linear theory shows that the front becomes spontaneously unstable above a critical velocity given by:

$$V_c = \frac{D_L G}{\Delta T}(1 + K\beta)$$
(C.X.11a)

where $\beta = D_S/D_L$. Measuring V_c thus yields $D_L + KD_S$.

An additional relation can be obtained by determining the melting velocity V^* (negative by convention) below which germs of isotropic liquid spontaneously nucleate ahead of the interface. If there is no nucleation delay, this velocity is given by the "constitutional superheating" criterion, expressed by:

$$V^* = -\frac{D_S G}{\Delta T}$$
(C.X.11b)

These two independent measurements give access to the two diffusion coefficients [2, 5]:

$$D_L \approx 1.4 \times 10^{-7}\,cm^2/s$$
(C.X.12a)

$$D_S \approx 0.85 \times 10^{-7}\,cm^2/s$$
(C.X.12b)

Note that these values are independent (within experimental precision) of the sample thickness (in contrast with the case of the nematic-isotropic interface, see chapter B.VI). In the following, we shall use the ratio β of the diffusion coefficients: $\beta = D_S/D_L \approx 0.6$.

These values are close to those measured by Daoud et al. for various dyes [6].

Finally, let us point out that similar directional growth experiments were performed using a lyotropic hexagonal phase of a non-ionic surfactant [7].

Another important parameter is the kinetic coefficient μ characterizing the molecular attachment at the interface. In section C.X.3, we shall see how it can be measured. Before that, let us describe the destabilization and nonlinear evolution of a circular germ in the diffusive regime, corresponding to low values of the growth velocity. In these conditions, the kinetic effects are negligible, meaning that the interface is locally very close to thermodynamical equilibrium.

X.2 Growth in the diffusive regime ($0 < \Delta < 1$)

The diagram in figure C.X.5 shows the evolution of an initially circular germ as a function of the imposed supersaturation. The sample is a thin layer (a few micrometers thick) of liquid crystal confined between two parallel glass plates. The growth is observed in the plane of the hexagonal lattice, as the columns orient spontaneously perpendicular to the glass plates. In the beginning, the germ nucleates on a dust particle (sometimes as small as to be barely visible under the microscope) then grows while maintaining a circular shape. Above a certain critical radius (depending on the supersaturation), it becomes unstable and takes a hexagonal shape. This deformation then becomes amplified and can lead to three different morphologies depending on the value of the supersaturation:

1. The first one corresponds to the **petal regime**. It can be observed at the beginning of the growth or at low supersaturation values, when the confinement effects related to the finite size of the sample block the growth (photograph taken at $\Delta \approx 0.1$ in figure C.X.5).

2. The second regime, observed in the range of intermediate supersaturation, is that of **dendrites** (photograph at $\Delta \approx 0.3$). In this case, each petal becomes a dendrite, growing at a constant rate.

3. At higher supersaturations, the side-branches invade the space between the dendrites and the primary dendrites become unstable. This is the **dense branching regime** (photograph at $\Delta \approx 0.8$).

To start with, let us analyze the destabilization of a germ and its weakly nonlinear evolution in the petal regime [8].

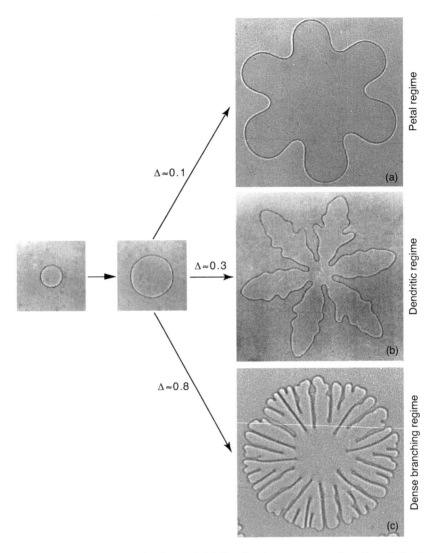

Fig. C.X.5 Possible evolution paths for an initially circular germ as a function of the imposed supersaturation. At a weak supersaturation ($\Delta \approx 0.1$), the germ remains in the petal regime (a) and its growth stops due to confinement effects related to the finite size of the sample; at an intermediate supersaturation ($\Delta \approx 0.3$), each petal turns into a dendrite (b); at high supersaturation ($\Delta \approx 0.8$), the "dense branching" regime (c) starts to develop.

X.2.a) Destabilization of a circular germ and weakly nonlinear evolution in the petal regime ($\Delta < 0.15$)

A quantitative study is only possible at a very low supersaturation Δ, otherwise the destabilization radius is too small and it becomes impossible to measure it.

One must also use very thin samples (a few μm) in order to avoid the destabilization of the germ in the thickness of the sample. If these two conditions are fulfilled, the three-dimensional effects are negligible [2] and the growth is essentially driven by the diffusion of impurities in the horizontal plane [8]. It is then possible to analyze the time evolution of the germ by expanding its radius $R(\theta,t)$ in a Fourier series. In keeping with its hexagonal symmetry, this amounts to:

$$R(\theta,t) = R_o(t) + \sum_{k=1}^{\infty} \delta_{6k} \cos(6k\theta) \qquad \text{(C.X.13)}$$

with

$$R_o(t) = \frac{1}{2\pi} \int_0^{\infty} R(\theta,t)\, d\theta \qquad \text{(C.X.14a)}$$

and

$$\delta_{6k}(t) = \frac{1}{\pi} \int_0^{\infty} R(\theta,t) \cos(6k\theta)\, d\theta \qquad \text{(C.X.14b)}$$

Note that such an expansion is only possible in the weakly nonlinear "petal" regime, since the $R(\theta,t)$ function must be univalued. Figure C.X.6 shows the variation of these functions with time for $\Delta \approx 0.1$, as well as the total surface area of the germ $A(t)$.

We see that R_o starts by increasing as $t^{1/2}$ (solid curve in figure C.X.6a), and then progressively drifts away from this law in the nonlinear regime. Nevertheless, the total surface area of the germ A remains proportional to t throughout the experiment (Fig. C.X.6b). This remarkable property is however not a general feature, holding only at very low supersaturation, as the Péclet number of the germ, based on its average radius $Pe = R_o \dot{R}_o / 2D_L$ is much smaller than 1. In this limit, $\Delta C = 0$ and $dA/dt = \text{Const}$. This property was confirmed numerically by Brush and Sekerka [9]. From the slope of this line, one can obtain the diffusion coefficient of the impurity using the theoretical law established for a circular germ [10]:

$$A = 4\pi \lambda^2 D_L t \qquad \text{(C.X.15)}$$

where the parameter λ depends only on the supersaturation:

$$\lambda^2 \ln(v^2 \lambda^2) + \Delta = 0 \qquad \text{(C.X.16)}$$

In this formula (only valid at a low supersaturation Δ) $v \approx 1.335$ is Euler's constant. Since in this experiment $\Delta \approx 0.1 \pm 0.01$, we obtain $\lambda^2 \approx 0.0366 \pm 0.07$, yielding $D_L \approx 1.1 \pm 0.3 \times 10^{-7}$ cm^2/s. This value is in good agreement with the one obtained from the critical velocity measurements in directional growth.

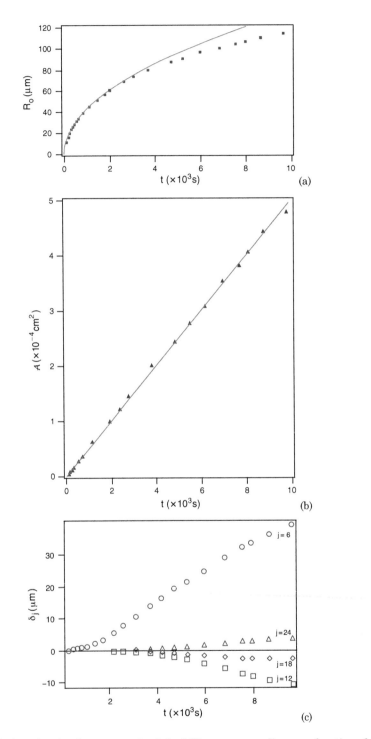

Fig. C.X.6 Evolution of a circular germ at $\Delta = 0.1$. a) The average radius as a function of time; b) the total area as a function of time; c) the harmonics as a function of time (from ref. [8]).

Finally, experiments show that the germ becomes unstable above a destabilization radius $R_6 \approx 15$ μm. Note that mode 6 is the first one to appear, certainly because it is already present due to the surface tension anisotropy. The following mode, 12, starts to develop for an average germ radius close to $R_{12} \approx 60$μm.

To compare these measurements with the theoretical predictions, we recall the results of the linear stability analysis for a circular germ. This calculation shows that a mode j develops above a radius R_j given by:

$$R_j = R_c \left[1 + j(j + 1)(1 + K\beta) \ln(1/\nu\lambda) \right]$$ (C.X.17)

where R_c is the critical nucleation radius:

$$R_c = \frac{d_0^c}{\Delta}$$ (C.X.18)

and d_0^c the chemical capillary length:

$$d_0^c = d_0 \frac{T_c}{mC_0(K - 1)} = d_0 \frac{T_c}{T_l - T}$$ (C.X.19)

In the experiment we described, the value of d_0^c is about 190 Å. As, on the other hand, $\Delta \approx 0.1$, $K \approx 0.33$ and $\beta \approx 0.6$, the theory predicts:

$R_6(\text{th}) \approx 13.5$ μm

$R_{12}(\text{th}) \approx 50$ μm

Once again, these values are in good agreement with those found in the experiments.

It is noteworthy that, at very low supersaturation values, the growth of a germ is eventually arrested in the "petal" regime, owing to the finite size of the sample or to the presence of other germs in its vicinity. In these conditions, the liquid becomes richer in impurities and the local supersaturation decreases. In practice, this confinement effect becomes important as the diffusion length $l_d = D_L/V$ is of the order of a few mm. At $\Delta \approx 0.1$, this problem appears as the radius of the germ reaches about 100 μm, since $l_d \approx R_0/2\lambda^2 \approx 14 R_0$. One must therefore increase the supersaturation Δ or the sample size (on the condition of being able to select a single germ) to reduce the influence of confinement effects. In these conditions, the "petals" turn into dendrites.

X.2.b) The dendritic regime

At a moderate supersaturation ($0.15 < \Delta < 0.5$, typically), experiment shows that the maximal radius R_{Max} of the germ starts by increasing as $t^{1/2}$, then as t when the germ becomes large enough (Fig. C.X.7). This regime change indicates the moment when the petals change into dendrites (Fig. C.X.8), a bifurcation that is not observed at very small supersaturation due to the confinement effects blocking the growth.

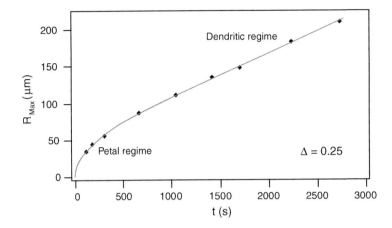

Fig. C.X.7 Radius of the germ envelope R_{Max} as a function of time. The two growth regimes are clearly visible in this graph (from ref. [12]).

We emphasize the difference between the two solutions: the petal slows down with time, while the dendrite grows at a constant velocity. This transition is also due to confinement effects [11, 12], but this time they are no longer artifacts due to the presence of other germs or to the finite size of the sample, but rather intrinsic effects related to the hexagonal morphology of the germ.

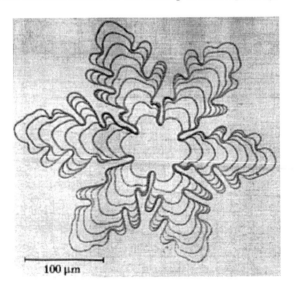

Fig. C.X.8 Growth of a germ in the dendritic regime at moderate supersaturation ($\Delta \approx 0.25$). The time interval between two images is constant, showing that each dendrite grows at a constant velocity (image sequence by J.C. Géminard, 1997).

Indeed, each branch of the germ is confined between its two neighbors, all three of them rejecting impurities in the liquid. Everything occurs as if each branch of the germ grew within a diverging channel of angular aperture $2\pi/6$. It turns out that theorists [13] showed that, in a channel of constant width L, no stationary solution exists below a certain critical supersaturation $\Delta_c(L)$; nevertheless, a stationary stable solution corresponding to the dendrite observed in an infinite medium does exist for $\Delta > \Delta_c(L)$. Conversely, one can calculate for each supersaturation Δ a critical channel width $L_c(\Delta)$ above which the dendrite exists and represents the unique stable solution (there is also another solution, of the type of the Saffman-Taylor finger, but it is unstable). On the other hand, there is no stationary solution for $L < L_c$. Let us assume that these results apply to the case of the divergent channel. One can see qualitatively that, at a particular value R_{Max}^c of the germ, the width of the effective channel $2\pi R_{Max}^c/6$ is equal to $L_c(\Delta)$. For this particular radius, the petals change into free dendrites.

Before describing the behavior of dendrites in experiments, let us briefly recall their properties. The theory tells us that a dendrite is characterized by its curvature radius at the tip ρ and by its growth velocity V. These two fundamental quantities are intimately related to the supersaturation Δ by the Ivantsov relation [14], and to the anisotropy ε_6 of the surface free energy through the marginal stability constant σ^* to be defined later.

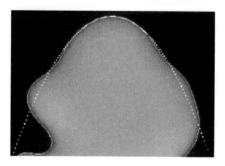

Fig. C.X.9 Parabolic fit of the tip of a dendrite. On can see that the side-branches are very close to the tip (from refs. [2, 11]).

Let us start by considering the first relation. In 1947, Ivantsov showed for the first time the existence of stationary solutions to the problem of growth in the diffusive regime [14]. These solutions, termed dendrites, have the shape of a parabola in two dimensions, and of a paraboloid (not necessarily a revolution one) in three dimensions. The Ivantsov relation, demonstrated rigorously for vanishing surface tension, reads

$$\Delta = \sqrt{\pi P}\exp(P)\,\mathrm{erfc}(\sqrt{P}) \qquad \text{in} \qquad 2D \qquad\qquad (C.X.20a)$$

and

$$\Delta = P\exp(P)\,E_1(P) \qquad \text{in} \qquad 3D \qquad\qquad (C.X.20b)$$

for a revolution paraboloid about its axis.

In these formulae

$$\text{erfc}(x) = \frac{2}{\sqrt{\pi}} \int_x^\infty \exp(-t^2)dt$$

is the complementary of the error function, and

$$E_1(x) = \int_x^\infty \frac{\exp(-t)}{t} dt$$

is the exponential integral function.

Eqs. C.X.20a,b give the Péclet number P of the dendrite as a function of Δ. This dimensionless number is defined by

$$P = \frac{\rho V}{2D_L} \qquad\qquad (C.X.21)$$

These relations can be shown to hold when the surface tension is taken into account, the capillary corrections being very small.

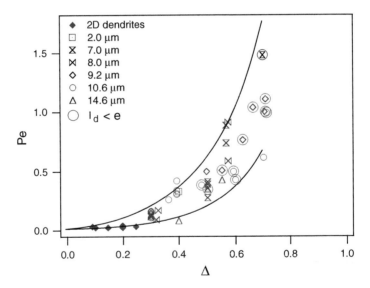

Fig. C.X.10 Péclet number measured at the tip of the dendrites as a function of the chemical supersaturation. These experiments clearly show that the dendrites follow the two-dimensional Ivantsov law (lower curve) for $\Delta < 0.3$. At higher supersaturation, the experimental points clearly move away from this law, without ever reaching (within experimental errors) the upper curve, computed for free axisymmetric dendrites in three dimensions (from refs. [2, 11]).

To check experimentally the Ivantsov relation, one must measure the curvature radius ρ at the tip of the dendrites and their growth velocity V. The first measurement is not easily performed, as the dendrites are always unstable at the sides and develop side branches, so that performing the fit with a parabola becomes difficult. An example of a fit is shown in figure C.X.9. In the following figure C.X.10, we show the variation of the Péclet number as a function of the supersaturation Δ. In the same graph, we traced the two theoretical laws C.X.20a and b corresponding to 2D (lower curve) and 3D (upper curve) dendrites. These measurements, performed on samples of different thickness, clearly show that the 2D Ivantsov law applies very well as long as $\Delta < 0.3$. On the other hand, the experimental points deviate considerably from the 2D law at higher supersaturation, without however reaching the law expected for three-dimensional dendrites. These measurements thus show that the dendrites are two-dimensional (i.e., they touch the two glass plates) at low supersaturation, a result fully confirmed by a direct measurement of their local thickness using the microscope (for more details on the optical method employed, see Géminard [2] and Géminard and Oswald [11]). At higher supersaturation, the dendrites become three-dimensional, a liquid film appearing between the glass plates and the dendrite. These dendrites are generally flattened, resembling the one depicted in figure C.X.11.

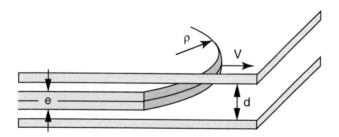

Fig. C.X.11 Schematic representation of a confined 3D dendrite. A thin liquid film appears between the dendrite and the two glass plates.

In this example, the curvature radius – measured at the tip in the horizontal plane – is larger than the sample thickness. Only at high supersaturation ($\Delta > 0.5$, typically) does ρ become comparable to or even smaller than the thickness d of the sample. Even in this case, however, the 3D Ivantsov law is not satisfied, due to the effects of confinement between the two plates.

The second theoretical prediction relative to the dendrites concerns their selection by the surface tension anisotropy. Indeed - the Ivantsov relation defining only the ρV product - one must choose among this continuum of solutions that are actually observed experimentally. The theory tells us (and experiment confirms this prediction) that the dendrites grow along favored crystallographic axes, those axes precisely for which the surface energy has a maximum. In the hexagonal columnar phase, there are six directions corresponding to 6 twofold axes of the hexagonal lattice, along which the surface energy is maximal. Therefore, it is not surprising that six dendrites develop. The second theoretical prediction concerns the value of the marginal stability constant, given by [15]

$$\sigma^* = \frac{2D_L d_o^c}{\rho^2 V} \tag{C.X.22}$$

This constant yields $\rho^2 V$, and then ρ and V, from the Ivantsov relation. The theory shows that σ^* is a universal function of the surface energy anisotropy ε_6, and that it varies as $\varepsilon_6^{7/6}$ for $\varepsilon_6 \to 0$ [16–21]:

$$\sigma^* = f(\varepsilon_6) \tag{C.X.23}$$

This function can be computed numerically (strictly speaking, it depends on the Péclet number of the dendrite and on the ratio β of the diffusion constants) and its value is [22]

$$\sigma^*_{2D}(\text{th}) \approx 0.025 \tag{C.X.24}$$

at low Péclet number, with $\beta = 0.6$ and $\varepsilon_6 = 1.5 \times 10^{-3}$.

Figure C.X.12 shows the experimental variation of the marginal stability constant σ^* as a function of the ratio between the diffusion length and the sample thickness l_d/d. As for the Péclet number, the experimental points separate into two distinct groups.

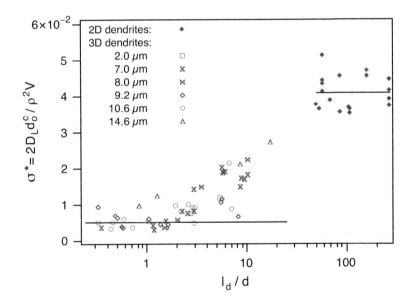

Fig. C.X.12 Stability constant measured as a function of the ratio between the diffusion length and the sample thickness (from refs. [2, 11]).

Those bunched in the upper right corner correspond to two-dimensional dendrites. For these dendrites, it is very important to point out that the diffusion length is always much larger than the sample thickness, as

$30 < l_d/d < 250$. On the other hand, σ^* is constant, with an average value [5,11]:

$$\sigma^*_{2D}(\text{exp}) \approx 0.041 \qquad\qquad\qquad (\text{C.X.25})$$

This value is larger than the theoretical one (0.025). This disagreement can stem from the experimental errors (ε_6 is difficult to measure precisely), but could also reflect the reality. Indeed, the measurements performed in other materials fairly consistently yield larger values than those calculated using the selection criterion of surface tension anisotropy. The theorists propose that this disagreement might be explained by the effect of noise, any perturbation leading to a systematic increase in σ^* [23].

The second group of points is more dispersed and corresponds to 3D dendrites. In this case, σ^* depends on the sample thickness and decreases as the l_d/d ratio decreases. For $l_d/d < 1$, σ^* saturates, tending toward a well-defined value:

$$\sigma^*_{3D}(\text{exp}) \approx 0.005 \qquad\qquad\qquad (\text{C.X.26})$$

Since the confinement effects are still important for these dendrites (as confirmed by measurements of the Péclet number), it would be useless to compare this value with any theoretical prediction (at this time, there is no theory for confined 3D dendrites).

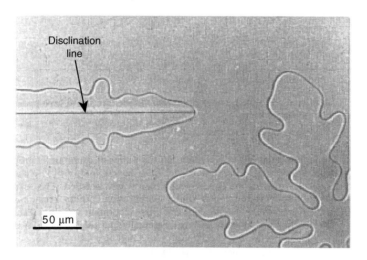

Fig. C.X.13 Abnormal dendrite growing along a disclination line (see figure C.IX.13). It is more pointed and grows faster than the two other ordinary dendrites. The supersaturation is close to 0.3 (from ref. [12]).

To conclude this section, let us mention the existence of abnormal dendrites (Fig. C.X.13) that propagate faster than usual dendrites, at the same supersaturation. Along the axis of these dendrites there is a disclination line, perfectly visible under the microscope. This line pierces the interface close to the tip and creates a local perturbation that modifies the selection mechanism of the dendrite. A highly simplified interpretation consists in assuming that this perturbation locally increases the surface tension anisotropy. The result is an increase in σ^*, and hence in $\rho^2 V$. Since ρV remains constant, according to the Ivantsov law, the radius of curvature ρ at the tip must decrease and its velocity V must increase. This effect is indeed observed experimentally.

X.2.c) Side branching and the dense branching regime

Experiment shows that the side branches of the dendrites are more and more invasive with increasing supersaturation (Fig. C.X.14).

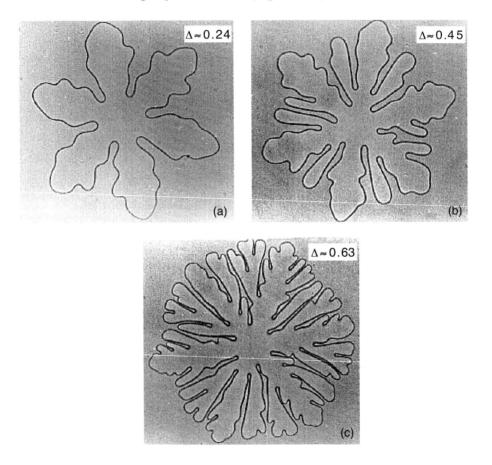

Fig. C.X.14 Evolution of the side branching of the dendrites (a–c) as a function of the imposed supersaturation (thin sample, a few μm thick). The last image (c) marks the beginning of the dense branching regime. The primary dendrites are still visible, pointing to the six vertices of the hexagonal envelope (from ref. [12]).

These branches appear close to the tip of the dendrite and are generally uncorrelated. This observation is consistent with a model of selective amplification of the experimental noise (the origin of the noise can be chemical, thermal, etc.) [19]. Above a certain supersaturation close to $\Delta = 0.7$, the branches fill the intercellular space almost completely, marking the beginning of the dense branching regime (Fig. C.X.5c). In this regime, the primary dendrites become unstable and often split close to the tip. The envelope of the germ is almost circular, with a slight hexagonal deformation along the direction of the primary dendrites. More important, the propagation velocity of the envelope is constant, meaning that no long-range impurity segregation occurs. Consequently, the impurities rejected by the solid concentrate in the inter-dendritic liquid, as well as in the two liquid layers appearing between the germ and the glass plates. This result was confirmed by concomitantly measuring the relative thickness of the germ (that does not touch the plates) and the apparent liquid fraction within the envelope. These two measurements yielded the actual bulk fraction of solid inside the envelope and it could be checked that it was indeed equal to the imposed supersaturation, in agreement with the law of total impurity conservation.

Let us now see what happens for a supersaturation larger than 1 (i.e., when the sample is quenched below the solidus temperature).

X.3 Growth in the kinetic regime ($\Delta > 1$)

At a supersaturation larger than 1 ($T < T_s$), the material crystallizes completely. This means that there is no long-range impurity segregation and that the average impurity concentration in the solid is equal to C_∞. The germ must therefore touch the two glass plates, a conclusion that is well verified by thickness measurements in optical microscopy. If the thermal effects due to latent heat rejection are negligible (and this is indeed the case [24]), the front must also propagate at a constant velocity, determined by the kinetics of molecular attachment at the interface. In this case, one speaks of the kinetic regime.

X.3.a) The quenching method

Before describing the experimental results, let us see how one can quench an initially molten sample below the solidus temperature rapidly, but in a controlled manner. To achieve this result, the sample (a few μm thick) is placed between two glass plates of thickness 1 mm. One plate is covered by an ITO (indium tin oxide) layer (in contact with the liquid crystal), through which an electrical current can pass. This setup is placed in an oven of large thermal inertia, regulated to a temperature T below the solidus temperature T_s. At this temperature (chosen for growth), the material crystallizes. To melt it, one sends through the conducting layer a very intense (50 V, typically) and short (0.1 s) electrical pulse. Immediately after the pulse, the heat diffuses across the glass plates and the temperature of the sample returns quickly to the setpoint temperature (the cooling rate is typically 10°C/s). If the imposed supersaturation is not too high, the setpoint temperature is reached before any germ can nucleate. At high supersaturation, on the other hand, a germ sometimes nucleates before the temperature is stabilized. In this case, one must determine the actual temperature of the sample during the growth of the germ, for instance by measuring the resistance of the conducting layer, which provides an excellent temperature probe. Using this quenching method, one can reach thermal supersaturations close to 7. It should also be noted that the measured velocities can be quite large (up to 500 µm/s). It is therefore imperative to use a video camera with a fast electronic shutter (capable of achieving one thousandth of a second) to obtain sharp images of the germ.

X.3.b) Measurements of the kinetic coefficient and of the absolute restabilization threshold

Figure C.X.15 shows the growth of two germs at two different supersaturations ($\Delta = 3$ and $\Delta = 5$). In both cases, the front is globally circular and propagates at a constant velocity (note that here the velocities are much larger than in the diffusive regime).

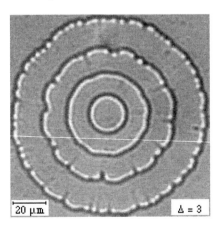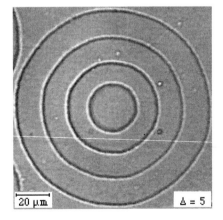

Fig. C.X.15 Growth of a germ in the kinetic regime for two different supersaturations. In the image sequence on the left ($\Delta \approx 3$), the time interval between two images is 0.08 s. In that on the right ($\Delta \approx 5$), the time interval is 0.04 s (from ref. [24]).

In figure C.X.16, we traced the front velocity as a function of temperature. It follows a linear law:

$$V = \mu \, (T_s - T) \tag{C.X.27}$$

with a slope $\mu \approx 130 \ \mu m/s/°C$. Experiment shows that μ is independent of the sample thickness, showing that the thermal effects related to latent heat rejection are negligible (there are some additional arguments for eliminating them [24]). The measured coefficient μ is thus indeed the kinetic coefficient of molecular attachment to the interface. Note that μ is very small (slow kinetics) with respect to the values expected in usual materials (in a metal, μ would be of the order of 1 m/s/°C, rendering its experimental determination extremely difficult). This effect is directly related to the large molecular size and to the high viscosity of the isotropic liquid [24]. Also note that the kinetic coefficient does not depend on ΔT (Fig. C.X.15), and hence on the impurity concentration in the dilute regime.

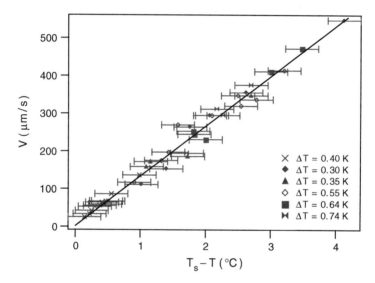

Fig. C.X.16 Growth velocity of a germ as a function of temperature in the kinetic regime (from ref. [24]).

Although the impurities do not fix the growth velocity, they are nevertheless responsible for the cell instability observed at moderate supersaturation (typically, for $\Delta < 4$). Indeed, a concentration jump occurs at the interface, imposed by the partition coefficient K of the impurities between the two phases. If the front velocity is not too large, meaning in practice that

$$V < \frac{D_L}{a} \approx 3 \ mm/s \tag{C.X.28}$$

where "a" is the distance between molecules, the impurities have enough time to diffuse across the interface, and the chemical equilibrium can be established. In these conditions (well satisfied in the experiments), K shifts very little with respect to its equilibrium value, and an impurity mound appears ahead of the front in the liquid. In this region, the concentration decreases exponentially from C_∞/K to C_∞ over a distance equal to the diffusion length $l_d = D_L/V$. This diffusion field is responsible for the Mullins-Sekerka instability observed (Fig. C.X.14a) for moderate supersaturation values. A linear stability analysis of the front shows that this instability disappears for $l_d \approx d_0^c = d_0 T_c / \Delta T$. This relation determines the threshold Δ^* for absolute restabilization of the front:

$$\Delta^* = 1 + \frac{D_L / d_0}{K \mu T_c} \qquad \text{(C.X.29)}$$

This phenomenon is observed experimentally for $\Delta > \Delta^*(\exp) \approx 4 \pm 2$ (see figure C.X.14). This value (independent of ΔT [24]) is lower than the one calculated using the theoretical expression C.X.29, yielding $\Delta^*(\text{th}) \approx 7 \pm 2$. Géminard [2] was able to show that this discrepancy can be easily explained by taking into account the finite resolution of the microscope and the size of the observed germs, smaller and smaller for increasing Δ (due to the nucleation).

X.3.c) Kinetic Wulff construction and anisotropy of the kinetic coefficient

In 1963, Chernov [25] showed that the growth shape of a germ in the kinetic regime is given by a geometrical construction similar to the one yielding the equilibrium shape of a crystal starting from the polar diagram of its surface energy (see chapter C.VII). Here, however, the relevant polar diagram is that of the kinetic coefficient of molecular attachment to the interface, $\mu(\theta)$. This kinetic Wulff construction shows that all germs have similar shapes and that, if the kinetic coefficient is of the form:

$$\mu(\theta) = \mu_0 [1 + \bar{\varepsilon}_6 \cos(6\theta + \Phi_6)] \qquad \text{(C.X.30)}$$

all germs are described by the polar equation:

$$R(\theta,t) = R_0(t) [1 + \bar{\varepsilon}_6 \cos(6\theta + \Phi_6)] \qquad \text{(C.X.31)}$$

The shape anisotropy of the germs is thus equal to the anisotropy of the kinetic coefficient.

To test these theoretical predictions, the radii $R(\theta,t)$ of several germs were measured and then expanded in a Fourier series:

$$R(\theta,t) = R_0(t) + \delta_6(t) \cos(6\theta + \Phi_6) + \dots \qquad \text{(C.X.32)}$$

The experiments, performed above the absolute restabilization threshold, show that all the germs exhibit the same shape anisotropy $\bar{\varepsilon}_6 = \delta_6(t)/R_0(t)$, independent of their size and growth velocity (Fig. C.X.17):

$$\bar{\varepsilon}_6(\exp) \approx 5.3 \pm 0.73 \times 10^{-3} \qquad \text{(C.X.33)}$$

These results show that the kinetic Wulff construction is well applicable to this experiment and that the measured anisotropy is indeed that of the kinetic coefficient.

In conclusion, let us find the directions along which the kinetic coefficient is maximal. To put it differently, we are looking for the value of the phase Φ_6, taking as origin a twofold axis of the hexagonal lattice along which the surface free energy is maximal (easy growth direction for the dendrites at low supersaturation). On symmetry grounds, $\mu(\theta)$ must be maximal along six twofold axes of the hexagonal lattice. The phase then is necessarily either 0 or 30°.

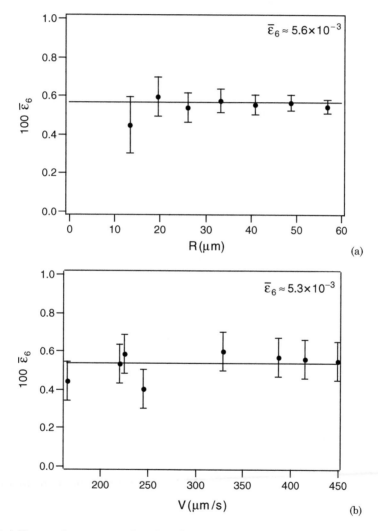

Fig. C.X.17 a) Shape anisotropy as a function of the radius for a given germ; b) shape anisotropy measured on germs growing at different velocities (from ref. [24]).

To remove this indetermination, we only need to grow the same germ, first in the kinetic regime, then in the dendritic regime (after partially melting it, in order to preserve its crystalline orientation), and to determine each time the easy (fast) growth directions. This procedure shows that the easy growth axes in the kinetic regime make a 30° angle with the axes of the dendrites in the diffusive regime (Fig. C.X.18). Thus, $\Phi_6 = 30°$.

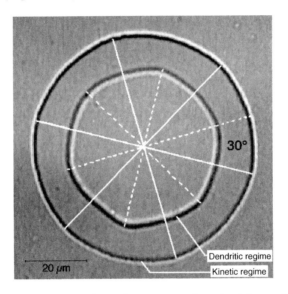

Fig. C.X.18 Two germs originating from the same nucleus (and hence identically oriented) photographed at different supersaturations; the larger one was grown in the kinetic regime at $\Delta \approx 5$, while the smaller one is observed in the dendritic regime at $\Delta \approx 0.3$. The solid lines mark the directions corresponding to the maxima of $\mu(\theta)$; $\gamma(\theta)$, on the other hand, is maximal along the dashed lines (growth directions of the dendrites at low supersaturation). These directions clearly make a 30° angle (from ref. [24]).

This angle between the easy growth directions at low and high velocity could play a role in the morphological transition observed for $\Delta \approx 0.7$ between the dendritic and the dense branching regime. Indeed, the theory of dendrite selection shows that the kinetic anisotropy plays at "high" velocity the same role as the surface tension anisotropy at "low" velocity [26]. It is then likely that the destabilization of the tips of primary dendrites observed at the threshold of the dense branching regime is partly due to this competition. This interpretation is supported by various numerical simulations, among which are those of Liu and Goldenfeld [27]. However, one still needs a theoretical elucidation of the role played by the thickness and confinement effects in this transition.

BIBLIOGRAPHY

[1] Destrade C., Mondon M.C., Malthête J., *J. Physique Coll. (France)*, **40** (1979) C3.17.

[2] Géminard J.-C., "Croissance libre de la phase colonnaire hexagonale d'un cristal liquide discotique," Ph.D. Thesis, Claude Bernard-Lyon I University, order number 128.93, 1993.

[3] Oswald P., *J. Physique (France)*, **49** (1988) 1083.

[4] Géminard J.-C., Oswald P., *Phys. Rev. E*, **55** (1997) 4442.

[5] Oswald P., *J. Physique Coll. (France)*, **50** (1989) C3-127.

[6] Daoud M., Gharbia M., Gharbi A., *J. Phys. II (France)*, **4** (1994) 989.

[7] Sallen L., Oswald P., *J. Phys. II (France)*, **7** (1997) 107.

[8] Oswald P., *J. Physique (France)*, **49** (1988) 2119.

[9] Brush L.N., Sekerka R.F., *J. Cryst. Growth*, **96** (1989) 419.

[10] Coriell S.R., Parker R.L., *J. Appl. Phys.*, **36** (1965) 632.

[11] Géminard J.-C., Oswald P., *J. Phys. II (France)*, **4** (1994) 959.

[12] Oswald P., Malthête J., Pelcé P., *J. Physique (France)*, **50** (1989) 2121.

[13] Brener E.A., Gellikman M.B., Temkin D.E., *Sov. Phys. JETP*, **67** (1988) 1003.

[14] Ivantsov G.P., *Doklady Akad. Nauk.*, **58** (1947) 567.

[15] Langer J.S., Müller-Krumbhaar H., *Acta Met.*, **26** (1977) 1681.

[16] Langer J.S., in *Chance and Matter*, Les Houches XLVI 1986, Ed. J. Souletie, J. Vanninemus, and R. Stora, North-Holland Publishing Company, Amsterdam, 1987.

[17] Barbieri A., Hong D.C., Langer J.S., *Phys. Rev. A*, **35** (1987) 1802.

[18] Saito Y., Goldbeck-Wood G., Muller-Krumbhaar G., *Phys. Rev. Lett.*, **58** (1987) 1541.

[19] Kessler D.A., Levine H., *Europhys. Lett.*, **4** (1987) 215.

[20] a) Pelcé P., in *Dynamics of Curved Fronts*, Ed. P. Pelcé, Academic Press, San Diego, 1988.
b) Pelcé P., *Theory of Growth Forms: Digitations, Dendrites, and Flames* (in French), EDP Sciences/CNRS Editions, Les Ulis, 2000.

[21] Kassner K., *Pattern Formation in Diffusion-Limited Crystal Growth*, World Scientific, Singapore, 1996.

[22] This value was calculated by M. Ben Amar, 1993.

[23] Ben Amar M., Brener E., *Phys. Rev. E*, **47** (1992) 534.

[24] Géminard J.-C. , Oswald P., Temkin D., Malthête J., *Europhys. Lett.*, **22** (1993) 69.

[25] Chernov A.A., *Sov. Phys. Crystallog.*, **7** (1963) 728.

[26] Brener E.A., *J. Cryst. Growth*, **99** (1990) 165.

[27] Liu F., Goldenfeld N., *Phys. Rev. A*, **42** (1990) 895.

INDEX

SUBJECT INDEX

When several page numbers are listed for an entry, the most important pages are boldface.

AUTHOR INDEX

Bibliography pages are marked in boldface.

D

E

F

G

H

M

Q

R

S